Plants and Microclimate

A Quantitative Approach to Environmental Plant Physiology

Third Edition

This rigorous yet accessible text introduces the key physical and biochemical processes involved in plant interactions with the aerial environment. It is designed to make the more numerical aspects of the subject accessible to plant and environmental science students, and will also provide a valuable reference source to practitioners and researchers in the field.

The third edition of this widely recognised text has been completely revised and updated to take account of key developments in the field. Approximately half of the references are new to this edition, and relevant online resources are also incorporated for the first time. The text shows how recent developments in molecular and genetic research on plants can be used to advance our understanding of the biophysical interactions between plants and the atmosphere, and how progress in molecular biology can itself be informed by an understanding of whole-plant physiology. Remote sensing technologies and their applications in the study of plant function are also covered in greater detail.

Hamlyn G. Jones is Emeritus Professor of Plant Ecology at the University of Dundee and Adjunct Professor in Plant Biology at the University of Western Australia. His research uses experimental approaches and mathematical modelling to investigate the characters that enable plants to be adapted to specific environments and to tolerate environmental stress.

Plants and Microclimate

A Quantitative Approach to Environmental Plant Physiology

Third Edition

HAMLYN G. JONES

Professor Emeritus, Division of Plant Science
University of Dundee at the James Hutton Institute
Invergowrie, Dundee DD2 5DA, UK
and
Adjunct Professor of Plant Biology
University of Western Australia

CAMBRIDGE
UNIVERSITY PRESS

CAMBRIDGE
UNIVERSITY PRESS

University Printing House, Cambridge CB2 8BS, United Kingdom

Published in the United States of America by Cambridge University Press, New York

Cambridge University Press is part of the University of Cambridge.

It furthers the University's mission by disseminating knowledge in the pursuit of
education, learning and research at the highest international levels of excellence.

www.cambridge.org
Information on this title: www.cambridge.org/9780521279598

Third edition © Hamlyn G. Jones 2014
First and second editions © Cambridge University Press 1983, 1992

First published 1983
Second edition 1992
Third edition 2014

Printed in the United Kingdom by TJ International Ltd. Padstow Cornwall

A catalogue record for this publication is available from the British Library

Library of Congress Cataloging-in-Publication Data
Jones, Hamlyn G.
Plants and microclimate : a quantitative approach to environmental plant physiology / Hamlyn G. Jones,
professor emeritus, Division of Plant Science University of Dundee at the James Hutton Institute, Invergowrie,
Dundee, DD2 5DA, UK. – Third edition.
   pages cm
Includes bibliographical references and index.
ISBN 978-0-521-27959-8 (Paperback)
1. Vegetation and climate–Mathematical models. 2. Plant-atmosphere relationships–Mathematical
models. 3. Plant ecophysiology–Mathematical models. I. Title.
QK754.5.J66 2013
581.7′22–dc23–dc23   2012045663

ISBN 978-0-521-27959-8 Paperback

CONTENTS

PREFACE

I have been delighted, and somewhat surprised, at the continued widespread use of this text, in spite of the fact that much has changed in associated fields since the previous edition was published around 20 years ago. Perhaps the major change in plant biology over this period has been the explosion of research on the molecular and genetic basis of plant responses to the environment, though there have been important advances in other relevant fields such as in remote sensing. Although I have not attempted to cover molecular aspects in any detail, as there are many suitable alternative texts, I have tried to relate recent advances in molecular sciences to our understanding of whole-plant responses to the environment. In this context I have aimed especially to indicate the ways in which the powerful new molecular tools and other 'omics' technologies can contribute to advancing our understanding of the biophysical interactions between plants and the atmosphere. As in the previous editions, however, I have continued the approach of describing only briefly the biochemical and molecular mechanisms involved in plant responses to the environment, so interested readers are referred to specialist reviews and books mentioned at appropriate places in the text.

For this third edition I have chosen largely to retain the general structure and aims of the successful previous editions. In particular the key aim remains to provide an authoritative introduction to environmental plant physiology suitable both as a text for upper undergraduate and postgraduate courses and as a reference for researchers in the field. As previously, the first half of the text concentrates on the general principles, with the later chapters going more into the physiology and practical applications. The emphasis throughout remains on the more quantitative and physical aspects of plant responses to the aerial environment as these topics tend to be relatively poorly treated by the standard plant physiology texts, yet our need to understand how the whole plant functions and responds to its environment has never been greater if we are to respond effectively to the challenges that face the world in the coming years. These include responding to the problems and opportunities raised by climate change and by the need to continue to feed the burgeoning and increasingly wealthy world population in the coming years.

As it is now 20 years since the previous edition was published it has proved necessary to completely revisit and revise the content throughout. Around half the publications referred to are new; nevertheless many of the references to earlier papers have been retained, especially for data. Even though many thousands of potentially relevant new papers and texts have appeared, often these only provide refinements rather than substantial improvements (for example I still quote the data using 1% O_2 to suppress photorespiration even though more recent papers use 2%). The citations included fall into one of several categories: general texts and reviews that give access to useful references and a limited number of key original references, while the majority of citations are simply useful examples selected from among many possible papers. I have also, where appropriate, included a limited number of key internet addresses, though it should be recognised that these change rapidly.

In revising the text I am indebted to many colleagues from around the world who have provided helpful and constructive comments on the previous editions, and to those who have read and commented on sections of the text. Particular thanks go to Abdellah Barakate, David Deery, Olga Grant, Anthony Hall, Amanda Jones, Ian MacKay, Barry Osmond, John Raven, Philip Smith and Bill Thomas.

ACKNOWLEDGEMENTS

I am very grateful to Dr M. L. Parker for permission to reproduce photographs in Figures 6.2 and 7.1 and to Dr B. M. Joshi for permission to use the electron micrographs in Figure 6.2. I am also grateful to the appropriate authors and to the following for permission to use previously published material: Carnegie Institution of Washington (Figures 7.18 and 9.19); British Ecological Society (Figure 7.21); The Royal Society (Figure 7.24); The Director, National Institute for Agricultural Botany, Cambridge (Figure 12.2(a)). I am also grateful to Michael Rosenthal for Figure 4.7(c) and for part of Figure 4.8; Manfred Stoll for the concept of Figure 10.14; Drs Bob Furbank, David Deery, Jose Jimenez-Berni and Xavier Sirault for the image data in Figure 12.3; Professor Ch. Körner for permission to use data in Figure 11.9; M. Rosenthal for part of Figure 4.8; Ian MacKay for part of Figure 12.2(b); and to Oxford University Press for permission to use Figures 2.1, 2.13, 2.31 and 7.1.

SYMBOLS

Where possible I have tried to use the most commonly accepted symbols for different quantities, though in some areas such as the treatment of radiation there appears to be no universal consensus. Where possible the use of the same symbol for different quantities has been minimised, though where duplication is unavoidable there should usually be little chance of confusion. Note that different fonts are used to distinguish quantities, for example molar and mass units for gas transfer. The individual superscripts and subscripts given for each main symbol may be combined to make compound symbols.

a	amplitude of oscillation (a′, modified amplitude)
a, A	constants
a_W	activity
	subscript: a_W water
A	amp (coulomb s^{-1})
A	area (m^2)
A	absorbance ($= \ln [\mathbf{I}_o/\mathbf{I}]$)
b, B	constants
b_i	control parameters (sensitivity analysis)
c	concentration (kg m^{-3} or mol m^{-3})
	subscripts: c_C carbon dioxide; c_H heat (J m^{-3} = $\rho c_p T$); c_M momentum (kg m^{-2} s^{-1} = ρu); c_o oxygen; c_W water; c_X concentration of pollutant X; c_a in the outside air; c_c at carboxylase; c_e in inlet air; c_i in intercellular spaces at surface of cell walls; c_x effective concentration internal to biochemistry (for CO_2); c_ℓ leaf; c_o in outlet air; c_{Pr} phytochrome concentration; c_s solute
	superscripts: c^m molar concentration; $c′$ carbon dioxide ($= c_C$)
c_D	total drag coefficient
c_f	form drag coefficient
c_p	specific heat of air (c_p* specific heat capacity of leaf tissue) (J kg^{-1} K^{-1})

c	speed of light (2.998 m s^{-1})
C	capacitance (m MPa^{-1} or m^3 MPa^{-1})
C_r	relative capacitance (MPa^{-1})
C	sensible heat exchange (W m^{-2})
	subscripts: $\mathbf{C}_{(d)}$ dry; $\mathbf{C}_{(w)}$ wet
C	control or sensitivity coefficients
d	days
d	zero plane displacement (m)
d	diameter (m) (e.g. d_p particle diameter)
d	characteristic dimension (m)
D	drainage (mm)
D	atmospheric vapour pressure deficit (kPa)
	modifiers: D^* integrated average daily vapour pressure deficit; D_ℓ leaf to air vapour pressure deficit; D_{cw} absolute humidity deficit of air; D_{r_W} water vapour mole fraction deficit of air
D	thermal time, growing degree days or temperature sum (°C day)
	subscript: D_{eff} effective day-degrees
D	diffusion coefficient (m^2 s^{-1})
	subscripts: $\mathbf{D}_A$ air; $\mathbf{D}_C$ CO_2, $\mathbf{D}_{CA}$ mutual diffusion coefficient for CO_2 in air; $\mathbf{D}_H$ heat; $\mathbf{D}_i$ the ith species; $\mathbf{D}_M$ momentum; $\mathbf{D}_O$ oxygen; $\mathbf{D}_W$ water; $\mathbf{D}_X$ pollutant X
	superscript: $\mathbf{D}^\circ$ reference value
$\mathscr{D}$	dielectric constant (dimensionless)
e	base for natural logarithm (2.71828)
e	water vapour pressure (Pa) (see also D)
	subscripts: e_a in bulk air; e_e in inlet; e_o in outlet; e_s surface; e_s saturation; $e_{s(T\ell)}$ saturation water vapour pressure at leaf temperature ($= e_\ell$); e_{ice} vapour pressure over pure ice
e	equation of time (min)
E	radiant energy (e.g. of a photon)
	subscript: E_λ radiant energy per unit wavelength
E_a	activation energy
E	evaporation (or transpiration) rate (kg m^{-2} s^{-1}, mol m^{-2} s^{-1} or mm h^{-1})

modifiers: E_ℓ transpiration; E^m evaporation or transpiration (molar units); E_o potential evaporation from free water surface; E_{eq} equilibrium evaporation; E_{imp} imposed evaporation

ET evapotranspiration

subscripts: ET_o reference evapotranspiration from short grass surface well supplied with water; ET_c expected value of **ET** for a specific crop and growth stage; $ET_{c,adj}$ actual evapotranspiration for any crop

f Pfr/Pr ratio

f fraction (e.g. fraction of O_2 unconsumed, fraction carbon allocated to leaves as compared with roots, f_{PSII} the fraction of absorbed PAR that is received by PSII, f_{open} fraction of reaction centres that are open)

f_{veg} fractional vegetation cover

f enhancement factor

f fraction of water in unfrozen state (dimensionless)

F fluorescence (arbitrary or mol m^{-2} s^{-1})

modifiers: F' fluorescence at any time ($=F_t$); F_m maximum value after equilibration in dark; F'_m fluorescence at any time obtained with saturating flash; F_o basal fluorescence with open reaction centres; F'_o basal fluorescence at any time; F_v variable fluorescence ($F_m - F_o$); F'_v ($F'_m - F'_o$)

F force (N)

g conductance (m s^{-1})

subscripts: g_A canopy boundary layer conductance; g_L canopy physiological conductance; g_C carbon dioxide; g_H heat; g_M momentum; g_O oxygen; g_R radiation ($= 4\varepsilon\sigma T_a^3/\rho c_p$); g_{HR} parallel heat and radiative transfer; g_W water; g_a boundary layer; g_c cuticle; g_g gas phase; g_ℓ leaf; g_i intercellular space; g_m diffusive component in mesophyll; g_o reference value; g_s stomatal; g_w wall; g_x internal biochemical conductance

superscripts: g^m molar conductance ($= g$); g' carbon dioxide ($= g_C$)

g molar conductance ($=g^m$) (mol m^{-2} s^{-1})

modifiers: as for conductance (g)

g acceleration due to gravity (m s$^{-2} = 9.8$ at sea level)

G Gibbs free energy (J)

G soil heat storage (W m^{-2})

h hour

h Planck's constant (6.626×10^{-34} J s)

h relative humidity (dimensionless)

h hour angle of the sun (the angular distance from the meridian of the observer; degree or radian)

h height or thickness (m)

HSAI hemi-surface area index (dimensionless)

I moment of inertia (kg m^{-2})

I electric current (A)

I thermal indices

subscripts: I_{CWSI} index analogous to CWSI; I_g conductance index

I irradiance (W m^{-2})

subscripts: I_S shortwave; $I_{S(diffuse)}$ diffuse shortwave; $I_{S(dir)}$ direct shortwave; I_L longwave; $I_{(PAR)}$ photosynthetically active radiation; I_c light compensation point; I_p photon irradiance; I_e irradiance in terms of energy; I_A solar irradiance at top of atmosphere; I_{pA} solar constant; I_o reference value

J joule (1 kg m^2 s^{-2})

J_{max} maximum electron transport rate (ETR)

J flux density or mass transfer rate per unit area (kg m^{-2} s^{-1})

subscripts: J_v volume flux density (m s^{-1}); for other modifiers see **D**

k thermal conductivity (W m^{-1} K^{-1})

k rate constant or other constant

subscripts: k_F, k_H, k_T and k_P, respectively, are the rate constants for de-excitation through fluorescence, thermal dissipation as heat, energy transfer to photosystem I, and PSII photochemistry with all reaction centres open; k_d rate of development ($= 1/t$)

k Boltzmann's constant

k extinction coefficient (dimensionless)

k von Karman's constant ($= 0.41$, dimensionless)

kg   kilogram

K   kelvin temperature

K_c crop coefficient (for evaporation from a well watered crop)

  modifiers: $K_{c\text{-adj}}$ adjusted crop coefficient allowing for drought effects; K_{cb} basal coefficient for crop; K_{stress} stress modifier; K_s soil coefficient

K hydraulic conductance (m MPa^{-1} s^{-1} or m^3 MPa^{-1} s^{-1}; see also L_p)

  modifiers: K_r root; K_{st} stem; K_{ℓ} leaf

$\boldsymbol{K}$ transfer coefficient (m^2 s^{-1})

  subscripts: as for $\boldsymbol{D}$

K_m Michaelis constant (dimensions as for concentration or irradiance)

  modifiers: K_m^C for carbon dioxide; K_m^I for light

ℓ length, thickness (m)

  superscript: ℓ^* thickness of leaf tissue

ℓ photosynthetic limitation (s m^{-1})

  modifiers: superscripts as for resistance (r); ℓ' relative limitation (dimensionless)

ln   natural logarithm

log   logarithm to base 10

L hydraulic conductivity (m^2 s^{-1} Pa^{-1})

L_p hydraulic conductance (m s^{-1} Pa^{-1})

L_v volumetric hydraulic conductance (s^{-1} Pa^{-1})

L radiance or intensity (W sr^{-1})

  subscripts: L_{in}, L_{out} for radiance within or outside a Fraunhofer line

L leaf area index

  modifier: L' leaf area index expressed per unit area of shaded ground (dimensionless)

m   metre

m mass fraction (dimensionless)

  modifiers: as for concentration (c)

m mass (kg)

m air mass (dimensionless)

M molecular mass

  subscripts: as for $\boldsymbol{D}$

$\boldsymbol{M}$ radiant exitance (W m^{-2})

$\boldsymbol{M}$ metabolic heat storage (W m^{-2})

mol   mole (amount of substance containing Avogadro's number of particles)

n hours bright sunshine; number

n number of moles

  subscripts: n_p photons; n_s solute; other subscripts as for $\boldsymbol{D}$

$n(E)$ number of moles with energy exceeding E (mol)

N   Newton

N daylength (h)

N reflectance in the near infrared ($= \rho_{NIR}$)

0   run-off (mm)

p partial pressure (Pa)

  modifiers: as for $\boldsymbol{D}$ and for concentration (c)

P   precipitation (mm)

P pressure (Pa)

  modifiers: P^o reference; P^* balance pressure

P period of oscillation (s)

P_e power output (W m^{-2})

$\boldsymbol{P}$ photosynthesis (mg m^{-2} s^{-1} or μmol m^{-2} s^{-1})

  subscripts: $\boldsymbol{P}_c$ rubisco-limited rate; $\boldsymbol{P}_g$ gross photosynthesis; $\boldsymbol{P}_j$ RuBP-limited rate; $\boldsymbol{P}_n$ net photosynthesis; $\boldsymbol{P}_{max}$ maximum value with either light of CO_2 saturating; $\boldsymbol{P}_t$ triose phosphate-limited rate

  superscripts: $\boldsymbol{P}^m$ molar; $\boldsymbol{P}^{max}$ maximum with light and CO_2 saturating; $\boldsymbol{P}^o$ reference value

q fluorescence quenching

  subscripts: q_I photo-inhibition; q_L estimate of the fraction of open PSII reaction centres using 'lake' model; q_N non-photochemical; q_o F$_o$ quenching; q_P photochemical quenching; q_T state transition

qr   quantum requirement

Q_{10} temperature coefficient: the ratio of the rate at one temperature to that at a temperature ten degrees lower (dimensionless)

$\boldsymbol{Q}$ radiant flux (J s^{-1} or W)

  modifiers: as for radiant flux density ($\boldsymbol{R}$)

r radius (m)

r resistance (s m^{-1} or m^2 s mol^{-1})

  modifiers: as for conductance (g); also $r'_* = \mathrm{d}r'_w/\mathrm{d}\boldsymbol{P}_n$ at normal ambient CO_2

r molar resistance ($= r^{m}$) (m^2 s mol^{-1})

modifiers: as for conductance (g)

R liquid phase hydraulic resistance (MPa s m^{-1} or MPa s m^{-3})

subscripts: R_ℓ leaf; R_p plant; R_s soil; R_{st} stem

R isotopic ratio (e.g. $^{18}O/^{16}O$)

subscript: R_0 reference value

R electrical resistance (ohm $=$ W A^{-2})

R reflectance in the red ($= \rho_R$)

$\mathbf{R}$ respiration rate (mg m^{-2} s^{-1} or μmol m^{-2} s^{-1})

subscripts: $\mathbf{R}_d$ dark; $\mathbf{R}_\ell$ light; $\mathbf{R}_g$ growth; $\mathbf{R}_m$ maintenance; $\mathbf{R}_{non-ps}$ from non-photosynthesising tissue

$\mathbf{R}$ radiant flux density (W m^{-2})

subscripts: subscripts 'e' and 'p' are used to distinguish radiant ($\mathbf{R}_e$) flux density (W m^{-2}) and photon ($\mathbf{R}_p$) flux density (mol m^{-2}) where necessary; $\mathbf{R}_{absorbed}$ absorbed; $\mathbf{R}_{emitted}$ emitted; $\mathbf{R}_{(d)}$ dry; $\mathbf{R}_{(w)}$ wet; $\mathbf{R}_d$ downward; $\mathbf{R}_u$ upward; $\mathbf{R}_n$ net radiation; $\mathbf{R}_{ni}$ net isothermal radiation; $\mathbf{R}_R$ radiative heat loss $= \mathbf{R}_n - \mathbf{R}_{ni}$

$\mathscr{R}$ gas constant (8.3144 J mol^{-1} K^{-1} or Pa m^3 mol^{-1} K^{-1})

s rate of change of saturation vapour pressure with temperature (Pa K^{-1})

s second

sr steradian

S stress (Pa)

$S_i(z)$ source density profile of entity i with height

$S(t)$ state of development at time t

$\mathbf{S}$ heat flux into storage (W m^{-2})

t time (s, h or day)

subscripts: t_o time at solar noon; t_d number of day in the year; $t_{1/2}$ half-time

T temperature (°C or K)

subscripts: T_a air; T_{dew} dew point; T_d dry; T_e equilibrium; T_ℓ leaf; T_w wet; T_{wb} wet-bulb temperature; T_{base} non-water-stressed-baseline temperature; T_h heated replica; T_m mean; T_o optimum; T_{max} maximum; T_s surface; T_s saturation; T_{sky} apparent radiative temperature of the sky; T_t threshold; T_u unheated replica; T_x observed temperature at given D

superscript: $T°$ reference temperature

$T(t)$ temperature as function of time

T growing season length (days)

$\mathbf{T}$ torque or turning moment (N m or J)

$\mathcal{T}$ transmission in discontinuous canopies (dimensionless)

subscript: $\mathcal{T}_f$ fraction that would reach the ground if all leaves non-transmitting

u molar air flow rate (mol s^{-1})

u wind velocity (m s^{-1})

subscripts: u_z wind velocity at height z; u_* friction velocity

v_s sedimentation velocity (m s^{-1})

v volume flow rate (m^3 s^{-1})

V_d deposition velocity (m s^{-1})

V rate of reaction

subscripts: V_c velocity of carboxylation; V_o velocity of oxygenation; $V_{c,max}$ maximum velocity of Rubisco for carboxylation

V volume (m^3)

modifiers: V_e expressed sap; V_o turgid volume of cell; $\overline{V}_W$ partial molal volume of water (18.048 $\times$ 10^{-6} m^3 mol^{-1} at 20°C)

V electrical potential difference (volt $=$ W A^{-1})

w mixing ratio (dimensionless)

subscripts: as for D

w vertical wind velocity (m s^{-1})

W watt (J s^{-1})

W water content (kg m^{-2} or kg m^{-3})

subscripts: W_ℓ leaf; W_{max} maximum

W leaf mass per unit area in CO_2 equivalents (g m^{-2})

x mole fraction

modifiers: as for D and concentration (c)

x distance or displacement (m)

Y yield threshold (Pa)

Y economic yield (tonne ha^{-1} or kg m^{-2})

subscript: Y_d dry matter yield

z distance, height or depth, or atmosphere thickness (m)

subscript: z_o roughness length

Z damping depth (m)

α contact angle (degree or radian)

α absorptivity, absorption coefficient or absorptance (dimensionless)

	modifiers: subscripts define waveband (e.g. α_{660} or α_S)
α	the azimuth or aspect of a surface (measured east from north)
α	the ratio between the woody tissue (hemispheric) area index and the total plant (hemispheric) area index
α	Priestley–Taylor coefficient
β	solar elevation (degree or radian; complement of θ)
$\boldsymbol{\beta}$	Bowen ratio (dimensionless $= \mathbf{C}/\lambda\mathbf{E}$)
γ	psychrometer constant (Pa K^{-1} $= Pc_p/0.622\lambda$) *superscript:* γ^* the modified psychrometer constant ($= \gamma g_H/g_W$)
γ_W	activity coefficient for water that measures departure from ideal behaviour
Γ	total soil heat flux ratio (dimensionless) *modifier:* Γ' energy partitioning at the soil surface
Γ	carbon dioxide compensation concentration (mg m^{-3} or mmol m^{-3}) *modifier:* Γ_* concentration at which CO_2 loss by oxygenation equals uptake by carboxylation
δ	deviation of isotope abundance from ratio in a standard sample (e.g. $\delta^{13}C$) *subscripts:* δ_a air; δ_p plant
δ	solar declination (degree)
δ	average thickness of laminar boundary layer (m) *subscript:* δ_{eq} thickness of equivalent boundary layer (m)
∂	partial differential
Δ	isotopic discrimination (dimensionless, ‰) *subscripts:* Δ_a AOX discrimination; Δ_c COX discrimination
Δ	finite difference
ΔF	difference between steady state and maximal fluorescence
ΔT_f	freezing point depression (K)
ε_i	elasticities of individual reactions in a pathway
ε	emissivity
ε	efficiency (dimensionless) *modifiers:* ε_p photon efficiency; ε_q quantum efficiency; $\varepsilon_{q(Pr)}$ quantum yield for phytochrome conversion

ε_Y	Young's modulus (Pa)
ε_B	bulk modulus of elasticity (Pa)
$\boldsymbol{\varepsilon}$	s/γ
ζ	ratio of the photon flux densities in the red (655–665 nm) and far-red (725–735 nm) portions of the spectrum
η	dynamic viscosity (N s m^{-2} or kg m^{-1} s^{-1})
θ	relative water content (dimensionless, %)
θ	angle from beam to normal; zenith angle of the sun (degree or radian)
$\Delta\theta$	change in soil moisture content
λ	wavelength (m) *subscript:* λ_m peak wavelength of Planck distribution
λ	latent heat of vaporisation of water (J kg^{-1})
λ	latitude (degree or radian)
λ_o	clumping index
$\boldsymbol{\lambda}$	constant; climate sensitivity parameter
μ_W	chemical potential (J mol^{-1}) *modifiers:* μ_W of water; μ° reference value
ν	frequency (hertz)
ν	frequency of stomata (mm^{-2})
υ	wavenumber (cm^{-1})
$\boldsymbol{\nu}$	kinematic viscosity (m^2 s^{-1} $= \boldsymbol{D}_M$)
π	pi, the ratio of circumference of a circle to its diameter (3.14159)
Π	osmotic pressure (MPa $= -\psi_\pi$)
ρ	density (often of dry air) (kg m^{-3}) *modifiers:* ρ_a dry air (sometimes abbreviated to ρ); ρ_{as} air saturated with water vapour; ρ_i ith component of mixture; ρ^* density of leaf or replica
ρ	reflectivity; reflection coefficient; reflectance or albedo (dimensionless) *subscripts:* ρ_{NIR} near infrared ($= N$); ρ_R red ($= R$); $\rho_{(\theta)}$ at any zenith angle θ; others as for α
σ	Stefan–Boltzmann constant (5.6703 × 10^{-8} W m^{-2} K^{-4})
σ	reflection coefficient (dimensionless)
$\boldsymbol{\sigma}$	surface tension (N m^{-1} $= 7.28 \times10^{-3}$ for water at 20°C)

$\sigma_{1,2}$ the fractional contribution of stomata to a change in limitation between two states $\mathbf{P}_{n1}$ and $\mathbf{P}_{n2}$

τ time constant (s)

τ transmissivity, transmission coefficient or transmittance (dimensionless)

subscripts: as for α

τ_a proportion of flow through AOX

$\boldsymbol{\tau}$ shearing stress (Pa = N m^{-2})

subscript: $\boldsymbol{\tau}_f$ form drag

ϕ evaporative fraction (dimensionless)

$\boldsymbol{\phi}_C$ losses of CO_2 not associated with losses through stomata

$\boldsymbol{\phi}_W$ losses of H_2O not associated with losses through stomata

ϕ phytochrome photoequilibrium Pfr/P$_{total} = f/(1+f)$

ϕ extensibility (s^{-1} Pa^{-1})

ϕ phase lag (s)

ϕ photorespiratory stoichiometry

ϕF quantum yield for fluorescence (dimensionless)

χ the zenith angle (= slope) of a surface (radian or degree)

ψ water potential (MPa)

subscripts: ψ_p, ψ_π, ψ_m and ψ_g respectively are components due to pressure, osmotic, matric and gravitational forces; ψ_ℓ leaf water potential, ψ_{st} stem water potential; $\psi_{\ell 0}$ initial value

ω angular frequency (= 2π × period)

ω angular velocity (rad sec^{-1})

Ω solid angle (sr)

$\boldsymbol{\Omega}$ decoupling coefficient (dimensionless)

MAIN ABBREVIATIONS AND ACRONYMS

ABA	abscisic acid
ADP	adenosine diphosphate
AOX	alternative oxidase
ATP	adenosine-5′-triphosphate
BRDF	bidirectional reflectance distribution function
BRF	bidirectional reflectance factor
BVOC	biogenic volatile organic compound
BWB	Ball–Woodrow–Berry model
C_3	three-carbon photosynthetic pathway
C_4	four-carbon photosynthetic pathway
CAM	crassulacean acid metabolism
CFCs	chlorofluorocarbons
CGR	crop growth rate (kg m^{-2} day^{-1})
COX	cytochrome oxidase
CRB	C-repeat binding proteins
CWSI	crop water stress index
DELLA	DELLA proteins are transcriptional regulators that regulate gibberellic acid signalling
DREB	deyhdration-responsive element binding proteins
DVI	difference vegetation index
ETR	electron transport rate through PSII
EUW	Effective use of water
FACE	free-air carbon dioxide enrichment experiment
FADH$_2$	reduced flavin adenine dinucleotide
fAPAR	fraction of absorbed photosynthetically active radiation
FR	far red
FSPM	functional-structural plant models
FvCB	the Farquhar–von Caemmerer–Berry model of photosynthesis
GCM	global circulation model
GDD	growing degree days (see *D*)
GNDVI	green normalised difference vegetation index
GWP	global warming potential
HFC	hydrofluorocarbons
HI	harvest index
HIR	high irradiance response
HU	'heat' units (should be avoided)
IAA	indole acetic acid
IR	infrared
IRGA	infrared gas analyser
LAD	leaf area duration
LAR	leaf area ratio
LD$_{50}$	minimum survival temperatures
LD	long-day (also LSD for long–short day)
LFR	low fluence rate response
LHC	light harvesting complex
LOV	light-oxygen-voltage domains
LUE	light use efficiency
MAS	marker-assisted selection
MIPs	major intrinsic proteins
MOST	Monin–Obukhov similarity theory
N$_2$	nitrogen
NAC	a superfamily of transcription factors
NADP$^+$	nicotinamide adenine dinucleotide phosphate
NADPH	reduced nicotinamide adenine dinucleotide phosphate
NAR	net assimilation rate
NDVI	normalised difference vegetation index
NPQ	non-photochemical quenching $((F_m/F_m') - 1)$
OAA	oxaloacetate
OTC	open-top chamber
P$_{680}$	reaction centre in PSII
P$_{700}$	reaction centre in PSI
Pa	pascal (N m^{-2} or kg m^{-1} s^{-2})
PAR	photosynthetically active radiation (400–700 nm)
PBM	process-based models
PCO	photorespiratory carbon cycle
PCR	photosynthetic carbon reduction cycle

PDB	Peedee belemnite formation in South Carolina		RNA	ribonucleic acid
PEP	phospho*enol*pyruvate		RNAi	RNA interference
Pfr	far-red absorbing form of phytochrome		ROS	reactive oxygen species
PGA	phosphoglyceric acid		Rubisco	ribulose-1,5-bisphosphate carboxylase-oxygenase
pH	negative logarithm of hydrogen ion activity		RuBP	ribulose-1,5-bisphosphate
phy	phytochrome		*RVI*	ratio vegetation index
PHY	PHY-A, PHY-B, etc. phytochrome proteins (genes in italic)		*SAVI*	soil adjusted vegetation index
phot1, phot2	phototropins		SD	short-day (also SLD for short–long day)
			SHAM	salicylhydroxamic acid
PIF	phytochrome interacting factor (e.g. PIF4 and PIF5)		SLA	specific leaf area
			SNP	single nucleotide polymorphism
PIPs	plasma membrane intrinsic proteins		TCA	tricarboxylic acid
ppb	volume parts per billion (10^9)		TE	transpiration efficiency
ppt	volume parts per trillion (10^{12})		TIPs	tonoplast intrinsic proteins
Pr	red light absorbing form of phytochrome		TPU	triose phosphate utilisation
PRI	photochemical reflectance index		UAV	unmanned aerial vehicles
PSI	photosystem I		UV	ultraviolet (UV-A: 315–400 nm, UV-B: 280–315 nm, UV-C: <280 nm)
PSII	photosystem II		VI	vegetation index
Q_A, Q_B	quinone acceptors		VLFR	very low fluence rate response
QTL	quantitative trait loci		vpm	volume parts per million
R	red light		WUE	water use efficiency
RF	radiative forcing (W m^{-2})		WUE*	water use efficiency integrated over the life of a plant
RGR	relative growth rate (day^{-1})		ZTL	ZEITLUPE family proteins

1 A quantitative approach to plant–environment interactions

Contents

Progress in environmental plant physiology, as in other scientific disciplines, involves repeated cycles of observation or experimentation followed by data analysis and the construction and refining of hypotheses concerning the behaviour of the plant–environment system. This process is illustrated in very simplified form in Figure 1.1. At any stage the information and hypotheses may be qualitative or quantitative, and there may be more or less emphasis on the use of controlled experiments for providing the necessary data.

The initial stages of an investigation tend to provide a more qualitative description of system behaviour: much early ecological research, for example, was concerned with the description and classification of vegetation types, with a relatively small proportion of effort being devoted to understanding the underlying processes determining plant distribution. Further improvements in the understanding of any system, however, require a more quantitative approach based on a knowledge of the underlying mechanisms.

It is at this second level that this book is aimed: I have attempted to provide an introduction to environmental biophysics and to the physiology of plant responses that can be used to provide a quantitative basis for the study of ecological and agricultural problems. Further information on specific topics may be found in specialised texts referred to throughout the book.

For the convenience of the reader I have included at the end of this book an extensive set of appendices that outline the SI system of units as used throughout the text (Appendix 1); tabulate important physical properties of air, water and other materials (Appendices 2, 3, 4 and 5); give values for a range of useful physical constants (Appendix 6); outline the calculation of solar geometry and leaf boundary layer conductance (Appendices 7, 8); provide the derivation of Eq. (9.9) (Appendix 9); and provide answers to the questions posed at the end of each chapter (Appendix 10).

1.1 Modelling

Mathematical modelling provides a particularly powerful tool for the formulation of hypotheses and the quantitative description of plant growth and function. As modelling techniques are being used increasingly in all areas of plant science, and because they are used throughout this book, it is necessary to start with a simple introduction to mathematical modelling. Mathematical models provide simplified

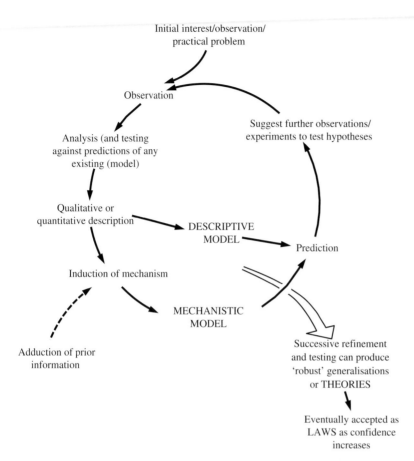

Figure 1.1 The role of models in scientific method.

descriptions of a real system (e.g. of a plant or of a process such as photosynthesis) by a series of equations that allow one to predict the development of the system over time or response of the system to external perturbations (e.g. changing temperature or water supply in the case of a plant). This predictive ability of mathematical models contrasts with other types of model, which include physical models (i.e. replicas such as model aeroplanes), conceptual models (verbal descriptions of systems) or pictorial models (illustrations and diagrams).

In the present context, a model is any representation of a real system, such as a plant, that can be used to simulate certain features of the more complex real system. For example light penetration in plant canopies is extremely complex (Chapter 2), but useful advances have been made by setting up the rather simple model where the actual canopy with

its individual leaves is replaced by a layer of homogeneous absorber. This model can be either an actual object (e.g. a solution of chlorophyll) whose properties can be studied empirically, or a mathematical abstraction that simulates those properties. A mathematical model, therefore, can constitute a concise formulation of a hypothesis (in this case that light penetration through a canopy is the same as through a homogeneous absorber). As such it can be readily used to generate testable predictions (e.g. of the effect of altering the angle of incident radiation). The results of these tests can then be used to refine, confirm or refute the initial hypothesis (Figure 1.1). In the present example, the accuracy with which the equations for a homogeneous absorber predict the penetration of light in a real canopy would be used to determine the adequacy of the model.

There are some areas of study where experiments are not possible; these include for example many studies of climate change, where it is not possible, or not ethical, to undertake the key experiments to test hypotheses. It would not be reasonable, for example, to attempt some of the major geo-engineering experiments that have been suggested as ways of combating global climate change. In such cases the only tools available are to model the system using our best understanding of the component processes and then to use predictions from the model to guide any management responses.

Because biological systems are so complex, one can rarely achieve complete mathematical descriptions of their behaviour. It is necessary, therefore, to make simplifying assumptions about the system behaviour and concerning the relevant components for inclusion in any study. This selection of variables is perhaps the most difficult task in the development of any mathematical model. An equally important step, however, in the development of useful models is their validation and testing. Some of the main advantages of mathematical models and the ways in which they can be used are summarised below.

1. They constitute precise statements of our hypotheses.
2. They are inherently testable.
3. They can 'explain' or describe a large number of separate observations in concise form.
4. They help to identify those areas where knowledge is lacking and further experiments or observations are required.
5. They can be used to predict system behaviour in untried combinations of conditions – this is particularly important in situations where experimentation is prohibitively expensive (large-scale field experiments) or inherently not possible (for example in the study of global climate or astronomical systems).
6. They can be used as management tools, for example in decision support systems for scheduling crops and management operations so as to maximise profit.

7. They can be used in diagnosis, for example in identifying crop diseases.

These last two applications have only been developed relatively recently with the advent of 'expert systems' and their use in 'decision support'. These attempt to encapsulate the knowledge of human experts into a set of rules that can be applied, among other things, to the diagnosis of disorders. A feature of this approach is that it can take account of uncertainty in any of the answers and weight them accordingly in coming to a conclusion. Although mathematical modelling has been widely used in the more physically based sciences, such as meteorology, it has, at least until recently, been underutilised in physiological and ecological studies.

Useful discussion of distinctions between different crop modelling approaches, their uses and misuses, and their relative advantages and disadvantages may be found in a series of reports arising from a symposium on crop simulation models (Boote *et al.*, 1996; Passioura, 1996; Sinclair & Seligman, 1996).

1.1.1 Types of model

Various types of mathematical model will be encountered throughout this book. These vary from relatively untested hypotheses (such as the models used in studies of 'optimum' stomatal behaviour – see Chapter 10), through partially tested models (i.e. theories), to well-tested models (i.e. laws – such as those dealing with well-known physical processes such as diffusion) where, given certain conditions, one can say with some certainty that a particular consequence will always ensue.

The majority of models can be separated into one of two groups: those that aim to improve our understanding of the physiology of crops and of their interactions with the environment; and those that aim to provide management advice to growers and farmers. The former approach requires a *scientific* or *mechanistic* approach, while the latter is usually based on more or less robust *empirical* relationships between plant responses and the main environmental variables. In empirical models no attempt is made to describe the mechanisms involved and minimal information is used a priori in their development. Mechanistic models, on the other hand, are developed using knowledge from previous work. A mechanistic model usually attempts to explain a phenomenon at a more detailed level of organisation. The choice of modelling approach depends on the particular research objective. Although both types of model may be used for predictions, the mechanistic approach probably has greater scope for generalised application and can lead to important advances in understanding. The large-scale dynamic crop simulation models discussed in Chapter 12 are examples where the model construction is usually based on mechanistic understanding, though they may involve a number of empirically fitted components. In the long run they are also likely to provide the more accurate predictions of system behaviour under a wide range of conditions. Nevertheless there is always a need for some care when attempting to use any model outside the range of conditions under which it was developed, and this should never be done for purely empirical models.

An example of the empirical approach is the use of relatively objective statistical regression techniques to describe and predict variation in crop yield in terms of weather variation from year to year (an example is given in Chapter 9 for hay yields in Iceland). This type of model can provide a useful description of the system by using routine techniques without the need for any physiological knowledge. The approach can, however, be made significantly more efficient with input of physiological knowledge to select the weather variates studied and to suggest appropriate forms for the relationships. It follows, therefore, that this approach is not completely distinct from the mechanistic approach, and indeed many empirical models tend to develop into more mechanistic ones as they are refined.

In addition to being empirical or mechanistic, models may be either *deterministic* or *stochastic*, and *dynamic* or *static*. In deterministic models, the output is defined once the inputs are known, while stochastic models incorporate an element of randomness as part of the model. Most models in physiological ecology are deterministic, mainly because of their greater simplicity and convenience, but some stochastic models have been used, for example to simulate random weather sequences, light penetration in canopies, spread of pathogens or ovule fertilisation (see Jones, 1981c).

Dynamic models include treatment of the time dependence of a process and are therefore particularly appropriate for simulating processes such as plant growth and yield production that integrate developmental and environmental changes over long periods. Many large-scale dynamic ecosystem and crop simulation models (see Chapter 12) have been developed, while the models used in climate modelling are also of this type. These complex computer simulations, however, can rarely be tested in the sense that physicists use the word, because of the large numbers of variables and assumptions used in their construction. They can, nevertheless, provide useful information on the sensitivity of crops or other systems to environmental variables.

Static models, in contrast to dynamic models, are used for steady-state systems or for simple descriptions of a final result. For example, many of the transport models described in this book consider only the steady state, so can be regarded as static models, as can those yield models where final yield is predicted by means of a simple regression equation between yield and certain weather variates during the season.

In addition to mathematical models, there are several examples where physical models can be used. For example, electrical circuits can be used to model diffusion and other transport processes, and with complex systems they may be easier to use than the corresponding mathematical abstractions.

Another class of models, which although not necessarily quantitative can contribute greatly to the development of understanding, are what might be termed conceptual models. These include general concepts such as the classification of plants into 'pessimistic' and 'optimistic' on the basis of their response to drought (Chapter 10), or more generally the development of what has been termed 'plant strategy theory' (Grime, 1979). This approach provides a valuable method for rationalising the vast array of evolutionary and ecological specialisations in plants and involves the assumption that there is a limited number of what have been called 'primary strategies' available to plants. In this case one type of specialisation for one type of existence and habitat condition tends to preclude success in other environments. The 'competitor–stress tolerator–ruderal' (CSR) model is a particularly powerful example of the application of this approach and can explain and predict stress responses very successfully (Grime, 1989). Although primarily a conceptual model it is amenable to quantitative analysis, since the equilibria between competition, stress and disturbance in vegetation may be readily quantified and represented graphically.

There has been particular interest in recent years in the development of 'virtual' plants in computers or what has become known as *functional–structural plant models* (FSPM; see Vos *et al.*, 2007). Functional–structural plant models combine architectural or structural models with process-based models (PBMs) to analyse problems where the three-dimensional spatial structure contributes essentially to any explanation of system behaviour. The process-based components include models of plant phenology, partitioning of carbohydrate between organs, and models of crop photosynthesis and growth. Effective simulation of crop photosynthesis, for example, requires not only the simple process-based relationship between photosynthesis and intercepted light, but also information on the geometric arrangement of leaves in space to allow calculation of the illumination on each leaf. Simulation of plant architecture was greatly stimulated by the development of L-systems by Lindenmayer (see Prusinkiewicz & Lindenmayer, 1990). This approach provided an iterative procedure for growing semi-realistic visualisations of plants based on a limited number of elements and sets of simple rules for their sequential addition. Functional–structural plant models are particularly useful for studies of phenomena such as competition between and within species where they offer opportunities to investigate interactions and feedbacks operating at the local (e.g. leaf) and global (e.g. canopy) scales. Similarly FSPMs provide tools for plant breeders to investigate plant ideotypes (see Chapter 12) that optimise photosynthesis and hence yield and growth.

In simple mathematical models a response is defined as a more or less complex function of a series of driving variables scaled by a number of fitted constants or *parameters*. In more complex models, however, the distinction between driving variables and responses may become blurred with complex feedbacks occurring and it may be the overall system response that is studied.

1.1.2 Fitting models and parameter estimation

Any observations that one makes need reducing to a simple framework, if they are to be of value in the development of a hypothesis or for predicting future behaviour of the system. Some form of curve-fitting or *calibration* procedure is necessary in order to derive a concise mathematical summary of the data. The summarising equation can be used to predict further values, as well as providing information to confirm or refute a theoretical model.

If, for example, a series of observations of photosynthetic rate at different irradiances has been made, a first step in the analysis might be to plot a graph of photosynthesis (on the ordinate, since it is likely to be the dependent variable) against irradiance (on the abscissa). One could then attempt to fit a line through the observations assuming that the points are particular examples of a general relationship. It is, however, unlikely that all points will fall on the line because some other factor (such as temperature) is also varying. The equation to the best-fit line (together with some description of the error) provides a useful mathematical summary of the observations. Multiple regression allows one to fit several x-variables at once.

For any particular set of points there may be an infinite number of equations that fit them and, although many may be far too complex for serious consideration, there may be several simple types that fit the observations satisfactorily. However, one must bear in mind Occam's razor (the principle that 'hypotheses must not be multiplied beyond necessity'). That is, when faced with the choice between two equally adequate models or hypotheses, one should take the simpler.

Useful introductions to the techniques for fitting curves may be found in appropriate statistical textbooks (e.g. Box et al., 2005; Sokal & Rohlf, 2012), while appropriate computer packages for performing the necessary analyses include GenStat (VSN International, Hemel Hempstead, UK), Minitab (Minitab Ltd., Coventry, UK) and SPSS (IBM, Armonk, New York).

Further details of modelling techniques and their application to plant physiological problems may be found in appropriate books and reviews (Rose & Charles-Edwards, 1981; Teh, 2006; Thornley & France, 2007; Vos et al., 2007). Specialised modelling platforms are available that facilitate the generation of FSPMs; these include the GroIMP platform (Kniemayer et al., 2007) and the GreenLab methodology (Kang & de Reffye, 2007) and their subsequent developments.

1.1.3 Validation of models

It is often argued that crop models and other models need to be *validated*, or *verified*, before use, but this is not strictly possible as they generally do not represent a single falsifiable hypothesis, but rather a collection of separate hypotheses. Therefore strict validation, as one might use for a physical law, is not possible – only their fit to a limited set of imprecise experimental data can be tested and quantified.

It is worth noting that there are two general components to the error in model predictions: the first arises from errors in estimating the necessary parameters during the model calibration process, and the second relates to those errors arising from a failure of the model itself (whether as a result of oversimplification or as a result of incorrect understanding of the process). The latter may frequently bias the result while the former largely affect the spread of predictions (Passioura, 1996).

1.2 Use of experiments

The observational and experimental phases of research are equally important as the modelling phase. Purely observational studies, of the type that has characterised much ecological research in the past, where one relies on natural variation in the environmental factors of interest, can be restrictive and difficult to interpret. This is because of the inherent complexity of the natural system and the tendency for correlations to occur between factors such as temperature and sunshine. For this reason it is usually necessary to be able to manipulate the various environmental factors independently in controlled experiments.

It is possible to perform experiments with either more or less interference with the natural environment (Table 1.1) and either more or less precise control of certain variables. In general there is a trade-off between good control of environment and minimal interference with the natural environment, with combinations nearer the top left in this table providing more precise, but not necessarily more accurate, information on plant response to individual factors. Field experiments may suffer from poor environmental control but, because the conditions are likely to be closer to natural than those in glasshouses or controlled environment chambers, any results obtained in the field are generally more likely to relate to the plant's behaviour in natural conditions. For this reason there has been increasing interest in recent years in conducting experiments where possible under conditions as near as possible to natural conditions; the best example of this has been the study of potential impacts of elevated atmospheric CO_2

Table 1.1 Differing degrees of experimental modification of root and aerial environments (modified from Evans, 1972). The symbol × indicates impractical combinations.

| | | Wholly artificial | | Aerial environment | | Wholly natural |
		Controlled environments	Daylit cabinets	Glasshouse compartments	Shelter, neutral screens	Field
Wholly artificial	Nutrient solution	✓	✓	✓	✓	×
	Inert base + nutrient solution	✓	✓	✓	✓	×
Root environment	Soil in pots	✓	✓	✓	✓	✓
	Field with fertilisation or irrigation	×	×	✓	✓	✓
	Transplant experiments	×	×	✓	✓	✓
Wholly natural	Natural	×	×	✓	✓	Observation only

concentrations using free-air carbon dioxide enrichment (FACE) systems (Long *et al.*, 2004). Unfortunately it is more difficult to modify air temperatures in natural systems, and although soil warming can be applied relatively easily, economic modification of the critical aerial environment while retaining natural variation in other environmental variables is more difficult (Aronson & McNulty, 2009).

In addition to varying degrees of modification of the physical environment, the results obtained depend to some degree on the biotic environment (competition, pathogens, etc.). Most of the studies described in this book can be classified as *autecological*, that is they consider the behaviour of one species in isolation. Although much valuable ecological information can be obtained from such studies, they can only go part way towards an 'explanation' of any ecological phenomenon. At least

in many important agricultural ecosystems, the most important type of biological competition is that from plants of the same species, while other biotic factors such as pests and diseases may be effectively controlled.

In practice, the choice of experimental system depends on the specific objectives. The more detailed a mechanistic explanation or model that is required for any phenomenon, the greater will be the need for controlled experiments. However, it then becomes important to minimise the interference with normal plant growth, or to become skilled in what Evans (1972) has called 'plant stalking'. It is usually necessary to carry out a range of types of experiment, from those in tightly controlled conditions to some in the field. The latter are necessary to confirm any model derived in controlled environments. Several examples of the dangers of relying too much on controlled environments will be encountered in what follows, with studies on antitranspirants (Section 10.3.7) providing a particularly good example of the problems. There is now extensive evidence that field-grown material behaves very differently from that grown in controlled environments and only in a few cases is the reason for this difference fully understood. One example is provided by the very different stomatal response to plant water potential shown by field- and controlled environment-grown plants (see Chapter 6).

The problems caused by the different 'coupling' of plants to their environment in controlled environments and in the field, and the consequences for studies of the control of evaporation by stomata, are discussed in Chapter 5. Plant morphology is also markedly different between these environments as a result of different irradiances and spectral distribution (see Chapter 8). Another example, as yet unresolved, is my own unpublished observation that certain genotypes of wheat showed marked leaf rolling in a dry season in the field. Attempts to investigate this phenomenon in controlled environments have not been successful, apparently because of differences in leaf morphology in the two environments.

Some features of the environment are easier to control than others. Field studies on plant nutrition and water status, for instance, have been conducted for over a hundred years, but it is only in the last 20 or so that any useful attempts have been made to control temperature in the field. But even now, temperature studies involve enclosing the plant canopy and altering a wide range of other factors at the same time. The use of reciprocal transplant experiments, such as those conducted at the Carnegie Institution's research gardens (Björkman et al., 1973) and those of Woodward and Pigot (1975) provide a powerful technique for studying the effect of the aerial environment without using controlled environments. Growing material at all sites in soil from the same source maximises the potential for studying the aerial environment in this type of experiment.

Whether, however, one should attempt to use controlled environments to mimic all the features of the natural environment is still controversial. Many elaborate systems have been set up to simulate the detailed daily trends of temperature and radiation (e.g. Rorison, 1981), but their advantages have not been demonstrated convincingly. The increased environmental complexity tends to negate the main advantage of a controlled environment.

2

Radiation

Contents

2.1 Introduction

There are four main ways in which radiation is important for plant life:

1. *Thermal effects.* Radiation is the major mode of energy exchange between plants and the aerial environment: solar radiation provides the main energy input to plants, with much of this energy being converted to heat and driving other radiation exchanges and processes such as transpiration, as well as being involved in determining tissue temperatures with consequences for rates of metabolic processes and the balance between them (see particularly Chapters 5 and 9).

2. *Photosynthesis.* Some of the solar radiation absorbed by plants is used to generate 'energy-rich' compounds that can drive energy-requiring (endergonic) biochemical reactions. These energy-rich compounds include those derived by dehydration (e.g. in the reaction of inorganic phosphate and ADP to form ATP) or reduction (e.g. of $NADP^+$ to NADPH). This harnessing of the energy in solar radiation in photosynthesis is characteristic of plants and provides the main input of free energy into the biosphere (see Chapter 7).

3. *Photomorphogenesis.* The amount, direction, timing and spectral distribution of shortwave radiation also plays an important role in the regulation of growth and development (see Chapter 8).

4. *Mutagenesis.* Very shortwave, highly energetic radiation, including the ultraviolet, as well as X- and γ-radiation, can have damaging effects on living cells, particularly affecting the structure of the genetic material and causing mutations.

This chapter introduces the basic principles of radiation physics that are needed for an understanding of environmental physiology, and describes various aspects of the radiation climate within plant stands. The latter part of the chapter goes on to describe in more detail the inversion of remote sensing observations of plants and plant canopies for the estimation of critical biophysical properties of plant stands such as leaf area index and leaf angle distribution.

The extreme complexity of the radiation climate means that inevitably much of the treatment is concerned with the derivation of useful simplifications or models that can be used by ecologists or crop scientists. More detailed discussion of aspects of radiation physics and the radiation climate may be found in Jones *et al.* (2003) and in texts such as those by Campbell and Norman (1998), Coulson (1975), Gates (1980), Liang (2004),

Monteith and Unsworth (2008), Jones and Vaughan (2010) and Rees (2001).

2.2 Radiation laws

2.2.1 Nature of radiation

Radiation has properties of both waves (e.g. it has a wavelength) and of particles (energy is transferred as discrete units termed quanta or photons). The wavelengths of radiation that are of primary concern in environmental plant physiology lie between about 300 nm and 100 μm and include some of the ultraviolet (UV), the photosynthetically active radiation (PAR, which is broadly similar to the visible) and the infrared (IR) (see Figure 2.1). The UV is conventionally split into the UV–C (<280 nm), the UV–B (280–315 nm) and the UV–A (315–400 nm), while the IR may be split into the near infrared

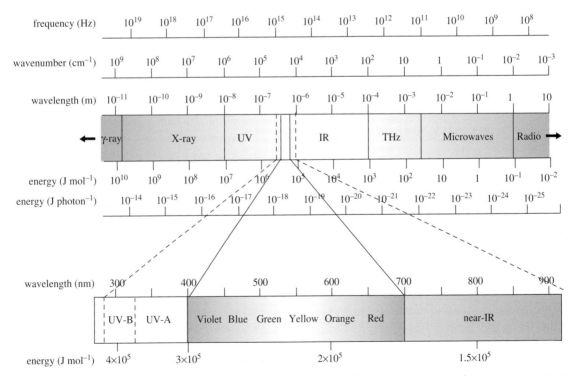

Figure 2.1 The electromagnetic spectrum: note that energy is given in J mol^{-1} (to convert to J photon^{-1} it is necessary to divide by Avogadro's number (6.022×10^{23}) so a photon of red light contains 2.84×10^{-19} J (from Jones & Vaughan, 2010).

(NIR, 700 nm–1 μm), the mid-infrared (MIR, 1–4 μm), the thermal infrared (TIR, 4–15 μm) and the far infrared (15–100 μm). The energy (E) of a photon is related to its wavelength (λ) or its frequency of oscillation (v) by:

$$E = hc/\lambda = hv \qquad (2.1)$$

where h is Planck's constant (approximately 6.63×10^{-34} J s), c is the speed of light (approximately 3×10^8 m s^{-1}) and v equals c/λ. An alternative measure of frequency that is commonly encountered is the wave number ($v = \lambda^{-1}$, usually expressed with units of cm^{-1}).

Using Eq. (2.1) it is apparent that shorter wavelength radiation has higher energy content than longer wavelengths. It can be calculated that a photon of red light (for example with $\lambda = 650$ nm) would have an energy, $E = 3.06 \times 10^{-19}$ J (i.e. $6.63 \times 10^{-34} \times 3 \times 10^8/(6.5 \times 10^{-7})$), while for a photon of blue light ($\lambda = 450$ nm), $E = 4.42 \times 10^{-19}$ J; that is 44% greater than the longer wavelength. It is often more convenient to refer to the energy in a mole (i.e. Avogadro's number, $= 6.022 \times 10^{23}$) of photons. Therefore for radiation of 650 nm, the energy per mole would be 1.84×10^5 J mol^{-1} (i.e. $3.06 \times 10^{-19} \times 6.022 \times 10^{23}$). The variation with wavelength of energy per mole is given in Figure 2.1. Although the Einstein is sometimes used to describe a mole of photons, it is an ambiguous term that is not part of the SI system, so should be avoided.

Some of the more useful measures of radiation for plant physiology and in remote sensing are summarised in Table 2.1. For further discussion see Bell and Rose (1981). The amount of radiant energy (in J) emitted, transmitted or received by a surface per unit time is called the *radiant flux* ($\mathbf{Q}_e$, which has units of power, i.e. J s^{-1} or W). The *net radiant flux* through a unit area of surface is the *radiant flux density* ($\mathbf{R}_e$, W m^{-2}). That component of the flux incident on a surface is termed *irradiance* ($\mathbf{I}_e$, W m^{-2}) while that emitted by a surface is termed the *radiant exitance* (or *emittance*) ($\mathbf{M}$, W m^{-2}). The subscript 'e' will be used where it is necessary to distinguish an energy flux from a flux of photons,

identified by subscript 'p' (see Table 2.1); for example a photon flux density ($\mathbf{R}_p$) has units mol m^{-2} s^{-1}.

Two other terms that occur in the plant physiological literature are also worth mentioning. The term radiation *intensity* is often used rather loosely as a synonym for flux density, but it is more correctly defined as the *radiance* (L), which is a flux per unit solid angle emitted from a source (and so has units of watts per steradian or W sr^{-1}) and its use should be restricted to that sense. Conversion from measurements of radiance to the corresponding radiant exitance for a surface involves multiplication by π because exitance refers to all radiation emitted into a hemisphere (π radians). The term *fluence rate* is also sometimes used as a synonym for flux density, though the two terms are not identical as fluence rate measures the flux per unit cross-sectional area incident from all directions on a spherical volume element, and therefore requires the use of a spherical detector. For some purposes, for example where one is concerned with light incident on a chloroplast, fluence rate may be the most appropriate measure of incident light, though the necessary spherical sensors are rare.

2.2.2 Black-body radiation

The process of radiation emission or absorption requires a corresponding change in the potential energy of the material. The wavelength of the radiation depends on the magnitude of the energy change (Eq. (2.1)), and thus on the possible transitions between available energy states. In atoms, the transitions occur between the limited number of allowed states for the orbital electrons thus giving rise to their characteristic spectra that consist of sharp lines at particular wavelengths that correspond to the particular electronic transitions. In molecules, transitions between different vibrational or rotational states are also allowed so that the spectra are much more complex and the vast numbers of possible transitions can give rise to broad bands of absorption or emission. A chemically complex body thus may have infinite possible energy transitions

Table 2.1 Terminology for radiation measurement. The various terms may be further qualified to include only radiation in certain wavebands (e.g. PAR, 'shortwave', etc.). Corresponding photometric terms based on luminous flux (lumen) are available, but these are weighted to the sensitivity of the human eye and should normally be avoided for plant work.

Term and symbol	Unit	Definition
Radiant energy (E)	$J = N\ m$	Total energy in the form of electromagnetic radiation radiated in all directions
Number of photons (n_p)	mol	Expressed as the number of moles of photons where one mole contains Avogadro's number (6.022×10^{23}) of photons
Radiant flux ($\mathbf{Q_e}$)	$J\ s^{-1} = W$	Radiant energy emitted or absorbed by a surface per unit time
Photon flux ($\mathbf{Q_p}$)	$mol\ s^{-1}$	Number of photons emitted or absorbed by a surface per unit time
Radiant flux (area)[a] density ($\mathbf{R_e}$)		Net radiant flux per unit area through a plane surface
Photon flux (area)[a] density ($\mathbf{R_p}$)	$mol\ m^{-2}\ s^{-1}$	Net photon flux per unit area through a plane surface
Irradiance[b] ($\mathbf{I}$)	$W\ m^{-2}$	Radiant flux incident on unit area of a plane surface
Spectral irradiance ($\mathbf{I_\lambda}$)	$W\ m^{-2}\ \mu m^{-1}$	Irradiance per unit wavelength (sometimes expressed per m)
Photon irradiance[c] ($\mathbf{I_p}$)	$mol\ m^{-2}\ s^{-1}$	Photon flux incident on unit area of a plane surface
Radiant exitance[d] ($\mathbf{M}$)	$W\ m^{-2}$	Total energy radiated in all directions in unit time per unit area (for isotropic radiation emitting into a hemisphere $\mathbf{M} = \pi L$)
Radiance (L)	$W\ m^{-2}\ sr^{-1}$	Radiant flux per unit area of surface per unit solid angle
Radiant intensity	$W\ sr^{-1}$	Radiant flux from a source into unit solid angle (Ω)
Fluence	$mol\ m^{-2}$	Number of photons across unit area (incident on a spherical surface)

[a] The qualifying term area is usually omitted.
[b, c] Also called energy (b) or photon (c) fluence rate, though more correctly fluence rate refers to the flux per unit cross-sectional area incident on a spherical volume element (therefore it requires a spherical detector). See Bell and Rose (1981) for detailed discussion of these terms.
[d] Also known as emittance.

covering all wavelengths, so that it should have a more or less continuous absorption or emission spectrum. An ideal material that is a perfect absorber or emitter of radiation at all wavelengths is termed a *black body*.

Because the energy transitions involved in emission and absorption of radiation are the same (but in opposite directions), it follows that absorption spectra correspond to emission spectra and that a good absorber at a particular wavelength will also be a good emitter at that wavelength. The *absorptivity* (or *absorptance*) (*a*) of a material is defined as the fraction of the incident radiation at a specified wavelength, or over a specified waveband, that is absorbed by a material. The appropriate wavelength, or wavelength interval, is usually indicated by a subscript. Similarly the *emissivity* (ε) at a particular wavelength is defined as the radiation emitted as a fraction of the maximum possible radiation at that wavelength that can be emitted by a body at that temperature. The maximum possible emittance is called the *black-body radiation*.

The energy distribution for emission from a true black body ($\varepsilon = 1$ at all wavelengths) is given by *Planck's distribution law*:

$$E_\lambda(\mathrm{d}\lambda) = \frac{2hc^2}{\lambda^5 (e^{hc/\lambda kT} - 1)} \mathrm{d}\lambda \qquad (2.2)$$

where $E_\lambda(\mathrm{d}\lambda)$ is the spectral radiant exitance (power emitted in all directions per unit area of a surface in a wavelength range $\mathrm{d}\lambda$) and k is Boltzmann's constant ($= 1.307 \times 10^{-23}$ J K^{-1}). Examples of the spectral distributions given by this law for black bodies at 6000 K (approximately equivalent to the sun) and 300 K (approximately equivalent to the Earth) are given in Figure 2.2.

The peak wavelength (λ_m, µm) of the Planck distribution for energy emission is a function of temperature and given by *Wien's law*:

$$\lambda_m = 2897.769/T \qquad (2.3)$$

As shown in Figure 2.2, λ_m for a black body at 6000 K is within the visible region of the spectrum at 483 nm, while λ_m for a radiator at a typical terrestrial temperature of 300 K is 9.65 µm (well into the IR). There is negligible overlap between the solar spectrum and the thermal radiation emitted by objects at normal terrestrial temperatures (Figure 2.2). It is convenient, therefore, to distinguish between shortwave radiation that mainly comprises radiation originating from the sun and falls between 0.3 µm and about 3 µm (or sometimes 4 µm), and longwave radiation (sometimes called terrestrial or thermal radiation) between 3 µm and 100 µm that is emitted by bodies at normal terrestrial temperatures. The symbols S and L will be used throughout as subscripts to **Q**, **R**, or **I** to distinguish shortwave and longwave radiation fluxes.

As well as being emitted or absorbed by a body, radiation can be reflected or transmitted. The *reflectivity* (or *reflectance*) (*ρ*) can be defined as the fraction of incident radiation of a specified wavelength that is reflected. Similarly the *transmissivity* (or *transmittance*) (*τ*) is the fraction

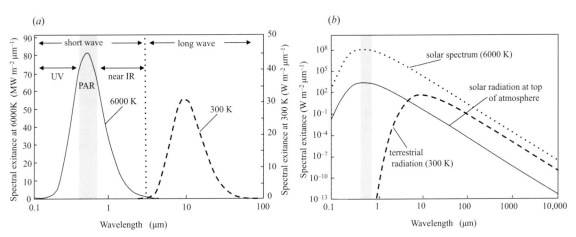

Figure 2.2 Spectral distribution of radiation emitted from black bodies at temperatures approximately equivalent to the sun (6000 K; giving shortwave radiation) and the Earth (300 K; giving longwave radiation): (*a*) expressed on linear scales and showing the UV, PAR and near IR regions within the shortwave, and the longwave region, (*b*) expressed on a logarithmic scale and also comparing the radiant exitance at the sun's surface and at the top of the atmosphere.

of incident radiation at a given wavelength that is transmitted by an object. The sum $\alpha + \rho + \tau$ at any wavelength equals 1, though the situation is complicated by the fact that a small proportion (usually <2–5%) of the radiation emitted from plant surfaces is not reflection but represents *fluorescence*, which is the rapid (c. 10^{-9} s) re-emission of absorbed radiative energy. Fluorescence always occurs at longer wavelengths (lower energy) than the exciting wavelengths with the difference in energy being lost as heat. As we shall see later, fluorescence from chlorophyll is widely used in monitoring photosynthetic function (Chapter 7), while fluorescence from other leaf pigments can be used in monitoring of responses to environmental stresses (Chapter 12).

When describing the absorption, reflection or transmission over a broad waveband, such as for solar radiation, the terms *absorption coefficient* (α_S), *reflection coefficient* (ρ_S) and *transmission coefficient* (τ_S) are used. Each of these represents an average absorptivity (or reflectivity or transmissivity) over the relevant wavelengths (0.3 to 3.0 μm for solar radiation) weighted by the distribution of radiation in the solar spectrum. The reflection coefficient for solar radiation of natural surfaces is sometimes called the *albedo*.

It is necessary to distinguish between radiation emitted by a body and that reflected. Snow, for example, is white because it reflects well in all the visible wavelengths but between 3 and 100 μm (the longwave region) it behaves almost like a black body being both a good absorber and radiator. In fact most natural objects (plants, soils, water) have emissivities close to one in the longwave part of the spectrum. Colour, however, depends on the wavelengths that are reflected in the visible, leaves being green because they reflect predominantly green light.

Stefan–Boltzmann law: The total amount of radiant energy emitted per unit area per unit time ($\mathbf{R}_e$) by a material is strongly dependent on temperature and is given by:

$$\mathbf{R}_e = \varepsilon \sigma T^4 \tag{2.4}$$

where σ is the Stefan–Boltzmann constant (= 5.67×10^{-8} W m^{-2} K^{-4}), and T is the Kelvin temperature. This equation gives the total area under the curves in Figure 2.2. For a black body $\varepsilon = 1$, but where $\varepsilon \neq 1$ the value of the exponent may not be exactly 4. Objects for which ε is constant at all wavelengths but less than unity are termed *grey bodies*.

2.2.3 Attenuation of radiation

Parallel monochromatic radiation passing through a homogeneous medium is attenuated according to *Beer's law:*

$$\mathbf{R}_\lambda = \mathbf{R}_{\lambda o}\, e^{-kx} \tag{2.5}$$

where $\mathbf{R}_{\lambda o}$ is the spectral flux density at the surface, x is the distance travelled in the medium and k is an extinction coefficient. This form of equation can be used to describe the attenuation of radiation in the atmosphere and in water, and approximates the light profile within leaves and in plant canopies. Although it is only precise for monochromatic radiation, it can be used with reasonable accuracy over any wavelength band where k is approximately constant.

It should be noted that similar terms are used with rather different meanings in chemistry texts concerned with absorption of radiation by solutions. In particular, it is common to take logarithms of Eq. (2.5) to give:

$$\ln(\mathbf{R}_\lambda / \mathbf{R}_{\lambda o}) = -kx \tag{2.6}$$

which, for convenience, is usually inverted to eliminate the negative sign, giving:

$$\ln(\mathbf{R}_{\lambda o}/\mathbf{R}_\lambda) = A = kx \tag{2.7}$$

where A is known as the *absorbance* (or *optical density*). The extinction coefficient in this equation is the product of the molar concentration and what is called a molar absorptivity, or a molar extinction coefficient.

2.2.4 Lambert's cosine law

The irradiance at a surface depends on its orientation relative to the radiant beam according to:

$$\mathbf{I} = \mathbf{I}_o \cos\theta = \mathbf{I}_o \sin\beta \tag{2.8}$$

where $\mathbf{I}$ is the flux density at the surface, $\mathbf{I}_o$ is the flux density normal to the beam, θ is the angle between

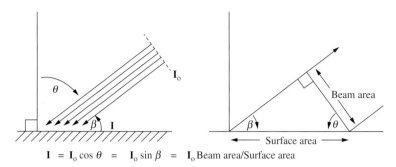

Figure 2.3 Lambert's cosine law.

$$I = I_o \cos \theta = I_o \sin \beta = I_o \, \text{Beam area/Surface area}$$

the beam and the normal to the surface (known as the *zenith angle*) and β is the complement of θ and is known as the beam *elevation* (see Figure 2.3). The more the beam is at an angle to the surface, the larger the area it is spread over, so the irradiance (which is expressed per unit area) decreases. The same relationship holds for the spatial distribution of the radiation emitted from a black body.

2.2.5 Spectral distribution and radiation units

There are several possible ways of expressing radiation that depend on the spectral response of the detector, each being of particular value for different purposes. For example, when the total energy exchange is of concern as in energy balance studies (see Chapters 5 and 9), measurements of total energy (made with detectors equally sensitive to all wavelengths), are most relevant with Q_e, R_e or I_e being summed over all wavelengths. In other cases, such as photosynthesis (Chapter 7) or morphogenesis (Chapter 8) only a limited range of wavelengths is effective, so it is usual to restrict measurements to the appropriate waveband. For photosynthesis, the photosynthetically active region (PAR) is usually defined as the waveband between 400 and 700 nm.

Most processes, however, including photosynthesis, are not equally responsive over the whole of the relevant waveband. The absorption spectra of some plant pigments important in photosynthesis and morphogenesis, together with an action spectrum describing the relative efficacy of different wavelengths for photosynthesis and the spectral sensitivity of the human eye are plotted in Figure 2.4.

An ideal detector would have the same spectral sensitivity as the process under consideration. An example where such detectors are used is in illumination measurement using photometric units (e.g. candela, lumen or lux), which are based on the spectral response of the human eye (Figure 2.4). Using such a detector, different sources with widely differing spectral distributions (e.g. sunlight or a fluorescent tube), would appear equally bright to a human observer when the luminous flux densities (in lux) are equal. In contrast, the radiant flux densities for the same sources in the same conditions would differ widely. Because the spectral sensitivity of the eye does not correspond to that of any plant process, it is best to avoid photometric units in plant studies.

For many biochemical processes, such as photosynthesis, the effect of radiation is more dependent on the total number of photons absorbed than on their energy. In these cases it is more appropriate to express radiation as a photon flux density (mol m^{-2} s^{-1}). The photon flux density within the PAR is commonly used in photosynthetic studies.

Table 2.2 shows how critically the relative values for different radiation measures depend on the light source and its spectral distribution. It can be seen that the photon irradiance gives the best measure of photosynthetic effectiveness of the different sources, being no more than 5% in error for any source, while irradiance in the PAR ($I_{e(PAR)}$) may be as much as 15% in error, incident luminous flux 53% in error and total shortwave irradiance (I_{eS}) even worse (particularly for quartz–iodine or tungsten lamps, which have a very high output in the near-IR).

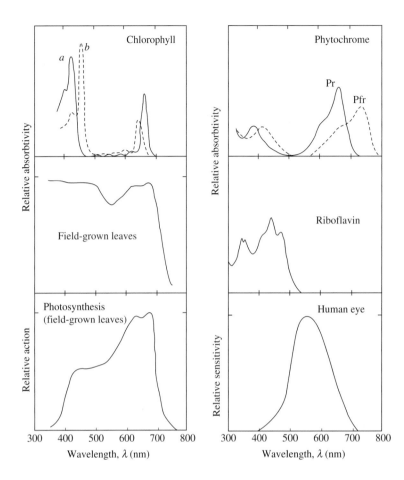

Figure 2.4 Absorption spectra of various plant pigments (chlorophylls *a* and *b*, the far-red light absorbing (Pfr) and red light absorbing (Pr) forms of phytochrome, and riboflavin – after Smith, 1981) and of field-grown leaves (data from McCree, 1972a). Also shown is the action spectrum (in terms of absorbed energy) for leaf photosynthesis and, for comparison, the sensitivity of the human eye.

2.3 Radiation measurement

Details of the instruments available for radiation measurement are well described elsewhere (Marshall & Woodward, 1986; Pearcy *et al.*, 1991; Vignola *et al.*, 2012). The purpose of this section is to emphasise some important considerations that influence the choice of method of radiation measurement. The commonest techniques involve either (i) photoelectric detectors (e.g. silicon cells, cadmium sulphide photoresistive cells, or selenium cells) or (ii) thermal detectors that measure the temperature difference between surfaces that differentially absorb the incident radiation (one at least is usually matt black and therefore a good absorber at all wavelengths). Photoelectric devices generally have a faster response than thermal detectors, but the latter tend to be preferable for energy balance studies and IR because of their wide wavelength response.

The usual way to measure leaf radiative properties including reflectance, transmittance and absorptance involves the use of an *integrating sphere* (Figure 2.5). An integrating sphere has a hollow spherical cavity lined with a highly reflective diffuse white reflector and has small inlet and exit ports so that all the entering radiation is eventually scattered to the detector port. For the measurement of absorptance of a leaf, the sample is placed inside the chamber and its effect on the radiation reaching the detector is a good measure of absorptance by the leaf. Similarly reflectance or transmittance of leaves can be measured by placing the leaf against one of the ports and either illuminating the sphere through the leaf

Table 2.2 A comparative table showing the values of different radiation measures for a range of different light sources, each providing an irradiance ($I_{e(PAR)}$) in the PAR (400–700 nm) of 100 W m^{-2}. In addition to giving the absolute values of the different measures for each light source (total energy in the shortwave, photon irradiance in the PAR and luminous flux), the table also gives the values for each measure expressed as a percentage of the value for sun + sky. Also shown in the second column are the relative photosynthetic efficiencies of the different sources, weighted for the average spectral response of photosynthesis. Data are from McCree (1972b), except for I_{eS} which was estimated for similar sources using a Kipp solarimeter with or without a Schott RG715 filter.

Light source	Irradiance in PAR ($I_{e(PAR)}$) (W m^{-2})	Relative photosynthetic effectiveness (%)	Corresponding I_{eS} (W m^{-2})	(%)	Corresponding $I_{p(PAR)}$ (µmol m^{-2} s^{-1})	(%)	Corresponding incident luminous flux (klux)	(%)
Sun + sky	100	100	200	100	457	100	25.2	100
Blue sky	100	93	152	76	425	93	20.4	81
Metal arc	100	114	210	105	498	109	36.0	143
Cool white fluorescent	100	101	146	73	466	102	38.8	154
Warm white fluorescent	100	97	n.a.	n.a.	457	100	36.5	145
Mercury vapour	100	102	208	(104)	471	(103)	27.7	(110)
Quartz-iodine	100	115	550	(275)	503	(110)	25.2	(100)

n.a. data not available

(transmission) or illuminating the leaf from inside the chamber (reflectance).

2.3.1 Spectral sensitivity

Because the spectral responses of various physiological processes differ, any measurement instrument must have an appropriate spectral sensitivity. This is particularly important when the spectral distribution of the radiation changes as it does, for example, when being filtered through vegetation (see below). The possible magnitude of the error that can occur when using inappropriate detectors is well illustrated by the data in Table 2.2.

It is possible to select combinations of photocells and filters (McPherson, 1969) that approximate the responses needed to measure either irradiance in the PAR or photon irradiance in the PAR. These are often called photosynthetic energy and quantum sensors, respectively. Many companies now produce commercial versions of sensors with tailored spectral sensitivities (see e.g. www.kippzonen.com,

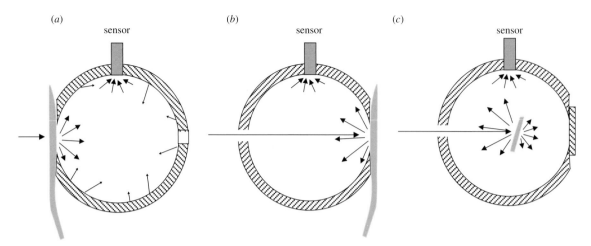

Figure 2.5 Use of an integrating sphere for the measurement of leaf optical properties. The inside of the sphere is coated with a highly reflective material (e.g. Spectralon or BaSO$_4$) so that all scattered radiation eventually enters the detector: (a) estimation of the directional-hemispherical leaf transmittance (τ), showing some of the primary and secondary reflections at the wall; (b) estimation of directional-hemispherical reflectance (ρ); and (c) estimation of directional leaf absorptance (α) (after Jones & Vaughan, 2010).

www.skyeinstruments.com, www.licor.com, www.delta-t.co.uk). For measurements of the total energy in the shortwave, near-IR or longwave, however, thermal detectors with appropriate filters are generally preferable. Instruments that measure radiant flux density are called radiometers; particular types include net radiometers that measure the difference between energy fluxes in opposite directions across a plane, and solarimeters (pyranometers) that measure total shortwave radiation incident upon a surface.

The importance of having a correct spectral response is particularly critical for studies of plant responses that have sharply defined action spectra (e.g. responses dependent on phytochrome absorption shown in Figure 2.4, and responses to UV radiation; see also Chapter 8). It follows that results obtained using broad-band UV sensors (e.g. www.apogeeinstruments.com, www.delta-t.co.uk, www.skyeinstruments.com) can be seriously misleading (Jones *et al.*, 2003). Approximate action spectra in the UV for various physiological processes and for a broad-band UV-B sensor are shown in Figure 2.6. The effectiveness of radiation at driving a

particular physiological response can be calculated as the integral over wavelength of action × irradiance (i.e. effect = $\int(\text{action}_\lambda \times \mathbf{I}_\lambda) \, d\lambda$). Using the action/sensitivity spectra shown in Figure 2.6 and the spectral irradiance of typical sunlight at the top of the atmosphere or at sea level, Jones *et al.* (2003) calculated that for one widely used action spectrum (plant action spectrum, PAS – Caldwell *et al.*, 1995) the true response would be 1.37-fold greater than that estimated by the SKU430 sensor for above atmosphere radiation but would be only 0.41 of the SKU430 value at sea level. The corresponding figures for DNA damage would be 3.45 and 0.05. It follows from these results that the most accurate studies can be made only where information on the full spectral irradiance is available, as may be obtained using a spectroradiometer with a cosine-corrected sensor.

2.3.2 Directional sensitivity

Conventionally radiation is usually measured on a horizontal surface, so that irradiance, for example, usually refers to the radiant flux incident per unit

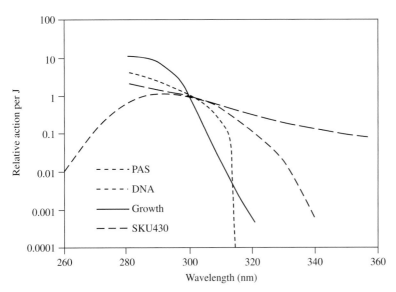

Figure 2.6 Estimated action spectra (normalised to 1 at 300 nm) for various physiological responses to UV-B radiation redrawn from Caldwell *et al.* (1995), where 'PAS' is Caldwell's generalised plant action spectrum, 'DNA' is Setlow's action spectrum for DNA damage in erythema (sunburn) and 'growth' is Steinmuller's response for seedling growth. These are plotted with the corresponding spectral response of a Skye SKU430 broadband UV-B sensor. (After Jones *et al.*, 2003.)

area of horizontal surface. In certain cases, however, other sensor orientations are needed. For example, in leaf photosynthesis studies the irradiance at the leaf surface may be more relevant than that on a horizontal surface.

For most purposes, it is important that the detector has a good cosine response. That is the irradiance measured for a beam at an angle θ to the normal should be proportional to $\cos \theta$ as expected from Lambert's law (Eq. (2.8)). Many sensors show a more severe reduction in sensitivity with grazing angles of incidence. The resulting errors can be particularly significant when the sun is low in the sky.

2.3.3 Averaging

Measurements of the radiation climate within canopies are often required as inputs to photosynthesis models or as a basis for estimation of canopy structure. The radiation climate within plant communities is extremely variable, both spatially and temporally. This heterogeneity can lead to large sampling errors so various techniques for minimising the problem have been suggested. One approach is to use large radiation sensors, such as the long tube solarimeters described by Szeicz *et al.* (1964) that average over a large area. Another is to use

multisensor arrays (e.g. Accupar 80, Decagon Devices Inc., Lincoln, Nebraska or the SunScan SS1 from Delta-T Devices, Burwell, Cambridge). These linear array instruments consist of a large number of small sensors on a linear probe that can be inserted in the canopy, and can be used to obtain a readout either of the average irradiance or of the fraction illuminated by high irradiance sunflecks. Alternatively one can move sensors through the crop taking readings at many positions using an instrument such as TRAC (Chen, 1996; Norman & Jarvis, 1974). It is necessary to remember, however, that although the mean irradiance may provide a general indication of light penetration into a canopy, more detailed information on the spatial and temporal variation is required for some purposes. For example photosynthesis does not respond linearly to light, with the light within sunflecks often not being used efficiently, so that the average irradiance could not be expected to predict CO_2 uptake well. The high irradiance within sunflecks may even cause photo-inhibitory damage to the choroplasts (see Chapter 7). It follows from this that for many purposes, especially in studies of photosynthesis, the average irradiance on a horizontal surface at any depth in the canopy is not the value of interest. Of more interest is the distribution of leaf area in different irradiance classes. It is necessary to integrate this over the whole diurnal cycle if one is, for example, to predict photosynthesis.

When irradiance measurements are being made in conjunction with photosynthesis measurements it may be appropriate to set the time constant of the light sensors similar to that of the photosynthetic system (seconds to minutes), but when relating irradiance to growth or phenology daily totals are likely to be adequate.

2.3.4 Estimation

Unfortunately radiation has only recently started to be measured at most meteorological sites so that it has in the past been frequently necessary to estimate solar or net radiation from other measurements, such as the duration in hours of bright sunshine (n). The conversion from sunshine duration to total solar or net radiation depends on site, type of cloud and time of year.

To a reasonable approximation, the average of I_S over periods of weeks or longer may be obtained from the Ångström equation:

$$I_S = I_A[a + (bn/N)] \tag{2.9}$$

where n/N is the fraction of the daylength with bright sunshine, I_A is the extraterrestrial irradiance on a horizontal surface appropriate for the time of year and latitude (see Appendix 7 for calculation of N and I_A), and a and b are constants that depend, for example, on site, atmospheric pollutants and time of year. Published values of these constants vary over a wide range (Martínez-Lozano et al., 1984; Shuttleworth, 1993). In England, for example, a is approximately 0.24 and b varies between 0.50 in winter and 0.55 in summer, with a small amount of additional variation around the country (Hough & Jones, 1997). Net radiation (see below) may also be estimated from sunshine duration (see Linacre, 1969 and Appendix 7).

The conversions between different radiation measures for different light sources can be obtained from Table 2.2. For example, this table shows that for average sun + sky light, $I_{PAR} \simeq 0.5\ I_S$. Some further approximations are presented in Appendix 7.

2.4 Radiation in natural environments

2.4.1 Shortwave radiation

The radiant flux density normal to the solar beam at the top of the atmosphere at the mean distance of the Earth from the sun is called the solar constant (I_{pA}) and is approximately 1370 W m^{-2}. The actual value of the flux density at the top of the atmosphere varies by about ±3.5% between July and January (when the sun is closest to the Earth). The radiation that actually reaches the Earth's surface is much modified in terms of quantity, spectral properties and angular distribution as a result of absorption or scattering by molecules in the atmosphere and by scattering or reflection from clouds and particulates. Reflection from and transmission through terrestrial objects such as leaves also modifies the radiation climate. A useful simplification when discussing solar radiation is to distinguish between the relatively unmodified parallel radiation in the direct beam – direct solar radiation ($I_{S(dir)}$), and the diffuse shortwave radiation ($I_{S(diff)}$) that includes reflected and scattered radiation from all portions of the sky. The sum of the direct and diffuse radiation incident on a horizontal surface is often called global radiation.

Scattering
Some of the direct solar beam is scattered by molecules or particles in the atmosphere. Rayleigh scattering is by molecules smaller than the wavelength of light, being most effective for shorter wavelengths (UV and blue) as it varies inversely to approximately the fourth power of λ; it is this phenomenon that makes the sky look blue because when you view from below you see mainly scattered radiation. The second main type of scattering is Mie scattering; this occurs mainly in the forward direction by larger particles such as

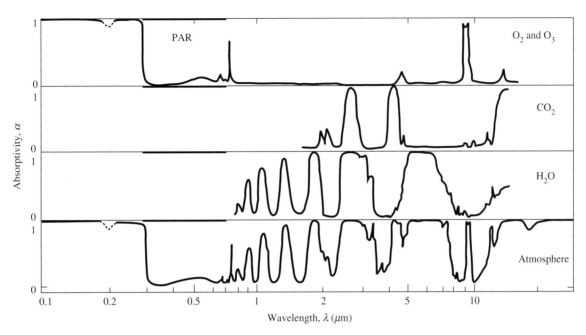

Figure 2.7 Atmospheric absorption spectra for the atmosphere and for some of its major components (after Fleagle & Businger, 1980). Details of transmission and absorption by the atmosphere and its components may be found on the HITRAN website (www.cfa.harvard.edu/HITRAN).

dust and water droplets and is relatively wavelength independent. *Non-selective scattering* occurs when the diameter of the particles is several times larger than the wavelengths involved; this includes the scattering of white light by suspended water droplets as in clouds.

Absorption in the atmosphere

Radiation absorption in the atmosphere is a function of the pathlength through the atmosphere and the content of absorbers, particularly water vapour. Absorption spectra for some of the more important absorbers, as well as the whole atmosphere, are given in Figure 2.7. This figure shows that there is a 'window' in the visible/PAR where the atmosphere is relatively transparent. Biologically important absorption bands include those in the UV (primarily ozone) that reduce the quantity of mutagenic UV radiation reaching the Earth's surface, and those in the IR (particularly water vapour and CO_2 – see Chapter 11).

Transmission through the atmosphere

Atmospheric transmission is a function of the optical air mass (m), which is the ratio of the mass of atmosphere traversed per unit cross-sectional area of the actual solar beam to that traversed for a site at sea level if the sun were overhead. The value of m therefore decreases with increasing altitude (in proportion to the atmospheric pressure, P) and with increasing solar elevation (β) or decreasing zenith angle (θ), approximately according to:

$$m \simeq (P/P_o)/\sin\beta = (P/P_o)/\cos\theta = (P/P_o)\,cosec\,\beta$$
(2.10)

where P_o is the atmospheric pressure at sea level (see Figure 2.8). If the transmissivity (τ) of the atmosphere when free of dust and clouds is defined as the fraction of incident solar radiation that is transmitted when $m = 1$, a combination of Beer's and Lambert's laws gives the direct irradiance on a horizontal surface ($\mathbf{I}_{S(dir)}$) as:

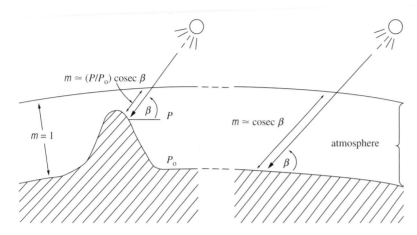

Figure 2.8 The calculation of optical airmass (m) in relation to solar elevation (β) and atmospheric pressure (P).

$$\mathbf{I}_{S(dir)} = \mathbf{I}_{pA}\tau^m \sin\beta \qquad (2.11)$$

where the first term represents attenuation by absorbers in the atmosphere and the second term represents the cosine correction (Eq. (2.8)) expressed in terms of the complement of the zenith angle. For 'clear sky' τ is commonly between about 0.55 and 0.70, though higher values can be obtained for very clear dry sky, particularly at higher elevations.

Diffuse radiation

Although equations such as (2.10) and (2.11) are useful for modelling the direct radiation environment, even on a clear day diffuse radiation contributes between about 10 and 30% of total solar irradiance. In England, for example, cloud cover is such that there is bright sunshine only about 34% of the time that the sun is above the horizon, with diffuse radiation contributing on average between 50 and 100% of the shortwave radiation depending on season and time of day (Figure 2.9). In drier climates, the proportion of diffuse radiation is much lower. For example in Yuma, Arizona, USA, the sunshine duration reaches 91% of the possible amount, so that the proportion of diffuse is correspondingly low. The proportion of diffuse radiation can be estimated with adequate precision for many purposes from the ratio of global radiation at the surface to the maximum potential (as indicated by the calculated irradiance at the top of the atmosphere, $\mathbf{I}_A$). This relationship is illustrated in Figure 2.10.

On a clear day, most diffuse radiation comes from the region near the sun (as a result of forward scattering). With an overcast sky, although it is significantly brighter near the zenith, it can be assumed, to a first approximation, that the sky is equally bright in all directions (called a uniform overcast sky). In practice, for an overcast sky, the radiance at the zenith is usually about 2.1 to 2.4 times the radiance of the sky at the horizon (see Monteith & Unsworth, 2008).

Although about 45% of the energy in the direct solar beam at the Earth's surface is in the photosynthetically active wavelengths, the average proportion of direct plus diffuse radiation in the PAR is approximately 50%. This is because diffuse radiation tends to be enriched in the visible wavelengths, particularly when the sun is low in the sky. This compensates for any depletion of the visible content of the direct beam at low solar elevations (as a result of Rayleigh scattering).

Radiation at different sites

The amount of radiation received on a horizontal surface at the top of the atmosphere ($\mathbf{I}_A$) is a simple function of latitude, time of day and time of year and can be calculated using readily available equations (see Appendix 7). Amounts received range from zero

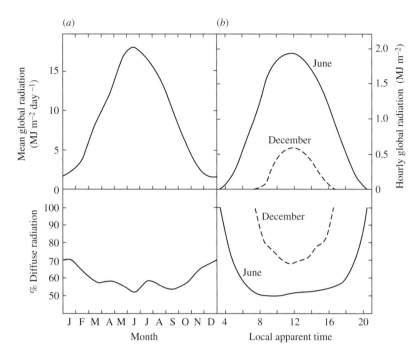

(a) *(b)*

Figure 2.9 Mean values (1959–1975) of shortwave irradiance (MJ m^{-2}) on a horizontal surface at Kew (51.5° N), and the proportion that is diffuse: *(a)* annual trend in mean daily global irradiance and *(b)* hourly values for June and December (data from Anon, 1980).

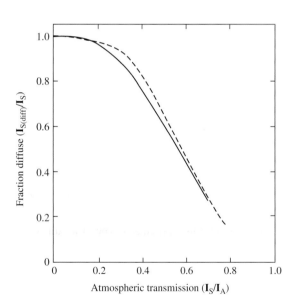

Figure 2.10 Average relationship between the fraction of total global radiation that is diffuse ($I_{S(diff)}/I_S$) and the atmospheric transmission (I_S/I_A), where I_A is the amount of radiation received on a horizontal surface at the top of the atmosphere (see Appendix 7), over a range of environments for hourly (– – – –) or daily data (——) (summarised from data collated by Spitters *et al.*, 1986).

in the polar regions in winter to more than 40 MJ m^{-2} day^{-1} in mid-summer for latitudes north of 40° N (June) or south of about 15° S (December). Similar geometrical considerations enable one to calculate the radiation receipts on slopes of different aspect or angle (Figure 2.11).

Annual and daily trends in actual total solar irradiance (on a horizontal surface) at a site in southeast England are shown in Figure 2.9. These values range from a mean of 1.8 MJ m^{-2} day^{-1} in December (approximately 25% of the extraterrestrial radiation), to 18 MJ m^{-2} day^{-1} in June (about 40% of the extraterrestrial radiation). In contrast, the mean daily radiation in June at China Lake, California, USA, is as high as 34.2 MJ m^{-2} day^{-1} and mean values greater than 25 MJ m^{-2} day^{-1} occur over most of the western and southern United States (Anon, 1964).

The maximum shortwave irradiance (at midday) commonly lies between about 800 and 1000 W m^{-2} during the growing season over much of the Earth's surface. For many purposes the diurnal trend in irradiance, at least on clear days, can be approximated by a sine curve:

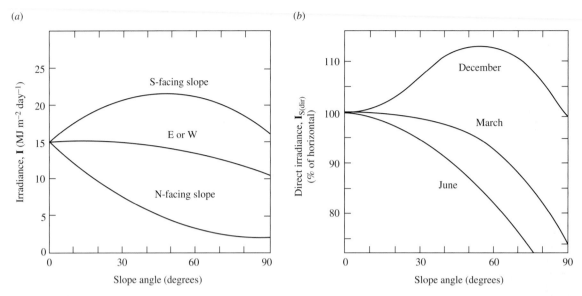

Figure 2.11 Influence of slope and aspect on direct solar radiation potentially received at a latitude of 53° 15′ N with $\tau = 0.7$ (after Pope & Lloyd, 1975). (a) Annual mean of potential daily direct irradiance (actual receipts will be considerably less than shown because of cloud). (b) Seasonal variation in potential irradiance relative to that on a horizontal surface, for east- or west-facing slopes.

$$\mathbf{I}_{St} = \mathbf{I}_{S(max)}\sin(\pi t/N) \qquad (2.12)$$

where $\mathbf{I}_{St}$ is the irradiance t hours after sunrise and N is the daylength in hours (see Appendix 7). Integration of Eq. (2.12) gives an estimate of the daily integral of irradiance as $(2N/\pi)\,\mathbf{I}_{S(max)}$. For example, with a maximum irradiance typical of a cloudless day in southern England in June of 900 W m^{-2} and a daylength of 16 h (5.8×10^4 s) this formula gives a daily insolation of 33 MJ m^{-2} compared with a measured maximum of about 30 MJ m^{-2}.

Variation of annual mean daily solar radiation over the Earth's land surface is shown in Figure 2.12. This illustrates that the greatest annual totals occur in the mid-latitudes.

2.4.2 Longwave and net radiation

In addition to shortwave radiation from the sun and sky, an important contribution to the radiation balance of plants is made by longwave (thermal) radiation. For example, the sky is an important source of longwave radiation: this is emitted by the gases (especially water vapour and CO_2) present in the lower atmosphere. However, as these atmospheric gases are not perfect emitters in the longwave (i.e. the emissivity of the atmosphere is <1; compare Figure 2.7), the apparent radiative temperature of the atmosphere (T_{sky}) is less than its actual temperature. Empirical results suggest that the sky behaves approximately as a black body with its temperature about 20 K below the temperature measured in a meteorological screen, so that the downward flux of longwave radiation ($\mathbf{R}_{Ld}$) under clear skies is given approximately by:

$$\mathbf{R}_{Ld} \simeq \sigma(T_a - 20)^4 \qquad (2.13)$$

where T_a is the air temperature (K). The difference between the downward longwave radiation and that emitted upwards by the surface ($\mathbf{R}_{Ld} \simeq \sigma\,T_a{}^4$) is the net longwave radiation ($\mathbf{R}_{Ln}$). With clear skies this net radiation ($\mathbf{R}_{Ld} - \mathbf{R}_{Lu}$) is approximately constant throughout the day and throughout the year at a net loss of 100 W m^{-2}, though, as shown in Figure 2.13, the average value taking

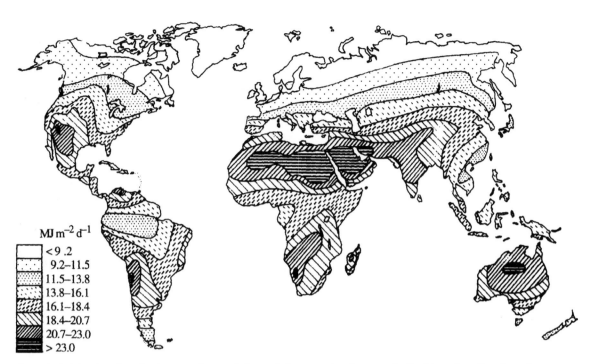

Figure 2.12 Historic values for the annual mean daily total shortwave irradiance at the Earth's surface (data from Anon,1964; Landsberg, 1961).

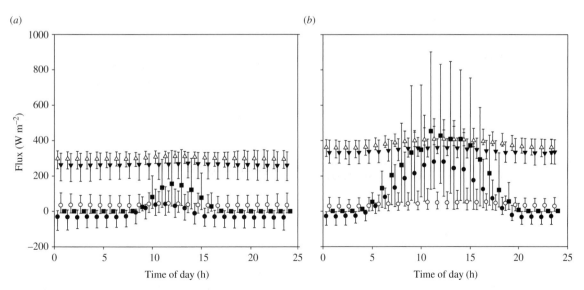

Figure 2.13 Diurnal trends in shortwave radiation (I_S; ■), net radiation (R_n; ●), downwelling longwave (R_{Ld}; ▼), upwelling longwave (R_{Lu}; △), and net longwave radiation (R_{nL}; ○) for grassland in the Netherlands (CABAUW experiment as described by Beljaars & Bosveld, 1997) for (a) January 1987 and (b) June 1987. The figure shows the mean values and ranges of the monthly data (from Jones & Vaughan, 2010).

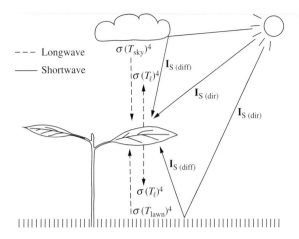

Longwave (dashed), Shortwave (solid)

$\sigma(T_{sky})^4$

$\sigma(T_\ell)^4$

$I_{S\,(diff)}$

$I_{S\,(dir)}$

$I_{S\,(dir)}$

$I_{S\,(diff)}$

$\sigma(T_\ell)^4$

$\sigma(T_{lawn})^4$

Figure 2.14 Schematic diagram of the longwave and shortwave radiation exchanges between a leaf and its environment.

account of cloudy days is substantially less than this.

The presence of clouds increases the downward flux of longwave radiation since clouds are more effective emitters and have a mean radiative temperature that averages only 2 K below the mean screen temperature. This corresponds to a net upward longwave flux of only 9 W m^{-2} in cloudy conditions (Figure 2.13). More precise methods for estimating longwave radiation fluxes are outlined by Sellers (1965) and Gates (1980).

The net flux of all radiation (shortwave and longwave) across unit area of a plane is called the *net radiation* ($\mathbf{R}_n$). Alternatively the net radiation absorbed by an object is the sum of all incoming radiation fluxes minus all outgoing radiation (see Figure 2.14). Incoming radiation includes all incident direct and diffuse solar radiation and that reflected from the surroundings, as well as incident longwave emitted by the sky and the surroundings. Radiation losses include thermal radiation emitted, as well as any incident radiation that is reflected or transmitted by the object.

Therefore, for a lawn (or a crop) that may be treated as a horizontal plane surface that does not transmit radiation, one can write:

$$\begin{aligned}\mathbf{R}_n &= \mathbf{I}_S + \varepsilon\mathbf{I}_{Ld} - \mathbf{I}_S\rho_{S(lawn)} - \varepsilon\sigma(T_{lawn})^4\\ &\simeq \mathbf{I}_S(\alpha_{S(lawn)}) + \mathbf{I}_{Ld} - \sigma(T_{lawn})^4\end{aligned} \quad (2.14)$$

where α_S and ρ_S, respectively, are the absorption coefficient and the reflection coefficient (of the lawn). All fluxes are referred to a horizontal surface and ε is usually assumed to equal 1.

For an object with two sides, such as an isolated horizontal leaf exposed above the lawn:

$$\begin{aligned}\mathbf{R}_n &\simeq \left(\mathbf{I}_S + \mathbf{I}_S\rho_{S(lawn)}\right)\alpha_{S(leaf)} + \mathbf{I}_{Ld}\\ &\quad + \sigma(T_{lawn})^4 - 2\sigma(T_{leaf})^4\end{aligned} \quad (2.15)$$

Note the additional terms for solar radiation reflected by the lawn and for longwave radiation from the lawn, and the factor of 2 in the term for longwave losses. Note also that all radiation fluxes are expressed per unit projected area. For a lawn, or any extended area of vegetation, that equals the ground area, but for a leaf it is the area of *one* side. This convention will be used throughout the book for treatment of heat, mass and radiation transfer.

The use of Eqs. (2.14) and (2.15) for one- and two-sided surfaces is best illustrated by the examples in Table 2.3, where the net radiation balance of a lawn and of an isolated leaf are calculated for different weather conditions. As might be expected $\mathbf{R}_n$ is lower on cloudy than on sunny days for both surfaces. At night, when there are no shortwave exchanges, $\mathbf{R}_n$ is negative. On a cloudy night, however, net radiation would be close to zero. The net radiation for the leaf is always smaller than that for the lawn. Note that the precise values for the different fluxes depend on the actual temperatures. These themselves depend on the complete heat balance including convection and evaporation (see Chapter 5).

For a non-horizontal leaf, calculations are more complex in that one has to take account of its angle relative to the solar beam, and of the different distribution of diffuse radiation from the ground and the sky. Net radiation is measured using net radiometers that detect the difference between upward and downward radiation fluxes. For a surface such as a lawn, $\mathbf{R}_n$, measured above the surface, equals the net radiation

Table 2.3 Comparison of net radiation balance for a lawn and an isolated horizontal leaf exposed above the lawn for different conditions. All fluxes are expressed per unit projected area (W m^{-2}) and it is assumed that $\alpha_{lawn} = 0.77$, $\alpha_{leaf} = 0.5$, $\rho_{lawn} = 0.23$ and that the emissivities of lawn and leaf are unity.

Assumed conditions	Sunny day	Cloudy day	Clear night
Shortwave irradiance on a horizontal surface (I_S, W m^{-2})	900	250	0
T_a (K)	293	291	283
T_{sky} (K)	273	289	263
T_{lawn} (K)	297	288	279
T_{leaf} (K)	297	288	277
Lawn (Eq. (2.14))			
$R_{S(absorbed)} = I_S(\alpha_{S(lawn)})$	693	193	0
$R_{L(absorbed)} = \sigma(T_{sky})^4$	309	389	266
$R_{L(emitted)} = \sigma(T_{lawn})^4$	433	383	337
R_n	569	199	-71
Leaf (Eq. (2.15))			
$R_{S(absorbed)} = I_S(\alpha_{S(leaf)})(1 + \rho_{S(lawn)})$	554	154	0
$R_{L(absorbed)} = \sigma(T_{sky})^4 + \sigma(T_{lawn})^4$	742	772	604
$R_{L(emitted)} = 2\sigma(T_{leaf})^4$	866	766	656
R_n	430	160	-52

absorbed by the surface. The net radiation absorbed by a layer of leaves in a plant canopy, however, must be measured as the difference between R_n above and below that layer. For more complex objects the net radiation absorbed can be obtained from net radiometers arranged around the object so as to integrate over all directions the net fluxes perpendicular to the surface.

2.5 Radiation in plant communities

2.5.1 Radiative properties of plants

Absorption, reflection and transmission spectra for typical plant leaves are shown in Figure 2.15. Although this general picture is true for most species, details vary with thickness, age, water content, surface morphology and orientation. The main features of the spectra are the high absorptance in most of the PAR except the green (hence the green colour of leaves) and the low absorptance in the near-IR. Leaves are good absorbers in the far-IR so that they behave approximately as black bodies in the longwave, with ε being between 0.94 and 0.99 for most species (Idso et al., 1969; Jones & Vaughan, 2010). A useful model for simulating the radiative transfer properties of plant leaves is PROSPECT (Jacquemoud & Baret, 1990); where one has spectral information on leaf reflectance this model can be inverted to give information on leaf biochemistry (Jacquemoud et al., 1996).

Table 2.4 presents some typical values for absorption and reflection coefficients of leaves,

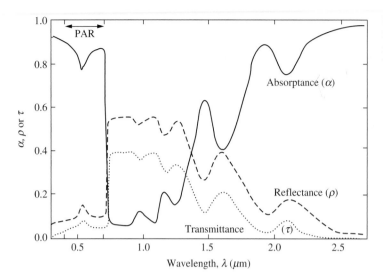

Figure 2.15 Absorption, transmission and reflection spectra for 'typical' leaves (the mean of various sources – see Jones & Vaughan, 2010).

vegetation and other natural surfaces. These results show that the solar reflection coefficient for leaves of temperate crop species is usually close to 0.30 with relatively little variation between species. Reflectance can be higher for white pubescent leaves, waxy leaves or for leaves with low moisture content. An example of the complex structure of leaf hairs on the lower surface of a heavily pubescent leaf is shown in Figure 2.16. The reflectance is strongly dependent on leaf moisture content, with changes in reflectance and absorption being greatest in the infrared and especially in the water absorption bands around 1200, 1450 and 1930 nm (Figure 2.17). The reflectivity at 550 nm (approximately proportional to ρ_S) for *Atriplex hymenelytra* leaves, for example, can vary between 0.35 in winter and 0.6 in summer as a function of leaf moisture content and leaf pubescence (Mooney *et al.*, 1977). Not only does the presence of leaf hairs tend to increase the reflectance of that surface, but there is also evidence that hairs on the lower surface of a leaf can increase reflection from the upper surface (Eller, 1977); this effect is particularly strong in the infrared. Reflectance in the PAR tends to be rather less than in the total shortwave.

As shown in Table 2.4, solar absorptance for many leaves is about 0.5, though this is quite variable, reaching as high as 0.88 for conifer needles. The

absorptance in the PAR is rather higher, averaging about 0.85. Absorptance is strongly dependent on leaf water content and pubescence, largely as a result of their effects on reflection. The effect of leaf hairs on α_{PAR} in closely related *Encelia* species is illustrated in Figure 2.18, where the extremely pubescent desert species *E. farinosa* has a reflectance in the PAR 50% greater than that of the glabrous coastal species *E. californica*. Leaf thickness is another factor determining absorptance, because thick leaves (e.g. succulent species) have a very low transmittance.

The reflectance of plant canopies tends to be rather lower than that of the component single leaves because the multiple reflections between adjacent leaves and between leaves and stems leads to trapping of radiation (see Campbell & Norman, 1998; Jones & Vaughan, 2010). This effect is particularly marked for taller crops such as forests where ρ_S may be as low as 0.10, while ρ_S for some dense short canopies may approach that of individual leaves (some typical values are given in Table 2.4). As with other surfaces, the crop reflection coefficient depends on solar elevation, increasing by a factor of about two as the solar elevation decreases from 60° to 10° (Ross, 1975).

Table 2.4 emphasises the different reflection and absorption behaviour in the PAR and in the near-IR. Although only about 50% of the

Table 2.4 Reflection (ρ_S) and absorption (α_S) coefficients for leaves, vegetation and other surfaces. All values are for typical shortwave radiation unless otherwise stated (Gates, 1980; Linacre, 1969; Monteith & Unsworth, 2008; Stanhill, 1981). Further useful information for specific surfaces may be found in various websites including the ASTER spectral library (http://speclib.jpl.nasa.gov), the LOPEX93 (http://ies.jrc.ec.europa.eu/index.php?page=data-portals) and the ORNL DAAC at Oak Ridge (http://daac.ornl.gov/holdings.html) databases and the USGS Digital Spectral Library 06 (http://speclab.cr.usgs.gov/spectral.lib06).

	ρ_S (%)	α_S (%)
Single leaves		
Crop species	29–33	40–60
Deciduous broad leaves (low sun)	26–32	34–44
Deciduous broad leaves (high sun)	20–26	48–56
Artemisia sp. (white pubescent, high sun)	39	55
Verbascum sp. (white pubescent, high sun)	36	52
Conifers	12	88
Typical mean values for total shortwave (ρ_S, α_S)	~30	~50
Typical mean values for PAR (ρ_{PAR}, α_{PAR})	~9	~85
Vegetation		
Grass	24	
Crops	15–26	
Forests	12–18	
Typical mean values for total shortwave (ρ_S)	~20	
Typical mean values for PAR (ρ_{PAR})	~5	
Other surfaces		
Snow	75–95	
Wet soil	9 ± 4	
Dry soil	19 ± 6	
Water	5–>20	

incident shortwave radiation is in the PAR, about 80 to 85% of all absorbed shortwave radiation is in this region. Therefore spectral properties in the visible dominate the leaf radiation balance. Transmission of different wavelengths in plant canopies is discussed in the next section, while the implications of leaf spectral properties to the energy balance and thermal regime of plant leaves are discussed in Chapter 9.

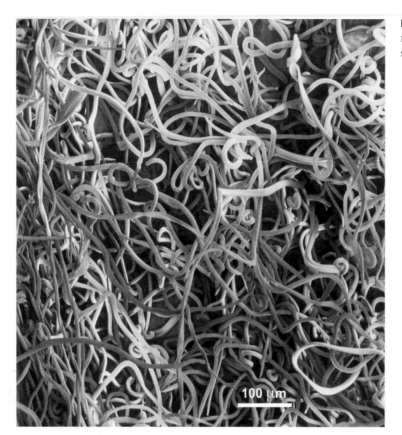

Figure 2.16 Scanning electron micrograph of leaf hairs on the abaxial surface of *Zizyphus mauritiana*.

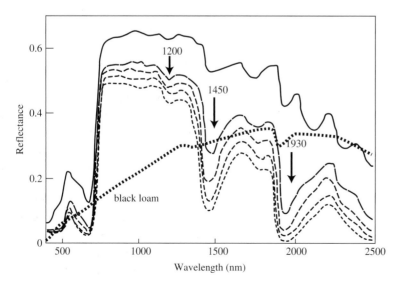

Figure 2.17 Reflectance spectra of *Magnolia grandiflora* leaves at relative water contents ranging from 5% (solid line) through 25%, 50%, 75% to 100% (short dashes) showing the increasing dips in reflectance with increasing water content at the water absorption peaks centred on 1200, 1450 and 1930 nm (data from Carter, 1991). The dotted line refers to black loam (from the ASTER spectral library; Baldridge *et al.*, 2009).

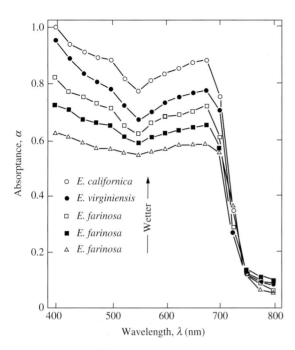

Figure 2.18 Absorption spectra of *Encelia californica,*
E. virginiensis and *E. farinosa* along an aridity gradient
during April (after Ehleringer, 1980).

2.6 Radiation distribution within plant canopies

A precise description of the pattern of radiation
distribution within any plant canopy is difficult
because of the necessity of taking account of detailed
canopy architecture, the angular distribution of the
incident radiation and the spectral properties of the
leaves. There are, however, useful simplifications that
give adequate precision for many purposes, including
the modelling of photosynthesis and productivity
(Campbell & Norman, 1998; Goudriaan, 1986; Krul,
1993; Ross, 1975; Wang & Jarvis, 1990). Only some of
the most useful approximations are reviewed below.

2.6.1 Horizontal leaves

A common simplifying assumption is that plant
canopies are horizontally uniform so that radiation
is constant in any horizontal layer, only changing
with height. In general the average irradiance at

any level tends to decrease exponentially with
increasing depth in a way similar to that predicted
by Beer's law (Eq. (2.5)) assuming that the canopy
is a homogeneous absorber.

One simple derivation assumes that the canopy
consists of randomly arranged horizontal leaves with
a canopy *leaf area index* (L, that is the 'one-sided'
leaf area per unit area of ground; this is often given
the symbol LAI)[1] that can be divided into a number
of horizontal layers each containing equal areas and
within which no leaves overlap. If one considers a
layer of canopy containing a small leaf area index, dL,
this will intercept an amount of radiation equal to
$\mathbf{I}_o\,dL$, where $\mathbf{I}_o$, is the irradiance at the top of the layer.
If the leaves are opaque, the change in irradiance ($d\mathbf{I}$)
on passing through this layer of canopy is equal to
$-\mathbf{I}_o\,dL$. Integration downwards through a total leaf
area index L, gives the average irradiance on a
horizontal surface below that leaf area index as:

$$\mathbf{I} = \mathbf{I}_o e^{-L} \qquad (2.16)$$

Although this is similar to Beer's law, in this case the
irradiance at any level is in fact the average of some
areas with unattenuated light (sunflecks) and some
completely shaded areas. This pattern of radiation
attenuation is illustrated for a hypothetical
horizontal-leaved canopy in Figure 2.19(*a*).

It follows from Eq. (2.16) that, for opaque leaves, $\mathbf{I}/\mathbf{I}_o$
$(= e^{-L})$ is the fraction of the area on a plane below a
leaf area index of L that is sunlit. Conversely, the sunlit
leaf area index (L_{sunlit}) in the canopy is given by:

$$L_{sunlit} = 1 - \mathbf{I}/\mathbf{I}_o = 1 - e^{-L} \qquad (2.17)$$

This function approaches 1 at high leaf area indices,
and indicates that the maximum leaf area index

---

[1] For thin flat leaves, this definition of leaf area index is
    equivalent to the projected leaf area per unit ground area,
    where the projected area is that defined by a beam normal to
    the largest leaf dimensions and projected on to a plane normal
    to the beam. This provides a convenient convention to treat
    other leaf shapes such as the cylindrical or hemi-cylindrical
    leaves as in conifers, though Campbell and Norman (1998)
    have argued that it can be better to use a *hemi-surface area
    index* (*HSAI*) that is half the total surface area of leaves per unit
    ground area. The definitions are equivalent for flat leaves.

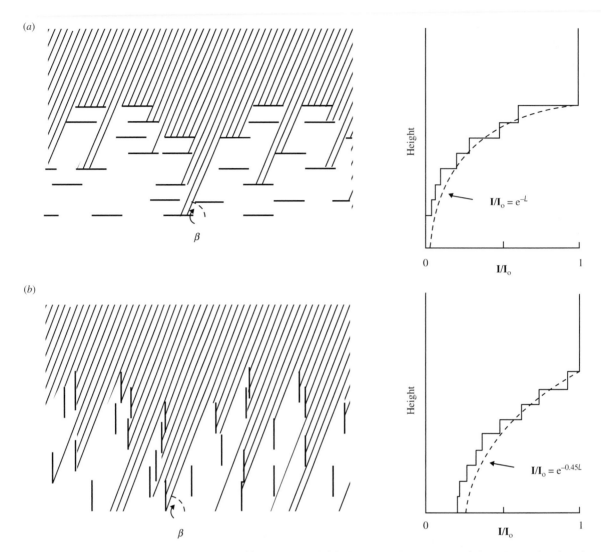

Figure 2.19 Light penetration at a solar elevation (β) of 66° through (a) a horizontal-leaved canopy (where k in Eq. (2.18) = 1) and (b) a vertical-leaved canopy (where $k = \cot 66° = 0.45$). This illustrates the more rapid extinction in the horizontal leaved canopy, though the converse would be true for $\beta < 45°$.

that can be in full sunlight, in a horizontal-leaved canopy, is equal to 1.

2.6.2 Other leaf angle distributions

The simple model for horizontal-leaved canopies may be extended to other patterns of leaf angle distribution. The general principle is to project the shadow of the leaves on to a horizontal plane,

and to use this area as the exponent in Eq. (2.16). If k is the ratio of shadow area to actual leaf area, Eq. (2.16) becomes:

$$I = I_0 e^{-kL} \tag{2.18}$$

where k is an extinction coefficient. The situation for vertical leaves oriented towards the sun is illustrated in Figure 2.19(b). In this case the ratio of shadow area to actual area is equal to $\cot \beta$ (where β is the solar elevation). In contrast to the horizontal-leaved situation, the radiation profile is dependent on β. With high sun, k is less than 1 and light penetrates further into the canopy than with a corresponding horizontal-leaved canopy. With low sun, however, the converse is true.

Using an argument similar to that above, it is possible to estimate the value of L_{sunlit}. In this case the maximum value of L_{sunlit} is equal to $1/k$, so:

$$L_{sunlit} = (1 - e^{-kL})/k \qquad (2.19)$$

though the actual irradiance per unit illuminated leaf area will be given by $k\mathbf{I}_o$, that is lower for high sun and vertical leaves. This has important implications for both leaf temperature regulation (Chapter 9) and canopy photosynthesis (Chapter 7).

In most real canopies leaves assume a range of orientations, with some canopies (*planophile*) having predominantly, but not exclusively, horizontal leaves, others (*erectophile*) having predominantly vertically oriented leaves, but many other distributions are also found. For example *plagiophile* (or plagiotropic) canopies have leaf angles clustered round an oblique angle. Some typical distribution functions of leaf inclination (the angle between the leaf and the horizontal) for different canopies are illustrated in Figure 2.20. It is often possible to approximate actual leaf angle distribution functions by simplified geometrical treatments that can be used to model light penetration through canopies.

One erectophile distribution that is of particular interest is the spherical leaf distribution. In this it is assumed that the leaves have an equal probability of any orientation so that they could be thought of as being capable of being rearranged on the surface of a sphere. Although on this assumption leaves have an equal probability of any azimuth (that is direction N, S, E, W, etc.), erect leaves (all around the equator) are more common than horizontal ones (only at the top or bottom). It follows that the extinction coefficient is related to the ratio of the projection of a sphere on to a horizontal plane to its surface area, which is $\pi r^2/(4\pi r^2 \sin \beta)$ ($= 0.25 \operatorname{cosec} \beta$), but because either side of the leaf can intercept radiation, the appropriate value for k is twice this value or $0.5 \operatorname{cosec} \beta$.

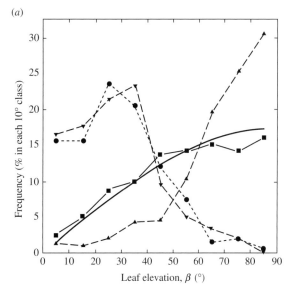

(a)

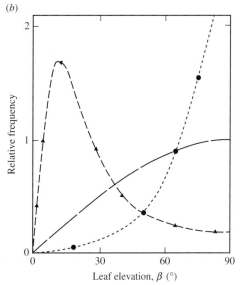

(b)

Figure 2.20 (a) Some examples of leaf inclination distribution functions (where β is the leaf elevation above the horizontal) for different canopies: Shamouti orange (—■—), perennial ryegrass (May 6 — — —▲— — —, August 21 — — —▼— — —) and flowering white clover (- - - -●- - - -). (Data from de Wit, 1965; Cohen & Fuchs, 1987). (b) Theoretical leaf inclination distribution functions, for ellipsoidal distributions with $x = 0.5$ (- - - - - -), $x = 1.0$ (spherical, ———) and $x = 3.0$ (— — — — —).

A more general function that is continuous over a range of leaf angles (like the spherical distribution) but can accommodate canopies with tendencies towards the erect or to the horizontal, as necessary, is the ellipsoidal distribution. A single parameter, x (the ratio of the horizontal semi-axis of the ellipsoid to the vertical semi-axis) is used to describe the shape of the distribution. The spherical distribution is a special case of the ellipsoidal distribution when $x = 1$. The leaf inclination distributions for several models are illustrated in Figure 2.20(b). Unfortunately the ellipsoidal function does not always closely approximate observed leaf angle distributions so several other distributions have been used. Among these are conical distributions that assume an average leaf angle (Monteith & Unsworth, 2008) and two-parameter beta distributions (Goel & Strebel, 1984), which are both rather better for plagiotropic canopies where leaf angles are distributed round 45°.

Extinction coefficients are more generally useful in modelling light climates than are leaf inclination distributions. The variation of extinction coefficients with angle of elevation for different canopy models are illustrated in Figure 2.21, and the functions tabulated in Table 2.5. Further information on radiation modelling in canopies may be found in Monteith and Unsworth (2008) and Campbell and Norman (1998) with more advanced discussion in Ross (1981) and Liang (2004).

2.6.3 Use in practice

In practice many real canopies cannot be approximated by one of these simple geometrical models and, furthermore, a proportion of intercepted radiation is either transmitted by the leaves or scattered downwards through the canopy. Therefore it is convenient for many purposes to replace the geometrically derived value of the extinction coefficient k by an empirically determined value. Equation (2.18), with an empirical k, was originally proposed by Monsi and Saeki (1953) in their classical work. Observed values of k have been found to vary from about 0.3 to 1.5 depending on species, stand type

Table 2.5 Dependence of the extinction coefficient (k) on beam elevation (β) for different leaf angle distribution functions that are commonly used in modelling canopy light climates, together with corresponding leaf angle distribution functions.

Leaf angle distribution	Extinction coefficient (k)
Horizontal	$k = 1$
Vertical	$k = (2 \cot \beta)/\pi$
Spherical	$k = 1/(2 \sin \beta)$
Ellipsoidal[a]	$k = (x^2 + \cot^2 \beta)^{1/2}/(Ax)$
Diaheliotropic	$k = 1/\sin \beta$

[a] where x is the ratio of the horizontal to the vertical axis of the ellipsoid and $A \simeq (x + 1.774 (x + 1.182)^{-0.733})/x$

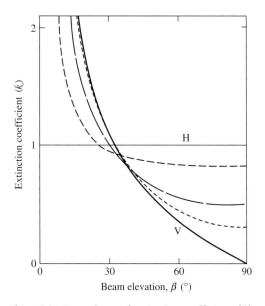

Figure 2.21 Dependence of extinction coefficients (k) for direct radiation as a function of beam elevation (β) for the ellipsoidal leaf angle distribution functions with $x = 0.5$ (- - - - -), $x = 1.0$ (spherical, ———) and $x = 3.0$ (– – – –), together with curves for horizontal (H) and vertical (V) distributions.

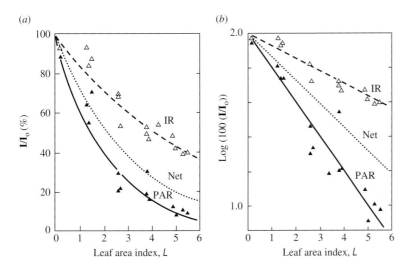

(a)

(b)

Figure 2.22 Daily means of transmission of near-IR (- - -△- - -), PAR (———▲———) and net (· · · · · ·) radiation in a wheat crop in early June (data from Szeicz, 1974). In (a), irradiance in each waveband is expressed as a percentage of that incident on the top of the crop, and in (b) the data are transformed to logarithms.

and so on. Values less than 1.0 are obtained for non-horizontal leaves or clumped-leaf distributions, while values greater than 1.0 occur with horizontal leaves or more regular arrangement in space.

There are several other complications that it is necessary to consider when modelling radiation in plant canopies. These include the following:

1. *Spectral distribution.* Because leaves are relatively transparent in the IR, the shortwave radiation deep in plant canopies is relatively enriched in the near-IR. Figure 2.22 illustrates the relatively greater attenuation of PAR than near-IR within a wheat crop. For this particular example, k for PAR was more than twice that for near-IR. The extinction profile for total shortwave radiation is between those for PAR and IR, being very close to the observed profile for net radiation (because the net radiation profile is dominated by the shortwave component).

2. *Heliotropism.* Radiation penetration models are further complicated in the many species, particularly among the Fabaceae, that show heliotropic leaf movements with the leaves tracking the sun during the day (Ehleringer & Forseth, 1980). Such movements can have large effects on the radiation received by the leaves, as illustrated in Figure 2.23 for isolated

leaves. This figure compares the diurnal trend of direct beam irradiance on a clear day, for a horizontal leaf, a vertical leaf oriented north–south and a diaheliotropic leaf that is continually oriented perpendicular to the solar beam. For these conditions the diaheliotropic leaf can receive nearly 50% more radiation over a day than the horizontal leaf. Conversely, paraheliotropic movements (orientation parallel to the solar beam) minimise interception of solar radiation, as shown for water-stressed *Vigna* (cowpea) leaflets in Figure 2.23(b). Paraheliotropic movements can help plants avoid damaging UV exposure (Grant, 1999) or photodamage in stressful environments (Pastenes *et al.*, 2004).

3. *Diffuse radiation.* Diffuse radiation is often an important component of incident shortwave radiation, but its penetration into canopies is not identical to that of direct radiation. In fact the leaf area index irradiated by diffuse radiation is greater than the corresponding L_{sunlit} for direct radiation. This is because a leaf that is shaded from the sky in one direction may be exposed to the sky in another. It can be shown that for horizontal leaves the irradiated leaf area is $\pi/2$ times L_{sunlit}. There is also a tendency for radiation to become more concentrated near the zenith as one goes down

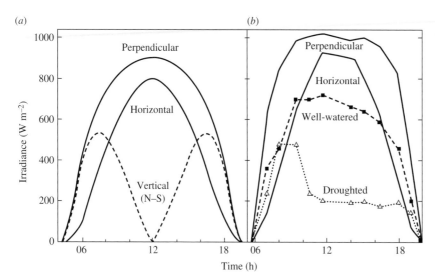

Figure 2.23 (*a*) Diurnal variation in direct beam irradiance on a clear day ($\tau = 0.7$) in early June at 50°N latitude for a horizontal leaf, a diaheliotropic leaf normal to the beam (perpendicular) and a vertical leaf oriented north–south. (*b*) Actual measurements for irradiance received by the upper surface of leaflets of water-stressed (· · · ·△· · ·) and well-watered (- - -■- - -) *Vigna* plants at Davis, California (after Shackel & Hall, 1979).

through a canopy. As the leaf area index increases, an increasing proportion of the leaf area is illuminated at low irradiances by diffuse light alone. Some data on the proportion of leaves in a mature sorghum canopy illuminated at different irradiances are presented in Figure 2.24. It is apparent from this figure, which is fairly typical, that much of the leaf area in this canopy receives only weak diffuse irradiation yet a large proportion of the total energy received is at relatively high irradiance.

4. *Discontinuous canopies.* An approach that is useful for calculating light interception or transmission (T) in discontinuous canopies such as orchards and widely spaced row crops is to divide the incident radiation into two components (see Jackson & Palmer, 1979). These are a fraction T_f that would have reached the ground even if the plants were completely opaque solid bodies, and a fraction ($1 - T_f$) that obeys the usual extinction law (Eq. (2.18)). Therefore one can write for the total transmission:

$$T = T_f + (1 - T_f)e^{-\hat{k}L'} \tag{2.20}$$

where L' is the leaf area index expressed per unit area of ground that would be shaded by non-transmitting structures of the same three-dimensional outline as the plants (i.e. L' = 'orchard' leaf area index divided by ($1 - T_f$).

5. *Penumbral effects.* When considering light penetration through canopies and the effects of sunflecks it can also be necessary to take account of the apparent size of the sun's disc (approximately 0.5° near the zenith). As indicated in Figure 2.25, the edges of shadows cast by leaves are not sharp and there is a region, called the penumbra, where the irradiance is less than that of full sunlight because direct light is coming from only part of the sun's disc. This effect commonly reduces the peak irradiance in sunflecks arising from small holes in canopies (see the photograph opposite p. 68 of Bainbridge *et al.*, 1968 for an illustration of this effect). From simple trigonometry it is apparent that for holes in a 10 m tall tree canopy, the sunflecks projected at ground level will be entirely penumbral if the holes are less than about 87 mm diameter (10 × tan 0.5 = 0.0873 m). Penumbral effects are particularly important in deep fine-leaved canopies.

2.7 Canopy reflectance and remote sensing

Remote sensing of canopy structure and function in the optical part of the electromagnetic spectrum, whether from satellite or airborne imagery, or from

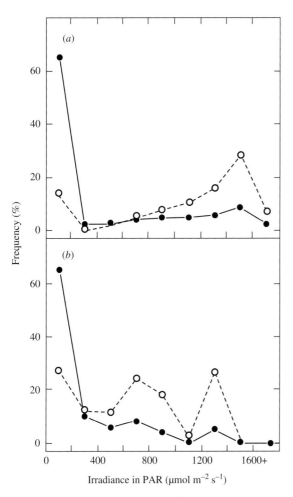

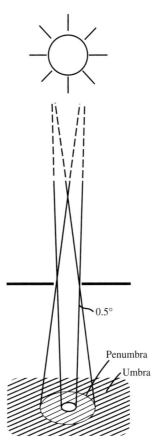

Figure 2.25 An illustration of the penumbral effect, showing the umbral and penumbral regions for a gap in the canopy slightly larger than the apparent diameter of the sun.

Figure 2.24 Frequency distributions of leaf area having different photon irradiances (in the PAR, in 200 μmol m^{-2} s^{-1} intervals) (————) for a sorghum leaf canopy under clear conditions in Rome, Italy, together with the corresponding distribution of total energy received in each irradiance interval (- - - - - -), (a) for a solar elevation of 25°, and (b) for a solar elevation of 60°. The leaf area index was approximately 6. Measurements were obtained by a canopy survey technique using a cosine-corrected quantum sensor and represent the higher of the abaxial or adaxial irradiances. (H. G. Jones, D. O. Hall & J. E. Corlett, unpublished data.)

closer range (e.g. from mobile platforms, 'cherry-pickers', balloons or unmanned aerial vehicles (UAVs)) largely depends on measurements of the spectral reflectance. Such information is increasingly

being used for monitoring canopy growth and function (e.g. photosynthesis and transpiration) and for diagnosis of biotic and abiotic stresses. In this section we consider some of the ways in which one can derive such information from remotely sensed reflectance.

2.7.1 Spectral indices

We have seen that the reflectance of plant leaves shows a very sudden increase at around the *red-edge* at 700 nm (Figures 2.15 and 2.17); this behaviour is very different from that of soil, which tends to show a steadily increasing reflectance with increasing wavelength (Figure 2.17). These characteristic spectral differences have provided the basis for the development of what are known as *spectral indices* (see Jones & Vaughan, 2010, for a detailed discussion). Spectral indices are new variables generated by mathematical combination of two or more spectral bands selected in such a way that the new indices are more closely related to biophysical parameters of interest (e.g. leaf area index, canopy chlorophyll content or canopy water content) than any of the original bands. The most widely used indices are the *vegetation indices* (*VI*) that were originally developed for studying vegetation cover in broad-band satellite images and are based on measurements of reflectance in the red (R) and near infrared (NIR).

Simple vegetation indices

The simplest approach to calculation of a *VI* is to take the difference between reflectances in the R and NIR: this difference would be large for areas that are predominantly vegetated, and smaller for bare soil. This gives a *difference vegetation index* ($DVI = \rho_{NIR} - \rho_R = N - R$, where N and R respectively, are used as shorthand for ρ_{NIR} and ρ_R). Note that one should use reflectances rather than the raw radiances reflected, as reflectances at least partially correct for variation in illumination. In order to improve the normalisation for environmental conditions one can instead calculate the ratio between NIR and R, to give a *ratio vegetation index* ($RVI = N/R$), or else, as is most commonly done, use what is known as the *normalised difference vegetation index* (*NDVI*), defined as:

$$NDVI = (\rho_{NIR} - \rho_R)/(\rho_{NIR} + \rho_R) \qquad (2.21)$$

It is often useful to further normalise *VIs* so that they scale between 0 and 1 by defining a scaled vegetation index (*VI**) as:

$$VI^* = (VI - VI_{min})/(VI_{max} - VI_{min})$$
$$= (VI - VI_{soil})/(VI_{veg} - VI_{soil}) \qquad (2.22)$$

where the subscripts max and min, respectively, refer to the values for dense vegetation and bare soil.

Many refinements of this basic concept have been proposed including indices such as the *green normalised difference VI* (*GNDVI*); this is similar to *NDVI* but with the reflectance in the red replaced by reflectance in the green), the *soil-adjusted VI* (*SAVI*, calculated as $(1 + L)(N - R)/(N + R + L)$, where L is a constant usually assumed $= 0.5$), and many others that have been found to be more linearly related to parameters of interest in specific situations than the original *NDVI* (see the useful tabulation in Jones & Vaughan, 2010).

Relationship of *VIs* to physiologically relevant quantities

Vegetation indices, especially *NDVI*, have been widely used by physiologists, agronomists and ecologists as proxies for quantities such as biomass, leaf chlorophyll content, leaf nitrogen, photosynthesis, leaf area index, fractional vegetation cover (f_{veg}) and fraction of photosynthetically active radiation absorbed (*fAPAR*). Unfortunately these relationships tend neither to be linear, nor to be consistent for different species and environments. In interpreting *VIs* we should note that there are three main mechanisms underlying variation in the *VI*. The most important is the direct comparison between soil and leaf, so that variation in *VI* is primarily determined (especially at a remote-sensing scale) by the fraction of leaf or vegetation in a pixel. Superimposed on this effect are influences of the spectral properties of leaves dependent on their biochemical composition (e.g. chlorophyll content and other aspects of leaf biochemistry) and any effects of canopy structure and view angle.

The basic assumption is that for any area viewed the spectral reflectance is a linear combination of the values for the background soil and the vegetation so that we can write, for example for red reflectance:

$$\rho_R = f_{veg}\,\rho_{R-veg} + (1 - f_{veg})\,\rho_{R-soil} \qquad (2.23)$$

where ρ_{R-veg} and ρ_{R-soil} are the red reflectances of pure vegetation and pure soil 'end members'. It is straightforward to show that substitution of these

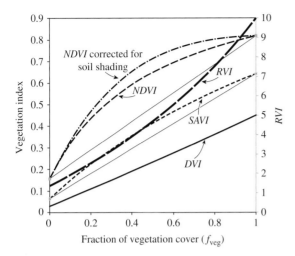

Figure 2.26 Calculated variation for vegetation indices ($NDVI = (N - R)/(N + R)$; $DVI = N - R$; $SAVI = (1 + L)$ $(N - R)/(N + R + L)$, and $RVI = N/R$; where N and R are the reflectances in the near infrared, and the red, respectively and L is a constant assumed $= 0.5$) as a function of fractional vegetation cover (f_{veg}), assuming typical component reflectances from Carlson and Ripley (1997) (vegetation: $\rho_R = 0.05$; $\rho_{NIR} = 0.5$, and soil: $\rho_R = 0.08$; $\rho_{NIR} = 0.11$) and neglecting any shading effect. Although DVI is linearly related to the vegetation fraction, $SAVI$ and to an increasing extent $NDVI$ and RVI are non-linearly related to f_{veg}. Also shown is the relationship for $NDVI$ corrected for soil shading. (Based on calculations by Jones & Vaughan, 2010).

values for the possible end member values that arise with different soil and plant combinations into various *VIs* gives substantial variation in the relationships between *VI* and f_{veg} as shown in Figure 2.26. For precise work the simple mixing model used here needs to be corrected for effects of shading and other radiative interactions within the crop canopy. Although *VIs* are commonly used as estimators of f_{veg}, Figure 2.26 shows that, except for the *DVI*, most indices are not strictly linearly related to f_{veg}. Where the view angle used to estimate the *VI* (and hence f_{veg}) is not from directly above, the true fraction of vegetation cover that would be seen from above ($f_{veg\text{-}nadir}$) can often be approximated by assuming a spherical leaf angle distribution and using simple geometry as $f_{veg}/\cos\theta$. For a typical canopy

absorptance for PAR of 0.85, the value of *fAPAR* can then be approximated as $0.85\,f_{veg}$.

Although *VIs* are most directly related to vegetation cover, they are often used to estimate leaf area index. The relationship between fractional cover or light interception and L can be derived using the Beer's law approximation (Eq. (2.18)) as:

$$f_{veg} \cong (1 - e^{-kL}) \qquad (2.24)$$

so that:

$$L \cong -\ln(1 - f_{veg})/k \qquad (2.25)$$

Rapid saturation of this expression as the vegetation becomes more dense means that estimation of L becomes very imprecise as L increases beyond about 3 to 4.

Narrowband indices, multiple wavelengths and derivative spectroscopy

There has been increasing interest in the use of high spectral resolution data to develop more specialised indices (usually expressed as normalised difference indices; Blackburn, 1998). Of particular interest to plant physiologists is the *photochemical reflectance index* ($PRI = (\rho_{570} - \rho_{531})/(\rho_{570} + \rho_{531})$). This measures the oxidation state of xanthophyll pigments in leaves (see Chapter 6) and normalises the small changes at 531 nm by values in a control region at 570 nm (Gamon *et al.*, 1992). Other approaches to the use of high spectral resolution data include the use of derivative spectroscopy to detect small changes in the position of the red-edge or combinations of multiple wavelengths. Other potentially useful spectral indices include *water indices* that estimate the water content of vegetation by focusing on the water absorption bands in the NIR (Figure 2.17).

The various approaches that can be used to select the appropriate wavebands to use in these more specialised vegetation indices have been reviewed by Jones and Vaughan (2010). Perhaps the most rational approach for wavelength selection is to make the selection based on knowledge of the absorption spectra of specific pigments. As a general rule, wavelengths where absorption is low are appropriate for distinguishing high pigment concentrations, while spectral regions with high absorption are more suitable where pigment concentrations are low. Alternatively, wavelengths can be selected empirically using statistical techniques such as *partial least squares regression* (Serbin *et al.*, 2012) or the construction of correllograms to identify the best wavelength combinations (e.g. Darvishzadeh *et al.*, 2008). These empirical techniques depend on the availability of a calibration data set having many samples with a wide range of pigment

concentration or of a functional property such as photosynthesis; even then results are not usually readily applicable to different species or even to different environments.

2.7.2 Multi-angular sensing and further development of radiation transfer models

The radiative properties of plant canopies are very dependent on both the angle of illumination and on the angle from which one views the canopy with the

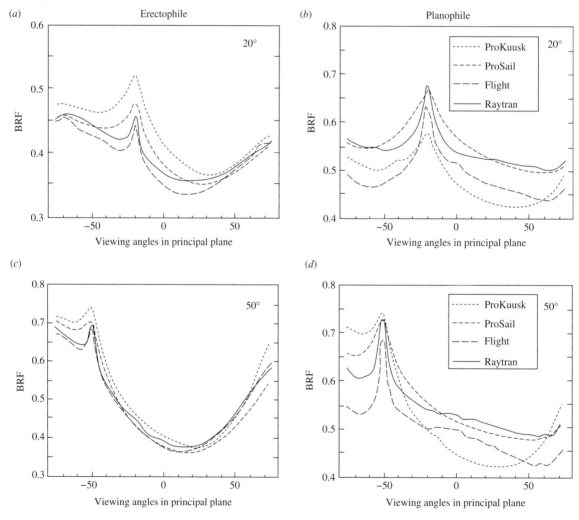

Figure 2.27 An illustration of the angular dependence of the BRF in the principal plane for an erectophile canopy (*a, c*) or a planophile canopy (*b, d*) in the near infrared as simulated by four different radiation transfer models at illumination zenith angles of (*a, b*) 20° and (*c, d*) 50° (data from Pinty *et al.*, 2001).

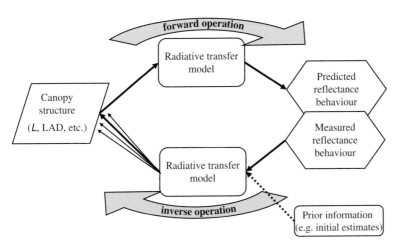

Figure 2.28 An illustration of the operation of forward and inverse modelling in the retrieval of canopy structure information in remote sensing. In the forward operation the radiative-transfer model is used to predict reflectance behaviour, while in the inverse mode measured reflectances are used to estimate canopy structure parameters. The extra outputs in running in the inverse mode indicate different equally possible combinations of the structure parameters. Prior information, such as knowledge of the crop and its leaf-angle distribution or an initial guess as to the value of L can be used.

observed reflectance showing substantial anisotropy. Many of us are familiar with the *hotspot phenomenon* observed when viewing a grass lawn, which often appears brighter when viewed with the sun directly behind the observer than when viewed towards the sun; this effect arises because there are no shadows visible when the sun is behind the observer.

Reflectance from structured vegetation contrasts markedly with the behaviour of smooth water bodies, which appear brightest when looking towards the sun as a result of direct *specular* reflection. A full description of the directional reflectance properties of surfaces is given by the *bidirectional reflectance distribution function* (BRDF), which defines the reflectance as a function of all possible illumination and view angles. In natural environments the BRDF is usually symmetrical about the solar plane and shows a peak (hotspot) at the solar angle (Figure 2.27). Because all sensors used to measure BRDF view finite angles, it is more convenient to refer to a *bidirectional reflectance factor* (BRF) that is defined as the ratio of the radiance reflected at any angle to what would be reflected by a perfectly diffuse, or *Lambertian*, reflector. Advanced treatments of multi-angular radiation modelling may be found elsewhere (Hapke, 1993; Liang, 2004) while details of the terminology are provided by Nicodemus *et al.* (1977).

Use of multi-angular observations of canopy reflectance can greatly improve our ability to extract information both about canopy structure and about biochemical and functional properties from remote sensing. The approach generally requires the inversion of canopy radiation transfer models that can effectively simulate observed patterns of BRF. Rather than using the models in a 'forward' mode, where they are used to estimate radiative properties on the basis of assumed structural parameters, their use for estimation of biophysical parameters involves operation in the 'inverse' mode (Figure 2.28). Essentially the inversion problem is to find the set of model parameters (L, leaf angle distribution (LAD), etc.) that best fits an observed BRF. Because most models cannot be inverted analytically it is usually necessary to resort to some statistical fitting routine, either by direct numerical optimisation or by the use of 'look-up tables' (LUTs). The latter method involves running the model in the forward mode many times to cover the full range of possible parameter values to generate a multidimensional table of results. This may then be searched to determine the parameters that best fit the data. Unfortunately the inversion problem is what is known as 'ill posed', in that several possible sets of fitted values may give equally good fits to the data so it is usual to provide some initial guesses to aid the selection of the appropriate optimum.

Some radiation transfer models

Various types of radiation transfer model have been used to predict the angular variation of radiation transfer and reflectance. These include simple *turbid medium* models that assume that canopies can be represented by a slab of infinitely small, homogeneously distributed scatters; these are particularly suitable for homogeneous agricultural crops. They include the Beer's law approach introduced earlier. A number of refinements have been introduced to this type of model to take account of finite leaf sizes, the angular distribution of radiant fluxes and the need to simulate the hotspot effect (e.g. more recent versions of the SAIL (scattering by arbitrarily inclined leaves) model (Verhoef & Bach, 2007). An alternative approach that is more applicable to non-homogeneous canopies such as forests assumes that the canopy can be described as an array of geometrical objects of defined shape and optical properties. In these *geometrical–optical* models the overall reflectance, $\rho_{(\theta)}$, at any zenith angle (θ) can be obtained as:

$$\rho_\theta = \rho_g f_g + \rho_c f_c + \rho_{c-sh} f_{c-sh} + \rho_{g-sh} f_{g-sh} \qquad (2.26)$$

where f is the fraction of the view area covered by sunlit ground (g), sunlit canopy (c), shaded ground (g-sh) and shaded canopy (c-sh). These fractions are obtained by simple geometric considerations as in the 'four-scale' model of Chen and Leblanc (1997).

For those interested in simulating the bidirectional reflectance of plant leaves and canopies the PROSPECT and SAIL models and a combined model (PROSAIL) are available at http://teledetection.ipgp. jussieu.fr/prosail/. Other useful models are available at www.npsg.uwaterloo.ca/models.php.

Another approach is to use *ray-tracing*; conceptually this technique involves analysis of the fate of a very large number of photons fired randomly into the canopy to generate the appropriate BRF. A number of strategies are available to speed up computation. A number of ray-tracers are available including the free POV-ray, which comes with a programming module to enable one to generate three-dimensional canopies for analysis of reflectance properties (www.povray.org). More rapid model inversion can be obtained using semi-empirical *kernel-driven* models (Nilson & Kuusk, 1989; Roujean et al., 1992; Walthall *et al.*, 1985); these express the BRF in terms of semi-empirical 'kernels' that represent the three main types of scattering (isotropic, volumetric scattering from homogeneous canopies and a geometric term representing scattering from objects that cast shadows).

Measurement of BRF

It follows from the discussion above that the most accurate estimates of canopy structural parameters may be obtained from multidirectional observations of reflectance. The necessary multi-angular data can be obtained from satellite sensors that have multi-angular capability (e.g. the Advanced Along-Track Scanning Radiometer (AATSR) or the Compact High Resolution Spectrometer (CHRIS) on the PROBA satellite). For in-field studies on agricultural crops various goniometers are available that allow measurement of reflectance at different angles (e.g. PARABOLA3 – Abdou *et al.* (2000) or simple rotatable field mounts – Casa & Jones (2005)). Further details of suitable approaches and their limitations may be found elsewhere (Jones & Vaughan, 2010).

2.8 Direct and subcanopy methods for determining canopy structure

Remote estimation of key canopy parameters such as leaf area index and leaf angle distribution requires calibration and validation on the basis of field measurements. These may involve direct sampling or the use of within-canopy measurements of radiation. These latter approaches are not entirely independent of the remote estimates as they usually depend on similar radiation transfer theory.

2.8.1 Direct measurement

The leaf area index (L) and its distribution through a canopy may be estimated from complete or stratified harvests of canopy above a known area of ground. The leaf area in the sample can be estimated either by direct measurement (e.g. tracing leaf outlines on squared paper and counting squares), by photographing the leaves on a contrasting background and using image analysis approaches (e.g. the 'magic wand' in Adobe Photoshop or similar software) or by using a commercial leaf area meter (e.g. Licor 3000C from Licor Inc., Lincoln, Nebraska, or WinDIAS from Delta-T, Burwell, Cambridge, UK). For trees, total harvest is even more difficult than for herbaceous canopies if not totally impractical, so scaling up using an appropriate sampling strategy is usual. For example Čermák et al. (2007) described an approach for olive trees based on counting leaves on individual branches and scaling up to whole trees. Direct estimation of leaf angle distribution is even more difficult. Although one can use an inclinometer to record angles and orientation of a small number of leaves, again this is extremely tedious. A range of instruments have been developed to facilitate such measurements (e.g. Lang, 1973; Sinoquet et al., 1998) with the modern instruments using sound propagation or magnetic fields to locate the necessary sensor head and facilitate digitisation. The use of stereoscopic images (Wang et al., 2009) or laser techniques (e.g. Omasa et al., 2007; Azzari et al., 2013) to derive detailed three-dimensional canopy structures is also becoming much more widely available. Because of the difficulty of direct measurement of leaf area and angle distributions it is most usual to derive information on these critical canopy biophysical properties by means of indirect methods based on inversion of radiative transfer models as outlined in the next section.

Within-canopy radiation transmission methods

These may be based on either (i) the attenuation of mean irradiance with depth in the canopy; or (ii) the 'gap fraction approach', which measures the proportion of area below the canopy that is illuminated by sunflecks and therefore only requires information on the direct beam transmission and does not need to consider scattered radiation. These methods involve the estimation of L (and potentially also k) from Eqs. (2.18) or (2.19), respectively. Unless one has pre-existing information on the canopy structure and the appropriate angular dependence of k (Table 2.5), it follows that one needs records of transmission at a number of different beam elevations to allow one to solve for both L and the leaf angle distribution. Several algorithms to accomplish the inversion are available, but they all make some assumptions about the expected canopy geometry. Figure 2.29 illustrates the variation of gap fraction with beam elevation and leaf area index. The gap-fraction approach tends to be the more accurate (or at least simpler to use) because the attenuation of total irradiance is complicated by the fact that it has a component of diffuse radiation coming from different angles; this needs to be considered when calculating k.

It is apparent from Figures 2.21 and 2.29 that the value of k (and hence $\mathcal{T}$) is nearly independent of canopy structure for an angle of elevation of about 33°, so that measurements at this angle tend to give the most robust estimates of L (Chapter 7).

The necessary information on gap-fraction can be obtained by a number of approaches including: (i) the use of multisensor arrays (as discussed above) with measurements made at different solar angles and set up with an appropriate threshold to avoid detection of diffuse or scattered radiation (Figure 2.30); (ii) the use of hemispherical photographs taken from the ground under overcast conditions to estimate the fraction of sky visible at different angles of elevation; (iii) use of automated versions of this approach (Figure 2.31).

The Licor LAI-2000 Plant Canopy Analyzer (Licor Inc., Lincoln, Nebraska) has an optical system that collects light from a whole hemisphere and splits it into fractions coming from different elevation classes. The contribution of radiation scattered by foliage is minimised by using filtered silicon light sensors that are sensitive only to radiation shorter than 490 nm.

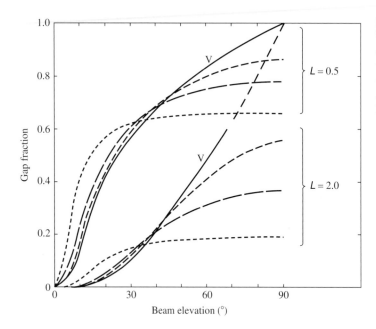

Figure 2.29 Variation of the gap fraction with beam inclination above the horizontal for different leaf angle distributions including the ellipsoidal with $x = 0.5$ (- - - - - -), $x = 1.0$ (spherical, —— ——) or $x = 3.0$ (------), and the vertical (V, ——) for canopies with leaf area indices (L) of 0.5 or 2.0.

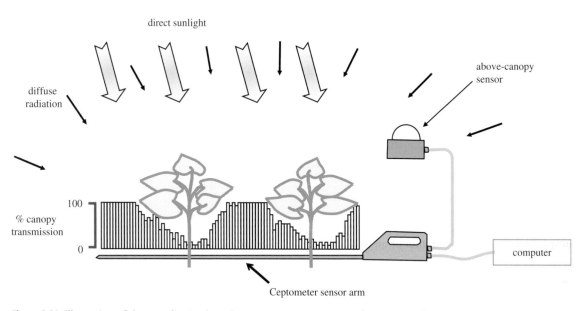

Figure 2.30 Illustration of the use of a simple multisensor transmission meter (SunScan SS1) for estimation of leaf-area index and leaf angle distribution. The above-canopy sensor is used to estimate the incident irradiance and the fraction of direct radiation. The instrument can be set up either to estimate the average fraction of radiation transmitted through the canopy to the level of the sensor, or else it can be set up to measure the sunfleck fraction for the direct component.

Among the uncertainties in application of these techniques include difficulties in distinguishing between radiation interception by leaves and by non-transpiring tissues such as branches and stems (leading to overestimates of true leaf area) and problems caused by clumping of leaves (leading to underestimation of leaf area). Kucharick *et al.* (1998a,b) have proposed the use of a multi-band

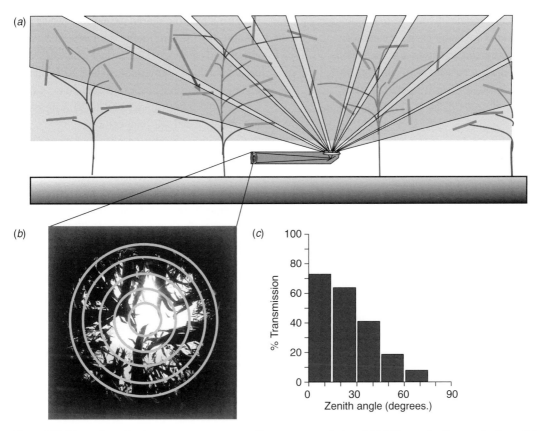

Figure 2.31 Illustration of subcanopy measurements of canopy transmission. (a) Multi-angular Canopy Analyzer (LAI-2000) under canopy showing view angle classes; (b) corresponding hemispherical photograph showing the angle classes; (c) histogram showing radiation transmission (or gap fraction) in each class.

vegetation analyser (MVA) that takes high resolution images of the canopy in the red and near infrared to allow separation of green and woody tissues and the correction of subcanopy images for the amount of woody tissue. Clumping errors can lead to underestimation of the true leaf area index by as much as 50%. One can convert the observed leaf area index obtained by any of the methods described above (L_{eff}) to give the true leaf area index (L) by defining a *clumping index* (λ_o) where:

$$L = L_{eff}/\lambda_o \qquad (2.27)$$

In this equation λ_o is normally <1.0 for clumped canopies and can be >1.0 for very regular canopies.

Leblanc *et al.* (2005) have proposed methods for estimating the clumping index from the frequency distribution of gap sizes. Combining the clumping correction with one for the ratio between the woody tissue area index and the total plant area index (α) gives the following estimate of the true leaf area index from observed values (Chen, 1996):

$$L = L_{eff}(1 - \alpha)/\lambda_o \qquad (2.28)$$

Useful discussions of these approaches and their limitations may be found in Norman and Campbell (1989) and Jones and Vaughan (2010). Detailed equations and the methods of application, together with essential precautions to ensure adequate sampling of the canopy of interest may be found in the appropriate instrument manufacturers' handbooks. There are many useful reviews of methods for extracting canopy parameters from subcanopy radiation measurements (Bréda, 2003; Garrigues *et al.*, 2008; Hyer & Goetz, 2004; Jonckheere *et al.*, 2004; Lang *et al.*, 2010; Leblanc *et al.*, 2005).

2.9 Concluding comments

Further detail on the radiative climate in plant canopies may be found in both plant environmental physiology texts (Campbell & Norman, 1998; Monteith & Unsworth, 2008) and the more vegetation-oriented remote sensing texts (Guyot & Phulpin, 1997; Jones & Vaughan, 2010; Liang, 2004; Rees, 2001). Microwave sensing (Radar) is another tool that can provide useful information on canopy structure and properties (Rees, 2001; Woodhouse, 2006), especially on soil and canopy water contents. The major advantage of microwave sensing from satellites is that microwave sensors have an all-weather capability, being insensitive to cloud.

2.10 Sample problems

2.1 Given the spectral properties of a leaf and the incoming solar radiation:

Wavelength interval (μm)	Average leaf absorptance	Total incident energy (W m^{-2})
0.3–0.7	0.85	450
0.7–1.5	0.20	380
1.5–3.0	0.65	70

(i) Calculate (a) the shortwave energy absorbed by the leaf; (b) the shortwave absorption coefficient; (c) the leaf temperature (assuming that the environment is at 20°C, there is no latent or sensible heat exchange and $\varepsilon = 1.0$ for wavelengths greater than 3 μm). (ii) Why do leaves not usually reach this temperature?

2.2 (i) Calculate net radiation absorbed by an isolated horizontal leaf ($\alpha = 0.5$) exposed above bare soil ($\rho_S = 0.3$), given that the total shortwave irradiance is 500 W m^{-2}, effective sky temperature is –5°C, soil temperature is 24°C and leaf temperature 20°C. (ii) What extra assumptions are required?

2.3 (i) What is the average energy per photon of green light ($\lambda = 500$ nm) and infrared ($\lambda = 2000$ nm)? (ii) What are the wavenumbers that correspond to these wavelengths? (iii) For a source that produces equal energy per nm centred on each of these wavelengths, calculate the ratio of the photon flux densities per nm at these wavelengths.

2.4 For a horizontal-leaved canopy with randomly distributed leaves: (i) What is the fraction of ground area sunlit if (a) $L = 1$ or (b) $L = 5$? (ii) What are the corresponding values of sunlit leaf area index? (iii) If instead, the leaves had random orientation and random inclination, what would be the corresponding values of sunlit leaf area index for a solar elevation of 40°?

2.5 Estimate the leaf area index for homogeneous canopies with either (i) a spherical leaf angle distribution; or (ii) a horizontal leaf distribution, given that the solar elevation is 60° and the transmitted fraction of direct radiation is 0.25.

3 Heat, mass and momentum transfer

Contents

Chapter 2 considered radiative energy exchange between plants and their environment. Other ways in which plants interact with their aerial environment include the transfer of matter, heat and momentum. The mechanisms involved in mass transfer processes, such as the exchanges of CO_2 and water vapour between plant leaves and the atmosphere, and in heat transfer, are very closely related so will be treated together. These can be broadly divided into those operating at a molecular level that do not involve mass movement of the medium (i.e. diffusion of matter and conduction of heat) and those processes, generally termed convection, where the entity is transported by mass movement of the fluid. The forces exerted on plants by the wind are a manifestation of momentum transfer.

Clear discussion of heat and mass transfer processes may be found in Campbell (1998) and in Monteith and Unsworth (2008) and a number of more advanced treatments (Cussler, 2007; Garratt, 1992; Kaimal & Finnigan, 1994; Monteith, 1975, 1976). The physical principles underlying these transfer processes and the analogies between them are outlined in this chapter,

and this information is used to analyse transfer between the atmosphere and both single leaves and whole canopies. Although the principles described are applicable to transfer in any fluid, the examples in this chapter will be confined to transfer in air.

3.1 Measures of concentration

Before going into details of the different mechanisms of heat and mass transfer it is necessary to define what is meant by concentration. In general the spontaneous transfer of mass, or other entities such as heat or momentum, occurs from a region of high 'concentration' to one of low 'concentration'. There are, however, many alternative ways in which one can specify the amount or concentration of an entity 'i' in a mixture, each of which may be appropriate for certain purposes, as can be seen in the following discussion.

3.1.1 Concentration

A widely used measure of composition of an entity 'i' is the (mass) concentration (c_i) or density (ρ_i) where:

$$c_i = \rho_i = \text{mass of i per unit volume of mixture} \quad (3.1)$$

Alternatively one can use the molar concentration (c_i^m) where:

$$c_i^m = \text{number of moles of i per unit volume of mixture}$$
$$= c_i/M_i \quad (3.2)$$

where M_i is the molecular mass. Although concentration is often used as a fundamental measure of gas composition, in a closed system concentration changes with temperature or pressure as these factors alter the volume according to the ideal gas law:

$$PV = n\mathscr{R}T \quad (3.3)$$

where n is the number of moles present, T is the absolute temperature, P is the pressure, V is the volume and $\mathscr{R}$ is the universal gas constant. Because liquids are not as compressible as gases, concentration is much less sensitive to pressure or temperature in solutions.

3.1.2 Mole fraction

A more conservative measure of composition is the mole fraction (x_i), which is the number of moles of i (n_i) as a fraction of the total number of moles present in the mixture (Σn):

$$x_i = n_i/\Sigma n \quad (3.4)$$

In this case alterations in temperature, pressure or volume do not affect the mole fraction as they affect all components equally. A related measure appropriate for gases is partial pressure (p_i), which for any component is the pressure that it would exert if allowed to occupy the whole volume available. The equivalence with mole fraction follows from combining the ideal gas law with *Dalton's law of partial pressures*, which states that, in a gas mixture of several components, the total pressure equals the sum of the partial pressures of the components, therefore:

$$x_i = p_i/P \quad (3.5)$$

Using the above relationships it may easily be shown that gas concentration is related to partial pressure by:

$$c_i = \text{mass}_i/V = n_iM_i/V = p_iM_i/\mathscr{R}T \quad (3.6)$$

3.1.3 Mass fraction

Another useful term is the mass fraction (m_i):

$$m_i = \text{mass of i per unit total mass of mixture} = c_i/\rho \quad (3.7)$$

where ρ is the density of the mixture. This is also independent of temperature and pressure. The mass fraction is related to mole fraction by:

$$m_i = x_i M_i/M \quad (3.8)$$

where M is the average molecular mass of the mixture.

3.1.4 Volume fraction

For a gas, the volume fraction (the volume of i per unit total volume of mixture) is identical to the mole fraction.

3.1.5 Mixing ratio

A term common in the meteorological literature to describe the composition of air is the mixing ratio (w_i) where:

$$w_i = \text{mass of i per(total mass} - \text{mass of i).} \quad (3.9)$$

3.2 Molecular transport processes

3.2.1 Diffusion: Fick's first law

The rapid thermal motions of the individual molecules in a fluid lead to random rearrangement of molecular position and, in an inhomogeneous fluid, to transfer of mass and heat. This process is called diffusion. For example, in a motionless fluid, mass transfer occurs as a result of the net movement of molecules of one species from any area of high concentration to one of lower concentration. (Quantities such as concentration of water vapour or temperature that are fully described by their magnitude are commonly referred to as scalars in fluid dynamics treatments.) In a one-dimensional system, the flux density or rate of mass transfer ($\mathbf{J}_i$) of a scalar entity i per unit area through a plane is directly related to the concentration gradient ($\partial c_i/\partial x$) of i across the plane by a constant called the diffusion coefficient (D_i). This can be written mathematically as:

$$\mathbf{J}_i = -\mathbf{D}_i(\partial c_i/\partial x) \tag{3.10}$$

This is the one-dimensional form of *Fick's first law of diffusion*. The minus sign is a mathematical convention to show that the flux is in the direction of decreasing concentration. Corresponding equations can be written for transfer in more than one dimension, but in what follows only the one-dimensional case will be treated.

Although it is common to use the concentration gradient as the driving force for diffusion as in Eq. (3.10), and this will be done in much of what follows, it can be inadequate for precise work when other factors are varying. For example, in solutions that depart significantly from ideal behaviour one needs to replace concentration by activity (Atkins & de Paula, 2009). Similarly in gases where there is a temperature gradient between the source and the sink, use of concentration can lead to significant errors (Cowan, 1977). This is because the rate of diffusion depends on the rate at which the individual molecules move (a function of temperature) as well as on their concentration. The use of any of mole fraction, partial pressure or mass fraction takes this effect into account. By using the appropriate substitutions for c_i (Eqs. 3.7, 3.6 and 3.5), Eq. (3.10) may be rewritten in any of the following forms that are more appropriate for non-isothermal gases:

$$\mathbf{J}_i = -\mathbf{D}_i\rho(\partial m_i/\partial x) \tag{3.11}$$

$$\mathbf{J}_i = -\mathbf{D}_i(M_i/\mathcal{R}T)(\partial p_i/\partial x) \tag{3.12}$$

$$\mathbf{J}_i = -\mathbf{D}_i(P\,M_i/\mathcal{R}T)(\partial x_i/\partial x) \tag{3.13}$$

These equations may appear similar to Eq. (3.10) since, for example, $\rho\,m_i = c_i$. However, $\rho\,\Delta m_i$ is not necessarily identical to Δc_i (where Δ represents a finite difference) so they can provide a significant improvement in non-isothermal systems. Unfortunately even these equations still involve some simplification.

3.2.2 Heat conduction

Heat transfer by conduction is analogous to diffusion. Conduction is the transfer of heat along a temperature gradient from a region of higher temperature (or kinetic energy) to one of lower temperature, without mass movement of the medium. In solids this energy transfer occurs as a result of molecular collisions transferring kinetic energy between molecules that are not themselves displaced, while in fluids the higher energy molecules themselves may diffuse.

Conductive heat transfer is described by *Fourier's law*, where the rate of sensible heat transfer per unit area (C, with units of $\text{W m}^{-2} = \text{J m}^{-2}\,\text{s}^{-1}$) is given by:

$$C = -k(\partial T/\partial x) \tag{3.14}$$

where k is the *thermal conductivity* ($\text{W m}^{-1}\,\text{K}^{-1}$). Although the driving force for heat transfer is the temperature gradient, it is convenient to make a simple mathematical manipulation so as to obtain the proportionality constant in the same units as were used for mass transfer (Monteith & Unsworth, 2008). If T is replaced by a 'heat concentration' $c_H = \rho c_p T$, where ρ is the density of the fluid and c_p is the specific heat capacity of the fluid (J kg^{-1}), one obtains an equation analogous to 3.10:

$$C = -\mathbf{D}_H\rho c_p(\partial T/\partial x) \tag{3.15}$$

where $\mathbf{D}_H$ is a *thermal diffusion coefficient* (often called a *thermal diffusivity*). Values of $\mathbf{D}_H$ and k for various fluids and solids are given in Appendices 2 and 5.

3.2.3 Momentum transfer

When a force is applied tangentially to a surface it tends to cause the surface layer to slide or shear in relation to the underlying material. A rigid solid transmits such a *shearing stress*, which is given the symbol τ and has units of force per unit area ($\text{kg m}^{-1}\,\text{s}^{-2}$), without undergoing deformation. In a fluid, however, adjacent layers slide relative to each other with any one layer being relatively ineffective at transmitting a shearing stress to the next layer of fluid, so that a velocity gradient develops when a fluid flows over a surface. The viscosity of a fluid is a measure of the internal frictional forces that arise from molecular interactions between adjacent layers, with viscous fluids being more effective at transmitting a shearing stress than non-viscous fluids. This process is described by *Newton's law of viscosity,* which states that the shearing stress at a plane in a fluid is directly proportional to the velocity gradient $(\partial u / \partial x)$:

$$\tau = \eta\,(\partial u / \partial x) \tag{3.16}$$

where η is called the *dynamic viscosity* (kg m^{-1} s^{-1}).

This equation is similar in form to those already introduced for heat and mass transfer. In this case the shearing stress has the dimensions of a momentum flux density, where momentum is mass × velocity (i.e. MLT^{-1}·L^{-2}·T^{-1} = ML^{-1}T^{-2}). As for heat transfer, the velocity gradient can be replaced by a gradient of momentum 'concentration' (c_M = mass × velocity/volume = ρu) thus giving a proportionality constant with the dimensions of a diffusion coefficient (L^2 T^{-1}). This diffusion coefficient for momentum (D_M) is also called the *kinematic viscosity* (ν).

The tangential force exerted on a surface by a fluid flowing over it is termed skin friction. In addition to this transfer of momentum to a body across the streamlines of flow, a moving fluid can also exert a force as a consequence of *form drag* (τ_f) where the pressure exerted on the front of an object is greater than that on the rear. This is the main force that causes the bending of trees and other plants in the wind. The magnitude of the form drag, that is the force per unit cross-sectional area normal to the flow (A) for any object, is given by:

$$\tau_f = 0.5\,c_f(\rho u^2) \tag{3.17}$$

where c_f is a form drag coefficient that relates the actual drag to the maximum potential force that could be exerted if all the air movement was completely stopped ($0.5\rho u^2$). Streamlined objects, such as aircraft, will have much lower drag coefficients than objects without streamlining (e.g. buildings). The value of c_f decreases dramatically if the airflow is turbulent.

Further discussion of turbulence and its importance may be found in the section on convective and turbulent transfer (Section 3.3), while drag and its significance is considered further in Section 11.1.4.

3.2.4 Diffusion coefficients

Choice of an appropriate 'concentration' gradient enables us to express the proportionality constant D for a wide range of transfer processes in common units. For example, dimensional analyses of Eqs. (3.10) to (3.13) for mass transfer and of (3.15) for heat transfer all give D with dimensions L^2 T^{-1}, while the same is true for momentum transfer.

The value of D in a binary mixture (e.g. CO_2 and air) is called a mutual diffusion coefficient, where D_{CA}, for CO_2 diffusing into air is the same as D_{AC} for air diffusing into CO_2, with very little effect of altering the proportions of air and CO_2. Values for diffusion coefficients for quantities including various gases, heat and momentum in both air and in water are listed in Appendix 2. When a substance is diffusing within itself, D is called a self-diffusion coefficient: this can be very different from the mutual diffusion coefficient. For example the self-diffusion coefficient for CO_2 (D_C) is 5.8 mm^2 s^{-1} compared with a D_{CA} of 14.7 mm^2 s^{-1} at 20°C and atmospheric pressure. Plant physiology is often concerned with ternary systems of air, CO_2 and H_2O, where CO_2 and H_2O fluxes may interfere; rigorous treatment of this effect can modify Eq. (3.10) (Jarman, 1974).

The relative values of D for different gases are approximately as predicted by *Graham's law,* which states that the diffusion coefficients of gases are inversely proportional to the square roots of their densities when pure (i.e. $D_i \propto (M_i)^{-1/2}$, since density is proportional to M_i). Effects of temperature and pressure on D are given by:

$$D = D^o(T/T^o)^n(P^o/P) \tag{3.18}$$

where the superscript 'o' refers to a reference value, which can be taken as the value at 20°C (293.15 K) and a pressure of 101.3 kPa (1013 mbar). Although the exponent n depends on the gas, a value of 1.75

predicts D over the normal range of environmental temperatures with less than 1% error. In addition D is modified (Knudsen diffusion) when diffusion occurs in a confined physical system where the average distance that a molecule travels between collisions (the mean free path, which for CO_2 in air at 20°C is approximately 54 nm) is of the same order as the size of the system. An example of a situation where this effect may be relevant is gaseous diffusion through nearly closed stomata. Diffusional transfer in plant systems is further complicated when temperature gradients exist; for example it has been suggested that a process termed thermal effusion, caused by solar heating of the leaves, drives air flow to the roots of water lilies (Dacey, 1980; see also Section 10.5.1). Advanced treatments of diffusion may be found in appropriate texts (e.g. Crank, 1979; Cussler, 2007).

3.2.5 Integrated form of the transport equation

The close similarity between the equations describing transport of a wide range of different entities including water vapour, CO_2, electric charge, heat and momentum, has led to them being sometimes referred to as particular examples of the *general transport equation*:

Flux density (or flux)
= proportionality constant × driving force (3.19)

In many practical situations it is more convenient to measure concentrations at two positions in a system, rather than to determine the concentration gradient at a point: therefore the transport equation is commonly applied in an integrated form. In the simple case where the flux is constant over the path being considered (i.e. there is no absorption or evolution of the transported species in that region) and where D does not change with position (generally true for molecular diffusion), integration of Eq. (3.10) between planes at x_1, and x_2, a distance ℓ apart, gives:

$$\mathbf{J}_i = \mathbf{D}_i(c_{i1} - c_{i2})/\ell \qquad (3.20)$$

where c_{i1} and c_{i2} are the concentrations at x_1, and x_2 (see Figure 3.1). In this equation, the driving force is the concentration difference across the path. The proportionality constant that relates any flux density to the appropriate concentration difference is equal to $\mathbf{D}_i/\ell$. In plant physiology this constant is conventionally called a *conductance* (and given the symbol g for diffusive transfer and for heat transfer). For many purposes, as will be seen later, it is more convenient to replace the conductance by its reciprocal, termed a *resistance* (given the symbol r), so that:

$$r_i = 1/g_i = \ell/\mathbf{D}_i \qquad (3.21)$$

Another transport process that fits the general transport equation is the transfer of electric charge as described by *Ohm's law* (when a steady current (I) is flowing through a conductor, the potential difference between its ends (V) is directly proportional to the current, with the constant of proportionality being called a resistance (R) – i.e. $V = IR$). The close analogy between Ohm's law and other transport processes has

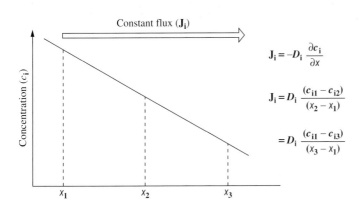

$$\mathbf{J}_i = -\mathbf{D}_i \frac{\partial c_i}{\partial x}$$

$$\mathbf{J}_i = \mathbf{D}_i \frac{(c_{i1} - c_{i2})}{(x_2 - x_1)}$$

$$= \mathbf{D}_i \frac{(c_{i1} - c_{i3})}{(x_3 - x_1)}$$

Figure 3.1 Diffusion down a concentration gradient showing that where D_i is constant, the flux is constant with distance and concentration drop is proportional to distance.

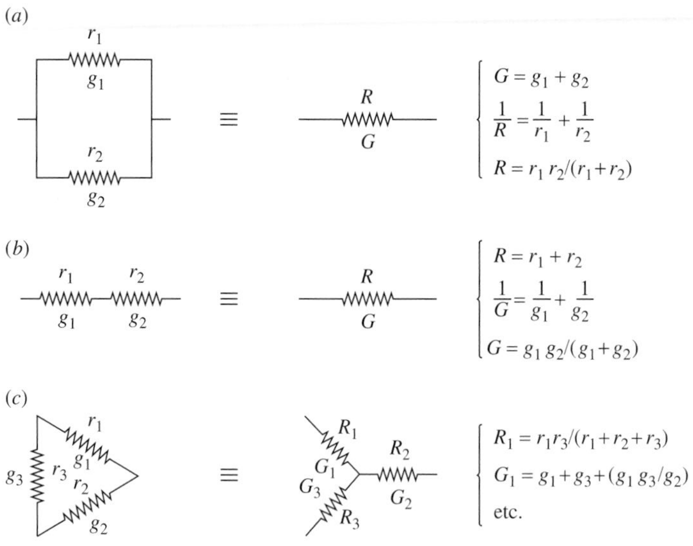

(a)

$$G = g_1 + g_2$$
$$\frac{1}{R} = \frac{1}{r_1} + \frac{1}{r_2}$$
$$R = r_1 r_2/(r_1+r_2)$$

(b)

$$R = r_1 + r_2$$
$$\frac{1}{G} = \frac{1}{g_1} + \frac{1}{g_2}$$
$$G = g_1 g_2/(g_1+g_2)$$

(c)

$$R_1 = r_1 r_3/(r_1+r_2+r_3)$$
$$G_1 = g_1 + g_3 + (g_1 g_3/g_2)$$
etc.

Figure 3.2 The rules for simplifying complex networks of resistors: (a) resistors in parallel; (b) resistors in series; and (c) the Delta–Wye transform.

proved extremely useful in the analysis of transfer processes in plants. This is because electrical circuit theory is well developed and is directly applicable to the analysis of the complex networks that occur in plant systems. As a simple example, for a leaf losing water by evaporation from both surfaces, the analogous electrical system is two resistors in parallel with the same potential difference across them. The rules for simplifying complex electrical networks are summarised in Figure 3.2.

It is often preferable to use conductances rather than resistances in transport studies because the flux across a path with a given driving force is directly proportional to its conductance, but inversely related to the resistance. This inverse relationship to resistance can be misleading in simple systems with only one dominant resistance (see e.g. Chapter 6). It is readily apparent from Figure 3.2, however, that when a system is predominantly composed of resistors in series, it is more convenient to work with resistances, particularly if one is concerned with the relative limitation imposed by each component (see Chapter 7). A system of resistors in parallel, on the other hand, is most easily treated using conductances. Either form will be used as appropriate in the following chapters so it is necessary to be familiar with both types of expression and their conversion.

The analogies between different transfer processes are summarised in Table 3.1. It is clear from this that the units for conductance depend on what is chosen as the driving force, it being to some extent arbitrary which factors are included in which term. In each case it is possible to manipulate units to give a conductance in m s^{-1} (or mm s^{-1}). Note that for electricity the current is a flux rather than a flux density, so that the analogy is not complete.

Diffusion coefficients are fundamental properties of the medium and of the material diffusing rather than of the particular system geometry. This is in direct contrast to conductances or resistances that are basically a property of the whole system in that they vary with geometry (e.g. the distance over which the transport occurs) as well as with the mechanism of transport (e.g. molecular diffusion or the rather more rapid turbulent transport).

3.2.6 Units for resistance and conductance

For many years it was normal practice among plant physiologists to express mass and heat transfer resistances in units of s m^{-1} (or s cm^{-1}) and conductances in m s^{-1} (or mm s^{-1}). These units arise (see Eq. (3.20)) if the flux is expressed as a mass flux density (e.g. kg m^{-2} s^{-1}) and the driving force is a

Table 3.1 Analogies between different molecular transfer processes.

General transport equation	Flux density	= apparent driving force	× conductance
Fick's law	J_i	$= \Delta c_i$	$\times D_i/\ell \ (= g_i)$
(mass transfer)	$(\text{kg m}^{-2}\,\text{s}^{-1})$	(kg m^{-3})	(m s^{-1})
	J_i^m	$= \Delta x_i$	$\times PD_i/\ell\mathcal{R}T \ (= g_i^m)$
	$(\text{mol m}^{-2}\,\text{s}^{-1})$	(dimensionless)	$(\text{mol m}^{-2}\,\text{s}^{-1})$
	J_i^m	$= (P/\mathcal{R}T)\,\Delta x_i$	$\times D_i/\ell \ (= g_i)$
	$(\text{mol m}^{-2}\,\text{s}^{-1})$	(mol m^{-3})	(m s^{-1})
Fourier's law	C	$= \Delta T$	$\times k/\ell$
(heat conduction)	$(\text{J m}^{-2}\,\text{s}^{-1})$	(K)	$(\text{W m}^{-2}\,\text{K}^{-1})$
	C	$= \rho c_p \Delta T \ (= \Delta c_H)$	$\times D_H/\ell \ (= g_H)$
	$(\text{J m}^{-2}\,\text{s}^{-1})$	(J m^{-3})	(m s^{-1})
Newton's law of viscosity	τ	$= \Delta u$	$\times \eta/\ell$
(momentum transfer)	$(\text{kg m}^{-1}\,\text{s}^{-2})$	(m s^{-1})	$(\text{kg m}^{-2}\,\text{s}^{-1})$
	τ	$= \rho\Delta u \ (= \Delta c_M)$	$\times D_M/\ell \ (= g_M)$
	$(\text{kg m}^{-1}\,\text{s}^{-2})$	$(\text{kg m}^{-2}\,\text{s}^{-1})$	(m s^{-1})
Poiseuille's law[a]	J_v	$= \Delta P$	$\times r^2/8\ell\eta \ (= L_p)$
(flow in pipes)	$(\text{m}^3\,\text{m}^{-2}\,\text{s}^{-1})$	$(\text{kg m}^{-1}\,\text{s}^{-2})$	$(\text{m}^2\,\text{s kg}^{-1})$
Ohm's law	I (flux)	V	$\times 1/R$
(electric charge)	(A)	(W A^{-1})	$(\text{A}^2\,\text{W}^{-1})$

[a] For details of Poiseuille's law see Chapter 4.

concentration (density) difference (kg m^{-3}). The same units arise for heat transfer when treated according to Eq. (3.15), and for momentum transfer (see Table 3.1).

It is, however, increasingly common, particularly in the biochemical literature, to express the flux as a molar flux density ($\mathbf{J}^m$, mol m^{-2} s^{-1}), because biochemical reactions concern numbers of molecules rather than the mass of material. (More correctly $\mathbf{J}^m$ should be termed a mole flux density because use of the term molar should strictly be limited to the meaning 'divided by moles', but we will retain the usage of molar here.) Similarly, concentrations (e.g. of water vapour and CO_2) are now usually measured as partial pressures (or the related volume fractions), as the appropriate driving force for diffusion is the gradient of partial pressure (p_i) or mole fraction (x_i) (rather than concentration). Therefore one can write the integrated form of the transport equation for a molar flux in either of the equivalent forms:

$$\mathbf{J}_i^m = (D_i P/\ell\,\mathcal{R}T)(x_{i1} - x_{i2})$$
$$= (D_i/\ell\,\mathcal{R}T)(p_{i1} - p_{i2}) \tag{3.22}$$

If one now follows general usage and defines a molar conductance (g^m) as $DP/\ell\mathcal{R}T$, it has dimensions mol L^{-2} T^{-1}, giving as appropriate units (mol m^{-2} s^{-1}) so the corresponding molar resistance (r^m) has units (m^2 s mol^{-1}). These molar units will be used frequently in much of the rest of this book, especially when considering gas exchange through stomata, so the superscript 'm' will often be omitted in what follows so as to simplify presentation of equations, and the type of units will be indicated by the choice of font (i.e. $g^m = g$ and $r^m = r$).

This alternative definition of conductance has some advantages. In the more usual definition where $g = D/\ell$, conductance is approximately proportional to the square of the temperature and inversely proportional to P (see Eq. (3.18)). Where, however, $g = PD/\ell\mathcal{R}T$, it is relatively independent of the properties of the air, being independent of P and approximately proportional to absolute temperature. The usual formulation is clearly less appropriate if one is considering effects of altitude (and hence total pressure) on gas exchange. A further advantage of using partial pressure is that it obviates the need for correcting for changing temperature and pressure that arises when using concentration. It is particularly important to use partial pressures rather than concentration gradients where the system is non-isothermal (Cowan, 1977).

It follows from Eqs. (3.20) and (3.22) that conversion between the two types of units is by means of:

$$g = g(P/\mathcal{R}T) \tag{3.23a}$$

and

$$r = r(\mathcal{R}T/P) \tag{3.23b}$$

At sea level and 25°C approximate conversions are, for resistance:

$$r(\text{m}^2\text{s mol}^{-1}) = 2.5r(\text{s cm}^{-1}) = 0.025r(\text{s m}^{-1}) \tag{3.24a}$$

and for conductance:

$$g(\text{mol m}^{-2}\text{s}^{-1}) = 0.04g \,(\text{mm s}{-1}) \tag{3.24b}$$

Conversions at other temperatures are given in Appendix 3.

In spite of the advantages of using a molar basis for expression of mass transfer resistances and conductances, the units s m^{-1} and mm s^{-1} will be retained for some purposes in the following treatment and especially for the analysis of heat and momentum transfer where there are no obvious analogies to molar fluxes. In addition, these units are still the most commonly used for many canopy-level studies, especially when considering evaporation, and will normally be used in that context. Of course it should be recognised that for any given conditions of temperature and pressure the two sets of units are always directly interconvertible using Eqs. (3.23) and (3.24) as appropriate.

3.2.7 Fick's second law of diffusion

In many situations where diffusion takes place the flux is not constant with distance because some of the material diffusing goes into changing the concentration at any position. Using the principle of conservation, that is, that matter cannot normally be created or destroyed, it is easy to show that, in a one-dimensional system where J_i is increasing with distance in the x direction, the extra material required must be obtained by decreasing c_i, so that:

$$\partial J_i/\partial x = -\partial c_i/\partial t \tag{3.25}$$

This is known as the continuity equation. Substituting for J_i using Fick's first law (Eq. (3.10)) leads to:

$$\partial c_i/\partial t = -\partial(-D_i(\partial c_i/\partial x))/\partial x = D_i(\partial^2 c_i/\partial x^2) \tag{3.26}$$

which is known as *Fick's second law*. This equation describes the time–distance relationships of concentration when diffusion occurs. The solution of this equation that is appropriate for any particular problem depends on the initial conditions and on the details of the system geometry. Solutions of this equation for a wide range of systems and boundary conditions have been presented by Crank (1979); see also Cussler (2007). Here I will discuss only one example that can be used to illustrate the scale of diffusive transport in plant systems. This is the case where a finite amount of material is released at time zero in a plane at the origin and allowed to spread by diffusion in one dimension. The shape of the resulting curve relating concentration to distance is that of the Gaussian or normal distribution. For this curve the distance from the origin at which the concentration drops to 37% (= 1/e) of that at the origin is given by:

$$x = \sqrt{(4Dt)} \qquad (3.27)$$

An alternative explanation of x is that 16% of the material initially placed at the origin will diffuse at least as far as x in time t. The distance over which diffusive transport occurs increases with the square root of time. Substituting a typical value for D in air of 20 mm s^{-1}, gives $x = 9$ mm for $t = 1$ s (i.e. $\sqrt{(4 \times 20 \times 10^{-6} \times 1)} \simeq 9 \times 10^{-3}$ m). This illustrates the sort of distance over which gaseous diffusion is an effective transport mechanism.

3.3 Convective and turbulent transfer

The transfer of mass or heat by diffusion is a consequence of the thermal movements of individual molecules and is the dominant mechanism in still fluids such as the air within the intercellular spaces of plant leaves. For surfaces that are exposed to the atmosphere, such as leaves, air movement over the surface can speed up heat and mass transfer considerably. There are two processes involved.

In the first, the air movement continuously replenishes the air close to the surface with unmodified air, thus maintaining a steep gradient of concentration (the driving force for diffusion) and therefore more rapid transport than obtains in still air. This is only important for isolated surfaces (such as isolated leaves or plants see Chapter 5). Where there is an extensive homogeneous surface, an equilibrium is achieved such that the air flowing close to the surface has already been modified by passage over an identical surface upwind. In this case the concentration gradient is the same as would occur in still air unless the airstream is turbulent. Turbulence or the existence of random eddies in the airstream provides the second way in which air movement can speed up transfer processes; in this case, materials are

transported directly by mass transfer in the moving air currents.

The air in the lower atmosphere is never completely still. Not only is there usually a net horizontal motion or wind, but there are also many random movements of small packets of air. The actual pattern of air movement depends on the type of convection regime that exists. This may be *free convection*, where the air movements are caused by changes in air density, as occur where the air adjacent to a heated surface expands and therefore rises, or where cold air sinks below a cool surface. Or it may be *forced convection*, where the air movement is determined by an external pressure gradient causing wind. Room heating with conventional 'radiators' relies largely on free convection, while fan-assisted radiators use forced convection to transfer heat to the room.

Forced convection may lead to the generation of eddies or turbulence as a result of the frictional forces acting between the wind and the surfaces over which it flows. The size and velocity of the individual eddies depend on a number of factors but they tend to decrease in magnitude as the surface is approached (Figure 3.3). Evidence for the random spatial distribution and the persistence of these eddies may easily be seen if one looks at the patterns on a field of waving barley. On a smaller scale they can be detected by instruments such as hot-wire anemometers that respond rapidly to changes in air velocity (see Chapter 11). Because the size of eddies in an airstream tends to

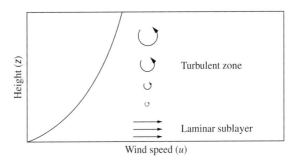

Figure 3.3 Profile of wind speed moving over a surface showing the laminar sublayer where wind speed changes rapidly with height and a turbulent zone with the eddy size increasing with distance from the surface.

be similar to the scale of surface irregularities, they are therefore several orders of magnitude larger than the average molecular movements giving rise to diffusion. For this reason turbulent transfer tends to be much faster than diffusion, typically by between three and seven orders of magnitude.

The relative importance of free and forced convection in heat and mass transfer depends on the balance between the buoyancy forces arising from temperature gradients and the inertial forces arising from air movements that cause turbulence. In most plant environments, heat and mass transfer are rarely determined by free convection alone, though it may be an important component of the transfer mechanism in very light wind or very bright sunlight.

3.3.1 Boundary layers

It was pointed out earlier that when a fluid flows over a surface the flow velocity decreases towards that surface as a consequence of the friction between the surface and the fluid and of the viscous forces within the fluid. The zone adjacent to a surface, where the mean velocity is reduced significantly below that of the free stream, is termed the boundary layer. In what follows, the transfer conductances and resistances in the leaf boundary layer will be distinguished by the use of the subscript 'a', so that the boundary layer conductance for heat transfer would be g_{aH}. One common arbitrary definition of the boundary layer defines its limit as that streamline where the velocity reaches 99% of that in the free airstream. Because the depth of the boundary layer in air tends to be about two orders of magnitude less than the size of the object, mass and heat transfer can be regarded as one-dimensional processes at right angles to the surface. For single leaves the boundary layer is of the order of millimetres deep, while for the atmospheric (planetary) boundary layer its depth ranges from tens to several hundreds of metres. The initial discussion here will concentrate on the behaviour of leaf boundary layers.

The pattern of fluid movement within a boundary layer may be either laminar, where all the fluid movement is parallel to the surface, or it may be turbulent. In a turbulent boundary layer the movements of individual molecules rather resemble the movements of commuters going to or from work in a large city: although the individual particles may be moving in a very irregular pattern, the overall motion is regular and predictable. Whether or not a particular boundary layer is laminar or turbulent depends on the balance between inertial forces in the fluid (because of its velocity) and the viscous forces that tend to produce stability and a laminar flow pattern.

Experimentally it has been found that, for a smooth plate, the transition from a laminar to a turbulent boundary layer generally occurs when the value of a group of terms called the *Reynolds number* exceeds a value between 10^4 and 10^5. The Reynolds number is a dimensionless group given by ud/v, where u is the free fluid velocity, d is a characteristic dimension of the object and v is the kinematic viscosity $(= D_M)$. For parallel-sided flat plates (approximately equivalent to grass leaves), d is the downwind width, while for circular plates (appropriate for certain other leaves) d is $0.9 \times$ diameter. For irregular plates d is the average downwind width, while for spheres or cylinders with their long axis normal to the flow d is equal to the diameter.

The build-up of a boundary layer in an airstream flowing over a flat plate such as a leaf is illustrated in Figure 3.4. Initially there is a laminar zone that gradually increases in thickness with increasing distance from the leading edge. The laminar layer may then break down to form a turbulent zone when d increases enough to make the local Reynolds number larger than the critical value. There is good evidence that this critical Reynolds number is achieved at values well below 10^4 (i.e. 400–3000; Grace, 1981) for plant leaves, because of the tendency of their surface irregularities, such as veins and hairs, to induce turbulence. Turbulence in the boundary layer is also encouraged by any turbulence in the free stream that might be caused by objects such as leaves and stems upwind. Even where the majority of the boundary layer is turbulent there remains a thin zone close to

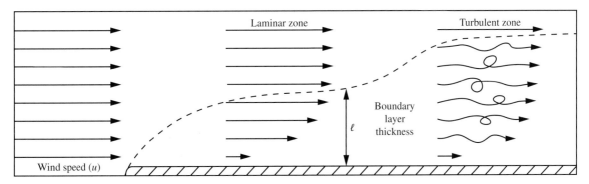

Figure 3.4 Diagrammatic representation (with much exaggerated vertical scale) of the development of the boundary layer over a smooth flat plate in a laminar airstream, showing the wind speed profiles, the initial laminar zone and the onset of turbulence.

the surface called the laminar sublayer where the flow is laminar, though this may be only a few tens of micrometres thick. As an indication of the sort of conditions under which turbulence may occur with real leaves, a leaf only 1 cm wide would achieve a possibly critical Reynolds number of 500 at a wind speed of only 0.76 m s^{-1} (i.e. 500 × 15.1 × 10^{-6}/0.01).

Mass and heat transfer through a boundary layer can be described by the general transport equation in the form already used for molecular diffusion in still air:

$$\mathbf{J}_i = g_i \left(c_{i1} - c_{i2} \right) \tag{3.28}$$

or for heat:

$$\mathbf{C} = g_H \rho c_p (T_1 - T_2) \tag{3.29}$$

where ρ_a is the density of air. As transport within a laminar boundary layer is by diffusion, the conductance of a laminar layer with a mean thickness δ is given by $\mathbf{D}_i/\delta$ (see Eq. (3.21)). The thickness of a laminar boundary layer over a flat surface increases in proportion to the square root of the distance from the leading edge and in proportion to the reciprocal of the square root of the free fluid velocity. The thickness, δ, is also weakly dependent on $\mathbf{D}_i$, being approximately proportional to $\mathbf{D}_i^{0.33}$, so that the boundary layer thickness is different for heat, mass and momentum.

Where the flow regime in the boundary layer is turbulent or mixed, the same form of equation applies but mass transfer is more rapid because the eddies

facilitate transfer. In this case the boundary layer conductance is increased because $\mathbf{D}$, the molecular diffusion coefficient, is replaced by a larger eddy transfer coefficient, $\mathbf{K}$. The value of this transfer coefficient varies with the size of the eddies and tends to increase with distance from the surface (Figure 3.3). The value of $\mathbf{K}$ may increase from around 10^{-5} m^2 s^{-1} near the leaves where the eddies are small to about 10^{-1} m^2 s^{-1} at the top of a plant canopy, reaching as much as 10^2 m^2 s^{-1} well above the canopy.

It has already been noted that the integration of Fick's first law (Eq. (3.10)) to obtain the integrated form (e.g. Eq. (3.29)) is easiest where the transfer coefficient does not alter with distance. Where it does vary, as in a turbulent boundary layer, the definitions of conductance and resistance in Eq. (3.21) must be replaced by:

$$r_i = g_i^{-1} = \int_{x1}^{x2} \frac{dx}{\mathbf{K}_i} \tag{3.30}$$

Figure 3.5 illustrates how the transfer coefficient and concentration gradient might vary across a typical mixed boundary layer having a laminar sublayer and a turbulent zone.

Because of the difficulties in integrating the transport equation in situations where the transfer coefficient varies, it is convenient to define an equivalent boundary layer of thickness (δ_{eq}). This is the thickness of still air that would have the same conductance or resistance as the turbulent boundary

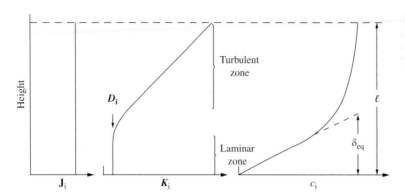

Figure 3.5 Profiles of J_i, K_i and c_i across a mixed boundary layer of total depth ℓ, illustrating the equivalent boundary layer thickness δ_{eq}.

layer of thickness ℓ (see Figure 3.5). Thus for a turbulent boundary layer where the value of the transfer coefficient, K, is say $10^3 \times D$, the thickness of the equivalent boundary layer (δ_{eq}) is $10^{-3} \times \ell$. Note that both δ and δ_{eq} are average thicknesses, since the actual thickness of the boundary layer is less near the leading than the trailing edge.

As it is often difficult to determine the boundary layer thickness it has been found convenient to express heat and mass transfer in terms of the characteristic dimension (d). The ratio d/δ_{eq} is often called the *Nusselt number* when studying heat transfer or the *Sherwood number* when referring to mass transfer. These two dimensionless groups are among those widely used in the fluid dynamics literature to summarise information on heat and mass transfer. For our purposes it is more convenient to express this information directly in terms of the dependence of boundary layer conductance or resistance on wind speed and leaf dimensions. The application of dimensionless groups is discussed by Monteith and Unsworth (2008) and in textbooks on heat and mass transfer (e.g. Kreith *et al.*, 2010).

3.3.2 Conductance of leaf boundary layers

The conversions between conductances for different entities depend on the nature of the boundary layer. Both when the air surrounding a plant organ is still and within the intercellular spaces of leaves transfer of heat or mass depends on molecular diffusion. Conductances for different entities (e.g. CO_2, water

vapour or heat) through such a layer of still air would be in the ratio of their molecular diffusion coefficients (Appendix 2). In a laminar boundary layer transport is still by diffusion so that one might expect the conductances to be in the same ratio, but as the effective boundary layer thicknesses for mass and heat transfer are proportional to $D^{1/3}$ it follows that conductances are approximately in the ratio of the $2/3$ power of the diffusion coefficients. As turbulence increases, transport in eddies becomes rapid in relation to molecular diffusion, so that in a fully turbulent boundary layer above a canopy, heat, water vapour and CO_2 are all transported equally efficiently and therefore the conductances approach equality. Appropriate factors for converting between conductances for other entities are given in Table 3.2.

The value of the boundary layer conductance for a leaf or other object depends mainly on its shape and size and on the wind speed. It is conveniently determined empirically for leaves of any given dimensions by measuring water loss from wet surfaces (e.g. blotting paper) of the same size under the same external conditions, or by energy balance measurements. These and other methods are outlined in Appendix 8. It has been found that conductance may be estimated with adequate precision for many purposes from the wind velocity (u) and the characteristic dimension (d), by making use of relationships that have been derived from a range of experiments and from heat transfer theory (Monteith & Unsworth, 2008). For flat plates in laminar forced convection conditions, the value of the

Table 3.2 Factors for converting conductances for different entities in different boundary layers relative to the heat transfer conductance (g_{aH}).

	Relationship	g_{aH}	g_{aW}	g_{aC}	g_{aM}
Still air	(D_i/D_H)	1.00	1.12	0.68	0.73
Laminar flow	$(D_i/D_H)^{0.67}$	1.00	1.08	0.76	0.80
Turbulent flow	(D_H)	1.00	1.00	1.00	1.00

boundary layer conductance to heat transfer (mm s^{-1}) is given by

$$g_{aH} = r_{aH}^{-1} = 6.62(u/d)^{0.5} \tag{3.31}$$

where d is the characteristic dimension (m) and u is the wind velocity (m s^{-1}). Note that this conductance refers to unit projected area of leaf (that is the area of *one* side) but includes heat transfer from both surfaces in parallel. Since mean boundary layer thickness is inversely related to g (i.e. $\delta = D_H/g_H$) it is easy to calculate the corresponding boundary layer thickness as ($\delta = 2 D_H(u/d)^{-0.5}/6.62$, where the factor 2 converts the conductance to that appropriate for exchange from one side of a leaf. For a 1 cm leaf in a wind of 1 m s^{-1}, therefore, $\delta = 0.65$ mm (i.e. $2 \times 0.215 \times 10^{-4} \times (1/0.01)^{-0.5}/6.62 \times 10^{-3}$ m).

Corresponding expressions for conductances of other shaped objects are:

for cylinders with their long axis normal to the flow:

$$g_{aH} = r_{aH}^{-1} = 4.03\,(u^{0.6}/d^{0.4}) \tag{3.32}$$

and for spheres:

$$g_{aH} = r_{aH}^{-1} = 5.71\,(u^{0.6}/d^{0.4}) \tag{3.33}$$

where d is the diameter of the cylinder or sphere. Equations (3.32) and (3.33) both refer to unit *surface* area.

Although Eqs. (3.31) to (3.33) are really only strictly applicable to smooth isothermal plates or other shapes in laminar flow, they are commonly used to estimate conductances for real leaves and other plant organs. In practice, however, surface temperatures are not uniform and some degree of turbulence in the leaf

boundary layer is common. Where turbulence occurs, Eqs. (3.31) to (3.33), which apply to laminar conditions, tend to underestimate the true conductance usually by a factor of between 1 and 2, though perhaps by as much as 3 in certain circumstances (see e.g. Grace, 1981; Monteith, 1981). In addition to factors such as leaf size and wind speed, the turbulent regime in the airstream can be important, with the altered turbulence pattern within dense plant canopies potentially affecting the conductance of rigid leaf models independently of wind speed.

The presence of leaf hairs also affects transfer in the leaf boundary layer (Johnson, 1975; Wuenscher, 1970). Sparse hairs may increase surface roughness and the tendency for turbulence. On the other hand, a dense mat of hairs is likely to increase the effective depth of the boundary layer (certainly for water vapour or CO_2 transfer) by up to the depth of the hair mat. A layer of still air trapped by hairs 1 mm long would have a resistance to water vapour diffusion of $\ell/D_W = 1 \times 10^{-3}/0.242 \times 10^{-4} = 41$ s m^{-1}. For momentum, however, the hairs would move the effective sink for momentum from the leaf surface to the surface of the hair mat, so hairs would affect the ratio between conductances for heat, mass and momentum.

Because of these complexities it is clear that it is difficult to estimate leaf boundary layer conductance accurately. Perhaps the best available generalisation is to increase the conductances calculated from Eqs. (3.31) to (3.33) by a factor of 1.5, giving the dependence of leaf conductance on wind speed and dimensions shown in Figure 3.6. The characteristic dimensions used in these calculations range from 1 mm (for narrow-leaved grasses and pine needles) to

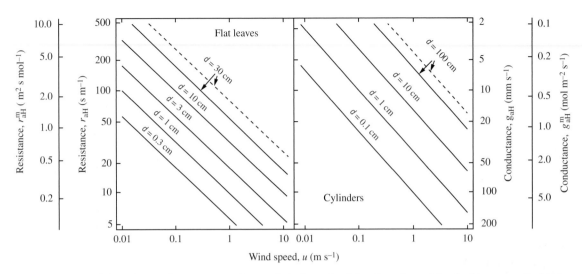

Figure 3.6 Estimated dependence of g_{aH} or r_{aH} on characteristic dimension (d) and wind speed for flat leaves or cylindrical leaves of stems in natural environments. The dashed lines give the value of g_{aH} predicted by Eqs. (3.31) or (3.32) for laminar flow, the solid lines ($g_{aH} = 1.5 \times$ value predicted by these equations) are for more typical flow conditions.

30 cm corresponding to very large leaves such as bananas. Variation of leaf size over this range changes g_a by more than an order of magnitude. Wind speeds at the top of plant canopies often exceed 1 m s^{-1}, but at times (e.g. at night) and deep in the canopy values may fall to 0.1 m s^{-1} or less.

Although forced convection is likely to dominate heat and mass transfer from leaves in natural environments, when large leaf-to-air temperature differentials occur, as with large leaves and high irradiances, there may be a significant contribution by free convection. With a 10°C leaf–air temperature differential, the free convection conductance for heat is likely to be about 3.2 mm s^{-1} (Monteith, 1981) so that it is comparable to that arising from forced convection only for the largest leaves at wind speeds less than 0.3 m s^{-1}.

3.4 Transfer processes within and above plant canopies

Several useful advanced treatments of the theory of transfer processes above plant canopies are available (e.g. Garratt, 1992; Kaimal & Finnigan, 1994) while the basic principles have been well summarised by Monteith and Unsworth (2008). Here we outline the

essentials necessary for interpreting modern studies of heat and mass transfer to and from plant canopies.

Many of the principles that have been applied to heat and mass transfer of individual leaves are also applicable to exchange by large areas of vegetation, but there are a number of important differences and complicating factors. First, the analysis is complicated by the fact that the 'surface' of a plant canopy (i.e. the sources or sinks of heat, water, CO_2 and momentum) is usually distributed over a significant depth of canopy; this distribution with depth is different for each entity. A second feature that has been of particular value in the development of micrometeorological techniques for the study of transfer processes between vegetation and the atmosphere is the difference in scale, with the boundary layer above a canopy being much deeper than that for a single leaf, so that it is possible to make measurements within the boundary layer; these measurements can be used to infer fluxes. A third feature of transfer within and above canopies is that the crop boundary layer is generally turbulent so that the transfer coefficients (K) for heat and mass transfer are usually assumed equal; this similarity assumption forms the basis of several of

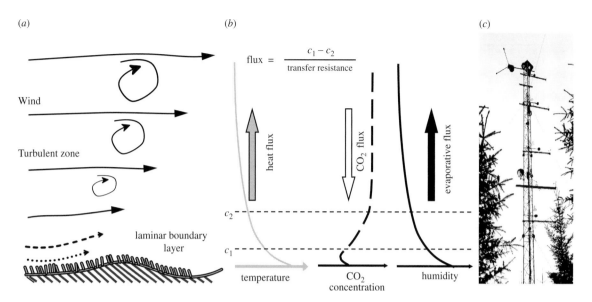

Figure 3.7 (*a*) Illustration of the wind-speed profile over vegetation, showing the transition from a laminar flow regime near the surface to a turbulent regime. Also shown in (*b*) are the gradients of air temperature, CO_2 concentration and atmospheric humidity showing the direction the corresponding fluxes of heat, CO_2 and water vapour. The flux of each entity can be derived from the concentration difference between any two heights and the transfer resistance over that zone. (*c*) A meteorological tower, showing environmental sensors at different levels.

the micrometeorological methods used to study canopy exchange processes. Perhaps most importantly, transfer processes within the boundary layer are usually very far from the traditional steady state that is often assumed as there is great temporal and spatial heterogeneity. For example, within the lowest part of the boundary layer (the interfacial or roughness sublayer – perhaps extending 1.5 to 3.5 times the height of roughness elements above the surface), there can be substantial horizontal variation relating to the detailed surface structure; this hinders effective sampling of representative surface fluxes in this zone. Above this roughness sublayer is a constant flux region known as the inertial sublayer; this is the preferred region within which micrometeorological studies of fluxes are conducted.

3.4.1 Flux measurement

Aerodynamic method for flux measurement
Although the aerodynamic approach to measurement of mass and heat fluxes in the canopy boundary layer

has largely been superseded by other methods (see below) it is still useful to introduce the relevant theory. These flux-gradient approaches (Figure 3.7) are based on Monin–Obukhov similarity theory (MOST) and require that there is no convergence or divergence of flux (i.e. there are no sources or sinks within the boundary layer for the entity being transported) and no advection (i.e. no lateral net import of energy or mass). In other words, the conservation equation applies and a one-dimensional vertical flux can be assumed, with the flux being constant at different heights, even though the transfer coefficient at any height z, $K_i(z)$, varies. The flux of mass or momentum above the crop, therefore, can be described by the standard gradient-diffusion assumption where the flux is proportional to a transfer coefficient multiplied by the driving concentration gradient:

$$\mathbf{J}_i = -\mathbf{K}_i(\partial c_i / \partial z) \tag{3.34}$$

Integrated between any two levels (1 and 2, or 2 and 3) this gives (cf. Figure 3.1):

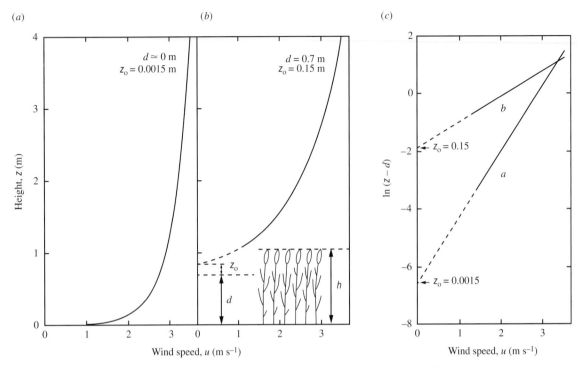

Figure 3.8 Hypothetical mean profiles of wind speed when wind speed at 4 m equals 3.5 m s^{-1} for (*a*) bare ground and (*b*) a cereal crop together with (*c*) corresponding linearising logarithmic transformations.

$$\mathbf{J}_i = \overline{\mathbf{K}}_i(c_{i1} - c_{i2})/(z_2 - z_1)$$
$$= \overline{\mathbf{K}}_i(c_{i2} - c_{i3})/(z_3 - z_2)$$
$$= g_{Ai}(c_{i2} - c_{i3}) \qquad (3.35)$$

where $\overline{\mathbf{K}}_i$ is the effective mean transfer coefficient over the distance between the sensor levels (dz) and g_{Ai} is the conductance.

For this equation to hold, measurements must be made entirely within the crop boundary layer that has developed from the 'leading edge' of the field or area of vegetation being studied. The depth of the boundary layer increases with distance or 'fetch' from the leading edge (Figure 3.7). In general it is assumed that measurements may be made with adequate precision up to a height above the canopy equal to about 0.01 × fetch. The 'footprint' of any sensor is the area of canopy that primarily determines the observed flux at a sensor and is usually assumed to refer to that upwind area of canopy within 100 × the height of the sensor above the canopy. It is also found

that measurements need to be made in the inertial sublayer, well above the underlying canopy, because the erratic turbulence structure within the roughness sublayer leads to such great variability in $\mathbf{K}$ that Eq. (3.34) has little practical value in this zone (Kaimal & Finnigan, 1994; Raupach, 1989b). It follows that micrometeorological studies of fluxes through the crop boundary layer require large areas of homogeneous vegetation, the size of which depends on the height above the canopy at which sensors are placed.

The importance of Eqs. (3.34) and (3.35) is that they allow one to estimate fluxes of heat or mass from measurements of the appropriate concentration gradients using the transfer coefficients or conductances obtained from analysis of the measured wind-speed profile. Wind speed increases with height above open ground or above plant communities, with the rate of increase being greatest near the ground, as shown in Figure 3.8. The shape of this wind-speed profile is such that over open ground the logarithm of height ($\ln z$) is linearly related to the wind speed at that height (u_z). Expressing u_z in terms of $\ln z$ gives:

$$u_z = A\,(\ln z - \ln z_0) = A \ln(z/z_0) \tag{3.36}$$

The intercept on the $\ln z$ axis is $\ln z_0$, where z_0 is called the *roughness length*, and is a measure of the aerodynamic roughness of the surface. The slope, A, is usually replaced by the term u_*/k, where u_* is called the *friction velocity* (having dimensions of velocity) and characterises the turbulent regime, and k is a dimensionless constant ($= 0.41$) named after von Karman.

Over vegetation, unlike over open ground, the wind-speed profile is no longer linear when u is related to $\ln z$. Instead u is linearly related to $\ln (z - d)$, where d is an apparent reference height, the *zero plane displacement* (Figure 3.8). As shown in Figure 3.8 wind speed extrapolates to zero at a height of $d + z_0$ (though actual wind speed at this height is still finite). Substituting $(z - d)$ for z in Eq. (3.36) gives:

$$u_z = (u_*/k)\ln\left[(z-d)/z_0\right] \tag{3.37}$$

as describing a wind-speed profile above vegetation. The plane at a height $(d + z_0)$ may be regarded as an apparent sink for momentum.

It has been found that reasonable approximations to d and z_0, for a range of relatively dense vegetation types are (Campbell & Norman, 1998):

$$d = 0.64\,h \tag{3.38}$$

and

$$z_0 = 0.13\,h \tag{3.39}$$

where h is the crop height. More appropriate values for coniferous forest are given by the following equations (Jarvis *et al.*, 1976):

$$d = 0.78\,h \tag{3.38a}$$

$$z_0 = 0.075\,h \tag{3.39a}$$

In practice, d and z_0 vary with wind speed and canopy structure in a fairly complex manner (Kaimal & Finnigan, 1994; Monteith, 1976).

It is possible to use the wind-speed profile to estimate transfer coefficients and conductances in the crop boundary layer and hence transfer of other entities. Remembering that the 'concentration' of momentum (c_M) equals ρu, one can write Eq. (3.34) as:

$$\tau = -\rho\,K_M(\partial u/\partial z) \tag{3.40}$$

where K_M is a turbulent transfer coefficient for momentum ($L^2\,T^{-1}$). It can also be shown (Monteith & Unsworth, 2008) that:

$$K_M = k\,u_*\,z \tag{3.41}$$

and that:

$$\tau = \rho u_*^2 \tag{3.42}$$

Assuming similarity of the transfer coefficients, fluxes for other entities such as water vapour (J_W or E), CO_2 (J_C) or heat (J_H or C) can therefore be obtained from the relevant concentration gradients:

$$\begin{aligned}
J_H &= C \simeq -(k\,u_*\,z)/(\partial(\rho c_p T)/\partial z)) \\
J_W &= E \simeq -(k\,u_*\,z)/(\partial c_W/\partial z) \\
J_C &\simeq -(k\,u_*\,z)/(\partial c_C/\partial z)
\end{aligned} \tag{3.43}$$

As momentum transfer is analogous to other transport processes one can also define a conductance for momentum transfer between height z and the reference plane ($z = d + z_0$) using the usual transport equation:

$$\tau = g_{AM}\rho[u_z - u_{(d+z_0)}] = g_{AM}\,\rho u_z \tag{3.44}$$

where g_{AM} is the canopy boundary layer conductance for momentum ($= r_{AM}^{-1}$).

Combining Eqs. (3.42) and (3.44) gives:

$$g_{AM} = u_*^2/u_z \tag{3.45}$$

which can be expressed in terms of the parameters of the wind profile equation (Eq. (3.36)) to give:

$$g_{AM} = \frac{k^2 u_z}{\{ln[(z-d)/z_0]\}^2} \tag{3.46}$$

Not only does this equation imply that g_{AM} increases with wind speed, but it also indicates that conductance tends to increase with crop height (as d and z_0 both increase with height). Substituting, for example, values of u, d and z_0 from Figure 3.8 into

Eq. (3.46) gives, for a wind speed of 3.5 m s^{-1} at 4 m, $g_{AM} = 9$ mm s^{-1} for the bare ground ($d \simeq 0$, $z_o = 0.0015$) and $g_{AM} = 62$ mm s^{-1} for the cereal crop ($d = 0.7$, $z_o = 0.15$).

Because the apparent sink for momentum in a canopy is above those for heat or mass exchange, there is a small extra resistance required when converting from r_{AM} to the corresponding resistances for heat or mass transfer. This extra resistance refers to transfer between the level of the momentum sink ($d + z_o$) and the alternate sink.

Making use of the similarity assumption for the turbulent transfer of different entities in the boundary layer, Eq. (3.46) can be used as an estimate for the canopy boundary layer conductances for other entities such as heat, CO_2 and water vapour so that fluxes may be estimated from appropriate concentration differences. Alternatively fluxes may be estimated directly from Eq. (3.43).

A problem with Eqs. (3.37) and (3.46) is that they hold only when the temperature profile in the atmosphere is close to neutrality. At neutrality the temperature decreases with height according to the dry adiabatic lapse rate (c. 0.01°C m^{-1} – see Chapter 11). If temperature decreases more rapidly with height there is a tendency for free convection to occur as a result of 'buoyancy' effects. This makes the atmosphere unstable and turbulent transfer is enhanced. Conversely, when temperature increases with height (a temperature inversion), the atmosphere is stable and transfer is suppressed because the less dense air is above the cooler denser air. In either case the normal profile equations need modification by an amount that depends on the balance between thermal buoyancy effects and mechanical turbulence effects (Garratt, 1992; Kaimal & Finnigan, 1994; Monteith, 1975; Monteith & Unsworth, 2008).

Bowen ratio

An alternative approach to the use of data from more than one measurement height is the *Bowen ratio* method; this energy balance has been widely used in the past to estimate evaporation fluxes from canopies. The canopy energy balance (see Chapter 5) is:

$$\mathbf{R}_n - \mathbf{G} = \lambda\mathbf{E} + \mathbf{C} \tag{3.47}$$

where $\mathbf{R}_n$ is the net radiation absorbed, $\mathbf{G}$ is the heat flux into the soil, λ is the latent heat of vaporisation of water, $\mathbf{E}$ is the evaporative flux and $\mathbf{C}$ is the sensible heat flux. This can be rewritten as:

$$\lambda\mathbf{E} = (\mathbf{R}_n - \mathbf{G})/(1 + \mathbf{C}/\lambda\mathbf{E}) = (\mathbf{R}_n - \mathbf{G})/(1 + \beta)$$
$$= (\mathbf{R}_n - \mathbf{G})/(1 + \gamma\partial T/\partial e) \tag{3.48}$$

where β is the Bowen ratio ($= \mathbf{C}/\lambda\mathbf{E}$), γ is the psychrometer constant and $\partial T/\partial e$ is the ratio between the gradients of temperature and humidity. This approach can be extrapolated to allow the estimation of fluxes of other scalars such as CO_2 (Monteith & Unsworth, 2008). An advantage of the Bowen ratio method is that it does not require as sophisticated instrumentation as do other methods.

Eddy covariance

Recent advances in sensor technology have allowed the widespread adoption of a more direct approach (*eddy covariance*) to the estimation of canopy fluxes that sums the vertical components of fluxes in the passing eddies at a single sensor position (see Aubinet *et al.*, 2012; Monteith & Unsworth, 2008). This approach makes use of the fact that atmospheric eddies are the main agents of transfer in the atmospheric boundary layer, and that the concentration of a scalar such as water vapour being transported upwards from a canopy will be on average higher in the upward moving arm of an eddy than in the downward moving side. The vertical flux density is the time-averaged product of the vertical component of wind speed (w') and the relevant scalar concentration (c_i') (Figure 3.9). Although conservation requirements indicate that the long-term average vertical wind speed must be zero, the covariance between the direction of flux and concentration of an entity being transported allows one to approximate the net vertical transfer of heat or mass as the time average of the product

$$\mathbf{J}_i \simeq \overline{c_i'w'} \tag{3.49}$$

The frequency response required of anemometers and other sensors used for eddy covariance depends

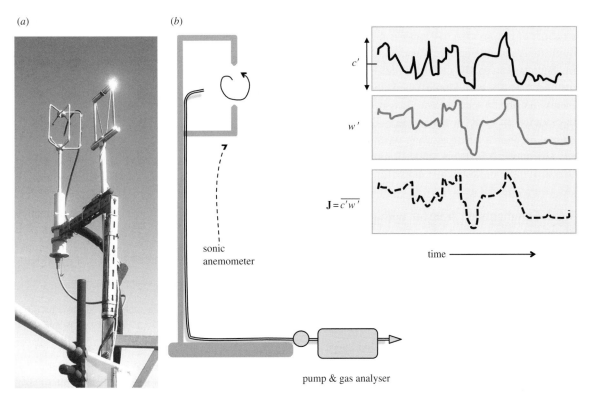

Figure 3.9 An illustration of (*a*) sensors for an eddy covariance system (showing on the left a three-dimensional sonic anemometer and on the right a krypton hygrometer) mounted on a meteorological mast and (*b*) the basic principles of operation of an eddy covariance system. The sonic anemometer measures the rapid changes in direction and velocity of air passing the sensor, infrared sensors measure the corresponding changes in concentrations of water vapour and CO_2, and a thermistor measures the corresponding rapid changes in air temperature. Together, these measurements allow calculation of the fluxes of heat, water vapour and CO_2 according to $J_i = \overline{c_i' w'}$, where $\overline{c_i' w'}$ is the instantaneous product of the vertical velocity (w') and the concentration (c_i') of the entity being transferred.

on the size of the eddies being used, which increases with height above the canopy. Frequency responses between 0.1 and 10 Hz are typically adequate for forest canopies, but a more rapid response is necessary for smooth surfaces such as grassland. Detailed precautions in the application of eddy covariance techniques are outlined in Aubinet *et al.* (2012).

Methods based on canopy temperature variation

A number of simpler and cheaper approaches to the estimation of above-canopy fluxes based on analysis of the dynamics of rapid temperature variation in the canopy have been proposed. These include *temperature variance* (Tillman, 1972) and

surface renewal. The latter approach is based on the observation that heat tends to build up steadily in a canopy and is only removed at intervals in sharp bursts in large parcels of air; the rate of heat loss can then be calculated from the rate and timing of the build up of temperature and the effective volume of the parcels of air ejected (Castellví & Snyder, 2009; Snyder *et al.*, 1996). Although these methods generally need calibration against eddy covariance, they can provide useful estimates of heat fluxes. An alternative approach involves the use of *scintillometry*: in this approach fluctuations in NIR radiation (e.g. at 0.94 μm) transmitted over a defined path within the boundary layer (which may be several km in length) are used to measure the turbulence structure of the air refractive index (caused especially by fluctuations in temperature and humidity) within the boundary layer. The derived turbulence structure parameter is used with Monin–Oblukhov similarity theory to estimate sensible heat flux, which, with information on available energy $(\mathbf{R}_n - \mathbf{G})$, allows estimation of area averaged water vapour fluxes (see e.g. Meijninger & De Bruin, 2000; Thiermann & Grassl, 1992; Van Kesteren *et al.*, 2013).

3.4.2 Comments on transfer within plant canopies

The erratic turbulence structure within plant canopies and the complexities introduced by the distribution of sources and sinks for heat, mass and momentum make the application of micrometeorological techniques to transfer processes within the canopy extremely difficult. Nevertheless clear wind-speed profiles can be observed within canopies (Figure 3.10), noting that in some canopies wind speed may be highest near the ground, particularly in forests that have little understorey vegetation.

The microclimate within a canopy depends on the source distributions and concentration fields of heat, water vapour and CO_2. The variation in source density for an entity i with height (the source density profile, $S_i(z)$, where a sink is a negative S), depends on physical and physiological processes particularly at the leaves, while the concentration profile $c_i(z)$ also depends on the turbulent wind flow and the way this distributes the entity under consideration. Since the law of conservation must apply, the source density in any horizontal plane is related to the change in vertical flux across that plane:

$$S_i(z) = d\mathbf{J}_i/dz \qquad (3.50)$$

The flux across the plane at height z is given by the integral of this equation from the ground to height z.

Analysis of the turbulence structure within and above canopies has demonstrated that the strongest turbulent events in a wide range of canopy types are gusts: energetic, downward incursions of air into the

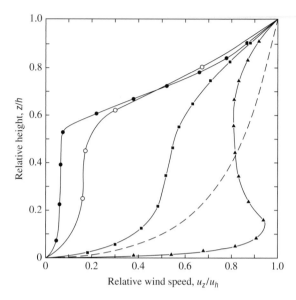

Figure 3.10 Normalised within canopy wind speed profiles for (●) a dense stand of cotton, (□) dense hardwood jungle with an understorey, (▲) isolated conifer stand with no understorey (Raupach, 1989a; Warland & Thurtell, 2000), (○) a corn crop CO_2 (see Businger, 1975 for original references) and (– – – – –) a logarithmic profile (Eq. (3.36)) with $z_0 = 0.001\, h$.

canopy space from the faster moving air above (see Raupach, 1989b). These gusts tend to be very intermittent but are responsible for most of the momentum transfer (more than 50% of energy may be transferred in events occupying less than 5% of the time). The resulting variability of $\mathbf{K}$ means that gradient analogies are often not helpful, with counter gradient fluxes (and hence apparently negative $\mathbf{K}$ values) sometimes observed (Denmead & Bradley, 1987). More sophisticated models such as the use of Lagrangian dispersion analysis, which involves a moving framework for analysis (rather than the usual fixed positions) and follows the trajectories of individual fluid particles (Raupach, 1989a; Warland & Thurtell, 2000), can be used to infer source–sink distributions within plant canopies. This approach is very dependent on the accurate characterisation of the turbulence statistics profiles within and above the canopy. The typical canopy eddies are coherent structures of similar dimensions to the canopy height that persist for long periods: the wind waves across cereal fields provide familiar visual evidence for the persistence of turbulent motions of this scale.

3.5 Sample problems

3.1 Water vapour is diffusing down a 10 cm isothermal tube at 20°C from a wet surface ($c_W = 17.3$ g m^{-3}) to a sink consisting of saturated salt solution (equilibrium $c_W = 11$ g m^{-3}). Calculate (i) J_W, (ii) g_W; and the equivalent molar values (iii) J_W^m and (iv) g_W (g_W^m).

3.2 For a 2 cm diameter circular leaf exposed in a laminar airstream moving at 1 m s^{-1} (i) what are (a) g_{aH}, (b) g_{aW}, (c) g_{aM} and (d) the mean boundary layer thickness for momentum? (ii) What would the values for these conductances be if the leaf were covered in a mat of hairs 1 mm deep? (iii) Is the assumption of a laminar boundary layer likely to be valid?

3.3 If the wind speed at 2 m is 4 m s^{-1} when blowing over a wheat canopy 80 cm tall, what are (i) u_*, (ii) the wind speed at the top of the canopy, (iii) τ and (iv) between the reference plane and 2 m?

4 Plant water relations

Contents

4.1 Physical and chemical properties of water

Water is an essential plant component, being a major constituent of plant cells, and ranges from about 10% of fresh weight in many dried seeds to more than 95% in some fruits and young leaves. Many of the morphological and physiological characteristics of land plants discussed in this book are adaptations permitting life on land by maintaining an adequate internal water status in spite of the typically rather dry aerial environment.

The unique properties of water (see Franks, 1972; Kramer & Boyer, 1995; Nobel, 2009; Slatyer, 1967 for details) form the basis of much environmental physiology. For example water is a liquid at normal temperatures and is a strong solvent, thus providing a good medium for biochemical reactions and for transport (both short-distance diffusion and long-distance movement in the xylem and phloem). Water is also involved as a reactant in processes such as photosynthesis and hydrolysis, while its thermal properties are important in temperature regulation and its incompressibility is important in support and growth.

The properties of water derive from its structure (Figure 4.1), and from the fact that it dissociates into hydrogen and hydroxyl ions that are always present in solution. The angle between the two covalent O–H bonds and the asymmetry of the charge distribution along the bonds gives rise to a marked polarity of charge so that water is a dipole. This polarity allows the development of so-called hydrogen bonds between adjacent water molecules, as illustrated in Figure 4.1, or between water and other charged surfaces or molecules. This hydrogen bonding, though weak (only having a bond energy of ~20 kJ mol^{-1}, compared with the bond energy of the covalent O–H bond of ~450 kJ mol^{-1}), can give rise to significant 'structure' even in liquid water.

It is a consequence of the intermolecular hydrogen bonds that water tends to be a liquid at much higher temperatures than other small molecules (e.g. NH_3, CH_4 or CO_2), which do not have the marked polarity and hydrogen bonding characteristic of water. The hydrogen bonds help to maintain water in

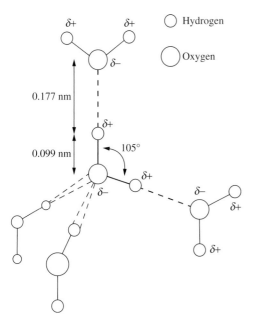

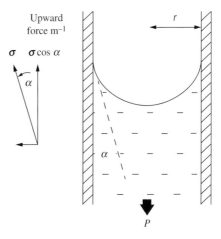

Figure 4.2 Capillary rise in liquids. At equilibrium the tension (P) is balanced by the vertical component of the adhesive force divided by the area over which it acts (πr^2). The vertical component of the adhesive force is given by $\sigma \cos\alpha$ multiplied by perimeter of the capillary ($2\pi r$), so that $P = (2\,\sigma \cos\alpha/r)$.

Figure 4.1 Schematic structure of water molecules showing the hydrogen bonding that results from the electrostatic attraction between net positive charge on hydrogen and net negative charge on oxygen atoms. In ice the tetrahedral structure shown is fairly rigid but there is ordering even in liquid water.

a semi-ordered liquid form. A measure of the high degree of ordering in liquid water can be obtained from the fact that the amount of energy required to convert solid ice to liquid water (the latent heat of fusion) is 6.01 kJ mol^{-1}, only about 15% of the total energy required to break all the hydrogen bonds in the ice.

Particularly important for its role in plants is the cohesive strength of water. The hydrogen bonding between water molecules gives rise to strong cohesive forces that tend to hold the molecules together and gives rise to its extremely high surface tension. The surface tension (σ) is defined as the force transmitted across a line in the surface and equals 7.28×10^{-2} N m^{-1} for water in air at 20°C. This property drives capillary rise phenomena including the retention of water within the interstices in soils and in the cellulose matrix of plant cell walls. Capillary rise involves both adhesive forces between the liquid water and the solid phase of the capillary wall and cohesive forces within the liquid water. The

phenomenon is illustrated in Figure 4.2, where it can be seen that as long as the contact angle (α) between the liquid and the surface is less than 90° there will be a vertical force acting on the liquid tending to draw it up a capillary. The magnitude of this force is given by the vertical component of the surface tension ($= \sigma \cos \alpha$) multiplied by the distance over which it acts (i.e. the perimeter of the capillary, which for a circular cross-section $= 2\pi r$), so that the total upward force $= 2\pi r\sigma \cos \alpha$. At equilibrium, this force is balanced by the gravitational force acting on the column of liquid. This gravitational force is approximately the product of the mass of liquid in the column ($\simeq \pi r^2 h\rho$, where h is the height of capillary rise and ρ is the liquid density $= 998$ kg m^{-3} for water at 20°C) and the acceleration due to gravity ($g = 9.8$ m s^{-2} at sea level). Equating these forces leads to:

$$h = 2\,\sigma \cos \alpha/r\rho g \qquad (4.1)$$

The contact angle is close to zero for wettable surfaces such as soils, cell walls and xylem vessels, which have many polar groups on their surfaces. Equation (4.1) shows that capillary rise is inversely proportional to the capillary radius.

Instead of calculating the capillary rise, it is often more convenient to have a measure of the suction or tension in the liquid (P) required to drain a capillary of given radius. The tension (with units of pressure, N m^{-2} or Pa) is given by the force divided by the area (πr^2) over which it acts, so that:

$$P = (2\pi r \,\sigma \cos\alpha)/\pi r^2 = (2\sigma \cos\alpha)/r \qquad (4.2)$$

which reduces to:

$$P = 2\sigma/r \qquad (4.3)$$

when α is zero. Applying this to the pores in the cell wall matrix of higher plants, which have a typical radius of approximately 5 nm, gives a suction of about 30 MPa ($(2 \times 7.28 \times 10^{-2})/(5 \times 10^{-9})$ Pa). This is the pressure required to force the air–water interface into the cell wall, i.e. to drain the cell wall. A suction of 30 MPa would support a column of water approximately 3 km tall! The static tension at the top of the tallest trees (c. 100 m) would only be 1 MPa. The following approximation of Eq. (4.3) is a useful 'rule-of-thumb' for calculating the pressure required to drain any pore: pressure (MPa) $\simeq 0.3$/diameter (in μm).

The strong cohesive forces between the molecules of water are also directly involved in drawing water up in transpiring plants. One can calculate from the energy in the hydrogen bonds that the maximum theoretical tensile strength in an unbroken column of pure water would be greater than 1000 MPa (Steudle, 2001). In practice, however, tensions greater than about 6.0 MPa can rarely be sustained in xylem vessels as is discussed below. Although the cohesive forces are normally strong enough to permit water to be drawn to the top of tall trees, on occasions the water columns in the xylem vessels rupture or cavitate. The role of capillarity and cohesion in water movement in plants and the significance of cavitation are discussed later in this chapter.

The high specific heat capacity of water ($c_p = 4182$ J kg^{-1} K^{-1}) (that is, the heat energy required to change the temperature of 1 kg by 1 K at constant pressure) contributes to temperature stability, both in aquatic environments and in plants with large water stores (e.g. cacti). Equally, the high latent heat of

vaporisation ($\lambda = 2.454$ MJ kg^{-1} at 20°C) (that is, the energy required to convert 1 kg of liquid water to vapour at constant temperature) has important consequences for temperature regulation of plant leaves (Chapter 10).

The dielectric constant is a measure of the impermeability of a medium to the attraction between electric charges, so that a material with high $\mathscr{D}$ reduces the strength of ionic attractions between different molecules. Water has a large dielectric constant ($\mathscr{D} = 80.2$ at 20°C); this also results from the polar structure of water. This value is over 40 times that for a non-polar liquid like hexane. This effect is involved in making water an extremely powerful solvent and hence a good medium for many biochemical reactions. Other properties, such as its dissociation into H$^+$ and OH$^-$, its spectral absorptance and the properties of its solid form, ice, also have important physiological and ecological consequences.

4.1.1 Water potential

The amount of water present in a system is a useful measure of plant or soil water status for some purposes. More commonly, however, the water status in plant systems is measured in terms of *water potential* (ψ, MPa), which is a measure of *free energy* available to do work. Water potential is defined in terms of the chemical potential of water (μ_W), which in turn is the amount by which the Gibbs free energy (G) in the system changes as water is added or removed while temperature, pressure and other constituents remain constant. In mathematical notation:

$$\mu_W = \left(\frac{\partial G}{\partial n_W}\right)_{T, P, n_i} \qquad (4.4)$$

where n_W is the number of moles of water added. Water moves spontaneously only from a region of higher chemical potential to one of a lower chemical potential. As water moves down its chemical potential gradient, it releases free energy so that such a flow has the potential to do work.

The chemical potential has units of energy content (i.e. J mol^{-1}). It is, however, the practice in plant physiology to express water status as water-potential using pressure units. This can be done by dividing the chemical potential by the partial molal volume of water ($\overline{V}_W = 18.05 \times 10^{-6}$ m^{-3} mol^{-1} at 20°C), and using the following definition for water potential:

$$\psi = \frac{\mu_W - \mu_W^\circ}{\overline{V}_W} \qquad (4.5)$$

where μ_W° is the chemical potential of water at a reference state consisting of pure free water at the same temperature, at atmospheric pressure and at a reference elevation. This definition has the consequence that ψ is zero when water is freely available decreasing to negative values as water becomes scarce, so 'higher' water potentials, at least in plant systems, are generally less negative. Although the bar has often been used as the unit of water potential, the appropriate SI unit is the pascal (1 Pa = 1 N m^{-2} = 10^{-5} bar) so water potentials will normally be expressed as MPa (1 MPa = 10 bar).

The total water potential may be partitioned into several components, one or more of which may be relevant in any particular system:

$$\psi = \psi_p + \psi_\pi + \psi_g \qquad (4.6)$$

where ψ_p, ψ_π and ψ_g respectively are components due to pressure, osmotic and gravitational forces. The pressure component (ψ_p) represents the difference in hydrostatic pressure from the reference and can be positive or negative. The osmotic component (ψ_π) results from dissolved solutes lowering the free energy of the water and is always negative. Rather than referring to a negative osmotic potential, many authors use the term osmotic pressure ($\Pi = -\psi_\pi$). This osmotic pressure is the pressure that can be generated within a solute-containing compartment (such as a cell) separated from pure water by a perfectly semipermeable membrane in the process of *osmosis*. The pressure arises because the solutes within the compartment act to impede outward diffusion of water through the water channel pores in the membrane as the solutes are repelled by the pores, so that more water molecules tend to flow in than out. It can be shown that the osmotic potential is

related to the mole fraction of water (x_W) or its activity (a_W) by:

$$\psi_W = \frac{\mathcal{R}T}{\overline{V}_W} \ln (\gamma_W x_W) = \frac{\mathcal{R}T}{\overline{V}_W} \ln (a_W) \qquad (4.7)$$

where γ_W is an activity coefficient that measures departure from ideal behaviour by the solution. As the concentration of solutes increases, x_W and ψ_π decrease. Although γ_W equals 1 in very dilute solutions, most plant systems show some departure from ideal behaviour (see Milburn, 1979 for some useful tabulations). A very useful approximation of Eq. (4.7) that is reasonably accurate for many biological solutions is the van't Hoff relation:

$$\psi_\pi = -\mathcal{R}T c_s = -\mathcal{R}T(n_s/V) \qquad (4.8)$$

where c_s is the concentration of solute expressed as mol m^{-3} solvent (or more precisely as mol per 10^3 kg solvent). Typical cell sap from many plants has an osmotic potential of about –1 MPa (Π = 1 MPa). Using Eq. (4.8) and substituting the value of $\mathcal{R}T$ at 20°C (2437 J mol^{-1}) gives a total cell sap solute concentration of $-(-10^6/2437) \simeq 410$ osmol m^{-3}. (An osmole is analogous to a mole in that it contains Avogadro's number of osmotically active particles – i.e. 1 mol of NaCl contains 2 osmol.)

Many treatments also include a matric potential (ψ_m) on the right-hand side of Eq. (4.6) to account for forces affecting water activity in solid matrices such as cell walls and soils, but it seems best to include this in a combined osmotic term (ψ_π) that incorporates all aspects of the reduction of a_W that result from interactions with solutes and solids (Passioura, 1980).

The gravitational component (ψ_g) results from differences in potential energy due to a difference in height from the reference level, being positive if above the reference, and negative below:

$$\psi_g = \rho_W gh \qquad (4.9)$$

where ρ_W is the density of water and h is the height above the reference plane. Although often neglected in plant systems, ψ_g increases by 0.01 MPa m^{-1} above the ground, so it must be included when considering water movement in tall trees. A tension of 1 MPa would therefore be required to raise water to the top of a 100 m tree.

4.2 Cell water relations

Plant cells behave as osmometers with an inner compartment, the protoplast, bounded by the semi-permeable plasma membrane, which is relatively permeable to water and impermeable to solutes. The degree of semi-permeability of a membrane to any solute is described by the *reflection coefficient* (σ), which varies between 0 for a completely permeable membrane to 1 for a perfectly semi-permeable membrane. Since water permeates the plasma membrane relatively easily, largely through the action of specialised membrane-spanning transporter proteins known as *aquaporins* (Maurel et al., 2008), the water potential within cells equilibrates with the immediate environment within seconds, though it takes longer for all cells in a tissue to equilibrate when the bathing solution is changed.

Although aquaporins have only been known in plant membranes since the early 1990s, their existence had been predicted in the 1960s (Dainty, 1963). It is now clear that they are important regulators not only of membrane water transport but also of the transport of small neutral molecules (glycerol, boric acid, lactic acid and urea) and gases such as NH_3 and CO_2. Aquaporins are members of the family of major intrinsic proteins (MIPs) and the many isoforms, each of which has specialised transport activity, fall into at least four subfamilies including the tonoplast intrinsic proteins (TIPs) and plasma membrane intrinsic proteins (PIPs). Different aquaporins are involved in the control of water and solute transport across many different internal and external membranes in plant cells. The opening or closing of the pores (gating) can be controlled by a range of effectors including H^+ and divalent cations. Changes in aquaporin amount and gating underlie the enormous differences in hydraulic permeabilities

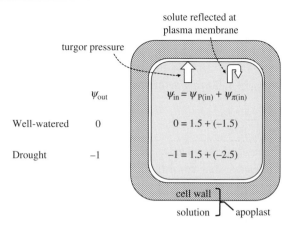

Figure 4.3 Illustration of a plant cell showing the cell wall and the plasma membrane, which is impermeable to solutes and leads to the generation of turgor pressure. Water potential and its components are shown at equilibrium when ψ_{in} equals ψ_{out}. Osmotic adjustment maintains turgor pressure as water potential falls.

of different membranes and at different stages in development.

Another important characteristic of plant cells is that they are encased in a relatively rigid cell wall that resists expansion, thus enabling the generation of an internal hydrostatic pressure (Figure 4.3). The main components of water potential that are relevant in plant cells, therefore, are the osmotic and pressure components so that:

$$\psi = \psi_p + \psi_\pi \tag{4.10}$$

The pressure difference between that inside and that outside the cell wall is usually positive and is commonly called the turgor pressure (P). For a given cell solute content the turgor pressure decreases as the total water potential falls. Water potentials in transpiring leaves are usually between −0.5 and −3.0 MPa.

The water relations of plant cells (and tissues) are conveniently described by the Höfler–Thoday diagram (Figure 4.4); this shows the interdependence of cell volume (V), ψ, ψ_π and ψ_p as a cell loses water. In a fully turgid cell $\psi = 0$, and ψ_p reaches a maximum value $= -\psi_\pi$. At this point the water content expressed as a fraction of the water content at full turgidity

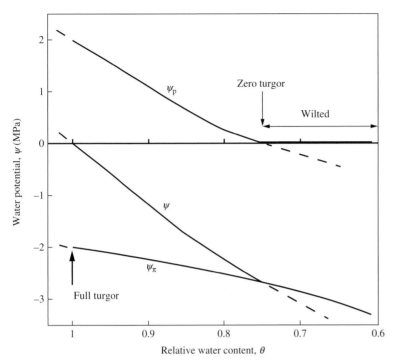

Figure 4.4 Höfler–Thoday diagram illustrating the relationships between total water potential (ψ), turgor potential (ψ_p), osmotic potential (ψ_π) and relative water content (θ) as a cell or tissue loses water from a fully turgid state. The dotted line below zero turgor pressure represents possible negative turgor in rigid cells.

(the *relative water content*, θ) is 1 (by definition). As water is lost, the cell volume decreases so that the turgor pressure generated as a result of elastic extension of the cell wall decreases approximately linearly with cell volume until the point of zero turgor (when ψ_p equals zero). In most plant cells, with any further decreases in water content turgor pressure remains close to zero (though see discussion of evidence for negative turgor by Acock & Grange, 1981). It is probable that negative pressures develop in certain rigid cells such as the ascospores of *Sordaria* (Milburn, 1979) driving explosive dispersal. As the cell volume decreases, the osmotic potential declines curvilinearly as expected from Eq. (4.8), which indicates that $-\psi_\pi$ is approximately inversely related to cell volume.

Many plants have an ability to at least partially maintain turgor by a process of *osmotic adjustment* as water potential decreases (whether as a result of soil drying or as a result of increasing soil salinity – see Chapter 10). This process of turgor maintenance involves increasing the solute content of the cells, largely by increasing the concentrations of salts or other solutes in the vacuoles. It is necessary, however,

that the corresponding increase of solute concentration in the cytoplasm is achieved by the synthesis of what are known as compatible solutes (e.g. proline, glycine-betaine, sugars and polyols) that protect the cytoplasm from potential damaging effects of high solute concentrations.

Visible wilting of leaves is usually observed when the point of zero turgor is reached. This corresponds with the point of limiting plasmolysis for cells immersed in solutions of lower water potential when the cells plasmolyse (i.e. the cell membrane separates from the cell wall), possibly doing irreversible damage. In normal aerial tissues, however, plasmolysis does not occur because the capillary forces at the air–water interface in the cell wall microcapillaries prevent them draining (see above), so that all the tension is supported by the wall rather than the membrane. Plasmolysis also does not occur when cells are immersed in solutions in which the solutes are too large to penetrate the cell walls because, like in normal aerial tissues, all the tension generated is supported by the cell wall. As solute concentrations increase, the tension eventually becomes insupportable so that the cells collapse – a condition known as cytorrhysis. In such a system, small negative turgor pressures of much less than 0.1 MPa cause collapse of mesophytic cells, but the small thickened cells of some desert plants can withstand in excess of 1.6 MPa (Oertli et al., 1990).

An important character determining the shape of the curves in Figure 4.4 is the elasticity of the cell walls. If the cell wall is very rigid, the water potential and its components change relatively rapidly for a given water loss. The wall rigidity as measured in a tissue is described by the *bulk modulus of elasticity of the cell* (ε_B), which can be defined by:

$$\varepsilon_B = dP/(dV/V) \qquad (4.11)$$

though some authors normalise it to V_o, the turgid volume of the cell, rather than to V. It should be noted that this bulk modulus of elasticity is distinct from the modulus of elasticity of the cell wall material itself, and in addition that it depends to some extent on the tissue structure and the nature of the interactions between the cells. Values of ε_B for plant cells are normally in the range of 1 to 50 MPa, where the larger values indicate relatively inelastic cells, or tissues with small cells. By analogy with Young's modulus for linear expansion (see Chapter 11), a bulk modulus of elasticity for a homogeneous solid would be expected to be constant with volume. Plant tissues are neither solid nor homogeneous and when they are compressed they lose water (i.e. matter is not conserved as it is in a solid), so it is not surprising that ε_B shows a marked non-linear behaviour, often increasing with turgor pressure in an approximately hyperbolic fashion from near zero at zero turgor.

Although the Höfler–Thoday diagram is appropriate for single cells, the various cells in any tissue are of different sizes and may have different wall elasticities and solute contents. In addition, in a tissue there is a pressure component caused by neighbouring cells pressing on each other. Therefore the properties of a tissue, although they can be represented by this type of diagram, can be very different from those of the component cells. For example, the individual cells in the mesocarp of a

cherry fruit are thin walled and elastic (low ε_B), but normally in the fruit these cells are constrained by a relatively rigid skin, so that the ε_B measured in intact tissue may be much larger than that for the component cells.

The solute concentrations in the *apoplast* (which refers to the solution in the cell walls) and in the major long-distance transport pathway, the xylem, are usually very low, giving rise to less than 0.1 MPa of water potential. In the xylem conduits, the dominant component of ψ is the pressure, which can reach very large negative values (below –6.0 MPa in some severely stressed desert plants) equating to large xylem tensions. The xylem vessels with their specialised thickenings are rigid enough to withstand these tensions without undergoing serious deformation. Note that water flow in the xylem depends only on the pressure component of ψ, with any gradient in ψ_π being irrelevant where there is no semipermeable membrane ($\sigma = 0$).

Further discussion of cell and tissue water relations may be found in texts such as Slatyer (1967) (still the best), Kramer & Boyer (1995) and Nobel (2009), and in papers and reviews by Maurel et al. (2008), Cheung et al. (1975), Steudle (2001) and Tomos (1987).

4.2.1 Growth and cell water relations

Volumetric growth depends mainly on cell expansion and is driven by water flow into the cell. A theoretical biophysical framework was developed by Lockhart (1965) and reviewed in more detail by Cosgrove (1986). The water flux into any cell and consequent growth depends on the driving force for water uptake, the hydraulic conductivity of the cell membrane and also on the rheological properties of the cell wall (which themselves depend on cell-wall biochemistry and enzymic activities: Cosgrove, 1999, 2005; see also https://homes.bio.psu.edu/expansins/index.htm).

The force for cell expansion is provided by the turgor pressure, which sets up large tensile forces in the wall; the mechanical stress (force per unit area) on cell walls can be as high as 30 to 100 MPa (because the turgor pressure in expanding cells is usually between about 0.3 and 1 MPa and this acts on the cell wall, which comprises perhaps 1% of the cell cross-sectional area). Cell expansion occurs through a highly regulated process of 'wall loosening' or controlled polymer creep that allows the cellulose microfibrils in the wall to gradually move apart through the breaking and reforming of non-covalent bonding between the cell wall polymers. Rapid changes in cell expansion are regulated by acid-stimulated wall proteins in the expansin superfamily that control wall loosening in a wide range of physiological processes. Although not directly coupled to the wall loosening, expansion also requires the synthesis and deposition of new polysaccharide wall materials.

In order to determine the driving force for water uptake by a cell it is necessary to take account of the degree of semi-permeability of the membrane. Where the membrane is perfectly semi-permeable, the driving force is the total water potential difference across the membrane ($\Delta\psi$, equal to the sum of the differences in pressure and osmotic potentials):

$$\Delta\psi = \Delta\psi_p + \Delta\psi_\pi = (\psi_{p(o)} - \psi_{p(i)}) + (\psi_{\pi(o)} - \psi_{\pi(i)}) \quad (4.12)$$

where the $\Delta\psi$ terms refer to differences between water potentials outside (o) and inside (i) the plasmalemma. Where the membrane is not perfectly semi-permeable and lets through some of the osmotica, then that component of $\Delta\psi$ that depends on solute gradients (the second term of Eq. (4.12)) becomes less effective at driving water flow. Remembering that the cell turgor pressure, P, equals $\psi_{p(i)} - \psi_{p(o)}$, the overall driving force for water flow can be written as:

$$\sigma\Delta\psi_\pi + \Delta\psi_p = \sigma\Delta\psi_\pi - P \quad (4.13)$$

When σ is 0 the driving force for water flux reduces to the hydrostatic pressure difference, and when σ equals 1, the driving force is given by the difference in total water potential ($\Delta\psi$).

From the general transport equation (Eq. (3.19); Table 3.1) the volume flux density of water ($\mathbf{J}_v$) is given by the product of the effective driving force and a proportionality constant:

$$\mathbf{J}_v = L_p(\sigma\Delta\psi_\pi - P) \quad (4.14)$$

In this case the proportionality constant L_p (m s^{-1} Pa^{-1}) is termed a *hydraulic conductance*. For plant cells, L_p ranges from about 10^{-13} to 2×10^{-12} m s^{-1} Pa^{-1} (Nobel, 2009). This water permeability is strongly dependent on the amount, isoform and gating of the aquaporin water channels in the membrane.

The rate of volume increase of a cell is therefore the product of $\mathbf{J}_v$ and the cell surface area (A). Dividing through by the cell volume (V) gives the relative rate of increase of cell volume as:

$$(1/V)(dV/dt) = (A/V)L_p(\sigma\Delta\psi_\pi - P)$$
$$= L_v(\sigma\Delta\psi_\pi - P) \quad (4.15)$$

where L_v (s^{-1} Pa^{-1}) is a volumetric hydraulic conductance obtained by multiplying L_p by the ratio of cell surface to volume (A/V).

As well as being dependent on cell water relations, the rate of volume increase is also dependent on wall rheological properties according to what is known as the Lockhart equation:

$$(1/V)(dV/dt) = \phi(P - Y) \quad (4.16)$$

where Y is the *yield threshold* (Pa) or turgor that must be exceeded before any extension occurs, and ϕ is the cell wall *extensibility* (s^{-1} Pa^{-1}), which describes the rate at which cells undergo irreversible expansion whenever Y is exceeded. (Extensibility is often termed plasticity in the cell wall literature and contrasts with elasticity (ε), which refers to reversible changes in cell dimensions.) In practice, however, it should be noted that the parameters in this equation (ϕ and Y) are under continual close regulation, for example by expansins and pH, so that growth may remain constant as water potential changes over even quite short time scales. Equations (4.15) and (4.16) can be equated to eliminate P, giving on rearrangement:

$$(1/V)(dV/dt) = (\phi L_v/(\phi + L_v))(\sigma\Delta\psi_\pi - Y) \quad (4.17)$$

When water transport is not limiting (i.e. $L_v \gg \phi$) turgor pressure approaches the maximum value and Eq. (4.17) reduces to (4.16) and growth is determined by the rheological properties Y and ϕ as determined by expansins; maintenance of growth also depends on continued synthesis and deposition of the necessary wall materials. Conversely, when water supply is limiting, turgor pressure approaches the yield threshold, and growth depends on the rate of water supply.

4.3 Measurement of soil or plant water status

In all studies of the effects of water deficits on plant functioning there is a need for an accurate and comprehensive definition both of the water deficit treatments and of their effects on plant water status; this is a prerequisite for the design of repeatable and interpretable experiments. Unfortunately in many recent, particularly molecular, studies of plant response to water deficits there has been a tendency to omit any detailed description of the drought treatments imposed or of their effects on plant water status (Jones, 2007); indeed many molecular studies have used such extreme drought treatments (e.g. desiccation of excised tissues) that the changes in gene expression observed have little or no relevance to plant responses in natural situations.

The choice of water status measurement in any study depends on the experimental objectives and on hypotheses about the mechanism of any response to the water deficit. In this context it is necessary to take account of any signalling mechanism involved in the response; for example root–shoot signalling has been shown to play an important role in plant adaptation to soil moisture deficits (see Davies & Zhang, 1991) so that measurements of water status of the leaves may be of doubtful relevance to stomatal or photosynthetic responses (Jones, 2007). Precise definition of the water status in different parts of the soil–plant system is required for useful studies of the mechanisms of plant response to water deficits.

Techniques for the measurement of soil and plant water status have been described in detail elsewhere (Barrs, 1968; Boyer, 1995; Jones, 2007; Kirkham, 2004); here we focus on some important questions

that must be addressed in the selection of an appropriate measurement of water status for any study. Measures available can be broadly classified into those based on water content and those based on the energy status of the water.

4.3.1 Water content-based measures

Simple measures of water content are usually expressed on either volumetric or mass bases ($m^3\ m^{-3}$, volume %, g g^{-1}, etc.) but tend to be of limited value for comparative studies. For soils it is most common to express the water content on a volumetric basis, whether measurements are made using gravimetric techniques or using indirect sensors such as neutron probes or capacitance or other electromagnetic sensors (Kirkham, 2004). Because different soils have very different moisture release curves, a given percentage water content may represent full saturation for a freely draining sand or may represent a rather dry clay soil. As a result it is useful to normalise results by expressing the soil moisture content as a fraction of either the total volume of pore space (giving a relative saturation) or as a fraction of the water-holding capacity of the soil. This is usually defined as the amount of water released in going between field capacity (measured after allowing natural drainage for 24 h) and what is often termed the wilting point (defined as the water content under a tension of 1.5 MPa). More useful measures of water status that directly relate to water availability are the energy-based measures described in Section 4.3.2 below.

For plant tissues a widely used normalisation is to use the *relative water content* (RWC, θ) where the water content is expressed as a fraction of the fully turgid water content of that tissue (i.e. when $\psi = 0$):

$$\theta = \frac{(\text{fresh mass} - \text{dry mass})}{(\text{turgid mass} - \text{dry mass})} \tag{4.18}$$

Relative water content is a particularly useful measure of plant water status since it is closely related to cell volume (Figure 4.4) and is therefore likely to be relevant in many studies of the metabolic effects of plant water deficits; indeed it is often likely to be more relevant than water potential. It should be noted, however, that where osmotic adjustment occurs, this can lead to artefacts in the calculation of θ. For example where 0.2 MPa of extra solutes have been synthesised to maintain P (and cell volume) constant at their full-turgor values as ψ falls to -0.2 MPa, when the tissue is rehydrated on water to obtain the turgid mass, extra water is absorbed until $\psi = 0$, increasing P and cell volume beyond their normal full-turgor values. This leads to the true turgid mass being overestimated and to Eq. (4.18) underestimating θ. The measurement of relative water content has been discussed in detail elsewhere (Barrs, 1968; Boyer et al., 2008).

4.3.2 Energy-based measures

Water potential is the most frequently used measure of plant water status and is particularly relevant in studies of water movement, though in certain situations, such as where there is no semi-permeable membrane in the flow path, a component (in this case pressure potential) may be more useful. Although ψ has the advantage of being a rigorous measure, there is strong evidence that it is often not directly involved in the control of physiological processes such as growth or photosynthesis (Jones, 1990). It is more usual for physiological processes to be related to turgor pressure, rather than to water activity (or ψ); this is not altogether surprising in view of the small changes in a_W that correspond to physiologically important water deficits (Eq. (4.7)). Although it is difficult to distinguish between turgor and cell volume (or relative water content) as major controlling factors, experiments using isolated cells or protoplasts (Jones, 1973d; Kaiser, 1982) have demonstrated that cell volume may be the important variable in some cases. How this might be sensed is not certain, though there is evidence for the existence of a number of membrane 'stretch' sensors (Árnadottír & Chalfie, 2010; Kacperska, 2004; Schroeder & Hedrich, 1989).

Although total leaf water potential (ψ_ℓ) is probably not as generally useful a measure of water status as had once been thought, it is still of value, especially in studies of water flow, or as a proxy indicator of a component such as turgor pressure. The two main instruments used for measurement of water potential or its components are the thermocouple psychrometer and the pressure chamber, though liquid-phase equilibrium techniques (Shardakov method) can also be used. A number of indirect methods, such as the use of the β-gauge for the measurement of leaf water content (Jones, 1973a), together with leaf, fruit or stem morphometry (Fereres & Goldhamer, 2003; Huguet et al., 1992) and leaf patch pressure sensors (Zimmermann et al., 2008) that detect morphological responses to turgor changes can also give useful information.

Psychrometer

The principle of the psychrometer is that a sample of tissue is allowed to come to water vapour equilibrium with a small volume of air in a chamber and the humidity of this air is measured using thermocouples set up to measure the wet bulb depression or the dewpoint of the air (see Chapter 5). These results can then be used to give the tissue water potential by calibration against solutions of known water potentials. Brown and van Haveren (1972) give details of the differences in procedure when using different instruments and include many tables with useful conversions. With growing tissues psychrometry may lead to underestimates of ψ because growth requires extra water if ψ is to be maintained at the original value.

Osmotic potentials of plant tissues may also be obtained using the psychrometer after first disrupting the cell membranes (e.g. by rapid freezing and thawing) so that there is no longer a turgor component of cell water potential. Osmotic potentials of extracted sap may also be determined with a psychrometer or else with an osmometer that measures the freezing point depression. It is, however, difficult to extract a representative sample of the cell sap since it is liable to be diluted by the extracellular (apoplastic) water during collection and an appropriate correction must be made (see Boyer, 1995). A further problem with psychrometers is that they tend to be laboratory instruments requiring long periods (often more than 8 h) of equilibration and carefully controlled temperature facilities, though fairly robust instruments that can be installed on stems in the field (Dixon & Tyree, 1984) are commercially available (www.ictinternational.com.au/stempsychrometer.htm).

Pressure chamber

A much more adaptable technique is the pressure chamber as popularised by Scholander *et al.* (1964), which is rapid and can be used for estimating leaf water potential in the field, though it is necessarily destructive. If a leaf is cut from a plant, the tension in the xylem (which is particularly large if the plant is transpiring rapidly or is under stress) causes the xylem sap to be withdrawn from the cut surface. This leaf may then be sealed into a pressure chamber (Figure 4.5) with the cut end exposed: the chamber is then pressurised until the sap just wets the cut surface thus restoring the sap to its position in vivo. At the time of collection the original average leaf water potential would be given by:

$$\psi_\ell(\text{original}) = \psi_p + \psi_\pi \qquad (4.19)$$

where ψ_p, and ψ_π are leaf averages of turgor and intracellular osmotic potentials. As the chamber is pressurised the water potential is raised by the amount of pressure applied so that, at the balance pressure (P^*), the water potential is zero (there is free water at the cut surface) so:

$$\psi_\ell(\text{original}) + P^* = 0 \qquad (4.20)$$

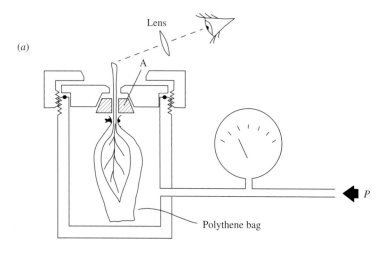

(a)

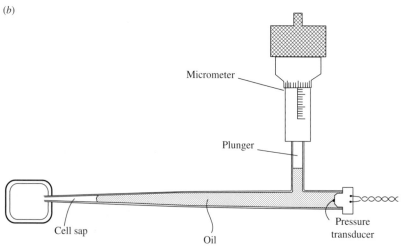

(b)

Figure 4.5 (*a*) Pressure chamber showing one possible way of sealing with a rubber bung (A). The leaf is sealed in a polyethylene bag to minimise evaporation during measurement. (*b*) Schematic diagram of a cell pressure probe: the intracellular hydrostatic pressure is transmitted to the pressure transducer via an oil-filled microcapillary inserted into the cell. Volume is adjusted using the micrometer and observing the interface between the cell sap and oil under a microscope. The elastic modulus is obtained from the initial pressure change on applying a step change in volume.

The negative of the balance pressure therefore equals the original ψ_ℓ. The method is rapid and accurate, generally giving results comparable to those given by the psychrometer. Although questions have been raised as to the validity of the pressure chamber measurements (see e.g. Zimmermann *et al.*, 2004) the consensus view strongly supports the interpretation given here (Angeles *et al.*, 2004; Steudle, 2001).

For precise work it is necessary to correct Eq. (4.20) for the osmotic potential of the apoplast (the water-filled space outside the cell membranes and including the xylem sap). Because the apoplast contains some solutes, at the balance pressure the water potential is still below zero by an amount equal to the osmotic potential in the apoplast. Since $|\psi_\pi|$ of the apoplast is usually less than 0.1 MPa, the consequent overestimation of ψ_ℓ is usually negligible.

It is also possible to dehydrate tissues by raising the pressure above P^* and, from measurements of the volume of sap expressed (V_e) for any pressure increment, it is possible to construct a Höfler–Thoday diagram as in Figure 4.4. This depends on the

assumption that turgor pressure falls to and remains at zero for all water potentials below that giving zero turgor. In this region, from Eq. (4.8):

$$P^* = -\psi_\pi = \mathscr{R}Tc_s = \mathscr{R}Tn_s/V \qquad (4.21)$$

where n_s is the total number of moles of solute present and V is the volume of water within the symplast of the cells in the tissue. Rearranging and substituting $V_o - V_e$, for V (where V_o is the original volume of the symplast when the tissue was cut) gives:

$$1/P^* = -1/\psi_\pi = (V_o/\mathscr{R}Tn_s) - (V_e/\mathscr{R}Tn_s) \qquad (4.22)$$

This equation shows that a plot of $1/P^*$ against V_e gives the curve shown in Figure 4.6. The intercept of the straight line portion ($V_o/\mathscr{R}Tn_s$) is equal to the inverse of the original osmotic potential, while the intercept on the V-axis gives the total osmotic volume of the system. The curved portion of the 'pressure-volume' curve occurs where turgor is positive.

A particularly interesting and useful version of the pressure chamber has been developed by Passioura and his colleagues (e.g. Gollan *et al.*, 1986), where by

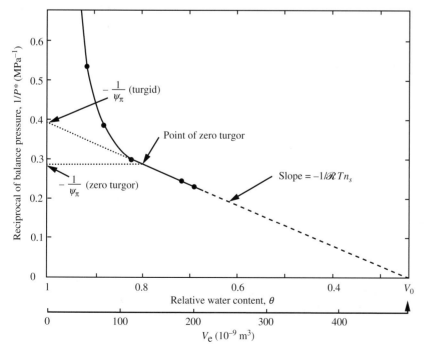

Figure 4.6 Pressure–volume curve for an apple leaf (cv. Golden Delicious) obtained by dehydration in a pressure chamber on 5 October 1978. The volume of sap expressed (V_e) may be obtained from the relative water content (θ) by multiplying by the turgid water content ($W = 652$ mg for this leaf).

use of a split lid it is possible to seal the roots of a growing plant into the chamber with the shoot emergent from the seal. By applying pressure to the roots it is possible to keep the shoot turgid irrespective of the water potential of the soil, and thus to investigate the response of plants to soil and leaf water potentials independently. An example of its use is presented in Chapter 6 (Figure 6.13).

Pressure probe

The pressure probe for direct measurement of the turgor of individual cells (Hüsken *et al.*, 1978; Figure 4.5*b*) allows one to make direct measurements of individual cell turgor in higher plants, and in combination with analysis of the extracted cell contents (Malone *et al.*, 1989) provides a powerful tool for studies of plant water relations, though it is not yet applicable to plants in the field. In essence the instrument consists of a microcapillary that can be inserted into individual cells: the pressure in the liquid in the capillary (in contact with the cell sap) can be monitored with a pressure transducer and the pressure changed by means of a motorised plunger. The pressure probe can be used to estimate a number of water relations quantities: the water potential of individual cells can be estimated (using Eq. (4.10)) from direct measurements of cell turgor pressure and the osmotic potential of extracted sap; the elastic modulus, ε, can be determined from the magnitude of the initial (before significant water flow through the cell membrane) pressure change on rapidly changing the volume of the cell by a small known amount (using Eq. (4.11)); while L_p can be obtained from the rate of relaxation of volume after a pressure change (from Eq. (4.14)). In certain circumstances it may even be possible to measure xylem tensions using a pressure probe if the oil is replaced by water (Balling & Zimmerman, 1990), though such an instrument does not seem able to detect pressures below about −0.8 MPa so is only of limited application to studies of xylem pressure (see discussion in Steudle, 2001). A detailed review of the potential uses of the cell pressure probe has been provided by Tomos and Leigh (1999).

4.3.3 Where should water status be measured?

Plant physiologists have concentrated on the measurement of plant water status, especially ψ_ℓ, on the grounds that it is the water status of the plant or, where one is concerned with a leaf process such as photosynthesis or stomatal closure, the leaf, that should determine any response. Unfortunately this simplistic view is complicated by the signalling that can be involved in growth coordination, with, for example, functioning of the leaves being partly dependent on transport of growth regulators from the roots. There are a number of lines of evidence that suggest that the use of leaf water status may often be inappropriate as a measure of plant water status (see Jones, 1990, 2007; Jones & Sutherland, 1991):

1. Leaf water status shows much short-term variability as a result of its dependence on environmental conditions with ψ_ℓ changing by as much as two-fold within minutes in response to the passage of clouds. In contrast ψ_ℓ at any one time is often maintained relatively constant over a wide range of soil water potential (ψ_s) (e.g. Bates & Hall, 1981). It is hard to envisage how the small differences in mean ψ_ℓ between soil water treatments can be distinguished from such large background fluctuations.

2. On occasion, ψ_ℓ may even be *higher* in droughted plants with more closed stomata than in well watered plants (Jones *et al.*, 1983); this stomatal closure cannot be explained by a feedback control of stomatal aperture acting through leaf water status, though it could be explained by a feed-forward response to ψ_s.

3. In a number of split-root experiments it has been possible to maintain leaf water status at control levels by keeping a proportion of the root system well watered while drying only part of the root zone: in such cases stomata may still close or leaf growth may be inhibited (Davies & Zhang, 1991; Gowing *et al.*, 1990) in response to soil drying even with no depression of ψ_ℓ. Analysis of xylem sap obtained from such split-root experiments has

provided evidence that abscisic acid (ABA) transported from the droughted roots may provide the root-sourced signal causing stomatal closure in such situations; this signal appears to be enhanced by coincident changes in xylem sap pH, with drought raising the apoplastic pH and facilitating partitioning of ABA into the guard cells (Wilkinson & Davies, 2002). Other experiments have implicated both reductions in xylem sap cytokinin content (Blackman & Davies, 1985; Ha *et al.*, 2012) and changes in inorganic ion transport in stomatal closure in response to soil drying (Wilkinson *et al.*, 2007) as important signals often interacting with ABA.

4. Theoretical modelling studies have shown that optimal stomatal control requires information on soil moisture availability to be signalled to the shoot (Jones & Sutherland, 1991).

5. Some other experiments that have shown stomatal behaviour to be more closely related to soil water content than to ψ_ℓ are presented in Section 6.4.5.

For these reasons, measures of water status that are closely related to soil water status are often more relevant to plant functioning than is midday ψ_ℓ. Unfortunately soil water content is very heterogeneous, so it is difficult to measure an effective mean ψ_s within the rooting zone weighted for the root distribution. The measurement of ψ_ℓ predawn (when ψ_ℓ approaches equilibrium with the effective ψ_s) therefore provides a particularly valuable measure of water status in drought studies, though it is inconvenient to measure and shows much smaller variation than midday water potential. An alternative method for calculating an effective mean ψ at the root surface of transpiring plants has been proposed by Jones (1983). This is based on the flow models described below (Eq. (4.26)) and uses simultaneous measurements of ψ_ℓ and transpiration at any time of day.

A number of studies have shown that *stem water potential* (ψ_{st}, estimated as the water potential on leaves that have been 'bagged' to stop transpiration

for at least 30 min before measurement and therefore eliminate the potential drop associated with the hydraulic resistance in the leaf) is a much superior estimate of tree water status than is leaf water potential, which tends to be excessively sensitive to the environment and relatively insensitive to soil moisture (McCutchan & Shackel, 1992; Patakas *et al.*, 2005).

Integration of water status over time to obtain a measure of the degree of stress to which plants have been subjected can be particularly useful for longer term studies. Perhaps the most convenient method is to sum predawn ψ_ℓ (Schultze & Hall, 1981). The use of a remotely sensed 'crop water stress index' as a measure of water stress that can readily be integrated over time is discussed in Chapter 10, though it should be remembered that this is a very indirect indicator of water status as it actually estimates stomatal closure.

4.4 Hydraulic flow

Mass flow of water can be described by the familiar transport equation. For flow in porous media and in capillaries, the appropriate driving force is the hydrostatic pressure gradient ($\partial P/\partial x$), so:

$$\mathbf{J}_v = -L\,\partial P/\partial x \qquad (4.23)$$

where $\mathbf{J}_v$ is the volume flux density ($m^3\ m^{-2}\ s^{-1}$), which is equal to an average velocity ($m\ s^{-1}$), and L is a *hydraulic conductivity coefficient* ($m^2\ s^{-1}\ Pa^{-1}$) corresponding to the diffusion coefficient in Fick's first law (see Eq. (3.10)). The choice of hydrostatic pressure rather than total water potential as the driving force in Eq. (4.23) can be understood if one realises that a gradient of ψ_π can only affect volume flow when there is a semi-permeable membrane present to generate a pressure. Equation (4.23) is known as *Darcy's law* when it is applied to flow in soils and can be used for any porous medium.

As with Fick's first law, it is often more convenient to apply Eq. (4.23) in the integrated form, and to include the pathlength in the coefficient to give:

$$\mathbf{J}_v = L\,\Delta P/\ell = L_p\,\Delta P \qquad (4.24)$$

where L_p is a *hydraulic conductance* (m s^{-1} Pa^{-1}), which for a uniform path is given by L/ℓ (where ℓ is the pathlength) so is analogous to diffusive conductance (g). L_p therefore depends on pathlength while L is a property of the material through which flow occurs and of fluid viscosity. Equation (4.24) is a special case of Eq. (4.14) for a system with $\sigma = 0$.

For hydraulic flow through cylindrical tubes, the hydraulic conductance can be expressed in terms of tube radius and fluid viscosity to give:

$$\mathbf{J}_v = (r^2/8\eta\ell)\Delta P \qquad (4.25)$$

where r is the radius of the tube (m) and η is the *dynamic viscosity* of the fluid (kg m^{-1} s^{-1} or Pa s $\simeq 1 \times 10^{-3}$ for water at 20°C). This equation is referred to as *Poiseuille's law*, and shows that the average flow rate through the unit cross-section of a tube increases as the square of the radius. The total flow *per conduit* will therefore increase by the fourth power of the radius; this conclusion is supported by much experimental data in plants but in real conducting systems needs adjusting for the restriction to flow imposed by the pits in the end walls between conduits (Sperry *et al.*, 2006). Although this relationship is only valid for laminar flow conditions, that is when the Reynolds number (see Chapter 3) – in this case given by $\mathbf{J}_v \rho r/\eta$ – is less than about 2000; this is generally true for liquid flow in plant systems. The frictional drag at the walls of any capillary leads to most rapid flow at the centre. This maximum flow velocity is twice the average velocity given by Eq. (4.25). The hydraulic conductance, L_p, for any capillary can be easily calculated as $r^2/8\eta\ell$. As with other transport processes electrical analogues can be used to analyse flow in complex systems, again using resistances ($R = 1/L_p$) in preference to conductance where the pathway consists primarily of several components in series. (A capital italic R is used to indicate hydraulic resistances.)

It is worth noting that application of Eq. (4.23) to flow in tubes and in porous media usually involves a different areal basis for the expression of $\mathbf{J}_v$. In the former it is the actual cross-section of the conducting tube, while in the latter it is the total cross-section including any non-conducting matrix.

4.4.1 Water flow in plants

The pathway for the transpiration flux through plants is illustrated in Figure 4.7 (see Milburn, 1979; Slatyer, 1967; Steudle, 2001). As was originally recognised by Hales (1727) the transpirational flow of water through the soil–plant system to the top of tall trees is driven by evaporation from the leaves.[1] This evaporation lowers the leaf water potential and in turn sets up a gradient of water potential between the soil and the leaf thus dragging water up the plant under tension. The cohesion–tension theory of sap flow recognises that this water flow depends on the tremendous cohesive forces within the water column and on the strong adhesion of water to the xylem cell walls to withstand the tension of at least 10 MPa that is required to draw water to the top of the tallest trees (Steudle, 2001). There have been a number of suggestions of alternative mechanisms, including the putative existence of distributed active pumps or one-way valves along the water-flow pathway. These have been partly based on suggestions that the pressure within xylem vessels may be less negative than expected from cohesion–tension theory, or than deduced from pressure chamber data, but the evidence strongly favours cohesion–tension (for useful citations, see Angeles *et al.*, 2004).

The main pathway for longitudinal flow is the xylem, in which the conducting elements are primarily the non-living and heavily thickened and lignified tracheids and xylem vessels. Vessels, which are found only in angiosperms, consist of files of cells whose end walls have broken down to form continuous tubes that may vary from a few centimetres to many metres in length, and from about 20 µm to as much as 500 µm in diameter. The tracheids, on the other hand, originate from single

---

[1] 'In animals it is the heart which sets the blood in motion, and makes it continually to circulate; but in vegetables we can find no other cause of the sap's motion, but the strong attraction of the capillary sap vessels, assisted by the brisk undulations and vibrations caused by the sun's warmth, whereby the sap is carried to the top of the tallest trees ... so the ascending velocity is principally accelerated by the plentiful perspiration of the leaves...' (Hales, 1727).

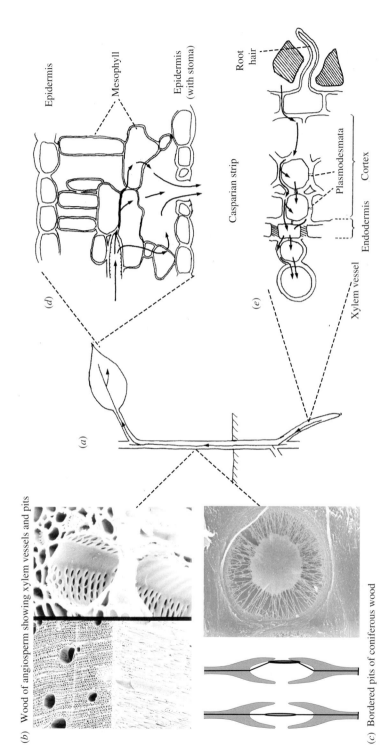

(b) Wood of angiosperm showing xylem vessels and pits

(c) Bordered pits of coniferous wood

(d)

Epidermis

Mesophyll

Epidermis
(with stoma)

(a)

(e)

Casparian strip

Xylem vessel

Root
hair

Plasmodesmata

Endodermis   Cortex

Figure 4.7 (a) The pathway of water flow from the soil to the leaves, showing (b) the xylem vessels and pits of the conducting tissue in the wood of an angiosperm tree, (c) cross-sections of the bordered pits in their normal and aspirated states in coniferous wood, together with a face-on image in *Picea abies* (kindly provided by Michael Rosenthal). (d, e) the parallel paths for flow in the cell walls (apoplast) and in the symplast within the roots and leaves.

cells and are found in all vascular plants so they constitute the main conduits in gymnosperms such as coniferous trees. Tracheids are typically only a few millimetres long and between about 10 μm and 60 μm in diameter. In woody plants the largest conduits tend to occur in the early formed wood of an individual growth ring and are much smaller in the 'late' wood. Typical examples of early and late wood diameters are 35 μm and 14 μm for tracheids of Scots pine, while the range of vessel size in an English oak was 268 μm down to 34 μm (Jane, 1970). Adjacent conduits are connected by pits through which sap can pass; the typical pore diameter in the pit membranes that separate angiosperm vessels is 5 to 20 nm while in the membrane in the margo of the specialised bordered pits found in the tracheids of conifers (Figure 4.7) they may be 0.1 to 0.2 μm. The bordered pits appear to act as safety valves that close under tension preventing water flow into embolised conduits, but allowing easy water flow when well hydrated.

The radial movement of water in the roots from the soil to the xylem occurs through the cortical tissue, partly in the water-filled free space of the cell wall (the apoplast) and partly within the symplasm (the connected protoplasm within the cell membrane). The separate cells forming the symplasm are interconnected by narrow cytoplasmic connections, the plasmodesmata. The symplasm is separated from the apoplast by the plasma membrane. Between the cortex and the vascular tissue in the root is a specialised cell layer, the endodermis, where the cell wall pathway for water movement is blocked by a band of suberised tissue, the Casparian strip (Figure 4.7). At this point all the water must pass through a membrane into the cytoplasm. Similarly, the path from the vascular tissue to the evaporating sites in the leaves may involve some movement in the symplast and some in the cell wall. The relative contribution of apoplastic and symplastic pathways is difficult to quantify because of the anatomical complexity, and because of lack of information on permeability of cell membranes and other components. Nevertheless roots are often observed to have a rather low or variable σ suggesting that

at least some flow does not go through the endodermal cells and that there can be some bypass flow that avoids the Casparian barrier, for example in young tissues before the Casparian strip forms (see Steudle, 2001).

It is easy to see why the xylem provides such an efficient long-distance transport pathway. If one assumes a typical pressure gradient in a transpiring plant of -0.1 MPa m^{-1}, it can be calculated from Poiseuille's law that the flow velocity in 100 μm radius conduits would be 125 mm s^{-1} (i.e. $(100 \times 10^{-6})^2 \times 10^5/(8 \times 10^{-3} \times 1)$ m s^{-1}): similarly, in 20 μm conduits it would be 5 mm s^{-1}, while, if the same law holds in the cell-wall interstices, which are only about 5 nm, flow velocity would be down to about 3.1×10^{-7} mm s^{-1}, or about seven orders of magnitude less than in an equal area of xylem vessel. Observed maximum flow velocities in stems of different trees actually range from about 0.3 to 0.8 mm s^{-1} in conifers and 0.2 to 1.7 mm s^{-1} in so-called diffuse-porous hardwoods (e.g. *Populus*, *Acer*) to 1.1 to 12.1 mm s^{-1} in ring-porous hardwoods (e.g. *Fraxinus*, *Ulmus*) (see Tyree & Zimmermann, 2002). In herbaceous species the xylem water stream may even reach velocities of 28 mm s^{-1} (100 m h^{-1}). These values are rather below those expected, partly as a result of the flow resistance in the pit membranes between vessels. In ring-porous trees, especially, the majority of flow occurs in the most recent wood. For example in *Ulmus* 90% of the flux has been observed in the most recent year's wood, even though some conduction occurred in wood up to four years old (Ellmore & Ewers, 1986).

Flow through the pits in the end-walls between angiosperm vessels or between conifer tracheids often provides c. 60% of the total flow resistance. Although angiosperm vessels tend to be an order of magnitude longer than corresponding diameter tracheids (and hence have fewer end walls) they are not necessarily more efficient because the tracheids have much lower end-wall resistances with their larger pores in the bordered (torus-margo) pits; this compensates for the greater number of walls along the path (Sperry *et al.*, 2006).

4.4.2 Hydraulic networks

There has been much discussion as to how the branched hydraulic and support systems of vascular plants scale from the smallest *Arabidopsis* to the largest *Sequoia* while maintaining effective support and water supply to all organs across such widely divergent life forms. There is clearly a need to optimise the distribution of resources between elements in the largest trunk to the smallest branch. West *et al.* (1997, 1999) proposed a general theory of resource distribution across hierarchical branching networks (the 'WBE' model) that explains much of the commonly observed 'allometric' scaling of a wide range of biological phenomena, where any biological variable (*y*) tends to vary with body mass (*m*) according to:

$$y = y_o\, m^b \tag{4.26}$$

where y_o is a constant that is characteristic of the type of organism and b is a scaling exponent (usually simple multiples of 1/4). For water flow in plants this rather simplistic model assumes that the vascular system comprises a parallel array of individual pipes of equal length and size running in parallel from the trunk to the petioles that are grouped together in increasing numbers on moving from smaller to larger branches in a fractal-like architecture (Figure 4.8). The basic assumption that energy dissipated in flow is minimised leads to a conclusion that flow resistance (and flow) should be independent of pathlength and hence plant size; this is achieved by allowing the tubes to taper at an appropriate rate.

Although the WBE model goes a long way towards explaining observed patterns of branching and hydraulic conductivity in vascular systems it does not appear to fully describe the trade-offs between efficiency and hydraulic safety that shaped the evolution of vascular networks (Savage *et al.*, 2010). These authors showed that a model based on optimal space filling and a general 'packing rule', where the frequency of xylem conduits varies approximately

(*a*)   (*b*)   (*c*)

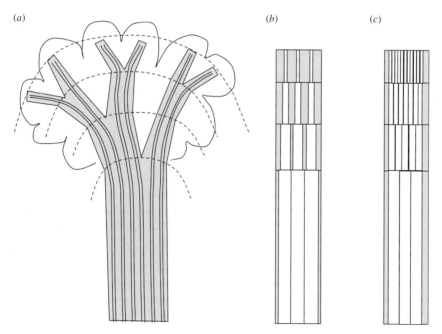

Figure 4.8 (*a*) A diagrammatic illustration of the branching structure of the vascular system in a tree arranged as a parallel array of pipes branching out to supply the remote branches and leaves, (*b*) an illustration of the WBE model showing a part of the structure demonstrating the taper of each of the tubes, and (*c*) modification to follow the generalised packing model (Savage *et al.*, 2010) where there are more of the smaller tubes filling the available stem area.

inversely with the square of conduit radius, accounts well for observed variation in a wide range of plants. This rule contradicts the WBE assumption that conduit frequency remains constant as conduits taper and leads to a prediction of a constant total conduit area at each level of branching and better fits the majority of anatomical observations in the literature. It appears that these space-filling principles may even apply at larger scales and allow predictions of the behaviour of whole forests and larger areas of land.

Limits to height of trees

The mechanisms that determine tree height and explain why it may vary for any one species with location and also explain why growth slows with increasing height and age have been surprisingly elusive. A number of possible hypotheses have been proposed to explain the limits to tree height: these include possibilities of increasing nutrient limitation (the nutrient limitation hypothesis) and the idea that respiration increases as trees grow until it finally balances assimilation (the respiration hypothesis). Ryan and Yoder (1997), however, have argued that the most likely, or at least the dominant, explanation depends on the increasing hydraulic resistance to transfer in the xylem as trees grow. In this *hydraulic limitation hypothesis*, the increasing hydraulic resistance as trees grow is proposed to lead to falling shoot water potential with consequent stomatal closure and reduced photosynthesis. Although some have argued that leaf:sap-wood ratios may be more important than the total axial resistance (Becker *et al.*, 2000) there is a general agreement that in spite of genetic differences, hydraulic limitation and the associated reductions in assimilation play the major role in limiting tree height.

Sap flow measurement

A number of methods have been developed for measuring in intact plants the velocity of xylem flow or the mass flow rate of water through the stem (see Pearcy *et al.*, 1991 for a useful summary). The most widely used techniques involve either the heat pulse technique (reviewed by Jones *et al.*, 1988) pioneered

by Huber in the 1930s, which is based on the measurement of the time between a pulse of heat being applied to a section of the stem and the detection of the temperature rise a certain distance downstream, or the continuous heating (e.g. Granier, 1987) or heat-balance (Čermák & Kucera, 1981; Čermák *et al.*, 2004) approaches. In the future it may even be possible to use approaches such as nuclear magnetic resonance imaging (MRI; Van As *et al.*, 2009). Because sap flow may be very heterogeneous within stems, substantial replication of sensors is usually required for good average data for plant stands.

4.4.3 Cavitation

In spite of its enormous tensile strength, the water in xylem can rupture (cavitate) under the extreme tensions that occur naturally, if suitable nucleation sites, either on the vessel wall or as a result of air entry through pit membranes, are available. Once initiated, the bubble then rapidly expands forming an embolism within the vessel or tracheid until it is stopped at the pit membranes (Figure 4.9). Further movement is usually prevented by the capillary effects in the narrow pores – for example 0.1 μm pores in the membrane will prevent passage of the vapour–water interface under as much as 1.5 MPa tension (Eq. (4.3)).

Soon after their formation the embolisms that block the vessels contain only water vapour (at a pressure equal to the vapour pressure of water), and can presumably refill readily as soon as xylem sap pressure recovers to zero or above, for example as a response to root pressure at night. This situation is unlikely to last long as air entry is likely to be complete within about 1000 s (Tyree & Sperry, 1989). Once air diffuses into the vessel and reaches equilibrium with atmospheric pressure, however, refilling will be much slower as the necessary gas dissolution requires a greater pressure in the gas bubble than at the surface of the stem and diffusion of the gas away from the bubble. As the pressure in the bubble depends on the net effect of the xylem tension and the capillary pressure generated by surface tension in the conduits, Yang and Tyree (1992) have

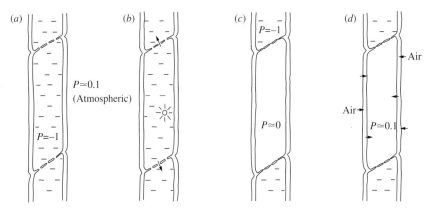

Figure 4.9 Schematic illustration of events during and after cavitation in a xylem vessel. The initial situation is shown in (a) where the absolute pressure (P) in the xylem sap is -1 MPa (the equivalent ψ_π is -1.1 MPa because it is referred to atmospheric pressure) and the air pressure outside is c.0.1 MPa. On the occurrence of a cavitation (b) within the vessel, the sap is rapidly withdrawn until the meniscus is held at the pit membranes (c). At this early stage the space is probably filled with water vapour at a few kilopascal absolute pressure ($P \simeq 0$ MPa); subsequently slow diffusion of air into the conduit probably raises the internal pressure to atmospheric ($P \simeq 0.1$ MPa). Refilling of a cavitated vessel is likely to be much easier in the early stages. (Modified after Milburn, 1979.)

shown that diffusion out of the stem of the dissolved gas in equilibrium with the gas in the bubble requires the xylem pressure potential (ψ_p) to be greater than $-2\sigma/r$, where σ is the surface tension and r is the conduit radius. For a 50 μm vessel this equates to −5.8 kPa. Although embolisms are frequently irreversible at this stage there are many reports of embolism repair even at xylem pressures well below the critical value (see e.g. Trifilò et al., 2003) and it has been suggested that in those plants where root pressure is not operating, an active solute loading process may facilitate osmotic water flow into the vessel. This xylem refilling forces dissolution of the entrapped gases and has been visualised by cryo-SEM microscopy (Canny, 1997) and more convincingly in real time by high resolution computed tomography (Brodersen et al., 2010) and. It seems likely that this dissolution process, which occurs primarily at night, is facilitated by the presence of night-time transpiration to remove the dissolved gases and does not occur in those species with no stomatal opening at night. Air entry can act as a stimulus for the formation of tyloses (ingrowths from the xylem parenchyma cells that completely block xylem vessels that have cavitated or been damaged by

infection). These cause the irreversible loss of xylem function in older wood.

The occurrence of cavitation events can be detected with a microphone as audible 'clicks' resulting from the shock waves created as the vessel walls relax after a cavitation (Milburn, 1979). The use of ultrasonic detectors that are sensitive in the region 0.1 to 1.0 MHz, appears to provide a better detection method because the lack of environmental noise in this frequency range means that such sensors can be used in the field (Tyree & Dixon, 1983). Unfortunately the profiles of audible and ultrasonic emissions produced as stems dry do not always coincide, indicating that they may represent cavitations or structural changes in different tissue elements. If the acoustic events are to have any value as indicators of xylem dysfunction, it is necessary to show that the events being detected occur within the main functional xylem elements, rather than predominantly in fibres or small tracheids. This has not yet been adequately demonstrated in all cases and there is good evidence that many acoustic events often originate in non-conducting elements. For example, when 5 cm lengths of apple wood were allowed to desiccate, high acoustic emission rates

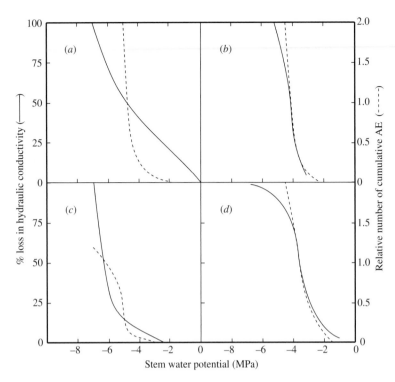

Figure 4.10 Some xylem vulnerability curves showing the loss in hydraulic conductivity (———) and acoustic emissions (AE, - - - - - -) with falling water potential for (a) *Cassipourea elliptica*, (b) *Acer saccharum*, (c) *Rhizophora mangle* and (d) *Thuja occidentalis*. Acoustic emissions are expressed relative to the number observed when ψ had fallen to a value giving a 50% reduction in hydraulic conductance. (Data recalculated from eye-fitted curves in Sperry *et al.*,1988b; Tyree & Dixon, 1986; Tyree & Sperry, 1989.)

were observed (Sandford & Grace, 1985), even though the number of complete vessel elements in such short pieces of wood would have been very small. In spite of such uncertainties the number of cavitations detected as a plant dehydrates is often related to the loss in stem hydraulic conductivity as measured on excised stems, though as indicated in Figure 4.10 the relationship is often not close. Analysis of the energy of the ultrasonic emissions in conifers has indicated that their energy may be related to tracheid diameter so that cumulative energy emitted may be a better indicator of conductivity loss than is the total number of emissions (Mayr & Rosner, 2010).

Because of the difficulties with acoustic detection as an indicator of xylem embolism it is better to use loss of hydraulic conductivity as the basic measure of embolism though results from the two approaches are often similar (e.g. Figure 4.10). Sperry *et al.* (1988a) described a simple pressure-flow apparatus that could be used to measure xylem conductivity on several stem segments simultaneously; their key advance was to express the conductivity at any sampled water

potential as a fraction of the conductivity measured after flushing the segment with an acidic perfusing solution at 175 kPa to remove any embolisms. This allows one to express the initial conductivity as a percentage of the maximum and is a measure of embolism. A *vulnerability curve* is then constructed from results for separate stems either sampled at different water potentials in the field, from stems allowed to dry out in the laboratory or from stems pressurised in a chamber (the air-injection technique: Cochard *et al.*, 1992). The air-injection technique assumes that excess pressure applied to a stem simulates tension in the xylem (by an amount equal but opposite in sign to the pressure applied) and forces air entry into the vessels. A convenient commercial pressure-flow apparatus that adopts these principles is available (www.bronkhorst.fr/fr/produits/xylem_embolie-metre).

An alternative approach, that allows one more rapidly to desiccate stem segments, is the rotor technique where stem segments are centred across the rotor of a centrifuge (Li *et al.*, 2008; Pockman *et al.*,

1995) and spun at different rotational velocities to generate a centrifugal force that causes tension in the xylem until cavitation occurs. The apparent outward force (centrifugal force) acting on the xylem water is balanced by the xylem tension. The maximum tension (P) occurs at the centre point of the rotor and integrates the centrifugal forces acting on water at radii from 0 to r_{max} (half the length of the segment) and is given by $P = -0.5 \rho \, r_{max}^2 \, \omega^2$, where ρ is the density of the sap and ω is the angular velocity (Alder et al., 1997). In this technique increasing rotor speed increases the tension in sap in any intact vessels in the stem segment according to the centrifugal force generated; the extent of cavitation is then determined from the difference in conductivity measured before and after spinning, though this approach may not work well in species such as vines where typical vessel lengths may be at least the size of the centrifuge rotor.

There is often a threshold water potential for cavitation; this varies both with the stress prehistory of the plant and with the species, and even the cultivar (Table 4.1). Vulnerability to cavitation is often closely related to the size and number of pores in the pit membranes, with those vessels having the largest pores being most vulnerable. This suggests that air entry provides the seeding mechanism that initiates cavitations (Tyree & Sperry, 1989). It has been suggested that the presence of small vessels favours avoidance of cavitation, but the relationship of cavitation threshold with size is rather weak when

Table 4.1 Variation in susceptibility of different plants to xylem cavitation. Approximate values of ψ giving different amounts of cavitation (the threshold, 50% cavitation and 90% cavitation) were read from published graphs. Data were obtained using either loss of hydraulic conductivity or acoustic emissions, with the values in parentheses representing acoustic data. See also Figure 4.9.

Plant	ψ (MPa)			Reference
	Threshold	50%	90%	
Acer saccharum	−3.0 (−2.5)	−4.1	−4.7	4
Acer pseudoplatanus	(−1.5)	(−1.8)	-	1
Cassipourea elliptica	−0.0 (−2.0)	−4.1	−6.6	3
Juniperus virginiana	−3.5 (−4.2)	−6.4	−8.8	4
Lycopersicon esculentum (tomato)	(−0.2)	(−0.4)	-	1
Malus × domestica (apple)				
on M9 rootstock (unstressed)	(−0.9)	-	-	2
on M9 rootstock (prestressed)	(−2.5)	-	-	2
Ricinus communis (castor oil)	(−0.5)	(−0.8)	-	1
Rhododendron ponticum				
mature leaf	(−1.7)	(−2.1)	-	1
immature leaf	(−0.8)	(−1.0)	-	1

References: 1. Crombie et al. (1985); 2. Jones & Peña (1987); 3. Tyree & Sperry (1989); 4. Tyree & Dixon (1986).

different species are compared. Although conifer tracheids are generally slightly more sensitive to cavitation at a given conduit diameter than are angiosperm vessels, this difference is much smaller than one would expect from the difference in pore sizes, because 'aspiration' of the bordered pits under tension seals their larger pores. At least in *Acer* species the mean cavitation pressure has been found to be more strongly correlated with the intervessel pit structure (pit membrane thickness and porosity, and chamber depth) rather than with numbers or areas of pits per vessel (Lens *et al.*, 2011). The smaller late-wood vessels within a species do tend to be less vulnerable than early wood. Vulnerability to cavitation is closely associated with the aridity of a plant's habitat and its drought tolerance (Pockman & Sperry, 2000). It can be argued that the chance that a pore may exceed the critical size for cavitation increases as the number of pits increases, purely as a result of the increasing chance of an extreme value as numbers increase.

Cavitations in conducting vessels decrease the hydraulic conductivity of the stem; this in turn tends to a decline in ψ_ℓ, which would itself favour further embolism. This is an unstable situation that might be expected to lead to what Tyree and Sperry (1988) termed 'runaway embolism' as in the absence of stomatal closure this cycle would continue until all the conducting tissue is lost. In practice, however, stomatal closure normally prevents such catastrophic xylem failure (Jones & Sutherland, 1991). These authors showed that optimal stomatal control could not be achieved on the basis of leaf water status alone and required the plant to utilise information on soil moisture status with a corresponding need for some root–shoot signalling. There is a tendency for cavitations to occur most in leaves and young twigs, which may help preserve the integrity of the main conducting system.

Steady-state flow

As was pointed out above, transpirational flow of water through the soil–plant system is driven by evaporation setting up a gradient of water potential down which water flows. In practice the actual driving force for hydraulic flow in the apoplast, the pressure gradient, $\partial \psi_p / \partial x \, (= -\partial P / \partial x)$ is similar to $\partial \psi / \partial x$, because matric and osmotic components are small in the xylem sap where ψ_π is generally above -0.1 MPa. (Note also that in a vertical vessel at equilibrium (hence ψ is constant and there is no flow) there is an additional gradient of P of -0.01 MPa m^{-1} to counteract the gravitational potential $\rho_w g h$.)

Because of the complexity of the flow pathway (Figures 4.7 and 4.11), steady-state flow is usually analysed in terms of rather simplified 'black-box' resistance models (Figure 4.11), rather than by Eq. (4.23). Most workers since van den Honert (1948) have based their analyses on the simple catenary series model in Figure 4.11(c). Using this simplification the following equation can be used to describe the relationships between steady-state flow and water potential within the system:

$$\mathbf{E} = \frac{\psi_s - \psi_\ell}{R_s + R_r + R_{st} + R_\ell} = \frac{\psi_s - \psi_r}{R_s}$$
$$= \frac{\psi_r - \psi_{st}}{R_r} = \frac{\psi_{st} - \psi_x}{R_{st}} = \frac{\psi_x - \psi_\ell}{R_\ell} \qquad (4.27)$$

where $\mathbf{E}$ is the water flux (transpiration rate) through the system, ψ_s, ψ_r, ψ_{st}, ψ_x and ψ_ℓ refer, respectively, to the water potentials in the bulk soil, at the surface of the roots, at the base of the stem, at the top of the stem and at the evaporating sites within the leaves. The hydraulic resistances in the soil (R_s), root (R_r), stem (R_{st}) and leaf (R_ℓ) refer to the flow resistances shown in Figure 4.11(c). Although resistances are most useful for the treatment of components in series, many workers use their inverse (hydraulic conductances, often given the symbols K_r, K_{st} and K_ℓ) when studying single components of the pathway. Since $\mathbf{E}$ is usually expressed as a volume flux density (m^3 m^{-2} s^{-1}), R has units of MPa s m^{-1} though the area basis used can be leaf area, stem cross-sectional area or ground area and flow may alternatively be measured in moles. If $\mathbf{E}$ is expressed as a flux per plant, then R has units MPa s m^{-3}. All these are valid alternative methods of expression.

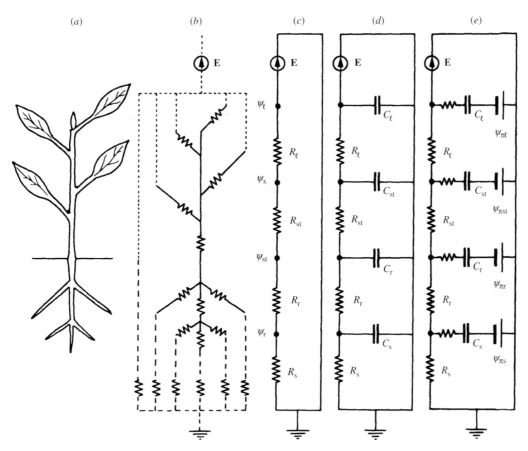

Figure 4.11 (*a*) Simplified representation of a plant; (*b*) the corresponding network of flow resistances in the soil, the roots, the stem and the leaves, with evaporation being driven by a constant current generator E; (*c*) simplified catenary model with the complex branched pathway of (*b*) represented as a linear series with the hydraulic resistances of the soil (R_s), roots (R_r), stem (R_{st}) and leaves (R_ℓ) each being represented by a single resistor; (*d*) the same as (*c*) but including capacitances (*C*) of the appropriate tissues; (*e*) as (*d*) but including resistances to and from storage and voltage sources (ψ_π) that represent the osmotic potentials of each component. In this case the voltage drop across each capacitor represents the turgor pressure.

The leaf resistance, of which approximately half relates to flow through petioles and veins and half to extra-xylem flow across the mesophyll tissues to the sites of evaporation, usually contributes the majority of the flow resistance in shoots and about 30% of R_p (Sack & Holbrook, 2006). The extra-xylem component of the pathway especially appears to involve at least some cell to cell transfer and the involvement of aquaporins. Reported values of K_ℓ vary over nearly two orders of magnitude between species (from 0.76 mmol m^{-2} s^{-1} MPa^{-1} (or 1.4 × 10^{-8} m s^{-1} MPa^{-1}) for the fern *Adiantum lunulatum* to 49 mmol m^{-2} s^{-1} (or 8.8 × 10^{-7} m s^{-1} MPa^{-1}) for the tree *Macaranga triloba*) and values can be sensitive to environment. In general, herbaceous plants tend to have higher conductances than trees with conifers and ferns being lowest (Sack & Holbrook, 2006).

The flux through the system is determined by the rate of evaporation, so leaf water potential can only

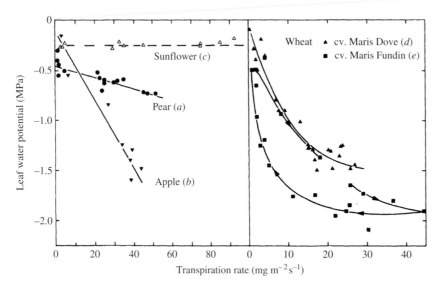

Figure 4.12 Examples of some of the types of relationship between leaf water potential and transpiration rate that have been observed. The sunflower and pear data are from Camacho-B *et al.* (1974), the apple data from Landsberg *et al.* (1975) and the wheat data from Jones (1978).

be regarded as indirectly *controlling* the flow through the plant by any effect it may have on the gas-phase (mainly stomatal) resistance. Using another analogy with electrical circuits, evaporation can be represented by a constant current generator (Figure 4.11) because the potential drop (and the resistance) across the vapour phase is typically more than an order of magnitude greater than that in the liquid phase. Therefore typical changes in the liquid-phase resistance have a negligible effect on the total resistance (and hence total flow), but they can affect the values of ψ at different points in the system according to Eq. (4.27). Typical values of ψ are between −1 and −2.5 MPa, giving $\Delta\psi$ in the liquid phase of 1 to 2 MPa, while water potential in the atmosphere is commonly lower than −50 MPa ($\simeq 69\%$ relative humidity – see Chapter 5), giving a $\Delta\psi$ in the vapour phase more than an order of magnitude greater. This comparison between gas- and liquid-phase resistances is only approximate because, as seen in Chapter 3, the driving force for gas-phase diffusion is partial vapour pressure, which is not linearly related to ψ Chapter 5).

Experimental results relating the water potential drop in the liquid phase in the soil–plant system to evaporation rates are illustrated in Figure 4.12. The water potential drop in the soil–plant system is sometimes linearly related to the evaporation rate (Figure 4.12(*a,b*)). Such results indicate that R_{sp} (the soil–plant resistance) is constant and independent of flow rate. In many cases, however, the relationship shows a marked curvature (Figure 4.12(*d,e*)) while, in extreme cases, water potential can be nearly constant over a wide range of evaporation rates (Figure 4.12 (*c*)). These results can be interpreted as R_{sp} decreasing at high flow rates. Both types of response have been reported in many species. The reason for this flow-rate dependence of R_{sp} is not known (see Fiscus, 1975; Kramer & Boyer, 1995; Passioura, 1984; Steudle, 2001), though the wide range of responses observed can perhaps best be explained on the basis of a composite model for water transport through roots where there is both an osmotically driven component of flow across membranes and a hydraulically driven component through apoplastic bypasses, with the contribution of each

Table 4.2 Flow resistances in apple trees calculated from data in Landsberg *et al.* (1976). All resistances were converted to a single plant basis and expressed as a percentage of $R_{st} + R_r$ (which included a soil component). Values in brackets represent the range of values.

	$R_{st} + R_r$ (MPa s m^{-3} × 10^{-7})	R_r (%)	R_{st} (%)	R_ℓ (%)
Potted trees (two-year old)				
Experiment 1	10.7 (8.2–12.9)	66 (48–74)	34 (26–52)	41
Experiment 2	30.5 (26.0–35.0)	52 (46–59)	48 (41–54)	21
Orchard tree (nine-year old)	0.8	60	50^a	35

^{a} Rounding error

depending on conditions and species (Steudle, 2001). The tendency for resistance to decrease with increasing flow rate is mechanistically slightly surprising, since it might be expected that the soil component of the resistance would increase, but it probably has an adaptive advantage in preventing the occurrence of severe plant water stress at high fluxes.

Measurements of water potential gradients up the stem of a transpiring plant are usually obtained with a pressure chamber using excised leaves. If attached leaves are sealed in plastic bags for approximately half an hour before the measurement, ψ_ℓ equilibrates with the stem xylem (ψ_x) and gives a true measure of the gradient up the stem. If the leaf is transpiring at the time of sampling, ψ_ℓ also depends on the resistance within the leaf petiole and can be used to estimate R_ℓ. The total plant resistance R_p (= $R_r + R_{st} + R_\ell$) is often obtained from measurements of ψ_ℓ in plants grown in solution culture as this eliminates any R_s. The gradient of ψ_ℓ for leaves at different insertion levels up a transpiring plant ranges from less than 0.03 MPa m^{-1} for branches of trees to about 0.1 MPa per node in a wheat plant (about 1 MPa m^{-1}) (Jones, 1977a). Table 4.2 gives some estimated values for flow resistances in different parts of the soil–plant system in apple trees, illustrating that the major proportion of the resistance tends to be in the roots, though there is also a large component in the petiole or leaf. This large leaf hydraulic resistance results in ψ_ℓ often being *c.* 0.2 MPa below ψ_{st} in transpiring trees.

Hydraulic redistribution

There is now substantial evidence that redistribution of water from wetter regions of soil to drier zones (usually nearer the surface) can be facilitated by transfer through plant roots – a process known as *hydraulic lift* or *hydraulic redistribution* (Caldwell *et al.*, 1998; Prieto *et al.*, 2012). As for other water transport processes this occurs down gradients of water potential with rates and directions dependent both on water potential gradients and on path resistances and can involve a reverse of the usual flow from the soil to the root; in some cases there are even reports of fog or dew interception by leaves being redistributed downwards to roots and to the soil. It is common for roots to have some 'rectifying' activity with a higher resistance for flow from roots to the soil than for the more usual water uptake, while the flow resistances through the system appear to be at least partly controlled by changes in aquaporin activity (Henzler *et al.*, 1999; McElrone *et al.*, 2007). Shared mycorrhizal networks may also contribute to redistribution while the improvements in soil water distribution can enhance root growth and function, and may even affect rhizosphere processes such

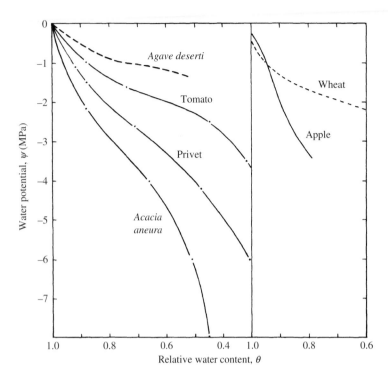

Figure 4.13 Relationships between water potential and relative water content for different species. (Data from Jones, 1978; Jones & Higgs, 1979; Nobel & Jordan, 1983; Slatyer, 1960.)

as decomposition and nutrient turnover and supply.

Dynamic responses

Figure 4.12(e) shows an example where the relationship between the water potential drop and **E** is not unique: that is, there is hysteresis. This indicates a failure of the steady-state model to simulate the true dynamic behaviour when **E** is varying. The hysteresis where liquid flow lags behind evaporative demand may be modelled by incorporating capacitors into the circuit analogue (Figure 4.12(d)).

The *capacitance* (C) of any part of the system may be defined as the ratio of the change in tissue water content (W) to the change in water potential, i.e.:

$$C = dW/d\psi = W_{max}d\theta/d\psi = W_{max}C_r \qquad (4.28)$$

where W_{max} is the maximum (turgid) tissue water content and θ is the relative water content. The term $d\theta/d\psi$ can be called a *relative capacitance* (C_r), which is an intrinsic property of the tissue so is useful when comparing tissues of very different shape or size.

When the water content is expressed per unit area (leaf, stem cross-section or ground) the units for C would be $m^3 m^{-1} MPa^{-1}$ ($= m MPa^{-1}$) but when expressed in absolute terms C would be in $m^3 MPa^{-1}$. For inclusion in flow models, the basis for C must be the same as that used for **E**.

Some examples of relationships between ψ and tissue water content are shown in Figure 4.13. Although these curves are in general non-linear, it is often possible to approximate C_r by a constant value (i.e. a straight line) over much of the relevant physiological range. Values of C_r taken from this figure range from about 5% MPa^{-1} for apple leaves to 33% MPa^{-1} for wheat and tomato leaves. Corresponding values for C_r for conifers have been estimated at 4.7% for *Larix* and 6.3% for *Picea* (Schultze *et al.*, 1985).

Inclusion of the capacitances of various tissues in the model of Figure 4.11(c) gives that shown in Figure 4.11(d). Although even this model is rather complex, a number of authors have extended the approach even further to include large networks of resistors and

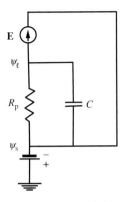

Figure 4.14 A simple lumped-parameter model of the plant hydraulic system, where flow is driven by the constant current generator **E**, and the soil water potential ψ_s is lowered below the reference value (ground) by a battery.

capacitors representing detailed anatomical and morphological measurements of hydraulic architecture. Another extension to this approach (Smith *et al.*, 1987) is illustrated in Figure 4.11(*e*). These authors showed that by including voltage sources it is possible to incorporate cell osmotic potential as an explicit variable. A consequence of this is that, since the voltage drop across the voltage source is ψ_π, and the voltage drop between the nodes (−●−) and ground is ψ, then the voltage drop across the capacitor represents the turgor pressure (ψ_p). These complex models can be used to simulate the dynamics of water relations of different tissues; examples where such an approach has been used include the dynamic simulation of water relations of plants as diverse as *Agave* (Smith *et al.*, 1987) and trees such as *Thuja* (Tyree, 1988). Although numerical solution of these models can be readily achieved by modern computer routines for electrical network analysis, their complexity often limits their value in understanding the underlying control processes. The general behaviour of such resistance–capacitance models is best illustrated using a simplified lumped parameter model (Figure 4.14) where the complex network in Figure 4.11(*d*) is approximated by a circuit with only one resistance and one capacitor. To analyse this, flow through the plant ($\mathbf{J}_p$) can be written (from Eq. (4.27)):

$$\mathbf{J}_p = (\psi_s - \psi_\ell)/R_p \qquad (4.29)$$

The rate of change of leaf water content (dW_ℓ/dt) is given by the difference between the flow of water into the leaf and that lost by evaporation, so:

$$dW_\ell/dt = \mathbf{J}_p - \mathbf{E} \qquad (4.30)$$

Substituting from Eq. (4.29) gives:

$$dW_\ell/dt = (\psi_s - \psi_\ell)/R_p - \mathbf{E} \qquad (4.31)$$

If one now assumes a constant capacitance one can substitute (from Eq. (4.28)) and rearrange to give:

$$d\psi_\ell/dt + \psi_\ell/R_p\, C = (\psi_s - \mathbf{E}\, R_p)/R_p\, C \qquad (4.32)$$

This is a first-order differential equation of the type commonly encountered in the analysis of electrical circuits. It can be solved using standard mathematical techniques to give the time of dependence of ψ_ℓ, after a step change in the equilibrium **E** from $\mathbf{E}_1$ to $\mathbf{E}_2$ to give:

$$\psi_\ell = A + B\exp(-t/\tau) \qquad (4.33)$$

where $A = (\psi_s - \mathbf{E}_2\, R_p)$, and $B = R_p\, (\mathbf{E}_2 - \mathbf{E}_1)$ and $\tau = R_p\, C$. The value $\tau\ (= R_p\, C)$ is called the time constant and is the time for 63% of the total change (see Figure 4.15). It can be easily shown that the half-time, or time for 50% of the total change to occur, is equal to $\tau \times \ln 0.5 = 0.693\tau$.

In practice, the assumptions of constant capacitance and resistance and the use of the simplified 'lumped parameter' model are adequate for many purposes even though realistic models may be much more complex (Bohrer *et al.*, 2005; Janott *et al.*, 2011; Jones, 1978). In particular, the relatively small number of parameters required in fitting this model to experimental data is a major advantage. Methods for solving Eq. (4.33) in the much more realistic cases where E is continuously varying are described for example by Jones (1978).

The dynamics of water exchange by single cells can also be treated in the same way as tissues. As for tissues (Eq. (4.33)), the time constant (τ) for equilibration of water potential is given by the product of the resistance to water uptake

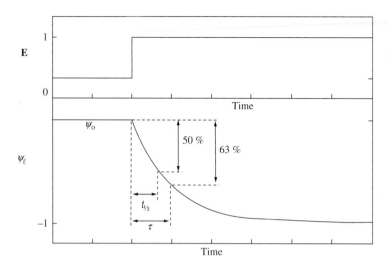

Figure 4.15 The time course of changes in ψ_ℓ for an instantaneous change in **E**, as predicted by the model in Figure 4.14, illustrating the half-time ($t_{1/2}$ – the time required for 50% of the total change) and the time constant (τ).

($R = 1/(A\ L_\text{p})$, where A is the cell surface area) and the cell capacitance ($C = V/(\varepsilon - \psi_\pi)$, where V is the cell volume and ε is the cell volumetric elastic modulus), so that:

$$\tau = V/(AL_\text{p}(\varepsilon - \psi_\pi))\qquad(4.34)$$

It follows from this equation that the rate of hydraulic equilibration of a cell increases with both L_p and ε.

4.5 Long-distance transport in the phloem

The movement of solutes from tissue to tissue or over shorter distances from cell to cell within plants is essential for normal growth and development. For example, minerals from the soil must reach the leaves and other aerial tissues while carbohydrates from the leaves must be transported downwards to build new roots. Details of the various transport processes may be found in Nobel (2009), as well as in other plant physiology texts and reviews on transport mechanisms (Taiz & Zeiger, 2010; Turgeon, 2010a). Here, only a brief outline will be presented to place the different processes in perspective.

In Chapter 3 it was stated that, when a finite quantity of material is released in a plane, the concentration profile is the shape of the Gaussian curve, and the distance from the origin (x) at which the concentration falls to 37% of that at the origin is given by:

$$x = \sqrt{4Dt}\qquad(4.35)$$

Because diffusion coefficients in solution are approximately four orders of magnitude smaller than in air (see Appendix 2), diffusion is much slower in solution and effective only over short distances. For example, solving Eq. (4.35) for a typical solute **D** of 1×10^{-9} m^2 s^{-1}, gives the time to diffuse 100 μm as 2.5 s (i.e. $(100 \times 10^{-6})^2/(4 \times 10^{-9})$ s). Corresponding times to diffuse 1 cm or 1 m are, respectively, 2.5×10^4 s ($\simeq$ 6.9 h) and 2.5×10^8 s ($\simeq$ 8 years). Clearly, therefore, simple diffusion cannot be an important mechanism for long-distance transport in plants, though it is important for cell-to-cell or within-cell transport. Within individual cells the cytoplasm is continually moving; this cytoplasmic streaming acts to speed transfer over short distances in a way analogous to turbulent transfer in a crop boundary layer. As long as there is no *net* cytoplasmic movement, the only effect is to lower the apparent value of **D**. Although the transport of solutes across plant cell membranes, whether by active or passive mechanisms, is important for plant functioning, a discussion of this topic is outside the scope of this book and details may be found in the texts referred to above.

The long-distance movement of solutes is primarily in the bulk flow of water in the xylem or in the other specialised conducting tissue, the phloem. The bulk flow of water in the xylem is important for the rapid movement of not only mineral nutrients from the roots to the shoots, but also metabolites and plant growth regulators such as cytokinins and ABA. However, because transpiration is largely unidirectional, it cannot be involved in basipetal transport so that another pathway is required for redistribution of solutes, particularly in a downward direction.

This alternative long-distance transport pathway, which is particularly important for carbohydrate movement from leaves to meristematic or storage tissues, is in the specialised (living) sieve tubes of the phloem. The mechanism of phloem transport is still incompletely understood but it appears to be based on an osmotically generated pressure-flow mechanism as proposed by Münch (1930). In this the pressure gradient that drives the flow is generated by the transfer of sugars from the mesophyll cells in the leaves into the phloem sieve elements. The high solute concentrations lower the water potential in the sieve elements so that water follows through the semi-permeable sieve-tube membrane, thus raising the hydrostatic pressure and causing mass flow out of the source tissue. There may also be active solute unloading at the sink tissues, further aiding the development of a pressure gradient.

It now appears that there are several mechanisms of phloem loading that can differ between species (Fu *et al.*, 2011; Turgeon, 2010b). One mechanism of phloem loading involves passive diffusion of sugars from the high concentrations in the mesophyll cells, through the companion cells and into the sieve elements. This seems to be particularly prevalent in tree species. Other plants, particularly herbaceous species, use active (energy-requiring) mechanisms that can concentrate the sugars in the phloem. In what has been called apoplastic loading, sucrose diffuses into the apoplast and is then pumped into the phloem companion cells by proton-coupled sucrose transporters (Sauer, 2007). A second active process,

though not strictly involving an active pump, is 'polymer trapping' where sugars diffuse through plasmodesmata into the companion cells where they are actively converted to the raffinose family of oligomers, which cannot leak out through the plasmodesmata. As a result they build up and diffuse into the sieve elements. Similarly there are also both active and passive loading strategies for sugar alcohols. A convenient simplified model that describes phloem transport and that can be used to provide quantitative predictions of the speed of translocation has been presented recently (Pickard, 2012).

For both xylem and phloem, the flux of the *i*th solute (J_i, kg m^{-2} s^{-1}) across a plane is given by the product of concentration and solution velocity:

$$J_i = c_i J_v \tag{4.36}$$

Typical concentrations of different solutes in phloem and xylem sap are presented in Table 4.3. This table shows that by far the major solute in the phloem is sugar while there is none in the xylem. Although this is true for many plants, some trees such as sugar maple do have significant amounts of sugar in the xylem, particularly in spring. The concentration of almost all substances, with the notable exception of calcium, are markedly higher in the phloem than the apoplast. Therefore, a given J_v in the phloem will transport more solute than an equivalent velocity in the xylem, though transport in the phloem does require expenditure of energy by the plant while the much greater bulk flow in the xylem is driven by energy from elsewhere.

Calcium is somewhat exceptional in that most workers have found fairly similar concentrations of calcium in the two types of sap, or else calcium is more concentrated in the xylem (e.g. Table 4.3). A consequence of the rather low calcium concentration in the phloem appears to be that low transpiring tissues, such as apple fruits or the enclosed leaves of lettuce, can suffer severe calcium deficiency, which gives rise to disorders such as bitter pit and breakdown in apple and tipburn in lettuce (see e.g. Bangerth, 1979).

Table 4.3 Comparison of concentrations of some solutes in xylem and phloem sap from *Nicotiana glauca* (data from Hocking, 1980).

	Xylem sap (g m^{-3})	Phloem sap (g m^{-3})
Chloride	64	486
Sulphur	43	139
Phosphorus	68	435
Ammonium	10	45
Calcium	189	83
Magnesium	34	104
Potassium	204	3673
Sodium	46	116
Amino compounds	283	11×10^3
Sucrose	0	155–168×10^3
Total dry matter	1.1–1.2×10^3	170–196×10^3

4.6 Sample problems

4.1 (i) To what height would water rise in
  (a) a wettable vertical capillary 1 mm diameter,
  (b) a similar capillary tilted at 45°, (c) a vertical capillary where the wall material has a contact angle of 50°, (d) a wettable capillary 1 μm in diameter? (ii) What pressure would need to be applied to the column in (d) to prevent any capillary rise?

4.2 A single cell has a ψ_π of -1.5 MPa when $\psi = -1$ MPa. The volume increases by 25% as ψ increases to -0.5 M Pa. What are (i) the original value of ψ_p; (ii) the new value of ψ_π; (iii) the new value of ψ_p; (iv) θ, the initial relative water content of the cell (assume ψ is linearly related to volume); (v) ε_B, the volumetric elastic modulus at full turgor?

4.3 (i) What would be the volume flow rate of water through a smooth cylindrical pipe 1 m long, 0.2 mm in diameter, with an applied pressure differential between the ends of 5 kPa? (ii) What are the values of L, L_p and R? (iii) What diameter tube would be required to maintain the same flow if the pressure differential decreased to 1 kPa?

4.4 If $\psi_{soil} = -0.1$ MPa, $\psi_\ell = -1.2$ MPa and transpiration rate $= 0.1 \times 10^{-6}$ m^3 H$_2$O (m^2 leaf)$^{-1}$ s^{-1}: (i) calculate total hydraulic resistance (a) on a leaf area basis, (b) on a plant basis (assuming a leaf area per plant of 0.1 m^2), (c) on a ground-area basis (assuming, 30 plants m^{-2}). (ii) Estimate ψ_ℓ if half the plants are removed without affecting total crop evaporation. (iii) Estimate ψ_ℓ if, instead, half the shoots on each plant are removed (assuming that half the hydraulic resistance is normally in the soil–root part of the flow path).

5 Energy balance and evaporation

Contents

The preceding chapters outlined the basic principles of mass transfer and described the application of electrical analogues. The application of these principles to evaporation requires the extension of the simple analogues to include the flow of heat as well as water vapour. This is because energy is required to supply the latent heat of evaporation. For this reason, as a preliminary to the treatment of evaporation, I first outline the energy balance equation and introduce the concepts of isothermal net radiation and the radiative heat transfer resistances, as well as define the various quantities that are used to specify the amount of water in air. The energy balance equation is then used to derive a general equation for the description of evaporation from single leaves and from plant communities, and this is used to investigate particular aspects of the environmental and physiological control of evaporation and of dewfall.

5.1 Energy balance

5.1.1 Component fluxes

The principle of the conservation of energy (the *first law of thermodynamics*) states that energy cannot be created or destroyed, but only changed from one form to another. Applying this to a plant leaf (or canopy) it can be seen that the difference between all the energy fluxes into and out of the system must equal the rate of storage, thus:

$$\mathbf{R}_n - \mathbf{C} - \lambda \mathbf{E} = \mathbf{M} + \mathbf{S} \tag{5.1}$$

where $\mathbf{R}_n$, is the net heat gain from radiation (shortwave plus longwave), $\mathbf{C}$ is the net 'sensible' heat *loss*, $\lambda \mathbf{E}$ is the net latent heat *loss*, $\mathbf{M}$ is the net heat stored in biochemical reactions, $\mathbf{S}$ is the net physical storage (causing a change of temperature). It is convenient to express all these fluxes per unit area (of leaf or ground) to give units of flux density (W m^{-2}). (Note that in meteorological applications it is conventional to express all vertical fluxes with downwards as positive.)

Net radiation

R_n is the dominant term in Eq. (5.1), not only because it is often the largest, but also it drives many of the other energy fluxes. Typical diurnal changes in R_n for different surfaces have been presented in Chapter 2 (Figure 2.13 and Table 2.3).

Sensible heat flux

The sensible heat loss, C, is the sum of all heat loss to the surroundings by conduction or convection. For example, whenever a leaf is warmer than the surrounding air, heat is lost and C is positive. The total sensible heat flux may be partitioned into that lost to the air (by conduction and convection), for which the symbol C is retained, and that lost by conduction to other surroundings, particularly the soil, which is given the symbol G. G is negligible for individual leaves. For plant canopies, however, G refers to the soil heat flux, which is positive during much of the day (representing a loss from the canopy and warming of the soil) and may range from 2% of R_n for a dense canopy to more than 30% of R_n in sparse canopies with little shading of the soil. The ratio G/R_n $(= \Gamma)$ can be approximated (Choudhury, 1994) by:

$$G/R_n = \Gamma \simeq \Gamma' \exp(-k L) \qquad (5.2)$$

where Γ is a total soil heat flux ratio and Γ' is the energy partitioning at the surface $(\simeq 0.4)$ and k is the extinction coefficient for radiation transfer through the canopy and L is the leaf area index (see Chapter 3). At night G is negative and of similar absolute magnitude to daytime values. This can be further approximated (Baret & Guyot, 1991) using the following vegetation index (see Chapter 2) as:

$$G/R_n \simeq \Gamma'(SAVI_{max} - SAVI)/(SAVI_{max} - SAVI_{min})$$
$$= \Gamma'\Gamma'' = \Gamma \qquad (5.3)$$

where $SAVI$ is the soil-adjusted vegetation index calculated as $1.5 \times (\rho_{NIR} - \rho_R)/(\rho_{NIR} + \rho_R + 0.5)$ and the subscripts represent maximum and minimum values for the relevant site. Therefore the available energy, $(R_n - G)$, can be estimated as $(1 - \Gamma)R_n$.

Latent heat flux

The rate of heat loss by evaporation (λE) is that required to convert all water evaporated from the liquid to the vapour state and is given by the product of the evaporation rate and the latent heat of vaporisation of water $(\lambda = 2.454$ MJ kg^{-1} at 20°C).

Storage

The rate of metabolic storage (M) represents the storage of heat energy as chemical bond energy and is dominated by photosynthesis and respiration. Typical maximum rates of net photosynthesis of 0.5 to 2.0 mg CO_2 m^{-2} s^{-1} correspond to M between 8 and 32 W m^{-2} (see Chapter 7), values that are usually less than 5% of R_n. At night, M takes smaller negative values associated with dark respiration, except in certain species of the Araceae where extremely high respiration rates can occur during the spring (see Chapter 9). The physical storage (S) includes energy used in heating the plant material as well as (for a canopy) heat used to raise the temperature of the air. In general the flux into physical storage is small except for massive leaves or stems (e.g. cacti) and forests. For example, a very dense cereal crop might contain 3 kg water m^{-2}, so that an estimate of S when canopy temperature is changing at 5°C h^{-1} would be only 17.5 W m^{-2} ($3 \times 4200 \times 5/3600 =$ mass $\times$ specific heat $\times$ d$T/$dt).

5.1.2 Isothermal net radiation

The value of the net radiation, R_n, is the difference between the total incoming radiation absorbed and the total longwave radiation emitted (from Eq. (2.4)):

$$R_n = R_{absorbed} - \varepsilon\sigma(T_s)^4 \qquad (5.4)$$

where fluxes are expressed per unit surface area, T_s is the Kelvin temperature of the surface, ε is the emissivity and σ is the Stefan–Boltzmann constant. Unfortunately the value of R_n is not an environmentally determined constant as it depends on processes such as transpiration and sensible heat exchange that affect the surface temperature. As was pointed out by Monteith (1973) it is therefore useful for predictive studies to define an 'environmental' net radiation that is independent of surface temperature. This *isothermal net radiation*, $\mathbf{R}_{ni}$, can be defined as the net radiation that would be received by an identical surface in an identical environment *if it were at air temperature*. Therefore:

$$\mathbf{R}_{ni} = \mathbf{R}_{absorbed} - \varepsilon\sigma(T_a)^4 \qquad (5.5)$$

Substituting from Eq. (5.4) gives the relationship between $\mathbf{R}_{ni}$ and $\mathbf{R}_n$ as:

$$\mathbf{R}_{ni} = \mathbf{R}_n + \varepsilon\sigma(T_s{}^4 - T_a{}^4) \qquad (5.6)$$

If one now makes the substitution $T_s = T_a + \Delta T$, and multiplies out, one gets:

$$\mathbf{R}_{ni} = \mathbf{R}_n + \varepsilon\sigma\left(T_a{}^4 + 4T_a{}^3(\Delta T) + 6T_a{}^2(\Delta T)^2 \right.$$
$$\left. + 4T_a(\Delta T)^3 + (\Delta T)^4 - T_a{}^4\right) \qquad (5.7)$$

Terms in $T_a{}^4$ cancel out and because $\Delta T << T_a$, all terms in $(\Delta T)^2$ and higher powers can be neglected, so that:

$$\mathbf{R}_{ni} \simeq \mathbf{R}_n + 4\varepsilon\sigma(T_a{}^3)\Delta T \qquad (5.8)$$

The second term in Eq. (5.8) represents a longwave radiative heat loss and can be put into a form analogous to the usual equation for sensible heat loss (see Table 3.1):

$$\text{radiative heat loss} \simeq (4\varepsilon\sigma T_a{}^3/\rho c_p)\rho c_p(T_s - T_a) \qquad (5.9)$$

where ρ is the density of air, c_p is the specific heat of air and $(4\varepsilon\sigma T_a{}^3/\rho c_p)$ is a 'conductance' to radiative heat transfer (g_R), so that:

$$\mathbf{R}_n = \mathbf{R}_{ni} - g_R\rho c_p(T_s - T_a) \qquad (5.10)$$

In addition to this radiative heat loss, an isolated leaf can also lose sensible heat by convection, a pathway that is in parallel with that for radiant heat loss (see Figure 5.1). Because radiant and sensible heat transfer are each proportional to the leaf-to-air temperature difference, we can now define a total thermal conductance (g_{HR}), as the parallel sum of g_H and g_R $(= g_H + g_R)$. The effect of temperature on g_H is shown in Table 5.1, illustrating that it has a small effect on the total thermal conductance, except when the boundary layer conductance for sensible heat transfer is rather small. As will be seen later, isothermal net radiation is particularly useful in modelling studies designed to determine the consequences for leaf temperature or evaporation rate of changing environmental or physiological factors.

5.1.3 Measures of water vapour concentration

The most commonly used measures of gas composition were described in Chapter 3; in addition there are several that are specifically used for describing the water vapour content of air.

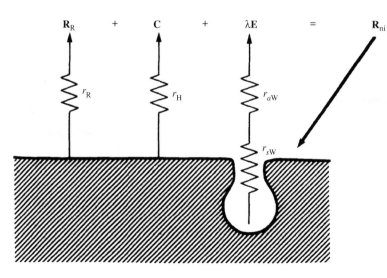

Figure 5.1 Energy exchanges for a leaf where the radiative heat loss $\mathbf{R}_R$ is the difference between the actual net radiation $(\mathbf{R}_n)$ and the isothermal net radiation $(\mathbf{R}_{ni})$.

Table 5.1 Temperature dependence of the 'radiative' conductance g_R, and some typical values of the total thermal conductance (g_{HR}) for a range of values of g_H. The value in brackets is g_H as a percentage of g_{HR}.

Temperature (°C)	g_R (mm s^{-1})	g_{HR} (mm s^{-1})		
		$g_H = 2$	$g_H = 20$	$g_H = 200$
0	3.54	5.5 (36%)	23.5 (85%)	204 (98%)
10	4.10	6.1 (33%)	24.1 (83%)	204 (98%)
20	4.69	6.7 (30%)	24.7 (81%)	205 (98%)
30	5.37	7.4 (27%)	25.4 (79%)	205 (97%)
40	6.10	8.1 (25%)	26.1 (77%)	206 (97%)

Water vapour pressure

We have already come across the use of the partial pressure of water vapour in air, usually given the symbol, e (Pa).

Absolute humidity

The mass concentration of water vapour (c_W, g m^{-3}) is called the *absolute humidity*. This is related to the water vapour partial pressure by Eq. (3.6), which simplifies to:

$$c_W = e M_W / \mathscr{R} T = (2.17/T)e \qquad (5.11)$$

An alternative expression relating c_W and e, that will be used in several subsequent derivations, may also be obtained from Eq. (3.6), remembering that the partial pressure of dry air $p_a = P - e$:

$$c_W = \rho_a(M_W/M_A)[e/(P-e)] \simeq \rho_a(M_W/M_A)(e/P) \qquad (5.12)$$

where p_a is the density of dry air and M_A, is the effective molecular mass of dry air ($\simeq 29$), so $M_W/M_A = 0.622$. The approximation in Eq. (5.12) usually introduces negligible error since $e << P$.

Water potential

The moisture status of air may also be described in terms of water potential (ψ). When air is allowed to equilibrate with liquid water at the same temperature, the equilibrium state is:

$$\psi_{liquid} = \psi_{vapour} \qquad (5.13)$$

so that vapour in equilibrium with free water has a water potential of zero. In this case the air is saturated with water vapour. When the liquid water potential falls below zero, the air humidity falls with the water activity, a_W, according to Eq. (4.7):

$$\psi = \frac{\mathscr{R}T}{V_W} \ln(e/e_s) \qquad (5.14)$$

where e_s is the saturation partial pressure of water vapour (or *saturation vapour pressure*; *svp*). The value of e_s over water is a function of temperature ($e_{s(T)}$), closely approximated by the following version of the Magnus equation (equation from the *CR-5 Users Manual 2009–12* from Buck Research – see www.hygrometers.com – modified from Buck, 1981; see Appendix 4 for further details and for tabulated values for $e_{s(T)}$, and for the corresponding concentration, c_{sW}):

$$e_{s(T)} = f(a \exp\left(bT/(c+T)\right) \qquad (5.15)$$

where T is in °C, $e_{s(T)}$ is in Pa, and the empirical coefficients are: a = 611.21, b = 18.678 – ($T/234.5$), c = 257.14, $f \simeq 1.0007 + 3.46 \times 10^{-8} P$ (in Pa).

Vapour pressure deficit

Another commonly used term expresses the difference between the saturation vapour pressure and the actual vapour pressure as the *vapour pressure deficit (D)* given by:

$$D = e_{s(T)} - e \qquad (5.16)$$

Relative humidity

Where air is not saturated with water vapour, the degree of saturation is often expressed as the *relative humidity* (*h*), which is the vapour pressure expressed as a fraction (or often percentage) of the saturation vapour pressure at that temperature:

$$h = e/e_{s(T)} \tag{5.17}$$

Dewpoint temperature

The *dewpoint temperature* (T_{dew}) is the temperature at which the water vapour pressure equals the saturation vapour pressure. As the air is cooled below its dewpoint, condensation occurs. The relationships between vapour pressure, vapour pressure deficit, dewpoint and relative humidity are illustrated in Figure 5.2.

Wet bulb temperature

Another useful term is the *wet bulb temperature* (T_{wb}), which is the temperature that a moist surface reaches when it evaporates *adiabatically* (i.e. without heat exchange) into an unsaturated atmosphere (see Figure 5.2). The value of the wet bulb temperature depends on vapour pressure, air temperature and on the rate of air movement over the evaporating surface. The wet bulb temperature is an experimentally measurable quantity that is commonly used for estimating atmospheric humidity using the relation:

$$e = e_{s(T_w)} - \gamma(T_a - T_{wb}) \tag{5.18}$$

where γ is the *thermodynamic psychrometer constant* ($Pc_p/0.622\,\lambda$). In spite of its name, γ varies with temperature and pressure, and experimental values also vary with the degree of ventilation (as this determines how closely the system approximates to adiabatic), and is 66.1 Pa K^{-1} for a well ventilated surface at 100 kPa pressure and 20°C; other values are given in Appendix 3). Psychrometric tables are widely available for determining *e* directly from dry bulb (T_a) and wet bulb (T_{wb}) temperatures measured either in normal meteorological screens or in aspirated enclosures. The theoretical derivation of γ is discussed by Monteith and Unsworth (2008) but see also Campbell and Norman (1998) for a discussion of its expression in molar units.

Measurement of humidity

There are many methods available for measuring air humidity. These include approaches that are based on gravimetric measurement or on the absorption of IR or UV radiation by water vapour, those that depend on the equilibrium adsorption of water onto solids with consequent changes in mechanical properties (e.g. length, as in the hair hygrometer) or electrical properties (e.g. capacitance sensors), and those that depend on measurement of the water potential of the water vapour in air using instruments such as psychrometers or dewpoint hygrometers (Bentley, 1998; Visscher, 1999; WMO, 2008).

5.2 Evaporation

5.2.1 Penman–Monteith combination equation

In Chapter 3 it was established that mass transfer is proportional to a concentration difference, so that for evaporation from a moist surface:

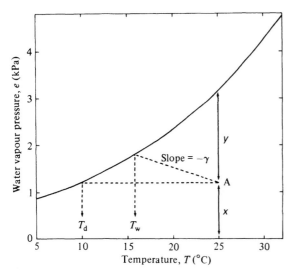

Figure 5.2 Curve of saturation water vapour pressure against temperature (solid line) illustrating the relationship between different measures of humidity. For the point marked A – vapour pressure (*e*) = *x*; vapour pressure deficit (*D*) = *y*; relative humidity = *x*/(*x* + *y*); dew point (T_d) = temperature where saturation vapour pressure = *x*; the wet bulb temperature (T_w) is also shown.

$$\mathbf{E} = g_W \Delta c_W \tag{5.19}$$

where g_W is the total conductance of the pathway between the evaporating sites and the bulk air (for plants it includes stomatal, cuticular and boundary layer components) and Δc_W is the water vapour concentration difference between the surface and the bulk air. Since vapour pressure is a more appropriate driving force where air temperature does not equal surface temperature, Eq. (5.12) can be used to replace Δc_W by $(\rho_a M_W/M_A P)\Delta e$, giving:

$$\mathbf{E} = g_W(0.622\rho_a/P)(e_{s(T_s)} - e_a) \tag{5.20}$$

where $e_{s(T_s)}$ is the saturation vapour pressure at surface temperature.

This equation provides the basis for measurements of leaf or boundary layer conductance in cuvettes (see Chapter 6) or, where independent estimates of g_W are available, it can be used to estimate $\mathbf{E}$. However, it does require a knowledge of surface temperature (for determining $e_{s(T_s)}$). If T_s is not measured, it can be determined from energy balance considerations (see Chapter 9), or else it is possible to eliminate the need for a knowledge of surface conditions by combining Eq. (5.20) with the energy balance Eq. (5.1) and using an approximation originally suggested by Penman (1948). In this, the surface–air vapour pressure difference $(e_{s(T_s)} - e_a)$ is replaced by the vapour pressure deficit of the ambient air $(D = e_{s(T_a)} - e_a)$ plus a term that depends on the temperature difference between the surface and the air (see Figure 5.3):

$$e_{s(T_s)} - e_a = (e_{s(T_a)} - e_a) - s(T_a - T_s)$$
$$= D + s(T_s - T_a) \tag{5.21}$$

where s is the slope of the curve relating saturation vapour pressure to temperature (which is assumed to be approximately constant over the range T_a to T_s), and D is the vapour pressure deficit of the ambient air. For values of s see Appendix 4.

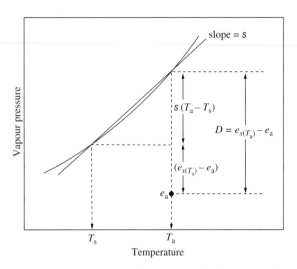

Figure 5.3 The Penman transformation. In this figure the solid curve represents the relationship between saturation vapour pressure and temperature. The slope of the curve over the range T_s to T_a is approximated by a straight line of slope s. The value of the surface to air vapour pressure difference $(e_{s(T_s)} - e_a)$ is given by the vapour pressure deficit of the ambient air (D) minus the difference between the saturation vapour pressures at T_a and T_s (approximated by $s(T_s - T_a)$).

Using this substitution in Eq. (5.20) gives:

$$\mathbf{E} = g_W(0.622\rho_a/P)[D + s(T_s - T_a)] \tag{5.22}$$

It is possible to write an equivalent expression for the sensible heat flux between the surface and the air (Eq. (3.29), with the volumetric heat capacity of dry air $(\rho_a c_p)$ used to approximate the heat capacity of air):

$$\mathbf{C} = g_H(\rho_a c_p)(T_s - T_a) \tag{5.23}$$

Eliminating $(T_s - T_a)$ from these two equations gives:

$$\mathbf{E} = g_W(0.622\rho_a/P)[D + (s\mathbf{C}/(g_H\rho_a c_p))] \tag{5.24}$$

In the steady state (and neglecting $\mathbf{M}$) the energy balance (Eq. (5.1)) reduces to:

$$\mathbf{C} = (\mathbf{R}_n - \mathbf{G}) - \lambda\mathbf{E} \tag{5.25}$$

substituting this for $\mathbf{C}$ in Eq. (5.25) and rearranging gives the following equation for $\mathbf{E}$:

$$E = \frac{[s(\mathbf{R}_n - \mathbf{G}) + \rho_a c_p g_H D]}{\lambda(s + (\gamma g_H / g_W))} \qquad (5.26a)$$

This equation can be expressed in a number of alternative forms. For example, by using the approximation in Eq. (5.11) and substituting the pyschrometer constant ($\gamma = Pc_p\, M_A/M_W\, \lambda$), it can be expressed in terms of the absolute humidity deficit of the ambient air (D_{cw}) as:

$$E = \frac{[\varepsilon(\mathbf{R}_n - \mathbf{G})/\lambda] + g_H D_{cw}}{\varepsilon + g_H / g_W} \qquad (5.26b)$$

where $\varepsilon = s/\gamma$. Where the boundary layer conductances for heat and water vapour are similar, as in a fully turbulent boundary layer, one can make use of the rules for adding conductances in series (Figure 3.2) to replace g_H/g_W by $(1 + g_a/g_\ell)$ where g_ℓ is the leaf conductance (otherwise referred to as a physiological or surface conductance) largely determined by stomatal aperture. Equation (5.26a) above, which was first applied to leaves by Penman (1953) and to plant canopies by Monteith (1965), is widely known as the *Penman–Monteith equation.* The predictive power of this equation for modelling studies can be improved by replacing $\mathbf{R}_n$ by the net isothermal equivalent, $\mathbf{R}_{ni}$, and by replacing g_H by g_{HR} to give:

$$E = \frac{[s(\mathbf{R}_{ni} - \mathbf{G}) + \rho_a c_p\, g_{HR} D]}{\lambda(s + (\gamma\, g_{HR} / g_W))} \qquad (5.26c)$$

The original derivation (Penman, 1948) was in a form to describe evaporation from a free water surface (known as *potential evaporation,* $\mathbf{E}_o$). For that surface there is no surface resistance to water loss, so that the term g_H/g_W in the denominator becomes equal to g_{aH}/g_{aW}. The term $\gamma g_H/g_W$ is sometimes referred to as the *modified psychrometer constant,* γ^*. For evaporation from free water surfaces in turbulent boundary layers these two conductances are approximately equal (see Chapter 3) so $\gamma^* \simeq \gamma$. The earliest use of the Penman equation was to estimate canopy evaporation on the basis of a series of empirically determined crop coefficients (see below) that converted the potential evaporation to crop

evaporation; these conversion factors commonly ranged from 0.6 to 0.8 in temperate climates (Doorenbos & Pruitt, 1984). Incorporation of the physiological conductance term (dependent on the stomatal conductance) in Eq. (5.26) allows one to estimate evaporation directly without the need for an empirical conversion where one has information on the leaf or canopy conductance (g_W).

When applied to single leaves, g_H is simply the boundary layer conductance for heat and g_W is the series sum of the leaf (mainly stomatal) and boundary layer conductances (i.e. $g_W = g_{\ell W}\, g_{aW}/(g_{\ell W} + g_{aW})$). Scaling up to canopies is not simple as the pathway of water loss is more complex. The simplest approximation is the 'big leaf' single source model (Figure 5.4(a)), where the physiological component of g_W (which actually includes both leaves and soil) can be estimated by the parallel sum of the individual leaf conductances. This physiological conductance (g_{LW} – note that we distinguish canopy conductances or resistances per unit ground area by capital italic subscripts) is reasonably well approximated by:

$$g_{LW} = \Sigma(\overline{g_{\ell i}} L_i) \qquad (5.27)$$

where $\overline{g_{\ell i}}$ is the mean leaf conductance per unit projected leaf area in a given stratum and L_i is the leaf area per unit ground area, or leaf area index, in that stratum of canopy (see Lhomme *et al.,* 1994 for a detailed discussion of this approach). The leaf conductance profile in the canopy may be measured using diffusion porometers (see Chapter 6) or else it may be estimated using stomatal models that include a response to the light profiles in the canopy (Irmak *et al.,* 2008) and the boundary layer resistance estimated by any of the methods outlined in Appendix 8. Some typical values for $g_{\ell W}$ are presented in Table 5.2. Low values can result from either stomatal closure or poor ground cover.

A problem with the simple 'big leaf' model is that it is not possible to separate completely the boundary layer from the 'physiological' or 'leaf' component (including any soil component). This is partly because the method assumes that the source and sink

(a) (b) (c)

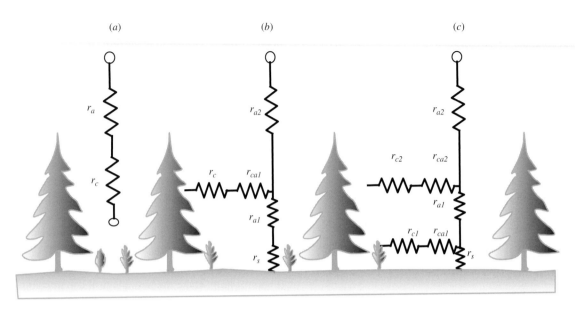

Figure 5.4 The pathways for water loss and the associated resistances for evaporation (**E**) and heat exchange between plant canopies and the atmosphere. (a) Simple 'big leaf' or single source model for canopy energy balance where the whole plant–soil system is lumped together, (b) two source model showing separate resistances for the soil and plant pathways, (c) a more complex, multilayer model that treats three or more sources separately. The various resistances involved include the leaf/canopy resistance (r_c), boundary layer resistances at different positions (r_a) and at the soil surface (r_s). (Modified from Jones & Vaughan, 2010.)

distributions within the canopy are identical for each entity (e.g. water vapour, heat, momentum: see Thom, 1975). Improved simulation of water and energy fluxes, especially in sparse canopies, may be achieved by the use of more complex two-component models (Figure 5.4(b); Shuttleworth & Wallace, 1985) that treat the energy balance and fluxes of the canopy and soil separately. Even more complex multilayer models have been proposed (Figure 5.4(c)) but their increased realism may be of only limited value (Raupach & Finnigan, 1988; Shuttleworth, 2007). It is useful to note that one can also invert Eq. (5.26) to estimate g_{LW} if **E** is known.

5.2.2 Environmental and plant control of evaporation

The dependence of **E** on environmental and physiological variables described by Eq. (5.26) is illustrated for a large flat leaf in Figure 5.5. As expected from inspection of the equation, **E** increases

linearly with both increasing net radiant energy (e.g. Figure 5.5(c)) and with increasing vapour pressure deficit of the ambient air (Figure 5.5(b)); these relations apply whatever the values of the other variables. The response to varying boundary layer conductance (resulting from altered wind speed), however, is rather more interesting, with increases in boundary layer conductance either increasing or decreasing **E** depending on radiation load (Figure 5.5(a)); this effect occurs because g_a occurs in both the numerator and denominator of Eq. (5.26). A similar effect is illustrated in Figure 5.5(c) where **E** is more sensitive to increasing radiation with low boundary layer conductances than under high g_a. It is easy to understand these initially surprising results once one realises that the crossover points arise where leaf temperature is at air temperature. Higher boundary layer conductance improves sensible heat transfer between leaf and air to bring T_ℓ closer to T_a. Similarly informative responses are shown when one plots **E** as a function of

Table 5.2 Some published values for canopy conductance (g_{LW}) for different crops and natural communities (Jarvis *et al.*, 1976; Miranda *et al.*, 1984; Wallace *et al.*, 1981).

Crop	Location	Canopy conductance (g_{LW}, mm s^{-1})
Alfalfa	Munich, Germany	14–40 (seasonal range)
Alfalfa	Arizona, USA	0.5–50 (seasonal range)
Barley	Nottingham, England	1–8 (diurnal range)
Barley	Nottingham, England	5–67 (seasonal range)
Grassland	Matador, Canada	1.7–10 (seasonal range)
Short grass surface	FAO standard	14.3
Heather moorland	Southern Scotland	4–20 (diurnal range)
Lentil	Madya Pradesh, India	1–17 (seasonal range)
Maple/beech	Montreal, Canada	0.8–10 (seasonal range)
Norway spruce	Munich, Germany	6.7–10 (seasonal range)
Rice	Kaudulla, Sri Lanka	16–117 (diurnal range)
Wheat	Madya Pradesh, India	1–10 (seasonal range)
Wheat	Rothamsted, England	2.5–40 (diurnal range)

physiological conductance and boundary layer conductance (Figure 5.5(*d*)). At low physiological conductances, **E** is greater where the boundary layer conductance is low, but when there is less stomatal control (open stomata or for wet leaves) **E** is greatest with high boundary layer conductances. Again this can be understood in terms of the crossover point where $T_\ell = T_a$. The same equation can also be used to predict dewfall (when **E** is negative – though note that there is no surface resistance in this case – see below).

Equation (5.26) is extremely flexible and of wide application, particularly in modelling studies where it is necessary to predict the water balance of leaves or plant communities in different conditions. The most accurate predictions are achieved by using the isothermal version (Eq. (5.26c)) though it is also possible to solve the energy balance equation using an iterative approach that does not require the approximations involved in Eq. (5.26) (Gates, 1980).

5.2.3 Coupling to the environment

For many years there was an interesting conflict between the usual meteorologists' view of the control of evaporation from canopies, where it is assumed that 'evaporation is independent of the character of plant cover, soil type and land utilisation to the extent that it varies under normal conditions' (Priestley & Taylor, 1972; Thornthwaite, 1944), and the longstanding experimental evidence for the critical role of stomata in controlling transpiration (see Jarvis, 1981; Jarvis & McNaughton, 1986). The fact that the total conductance to water vapour loss (of which $g_{\ell W}$ is the major variable component)

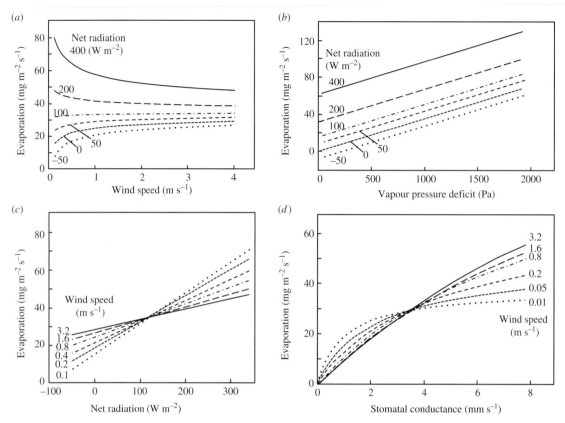

Figure 5.5 Dependence of evaporation rate from leaves on environmental and plant factors; unless otherwise indicated, leaf size = 0.1 m, air temperature = 20°C, g_s = 4 mm s^{-1}, relative humidity = 50%, wind speed = 0.8 m s^{-1} and $\mathbf{R}_n$ = 100 W m^{-2}. (a) Evaporation as a function of vapour pressure deficit and net radiation absorbed, (b) evaporation as a function of wind speed and net radiation absorbed, (c) evaporation as a function of net radiation absorbed and wind speed and (d) evaporation as a function of stomatal conductance and wind speed.

occurs as a proportionality constant in Eq. (5.19) has led to the belief that stomata play a dominant role in controlling transpiration. Inspection of Eq. (5.26), however, shows that other factors (e.g. boundary layer conductance, radiation and humidity) are also important. The simple approach of Eq. (5.19) is misleading because the driving force for evaporation (D_{cw}) is itself dependent on the evaporation rate (and hence on a wide range of these other factors), and this is taken into account in the Penman–Monteith equation. A major step forward in quantifying the degree of stomatal control of evaporation from different sized leaves and from crops was made by McNaughton and Jarvis (1983), who rewrote

Eq. (5.26) in a form that partitioned crop evaporation into two components: a so-called *equilibrium evaporation rate* ($\mathbf{E}_{eq}$) that depends only on the energy supply (radiation), and an *imposed evaporation rate* ($\mathbf{E}_{imp}$). The relative importance of these two depends on the degree of coupling of the evaporating surface (leaf or crop) to the environment.

Imposed evaporation

Where the boundary layer conductance is large (for example with isolated plants or thin leaves), heat and mass transfer are very efficient so that leaf temperature approaches air temperature whatever the input radiation, and the surface is said to be well coupled to the environment (Figure 5.5(*a*)). In this case, as the boundary layer conductance tends to infinity, leaf temperature tends to air temperature, and Eq. (5.26) reduces to a form similar to that of Eq. (5.20):

$$\mathbf{E}_{imp} = (\rho_a c_p/\lambda\gamma)g_\ell D \tag{5.28}$$

where g_ℓ (or g_{LW} for a canopy) is the physiological or 'surface' conductance. The efficient transfer between the surface and the atmosphere means that the conditions of the bulk atmosphere are 'imposed' at the leaf surface, so that evaporation is proportional to the leaf conductance as expected from Eq. (5.20).

Equilibrium evaporation

At the other extreme, when the boundary layer conductance is very small, heat and mass transfer between the surface and the atmosphere is extremely poor and the surface is said to be poorly coupled to the environment. In this case, evaporation tends to the equilibrium rate. In the extreme of complete isolation, as the boundary layer conductance tends to zero, $\mathbf{E}_{imp}$ decreases to zero and Eq. (5.26) reduces to:

$$\mathbf{E}_{eq} = \varepsilon\,\mathbf{R}_n/\lambda(\varepsilon+1) = s\,\mathbf{R}_n/\lambda(s+\gamma) \tag{5.29}$$

Although the lack of any effect of stomatal aperture or atmospheric humidity on evaporation in this case is somewhat surprising, it can be explained by considering what might happen to evaporation from a crop growing in a sealed glasshouse. In this case the crop and its adjacent air is completely decoupled from the outside air, with no transfer of water vapour through the glass and little or no heat loss. When there is no radiation input into the glasshouse, evaporation from the crop raises the humidity until, eventually, when the air is saturated, the leaf-to-air vapour difference falls to zero and evaporation rate eventually falls to zero, as one would expect. In contrast, when there is an input of radiation, this raises the temperature inside the glasshouse at a rate that is proportional to the irradiance. As the temperature increases, the water vapour holding capacity of the air also increases (Figure 5.2), so that

it is possible to maintain a nearly constant leaf–air vapour pressure gradient even though the vapour pressure of the air in the glasshouse may be continually increasing. This steady-state rate of evaporation depends on the rate of energy input into the system according to Eq. (5.29).

For large areas of crop it has been found that $\mathbf{E}_{eq}$ tends to a value about 26% greater than that given in Eq. (5.29), i.e.:

$$\mathbf{E}_{eq} = \alpha\,\varepsilon\,\mathbf{R}_n/\lambda(\varepsilon+1) = \alpha\,s\,\mathbf{R}_n/\lambda(s+\gamma) \tag{5.30}$$

where α is known as the Priestley–Taylor coefficient (Priestley & Taylor, 1972) and is usually approximately equal to 1.26 for most situations. This enhanced evaporation above the equilibrium rate in Eq. (5.29) arises because mixing in the atmospheric boundary layer prevents the air saturation required for Eq. (5.29) to hold (Eichinger *et al.*, 1996).

In practice, evaporation from a leaf, a crop or any area of vegetation operates between these limits and can be expressed as the sum of an imposed component and an equilibrium component by rewriting Eq. (5.26) as:

$$\mathbf{E} = \mathbf{\Omega}\mathbf{E}_{eq} + (1-\mathbf{\Omega})\mathbf{E}_{imp} \tag{5.31}$$

where:

$$\begin{aligned}\mathbf{\Omega} &= (\varepsilon+1)/(\varepsilon+g_H/g_W) \\ &\simeq (\varepsilon+1)/(\varepsilon+1+g_a/g_\ell)\end{aligned} \tag{5.32}$$

(if one assumes that $g_{aH} \simeq g_{aW}$). The $\mathbf{\Omega}$, or *decoupling coefficient*, is a measure of the coupling between conditions at the surface and in the free airstream and can vary between 0 (for perfect coupling) and 1 (for complete isolation). It is apparent from Eq. (5.32) that $\mathbf{\Omega}$ depends on the ratio between the surface conductance and that of the boundary layer, rather than on their absolute values.

5.2.4 Implications of coupling for water relations experimentation

It is important to recognise the difficulties involved in extrapolating results from physiological studies on single leaves, potted plants or even single plants or small plots in the field to the prediction of the behaviour of large areas of vegetation in the field. This is especially true for studies of evaporation. Some examples of the importance of coupling to the design of experiments will be presented to illustrate the pitfalls.

Sensitivity of evaporation to stomatal closure

A particular value of the coupling approach is that it enables one to estimate the degree to which stomata control evaporation. Because g_ℓ is largely determined by stomatal conductance, a measure of the stomatal control of evaporation is the relative sensitivity of $\mathbf{E}$ to a small change in g_ℓ (i.e. $(d\mathbf{E}/\mathbf{E})/(dg_\ell/g_\ell)$). This can be achieved by differentiating Eq. (5.26), which, after some rearrangement (Jarvis & McNaughton, 1986) gives:

$$(d\mathbf{E}/\mathbf{E})/(dg_\ell/g_\ell) = (1 - \mathbf{\Omega}) \qquad (5.33)$$

The sensitivity of evaporation to stomatal or surface conductance for different situations is summarised in Table 5.3. The differing sensitivity of $\mathbf{E}$ to surface resistance between extensive canopies and single plants whether outside or in controlled environment chambers (where there is no feedback effect of any water evaporated) also partly explains the failure of chemical anti-transpirants, which are effective at saving water in controlled environments, to have such large beneficial effects in the field (Jarvis & McNaughton, 1986).

Plant breeding

A plant breeder may wish to select plants that use water sparingly. Two lines may be found that differ in stomatal conductance ($\simeq g_\ell$) by, say, 20%, but it is not appropriate to infer that one would transpire at a rate 20% lower than the other. If they are compared in a controlled environment, or even in typical small field plots ($\mathbf{\Omega} \simeq 0.05$; see Table 5.3), the difference in transpiration would be about 19% (from Eq. (5.33), $d\mathbf{E}/\mathbf{E}$ would be equal to $(1 - \mathbf{\Omega})\, dg_\ell/g_\ell$ or 19%). If instead large field plantings of several hectares ($\mathbf{\Omega} \simeq 0.7$) of each line are compared, $d\mathbf{E}/\mathbf{E}$ would be

0.3 dg_ℓ/g_ℓ or only 6%. The advantage of the better selection would decrease with increasing area of planting.

Sensitivity to soil water

An understanding of coupling has also helped to resolve the controversy that surrounded the sensitivity of crop evaporation to soil water. For example, Denmead and Shaw (1962) reported that crop evaporation started to fall below the potential rate when only 10% of soil water was removed, while others (see Ritchie, 1973) have found that as much as 80% of 'available' soil water may be removed before crop evaporation falls significantly. It is worth noting that Denmead and Shaw's experiment was conducted on maize plants grown in large pots in the field among a continuous canopy of unstressed plants. This experimental design resulted in the vapour pressure deficit around each individual plant being determined by the weather and by the evaporation from the *unstressed* plants, with there being no opportunity for feedback from stomatal conductances of the stressed plants on the canopy humidity. Although grown in the field the droughted plants would have behaved as isolated plants (small $\mathbf{\Omega}$) with a much greater sensitivity of $\mathbf{E}$ to stomatal closure than would have been the case for a canopy of identical stressed plants.

5.3 Measurement of evaporation rates

Many of the methods used for estimating the actual evaporation from crops or other vegetation are based on first estimating the evaporation from a standard surface, which would then be a measure of the environmental demand, and then expressing the actual evaporation as a fraction of the reference value. Although potential evaporation from a free water surface ($\mathbf{E_o}$) as calculated from the Penman equation has often been used as the reference, the Penman–Monteith equation (Eq. (5.26)) with its strong underlying physical basis is now the most widely

Table 5.3 Likely values for the decoupling coefficient (Ω) and for the sensitivity of evaporation to leaf or canopy physiological conductance $(\mathrm{d}\mathbf{E}/\mathbf{E})/(\mathrm{d}g_\ell/g_\ell)$ for different leaves and crops (Jarvis, 1985; Jarvis & McNaughton, 1986; Jones, 1990).

Single leaves	Leaf width (mm)	Ω	$(\mathrm{d}\mathbf{E}/\mathbf{E})/(\mathrm{d}g_\ell/g_\ell)$
Rhubarb	500	0.8	0.2
Cucumber	250	0.7	0.3
Bean	60	0.5	0.5
Onion	8	0.3	0.7
Asparagus	1	0.1	0.9
Open field crops	Crop height (m)	Ω	$(\mathrm{d}\mathbf{E}/\mathbf{E})/(\mathrm{d}g_\ell/g_\ell)$
Grass	0.1	0.9	0.1
Strawberry	0.2	0.85	0.15
Tomato	0.4	0.7	0.3
Wheat	1.0	0.5	0.5
Raspberry	1.5	0.4	0.6
Citrus orchard	5.0	0.3	0.7
Forest	30	0.1	0.9
Other situations		Ω	$(\mathrm{d}\mathbf{E}/\mathbf{E})/(\mathrm{d}g_\ell/g_\ell)$
Uncontrolled glasshouse		>0.9	<0.1
Lysimeters or 1 m^2 plots		<0.1	>0.9
Controlled environment chambers		<0.1	>0.9

recommended (Allen *et al.*, 1998; Shuttleworth, 2007). This equation is used to estimate from a standard set of meteorological observations (radiation, air temperature, humidity and wind speed) the total evapotranspiration from a standardised well-watered short 'grass' surface that fully covers the ground to a height of 0.12 m, has a surface resistance of 70 s m^{-1} ($g_{LW} = 14.3$ mm s^{-1}) and an albedo of 0.23. This quantity is defined as the *potential evapotranspiration* or the *reference evapotranspiration*, $\mathbf{ET}_0$. The actual evapotranspiration for any specific crop or vegetation ($\mathbf{ET}_{c-adj}$) is then estimated from the reference $\mathbf{ET}_0$ using a series of empirically determined crop coefficients (K_c, K_{stress}) according to:

$$\mathbf{ET}_{c-adj} = \mathbf{ET}_0 K_c K_{stress} \qquad (5.34)$$

where K_c is a *crop coefficient* that corrects $\mathbf{ET}_0$ to the value expected for a specific well-watered crop at a particular growth stage ($\mathbf{ET}_c$; see Figure 5.6) and K_{stress} is a further coefficient to account for any suppression of $\mathbf{ET}$ below the standard value as a result of water stress or even other management practices (Allen *et al.*, 1998). The overall crop coefficient K_c is frequently partitioned into a component representing soil evaporation (K_s), which can be adjusted according to the frequency of rainfall, and a basal coefficient relating to the crop alone (K_{cb}) as shown in Figure 5.6. Values of K_c for different crops are tabulated in FAO 56 (Allen *et al.*, 1998).

The various approaches available for estimation of the reference $\mathbf{ET}_0$ and for the direct estimation of $\mathbf{ET}$ are outlined below. Further details and precautions to be adopted can be found in Allen *et al.* (2011).

5.3.1 From meteorological data

Penman–Monteith equation

As indicated above, the FAO recommended approach uses the Penman–Monteith $\mathbf{ET}_0$ and an associated set of crop coefficients (Allen *et al.*, 1998, 2011). It is, however, possible to estimate $\mathbf{ET}$ directly from the Penman–Monteith equation where one has information on the value of the required physiological conductance, as might be obtained from measurements or predictions of stomatal conductance. Various refinements of the Penman–Monteith approach are available; for example a system (MORECS) used in the UK incorporates a

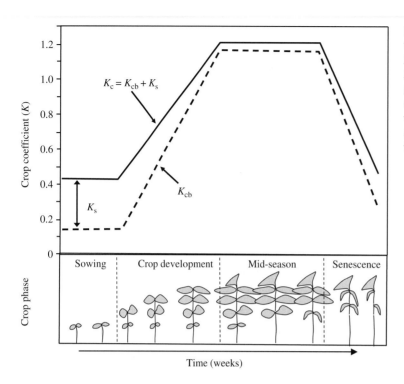

Figure 5.6 Illustration of the seasonal variation in crop coefficients ($K_c = K_s + K_{cb}$); note that the seasonal variation in K_c primarily relates to the changing ground cover, while the soil evaporation represents an average value as it fluctuates with irrigation and rainfall. (Values for K_c are tabulated in Allen *et al.*, 1998.)

number of corrections to the Penman–Monteith equation (Hough & Jones, 1997; Thompson *et al.*, 1982): these include both the isothermal radiation correction (Eq. (5.9)) and a water balance calculation (to modify evaporation via changes in the physiological conductance that result from stomatal closure as soil water is depleted).

Simple meteorological methods

Many other empirical formulae for estimating evaporation rates have been used over the years: these usually have a less rigorous physical basis than the Penman equation (see Doorenbos & Pruitt, 1984 for a summary of these other methods), but can have the advantage of correspondingly smaller data requirements. For example, readily available climatic data such as temperature, and sometimes also humidity, duration of bright sunshine or solar radiation, may be all that are required.

An important example that only requires radiation and air temperature data is the Priestley–Taylor relationship (Eq. (5.30)). This approximates quite

successfully the evaporation from large areas of many types of vegetation, at least when well supplied with water. Although the constant α is usually ≃ 1.26, there can be significant deviations from this, particularly in dry conditions and for very rough canopies. Where only air temperature data are available a very approximate estimate of reference evapotranspiration may be obtained from a simple function of daily temperature and daylength using the Blaney–Criddle equation (see Doorenbos & Pruitt, 1984 for details of the calculations involved).

Evaporation pans

An alternative method that is still widely used for estimating a potential evaporation rate is the use of evaporimeters or evaporation pans. These consist of pans of open water whose evaporation rate can be monitored. Unfortunately there are several types of pan in use with differences in size, shape and recommended exposure. Evaporation rates from such instruments tend to be 10 to 45% higher than even the Penman E_o, because they are too small to achieve an equilibrium evaporation estimate. Where pans are used it is usual to add a 'pan correction' into Eq. (5.34). With the increasing availability of accurate meteorological stations, however, it is generally more convenient and reliable to estimate $\mathbf{E_o}$ or $\mathbf{ET_o}$ directly by means of the standard Penman equation or its modifications.

5.3.2 Micrometeorological methods

In contrast to the meteorological methods outlined above, which can be suitable for regional studies, micrometeorological approaches indicate fluxes from the crop or vegetation within the 'footprint' of the measurement instruments so can provide useful local data. Flux gradient methods are rather little used these days, so eddy covariance (see Chapter 3) tends to be the method of choice, even though it can be rather expensive as it requires sophisticated instrumentation and careful maintenance. There is also increasing interest in the use of turbulent transfer theory-based methods such as temperature variance, scintillometry and surface renewal (see Chapter 3), as some of these can be cheaper to implement than eddy covariance. A distinct advantage of scintillometry is that it allows integration of fluxes over a long (and known) transect.

One older micrometeorological method that is still widely used, however, is the *Bowen ratio energy balance approach* (where the Bowen ratio, β, is the ratio of sensible to latent heat losses, $\mathbf{C}/\lambda\mathbf{E}$). Substituting this in Eq. (5.25) gives:

$$\lambda\mathbf{E} = (\mathbf{R_n} - \mathbf{G})/(1 - \beta) \tag{5.35}$$

Assuming that the transfer coefficients for heat and water vapour in the turbulent boundary layer above a crop are equal (Chapter 3) and substituting the appropriate driving forces for $\mathbf{C}$ and $\mathbf{E}$ gives:

$$\beta = \frac{\mathbf{C}}{\lambda\mathbf{E}} = \left(\frac{P\rho_a c_p}{0.622\rho_a}\right)\frac{\triangle T}{\lambda\triangle e} = \gamma\frac{\triangle T}{\triangle e} \tag{5.36}$$

where $\gamma\ (= Pc_p/0.622\lambda)$ is the psychrometer constant. Therefore all that is needed to estimate $\mathbf{E}$ is a knowledge of net radiation and soil heat flux together

with the gradients of T and e in the boundary layer. When β is small this method is particularly insensitive to errors in measurement of T or e, because all energy is then being used for evaporation and the error is just that in determining $(\mathbf{R_n} - \mathbf{G})$ (see Eq.(5.35)). As well as not requiring information on surface conditions, this technique does not even require knowledge of the actual physiological or boundary layer conductances.

5.3.3 Remote sensing

Remote sensing is not particularly well suited to estimation of surface energy fluxes, because of problems including: (i) most satellites only provide infrequent 'snapshots' of surface conditions, which are difficult to integrate over time; (ii) it is usually difficult to estimate remotely surface roughness and the necessary transfer coefficients or conductances; and (iii) cloud cover restricts the ability to determine surface properties including temperatures. Nevertheless, remote imagery has some important advantages including the ability to provide regular information on spatial variation which cannot be obtained from ground stations, and the information on spatial variation within a scene that can be used for internal calibration of calculations. Evaporation is particularly difficult to estimate by remote sensing, so it is often estimated as a residual from the energy balance equation (Eq. (5.25)). Although it is not easy to estimate the reference $\mathbf{ET_o}$ remotely, the Priestley–Taylor relationship (Eq. (5.30)) can give a useful first approximation. Remote sensing can then potentially provide good estimates of the crop coefficients (K_c and K_{stress}) that respectively correct $\mathbf{ET_o}$ for ground cover (on the basis of observed spectral vegetation indices) and stomatal closure under water stress (on the basis of surface temperature measurements). In many cases, remote estimates are improved if they can be supplemented by ground-level meteorological data such as for $\mathbf{R_S}$.

A more detailed discussion of the remote estimation of evaporation may be found in Jones and Vaughan (2010) and in key papers (Allen et al., 2007; Bastiaanssen et al., 1998); here we just outline the principles underlying the main approaches and refer the reader to remote sensing texts for further details on the actual remote sensing technologies (Campbell, 2007; Jensen, 2007; Liang, 2004). Slightly different approaches are suitable for satellite remote sensing and for in-field proximal sensing.

Estimation as a residual in the energy balance

From Eq. (5.25) and using Eq. (5.3) to eliminate $\mathbf{G}$, we see that:

$$\lambda \mathbf{E} = (\mathbf{R}_n - \mathbf{G}) - \mathbf{C} = (1 - \Gamma)\mathbf{R}_n - \mathbf{C} \qquad (5.37)$$

where we use the convention that $\mathbf{R}_n$ is positive for fluxes to the surface and the other fluxes are positive for losses from the surface. Estimation of the net radiation term requires information on the

incoming and reflected shortwave radiation and on the incoming and reradiated longwave radiation (Figure 5.7). Incoming shortwave radiation ($\mathbf{R}_S$) is estimated from the calculated above-atmosphere incoming radiance (see Appendix 7) corrected using a standard radiative transfer model such as MODTRAN or 6S (Berk et al., 1998; Vermote et al., 1997) parameterised according to satellite measurements of cloud cover and atmospheric composition, while the reflected shortwave at the surface is calculated from the at-satellite measurements of reflected radiance. The upwards emitted component of the net longwave can be estimated using the thermal bands on satellites by applying appropriate radiative transfer models (Norman et al., 1995), especially where there are several distinct thermal bands that might allow the use of 'split window' algorithms for the extraction of surface temperature. The downward component of

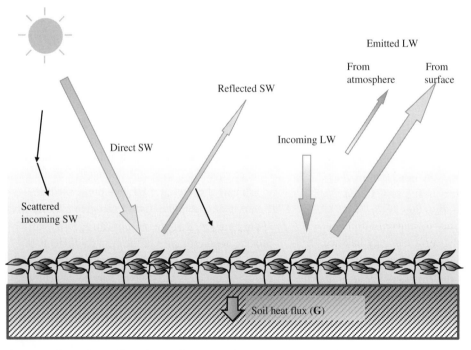

Figure 5.7 Illustration of the radiative fluxes to and from a vegetated surface, showing that the net absorbed shortwave radiation is the sum of the direct and scattered incoming minus that reflected, while the net longwave radiation is the difference between the incoming from the atmosphere and that emitted by the surface. When estimating fluxes from satellite observations it is necessary to allow for both absorption and scattering in the atmosphere and, for longwave, emission by the atmosphere.

longwave radiation depends on the effective sky temperature, which again can be estimated remotely. In the absence of accurate remote estimates of R_{Ln} it is possible to get a reasonable estimate as $100 \times (\mathbf{I}_{S(24h)}/\mathbf{I}^*_S)$, where $\mathbf{I}_{S(24h)}$ is the 24 h total of $\mathbf{I}_S$ and $\mathbf{I}^*_S$ is the maximum value of $\mathbf{I}_{S(24h)}$ expected for that time of year (Jones & Vaughan, 2010).

The sensible heat transfer term ($\mathbf{C}$) can be estimated from Eq. (5.23), though this requires not only estimates of T_s and T_a but also an estimate of g_H; of these only T_s is readily accessible remotely. As a result it is common to estimate the daily total of $\mathbf{C}$ from midday temperatures using a simple empirical approximation based on a form initially proposed by Jackson *et al.* (1977):

$$\mathbf{C}_{24h} = A + B(T_s - T_a)^n \tag{5.38}$$

where A, B and n are constants. It should be noted that remotely sensed surface temperatures are weighted to the upper layers of the canopy and may not accurately represent the true aerodynamic surface temperature.

SEBAL and METRIC

The residual energy balance approach has been incorporated into a number of operational algorithms for estimation of regional evaporation. Particularly important is the Surface Energy Balance for Land (SEBAL) algorithm (Bastiaanssen *et al.*, 1998). This essentially uses a physically based approach to the estimation of **ET** from remotely sensed data on land surface temperature, albedo and vegetation index to derive a land surface parameterisation which is then used to solve the surface energy balance. An important step in this algorithm is the estimation of the boundary layer conductance from those dry pixels where it can be argued that differences in T_s are driven by sensible heat transfer only; in this case the slope of the relationship between albedo (ρ) and surface temperature ($\partial \rho / \partial T_s$) can be shown to be approximately proportional to $-\partial \mathbf{C}/\partial T_s$, from which the effective heat transfer conductance can be derived using Eq. (5.23). One can define an evaporative fraction (ϕ) as equal to $\lambda \mathbf{E}/(\mathbf{R}_{ni} - \mathbf{G})$ so that for the

wettest (coolest) extreme pixels $\phi = 1$ and $\lambda \mathbf{E} = (\mathbf{R}_{ni} - \mathbf{G})$, while for the driest (hottest) pixels $\phi = 0$ and $\lambda \mathbf{E} = 0$. Values of ϕ, and hence $\lambda \mathbf{E}$, can then be estimated from the temperatures of all intermediate pixels. The effectiveness of remote sensing algorithms for estimation of **ET** can be improved by the incorporation of selected ground data, as in METRIC (Allen *et al.*, 2007).

Estimation from the relationship between surface temperature and evaporation

It is readily shown that for similar surfaces in a given environment, surface temperature decreases linearly with increasing evaporation. Using net isothermal radiation (Eq. (5.10)) and Eq. (5.37) we can write for a dry surface:

$$(1 - \Gamma)\mathbf{R}_{ni} = g_{HR}(\rho_a c_p)(T_{dry} - T_a) \tag{5.39}$$

and for any evaporating surface (s):

$$(1 - \Gamma)\mathbf{R}_{ni} = \lambda \mathbf{E} + g_{HR}(\rho_a c_p)(T_s - T_a) \tag{5.40}$$

Substituting 5.39 into 5.40 gives:

$$\lambda \mathbf{E} = g_{HR}(\rho_a c_p)(T_{dry} - T_s) \tag{5.41}$$

or else one can eliminate g_{HR} from Eqs. (5.39) and (5.41) to give for any surface:

$$\lambda \mathbf{E} = (1 - \Gamma)\mathbf{R}_{ni}[1 - (T_s - T_a)/(T_d - T_a)]$$
$$= (1 - \Gamma)\mathbf{R}_{ni}\phi \tag{5.42}$$

where the term in square brackets is equivalent to the evaporative fraction and is a simple multiplier that can be used to estimate **E** from the available energy, T_s and the temperatures of the extreme pixels (T_d, and $T_{wet} \simeq T_a$). This approach has been widely used in satellite imagery to study variation over heterogeneous areas.

5.3.4 Other methods

Water budget approaches

It is also possible to estimate evapotranspiration from vegetation by means of soil water budget approaches:

$$\mathbf{ET} = P - 0 - D - \Delta\theta \tag{5.43}$$

where P is precipitation, O is run-off, D is deep drainage and $\Delta\theta$ is the change in soil water content. Changes in soil water content can be measured using any of the standard methods (e.g. neutron scattering, time-domain reflectometry or capacitance sensing; see Allen *et al.*, 2011; Boyer, 1995; Kirkham, 2004). Evaporation from vegetation can also be measured directly using weighing lysimeters, but unfortunately plants grown in lysimeters do not necessarily respond identically to those grown in a normal soil profile.

Sap flow

Where sap flow rates are measured on main stems these can readily be scaled up to estimate stand evaporation, especially for forests, on the basis of a knowledge of stem density (see Steppe *et al.*, 2010). Three main methods are available to estimate sap flow rates through stems: the heat pulse-velocity method (Green *et al.*, 2003; Jones *et al.*, 1988), the Granier heat dissipation method (Granier, 1987) and the tissue heat balance method (Kjelgaard *et al.*, 1997). Although the last of these methods has been said to give absolute sap flow rates without calibration, it is not readily applicable to large tree trunks, and in practice calibration is required for all methods.

Further methods

Information on evaporation rates and even on sources of water (for example the depth in the soil profile from which it is extracted) can be obtained from studies of the natural deuterium/hydrogen (D/H) ratios in water (Pearcy *et al.*, 1991). Other techniques available include the use of crop enclosures (see Chapter 7, though the unnatural environmental coupling in such devices means that very great care needs to be taken in interpreting any data), or even silicon accumulation in leaves (Hutton & Norrish, 1974).

5.4 Evaporation from plant communities

In recent years there have been major advances in the measurement of energy balance, evaporation rates and CO_2 fluxes from different plant communities with the widespread adoption of the eddy correlation technique with many regional networks of sensors now being installed and coordinated through the international FLUXNET programme (http://fluxnet. ornl.gov). Some typical rates of evaporation from different plant canopies in a range of climates are presented in Table 5.4; in arid environments rates of evaporation as high as 10 mm day^{-1} or 1 mm h^{-1} (which is equivalent to an energy requirement of 680 W m^{-2}) are reasonably common for well-irrigated crops, while in temperate climates in winter, rates fall to less than 0.3 mm day^{-1}. Factors affecting crop evaporation are discussed in relation to the principles outlined above.

5.4.1 Canopy type and area

Considering first a tall isolated plant in an otherwise non-vegetated area (Figure 5.8), it is apparent that the appropriate air vapour pressure deficit (D) in Eq. (5.26) will be that of the airstream in the immediate vicinity of the plant, and the appropriate value for g_a is that for transfer from the individual leaves to the bulk airstream (perhaps 100 mm s^{-1} when not sheltered within a canopy: see Chapter 3). By substituting in Eq. (5.31), this gives Ω between 0.14 and 0.24 for typical unsheltered leaves of well-watered plants (assuming a typical range of g_ℓ of 5 to 10 mm s^{-1} and a temperature of 20°C – i.e. $(2.2 + 1)/(2.2 + 1 + 100/5)$).

As the airstream moves over an extensive area of homogeneous vegetation (Figure 5.8; see also Figure 3.4), transpiration will raise the humidity of the air near the crop surface, with the upper limit of the modified boundary layer increasing with the area of vegetation and hence the fetch. This has a feedback effect on **E**, tending to reduce it. It follows that the relevant g_A which here needs to describe transfer from the leaves right up to the mixed layer where D is unaffected by the crop, is correspondingly smaller than g_a of the individual leaves and decreases with increasing area of uniform vegetation. The value of Ω is therefore much larger than for an isolated plant (Table 5.3). Figure 5.8 also shows that Ω is strongly dependent

Table 5.4 Typical evaporation rates for reasonably dense plant canopies well supplied with water. Values in brackets represent maximum rates (Denmead, 1969; Denmead & McIlroy, 1970; Grant, 1970; Kanemasu et al., 1976; Körner, 1999; Kowal & Kassam, 1973; Lang et al., 1974; Lewis & Callaghan, 1976; Monteith, 1965; Rauner, 1976; Ritchie, 1972).

		Daily E (mm day^{-1})	Hourly E (mm h^{-1})
Central England	January	0.2	–
	April	1.3 (3)	–
	July	3 (6)	0.25
Continental Europe	Summer	5	–
European Alps – 500 m elevation[a]	Summer	4.5 (9)	–
European Alps – 2500 m elevation[a]	Summer	3.8 (8.2)	–
Arctic tundra	July	2 (3)	–
USA Great Plains	July	6 (10–12)	–(1.0)
North Nigeria	July	5	–
South East Australia	Summer	5.6 (9–10)	0.4 (0.9)

[a] bright days only

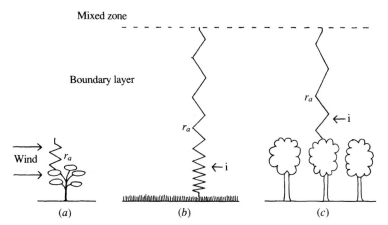

Mixed zone

Boundary layer

Wind

r_a

r_a ←i

r_a ←i

(a)   (b)   (c)

Figure 5.8 An illustration of the relative magnitudes of the boundary layer resistances appropriate for (a) an isolated plant; (b) an extensive short, smooth canopy such as mown grass; and (c) an extensive tall, rough canopy. Also shown is the normal height for micrometeorological measurements in the inertial sublayer (i), indicating that there is a significant resistance between this height and the mixed layer that is not normally included in micrometeorological estimates of the boundary layer resistance or conductance.

on the aerodynamic characteristics of the vegetation, with the improved transfer (higher g_A) above a rough canopy such as a forest (typical g_A 100 to 300 mm s^{-1}: see Chapter 3) as compared with a smooth canopy such as grassland (typical g_A perhaps 10 mm s^{-1}), tending to decrease Ω.

Some estimated values of Ω for a range of leaves and crops are shown in Table 5.3. It is clear from this that the degree of coupling increases (Ω decreases and the sensitivity of E to g_ℓ increases) as leaf size decreases, as canopy height increases and as the area of vegetation decreases.

The dependence of transpiration on canopy type and on canopy conductance is illustrated in Figure 5.9. Transpiration from forests is often rather less than from field crops because of the relatively low values of canopy conductances (especially in coniferous forest). Figure 5.9 confirms the result

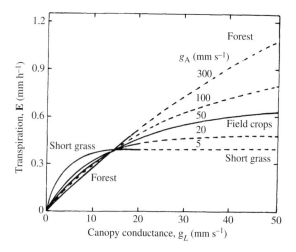

Figure 5.9 Calculated relationships between transpiration rate and canopy conductance at different boundary layer conductances, for 400 W m^{-2} available energy, 1 kPa vapour pressure deficit and 15°C. The solid lines represent the probable range of values of canopy conductance for different crops, being up to 50 mm s^{-1} for some field crops (after Jarvis, 1981).

in Table 5.3 that evaporation from short crops would be expected to be particularly insensitive to g_L as long as g_L remains above about 10 mm s^{-1}.

Another effect of canopy boundary layer conductance is that the vapour pressure term $(\rho_a c_p g_H D)$ in the numerator of Eq. (5.26) is greater than the available energy term $(s(R_n - G))$ in forests but smaller in short grass. Figure 5.10 shows that **E** for a forest would be much more sensitive to D than to R_n, while the converse is true for grass.

A further important factor in the water loss from plant canopies is the evaporation of intercepted rainfall. The difference between tall and short crops is particularly apparent when one considers the *total* evaporation from the canopy including that occurring directly from the surface of leaves wetted by rainfall. This loss of intercepted rainfall can be greater than total transpiration for forests in humid areas (Calder, 1976), and may lead to total forest evaporation being as much as twice that from grass. The reason for the different behaviour can be seen in Figure 5.9 where, as g_ℓ increases (as happens when the surface gets wet and stomatal control is eliminated) forest evaporation may increase three-fold while grass evaporation changes relatively little.

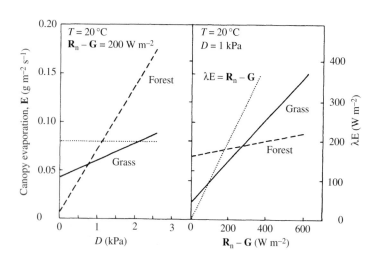

Figure 5.10 Dependence of canopy evaporation on humidity deficit (D) and available energy ($R_n - G$) for short grass ($g_{AH} = 10$ mm s^{-1}) or for forest ($g_{AH} = 200$ mm s^{-1}) with a constant physiological conductance (g_{LW}) of 10 mm s^{-1}. The line where evaporation consumes all available radiant energy is shown dotted.

5.4.2 Advection

In arid regions, where crops may be surrounded by relatively unvegetated areas, it is possible for the air to be significantly hotter and drier than in the crop, thus providing an extra source of energy so that the latent heat loss can significantly exceed any net radiative gain. This advection, or transfer of heat from the surroundings, is often called the 'oasis' effect and is adequately treated by the Penman equation, as long as measurements are restricted to the crop boundary layer. The magnitude of this advection effect is illustrated by results for rice growing at Griffith, New South Wales, Australia (Lang *et al.*, 1974), where $\lambda\mathbf{E}$ could be as much as 170% of $\mathbf{R}_n$, near the leading edge of the crop.

5.4.3 Soil evaporation

The evaporation from the soil is often an important component of canopy water loss, but is strongly dependent on soil wetness and on plant cover. Direct measurements of soil evaporation are few and far between, partly because of the difficulty of getting representative values (the conditions in microlysimeters and soil gas-exchange cuvettes, for example, are both unlikely to be representative of natural soil evaporation).

Although there have been relatively few direct measurements of the fraction of evapotranspiration attributable to soil evaporation (see Herbst *et al.*, 1996; Wallace *et al.*, 1993), it is clear that even with wet soil, soil evaporation may be only about 5% of the total when the leaf area index (*L*) reaches 4. When *L* falls to 2 or less, however, wet soil evaporation can be half the total, while for bare soil, when wet after rain, the surface resistance is small or zero and evaporation may approach $\mathbf{E}_o$ as calculated by the Penman equation. This high rate of soil evaporation may be maintained for about 10 mm of soil evaporation (Ritchie, 1972) but declines to near zero for dry soil. Equations for estimating the decline of soil evaporation from bare surfaces with time after rainfall (expressed as a function of reference $\mathbf{ET}_o$) are available (Allen *et al.*, 2005). Soil evaporation is discussed further by van Bavel and Hillel (1976) among others.

One interesting approach to the estimation of soil evaporation is to use stable isotope methods. Although water is principally composed of $^1H_2^{16}O$, a small fraction of the hydrogen and oxygen atoms are the heavier isotopes deuterium (D or ^2H) and ^{18}O (where $D/^1H = 1/6420$, and $^{18}O/^{16}O = 1/500$). The slightly greater density of water containing these heavier isotopes means that they evaporate less readily than standard water, so that the remaining water is enriched in the heavier isotopes. If it is assumed that no fractionation of water occurs during uptake by plant roots, the difference between the isotopic composition of rain water throughfall in the canopy and the composition of stream water must result from enrichment during evaporation from the soil. The actual rate of evaporation can therefore be estimated from the measured isotope fractionation and the amounts of throughfall and total evaporation by applying the known (temperature dependent) equilibrium fractionation factor during Rayleigh distillation (Kubota & Tsuboyama, 2004; Tsujimura & Tanaka, 1998).

5.4.4 Soil water availability

Soil drying also eventually leads to physiological stresses that decrease transpiration from plant canopies by effects such as stomatal closure. As has been pointed out above, the sensitivity of canopy evaporation to stomatal closure as the soil dries is very dependent on canopy and environmental characteristics that affect the degree of coupling ($\boldsymbol{\Omega}$), while the stomatal response to soil drying is itself very species dependent (see Chapter 6).

5.5 Dew

The condensation of water vapour on plant leaves or other surfaces is governed by the same physical principles as evaporation, so that Eq. (5.26) is applicable, with dew forming whenever **E** is negative. However, as there is no equivalent to the physiological resistance (i.e. $g_{LW} = \infty$), Eq. (5.26c) simplifies to:

$$\mathbf{E} = \frac{s(\mathbf{R}_{ni} - \mathbf{G}) + \rho_a c_p g_{HR} D}{\lambda(s + \gamma)} \qquad (5.44)$$

This predicts that dew occurs when $-s(\mathbf{R}_{ni} - \mathbf{G}) > \rho_a c_p g_{HR} D$.

If one assumes that $(\mathbf{R}_{ni} - \mathbf{G})$ falls to a minimum of about -100 W m^{-2} on a clear night, substitution of this in Eq. (5.44) predicts maximum rates of dewfall of almost 0.1 mm h^{-1} when the air is saturated. This compares with maximum observed rates of dewfall of up to about 0.5 mm per night, while more typical amounts on clear nights may be 0.1 to 0.2 mm per night (Monteith, 1957). There are several reasons why actual rates of dewfall tend to be well below the theoretical maximum. Most important is the fact that air is often not completely saturated, so that some of the heat required to satisfy the radiation loss is obtained by cooling the air to the dewpoint temperature before any condensation can occur. In addition, as the humidity deficit increases, dewfall becomes more sensitive to increasing wind speed (Eq. (5.44) and Figure 5.11). This is because a high wind speed increases sensible heat exchange, thus reducing the amount that the temperature of a leaf can fall below air temperature for a given radiation loss. In the extreme case where g_A tends

to infinity, $T_\ell \simeq T_a$, so that condensation can only occur when the air is saturated.

There is, however, a conflict between the requirement for calm conditions for maximum condensation according to Eq. (5.44), and the requirement for adequate transfer of water vapour down through the atmosphere to the surface. The magnitude of this downward flux can be appreciated from the fact that 0.4 mm of liquid water represents the *entire* water vapour content of the bottom 30 m of a saturated atmosphere at 15°C. Therefore significant wind is required to maintain the humidity near the surface high enough for condensation to continue at high rates. At very low wind speeds most of the condensation probably originates by distillation from the soil, rather than as dewfall from the atmosphere, though dewfall usually predominates in England (Monteith, 1957). Both types of condensation must be distinguished from guttation – the droplets of water exuded from hydathodes on plant leaves.

Quantitatively, dewfall is usually at least an order of magnitude less than potential rates of evaporation and so rarely contributes significantly to the water balance of plant communities. However, there are areas such as the dew deserts of Chile where condensation can equal or exceed rainfall. Taken together with evidence that dew can be absorbed by mosses, lichens and leaf tissues and raise leaf water potentials (Kerr & Beardsell, 1975; Lakatos

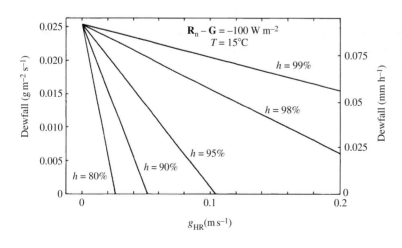

Figure 5.11 The effect of relative humidity (h) and boundary layer conductance (g_{HR}) on the rate of dewfall as predicted by Eq. (5.44) for a temperature of 15°C and an available energy of -100 W m^{-2}.

et al., 2011), it is probable that dew is a factor contributing to plant survival in such climates. A more important ecological effect results from the favourable microenvironment that wet leaf surfaces provide for fungal infection. In this case, the duration of surface wetness is likely to be more significant than the total amount. At least during summer in England, dew typically persists for periods of 6 to 12 h.

There is substantial interest in the development of dew collectors as a source of extra water in semi-arid regions (see the website of the International Organization for Dew Utilization (OPUR; http://www.opur.fr). Maximisation of dew collection can be achieved by ensuring that the collector surface has as high as possible emissivity so that radiative cooling is maximised (Maestre-Valero *et al.*, 2011).

There is evidence that some plants can condense water from a non-saturated atmosphere by excretion of hygroscopic salts onto the leaf surface. Even when the ambient relative humidity was no higher than 82%, significant water was found to accumulate on the leaves of *Nolana mollis* in the Atacama desert of Chile, though the quantities were only of the order of 0.1 mm per night (Mooney *et al.*, 1980).

The most direct methods for dew estimation include the use of weighing lysimeters or actual collection using absorbent tissue. Alternatively, dew may be estimated using an artificial surface positioned on a recording balance, but it is necessary to ensure that the radiative and aerodynamic properties of the collector resemble those of real leaves. Another very useful technique for estimation of dew duration is to measure the electrical resistance between two electrodes (e.g. strips of copper wire) attached to the surface of a leaf. When the leaf is wet the resistance decreases dramatically. These and other techniques are outlined by Agam and Berliner (2006).

5.6 Sample problems

5.1 Assuming an air temperature (T_a) of 30°C and a relative humidity, h, of 40%, estimate (i) e_s, (ii) e, (iii) c_W, (iv) D, (v) T_{wb}, (vi) T_{dew}, (vii) m_W, (viii) x_W, (ix) ψ.

5.2 The net radiation absorbed by a given leaf at 22°C is 430 W m^{-2}. What is (i) the corresponding isothermal net radiation ($\mathbf{R}_{ni}$) if $T_a = 19$°C, (ii) the radiative conductance (g_R)?

5.3 (i) Using the Penman–Monteith equation, calculate the evaporation rate from a forest ($g_A = 200$ mm s^{-1}) or from short grass ($g_A = 10$ mm s^{-1}) for ($\mathbf{R}_n - \mathbf{G}) = 400$ W m^{-2}, $T_a = 20$°C, $D = 1$ kPa; (a) when the surface is wet, (b) when $g_L = 30$ mm s^{-1}. (ii) What are the corresponding values of the Bowen ratio?

5.4 Assuming an area of homogeneous crop where the total boundary layer conductance is 15 mm s^{-1}, the crop physiological conductance is 5 mm s^{-1} and the air temperature is 20°C, calculate (i) the value of the decoupling coefficient ($\mathbf{\Omega}$), and (ii) the relative reduction in crop transpiration rate if the stomata close to such an extent that the physiological conductance decreases by 50%. (iii) What particular assumption have you had to make in the calculation of (ii)?

6

Stomata

Contents

The evolution of the stomatal apparatus was one of the most important steps in the early colonisation of the terrestrial environment. Even though the stomatal pores when fully open occupy between about 0.5 and 5% of the leaf surface, almost all the water transpired by plants, as well as the CO_2 absorbed in photosynthesis, passes through these pores. It is only in rare cases, such as in the fern ally *Stylites* from the Peruvian Andes, that significant CO_2 may be absorbed through the roots (Keeley *et al.*, 1984). The central role of the stomata in regulating water vapour and CO_2 exchange by plant leaves is illustrated in Figure 6.1. This figure also shows some of the complex feedback and feedforward control loops that are involved in the control of stomatal apertures and hence of diffusive conductance; these are discussed in Section 6.6.1. It is the extreme sensitivity of the stomata to both environmental and internal physiological factors that enables them to operate in a manner that optimises the balance between water loss and CO_2 uptake.

This chapter outlines the fundamental aspects of stomatal physiology, their occurrence in plants, their morphology, their response to environmental factors and mechanics of operation, including a description of the various control loops illustrated in Figure 6.1. The role of the stomata in the control of photosynthesis and of water loss is discussed in more detail in Chapters 7 and 10.

Further information on stomata, their responses and mechanism of operation, may be found in older texts or symposia such as those by Meidner and Mansfield (1968), Jarvis and Mansfield (1981), Zeiger *et al.* (1987), Weyers and Meidner (1990), Willmer and Fricker (1996) and in a number of more recent reviews (Bergmann & Sack, 2007; Buckley, 2005; Davies *et al.*, 2002; Kim *et al.*, 2010; Roelfsema & Hedrich, 2005; Schroeder *et al.*, 2001) and a whole issue of *New Phytologist* (issue no. 3, volume 153, 2002).

6.1 Distribution of stomata

True stomata are distinguished from other epidermal pores, such as hydathodes, lenticels and the pores in the thalli of some liverworts, by their marked

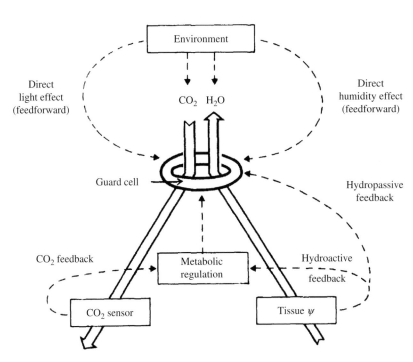

Figure 6.1 Simplified diagram illustrating the role of stomata in regulating CO_2 and H_2O fluxes, showing the feedback and feedforward control pathways (dashed lines). For details see text (modified after Raschke, 1975).

capacity for opening and closing movements. These changes in aperture depend on alterations in the size and shape of specialised epidermal cells, the guard cells (Figure 6.2). Stomata are present in the aerial parts of practically all the land flora, being found in the sporophytes of mosses, in ferns and in both gymnosperms and angiosperms. Functioning stomata appear to have evolved at least 400 million years ago and it may be that all existing plants with stomata are descended from a single ancestor (Raven, 2002). Although they are most frequent on leaves, they also occur in other green tissues such as stems, fruits and parts of inflorescences (e.g. awns of grasses and sepals of angiosperms). They tend to be most frequent on the lower surface of plant leaves, while in some species, especially trees, they occur only on the lower epidermis. Leaves with stomata on both sides are called *amphistomatous*, and those with stomata restricted to the lower epidermis are *hypostomatous*.

The two main types of stomata found in higher plants (Figure 6.2) are (a) the elliptical type and (b) the graminaceous type, which is found in the Gramineae and Cyperaceae and has distinctive dumbbell-shaped guard cells arranged in rows. In many species, the stomata have antechambers outside the pore or special protective structures such as outer lips or even membrane 'chimneys' (see Figure 6.2(d)–(h)), or else they are partially occluded by wax. All these features increase the effective diffusive resistance of the pore.

Representative examples of dimensions and frequencies of stomata in different species and on different leaves are presented in Table 6.1. It is clear that frequency and size vary as a function of leaf position and growth conditions. Even within one species there may be a large genetic component of the variation between different cultivars or ecotypes. It is also worth noting that the *stomatal index* (the number of stomata per unit area of leaf expressed as a percentage of the total number of epidermal cells plus stomata per unit area) can vary widely with species and with growing conditions. As a typical response, the stomatal index for wheat leaves has been shown to decrease by about 13% (from 17.3%) in response to water deficit at the same time as the leaf area decreased to about 50% of the control size (Quarrie & Jones, 1977).

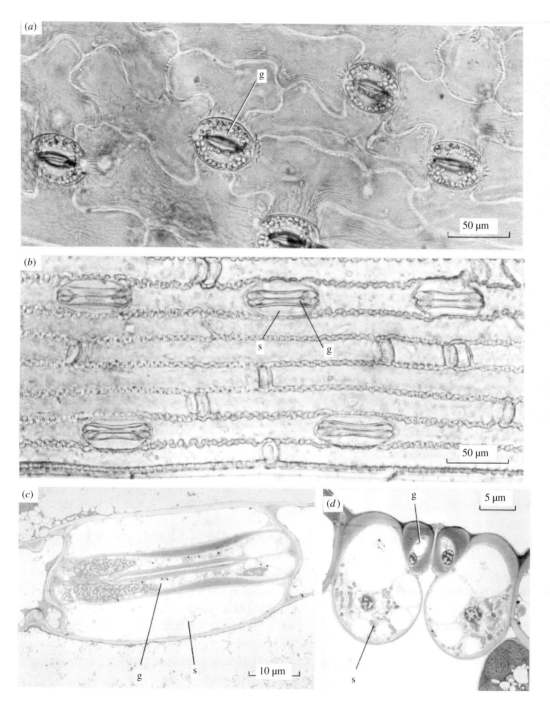

Figure 6.2 Representative stomata from the leaves of different species: (*a*) elliptical type (abaxial surface of *Gloriosa superba*); (*b*) graminaceous type (abaxial surface of wheat, *Triticum aestivum*) (light micrographs of stripped epidermis by M. Brookfield); (*c*) and (*d*) paradermal and transverse sections, respectively, of wheat stomata showing the thickened guard cells (g) and the subsidiary cells (s) (Courtesy of Dr M. L. Parker, Plant Breeding Institute Cambridge). (*e*) to (*h*) Scanning electron micrographs of the lower epidermis of some arid zone species (courtesy of Dr M. B. Joshi, University of Jodhpur) showing surface structure: (*e*) *Acacia senegal* (note the wax platelets); (*f*) *Echinops echinatus*; (*g*) *Tribulus terrestris*; (*h*) *Chorchorus tridens*.

(e)

(f)

(g)

(h)

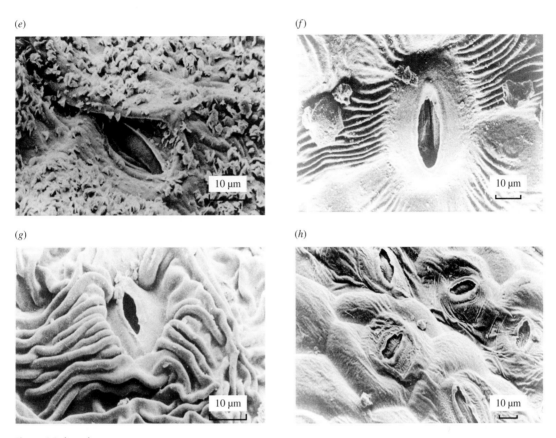

Figure 6.2 (*cont.*)

6.2 Stomatal mechanics and mechanisms

It has long been recognised that stomatal movements depend on changes in turgor pressure inside the guard cells and in the adjacent epidermal cells (which are sometimes modified to form distinct subsidiary cells) with the balance between pressures in these two cell types being critical. The changes in turgor can result either from a change in total water potential (ψ) of the guard cells as the supply or loss of water changes, or from active changes in osmotic potential (ψ_π). The former mechanism, relying on changes *outside* the guard cells has been termed 'hydropassive' (Stålfelt, 1955), while the latter has been termed 'hydroactive'. Both involve movement of water into or out of the guard cells.

Although purely hydraulic mechanisms can be involved in natural stomatal movements in response to water deficits and humidity changes, even in these cases some active ion pumping in response to the hydraulic changes is probably commonly involved.

Changes in guard cell turgor cause alteration in pore aperture as a consequence of the specialised structure and geometry of the stomatal complex. Two significant features of the commonest types of stomata are the presence of inelastic radially oriented micellae in the cell walls and a markedly thickened ventral wall (adjacent to the pore). Although it appears that neither of these characters is essential to the operation of elliptical stomata, they do influence the details of their movement (see Sharpe *et al.*, 1987 for a detailed analysis of stomatal mechanics).

Table 6.1 Examples of the range of values for stomatal density (v, mm^{-2}) and pore length (ℓ, μm) for different species, leaf position and growing conditions.

	Adaxial surface		Abaxial surface		Reference
	v (mm^{-2})	ℓ (μm)	v (mm^{-2})	ℓ (μm)	
Trees					
Carpinus betulus	0	n.d.	170	13	1
Malus pumila (cv. Cox)	0	n.d.	390	21	2
Malus pumila (variation within season)	0	n.d.	230–430	n.d.	2
Malus pumila (different cultivars)	0	n.d.	350–600	n.d.	3
Pinus sylvestris[a]	120	20	120	20	1
Picea pungens[a]	39	12	39	12	1
Other dicots					
Beta vulgaris	111	14.6	131	15.3	4
Tomato (low-high light range)	2–28	n.d.	83–105	n.d.	5
Soybean (range for 43 cultivars)	81–174	21–23	242–385	19.5–21.7	6
Soybean (well-watered – stressed range)	149–158	n.d.	357–418	n.d.	6
Ricinus communis	182	12	270	24	1
Tradescantia viginiana	7	49	23	52	1
Grasses					
Sorghum bicolor (mean of 6 cultivars)	n.d.	22.6	135	23	7
Hordeum vulgare (flag leaf)	54–98	17–24	60–89	17	1, 8, 9
Hordeum vulgare (5th below flag leaf)	n.d.	n.d.	27–42	n.d.	9

[a] Note that needles have no identifiable upper or lower surface

n.d. = no data

References: 1. Meidner and Mansfield (1968); 2. Slack (1964); 3. Beakbane and Mujamder (1975); 4. Brown and Rosenberg (1970); 5. Gay and Hurd (1975); 6. Ciha and Brun (1975); 7. Liang *et al.* (1975); 8. Miskin and Rasmusson (1970); 9. Jones (1977b).

At least for elliptical stomata, movement probably involves deformation of the guard cells out of the plane of the epidermis, though some bulging into the subsidiary cells may also occur. The changes in guard cell volume that occur during opening are not well documented but, at least in *Vicia faba*, anatomical studies indicate that the lumen volume may double when opening from closed to an aperture of 18 μm (Raschke, 1975). Stomatal apertures are approximately linearly related both to guard cell volume and, more approximately, to guard cell turgor pressure, though opening tends to saturate as pressure increases (Buckley, 2005).

An important feature of stomatal operation is the role of the subsidiary cells. Analysis of stomatal mechanics has shown that in many cases the subsidiary cells have a mechanical advantage over the guard cells, such that equal increases in pressure in guard and subsidiary cells cause some closure. This implies that closure cannot normally occur as a simple hydraulic response to declining bulk leaf water status, and that all stomatal movements normally result from an active process. This conclusion is supported by the well-known observation of transient stomatal opening on excising leaves (Iwanoff, 1928). Though this could partly be explained in terms of the epidermal cells losing turgor before guard cells as bulk leaf turgor falls, it seems more likely that such a transient wrong-way closure could arise as a result of the turgor in both guard and epidermal cells declining and the mechanical advantage of the subsidiary cells leading to transient opening – subsequent closure would then be an active metabolic process. The antagonism between guard and subsidiary cells has been recognised for over a century (von Mohl, 1856), while recent work has been reviewed by Buckley (2005).

Stomatal guard cells generally contain chloroplasts, though they are commonly less frequent, smaller and of different morphology to those in mesophyll cells. In spite of this widespread occurrence of chloroplasts in guard cells, however, there is little evidence that photosynthetic carbon reduction occurs or is involved in stomatal opening. Nevertheless it is likely that photophosphorylation and $NADP^+$ reduction provide energy for stomatal opening. Certain orchids (*Paphiopedilum* spp.) are exceptional in that their guard cells lack chlorophyll yet their stomata are functional (at least in intact leaves).

As we have seen, most stomatal movements, including those in response to changes in water status, involve closely regulated active changes in guard cell osmotic potential. A general feature of active stomatal opening movements is that a large proportion of the osmotic material consists of potassium ions. In *Vicia faba*, for example, the K^+ content increases from about 0.3 to 2.4 pmol per guard cell (equivalent to concentrations of 90×10^{-6} to 680×10^{-6} mol m^{-3}) as stomata open (MacRobbie, 1987). It is now generally accepted that this K^+ uptake is driven by ATP-powered primary proton extrusion at the plasmalemma. This sets up an electrical driving force that drives K^+ uptake via inward-rectifying K^+ channels and a pH gradient (cytoplasm alkaline). The pH gradient may be dissipated by synthesis of malic acid in the cytoplasm or by chloride uptake by co-transport with protons. The malate is generated within the guard cells from storage carbohydrates such as starch, though in *Allium* species and others that lack starch in their guard cells, Cl^- provides the counterion for K^+. There is some evidence that rapid changes in stomatal aperture are driven by K^+ fluxes, but that longer term maintenance of stomatal opening includes a substantial osmotic contribution from sucrose synthesised by starch hydrolysis partially replacing K^+ as the osmoticum (Talbott & Zeiger, 1998). This observation goes some way to unifying the classical starch–sucrose hypothesis for stomatal function and the currently favoured K^+/anion mechanisms.

It has been estimated for *Vicia faba* (see MacRobbie, 1987) that an increase in pore width of 10 μm requires a 1.25 MPa decrease in osmotic potential, which can be largely accounted for by the increase in potassium malate. The well-known decrease in starch content that occurs in the guard cells of many species as stomata open may provide both organic anions and energy for the ion pumps. Stomatal closure, for example in response to water stress, is also usually an active metabolic process involving active extrusion of ions, with anion extrusion at the plasmalemma being a critical early step, followed by conversion of sucrose or malate to starch. The membrane depolarisation caused by anion efflux subsequently drives K^+ efflux.

Among the various sensing pathways that are known to regulate the osmoregulation involved in stomatal movements are the responses to red and blue light (discussed below), to CO_2 and to water deficits, where the plant growth regulator abscisic acid (ABA) plays a major role. The importance of ABA was initially demonstrated by the fact that externally applied ABA closes stomata, concentrations of endogenous ABA increase rapidly in stress (often in parallel to stomatal closure), and mutants that are deficient in the capacity for ABA synthesis (for example *sitiens* and *flacca* in tomato, and *droopy* in potato) are not able to close their stomata, though the lesion can be reverted by the supply of exogenous ABA (see e.g. Addicott, 1983). Increased binding of ABA to receptors in the guard cells is now known to trigger multiple regulatory cascades of cellular biochemistry including activation of G-proteins, generation of nitrous oxide, the production of reactive oxygen species, increases in cytosolic Ca^{2+} concentrations (both from intercellular stores and by enhanced uptake) and increased sensitivity to Ca^{2+}; these then stimulate the outward anion channels and inhibit the outward H^+-ATPase and inward K^+-channels thus leading to decreased guard cell turgor and closing (Acharya & Assmann, 2009; Kim *et al.*, 2010). Although endogenous ABA concentrations do not always correlate well with stomatal aperture, particularly during recovery from stress, these observations may be often explained in terms of compartmentation into physiologically active and inactive pools, or that other compounds are involved. Although ABA is the primary controller of stomatal closure in response to water stress,

auxins and cytokinins can enhance opening, while ethylene, brassinosteroids, jasmonates and salicylic acid tend to favour stomatal closure, though the interactions can be complex (Acharya & Assmann, 2009).

6.3 Methods of study

The stomatal pathway and the corresponding resistance (r_s) or conductance (g_s) to mass transfer is only one component of the total leaf resistance (r_ℓ) (Figure 6.3). The cuticular transfer pathway (r_c) is in parallel with the stomata, while there is also a transfer resistance within the intercellular spaces (r_i) and, in series with that, a possible wall resistance (r_w) at the surface of the mesophyll cells. Many methods for determining r_s include these other resistances so more correctly provide a measure of r_ℓ; furthermore they often also include a boundary layer component. Throughout this book r_ℓ will be used as a measure of the true stomatal resistance, r_s, even though these terms are not exactly equivalent. The relative values of stomatal and other components of r_ℓ are discussed in detail below. It follows from discussion in Chapter 3 that stomatal resistances for different gases are inversely proportional to their diffusion coefficients.

A detailed description of the various methods available for studying stomata may be found in the book by Weyers and Meidner (1990). We outline

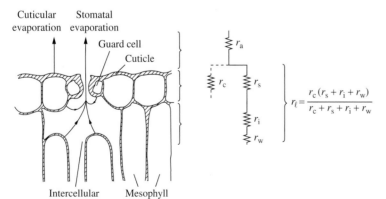

Cuticular evaporation   Stomatal evaporation

Guard cell
Cuticle

$$r_\ell = \frac{r_c (r_s + r_i + r_w)}{r_c + r_s + r_i + r_w}$$

Intercellular spaces   Mesophyll cells

Figure 6.3 Pathways for water loss from one surface of a leaf, showing the boundary layer (r_a), cuticular (r_c), stomatal (r_s), intercellular space (r_i), wall (r_w) and leaf (r_ℓ) resistances. The total leaf resistance is the parallel sum of r_ℓ for the upper and lower surfaces.

below the most important techniques for the study of stomatal functioning.

6.3.1 Microscopic measurement

Anatomical measurements can be made in vivo, as well as on fixed and cleared leaves or on epidermal imprints. These allow one to quantify the size of the stomatal complexes and their frequency on leaf surfaces. A convenient method for obtaining epidermal imprints is to spread a solution such as nail varnish over the epidermis, allow it to dry, peel it off and store in small envelopes for subsequent microscopic examination. Although useful for counting stomata and for measuring their dimensions, it is difficult to determine apertures from imprints as the impression material may not rupture at the narrowest part of the pore. It is always necessary to demonstrate that microscopic measurements, whether on living or prepared material, isolated epidermes or replicas, are representative of the situation in vivo, though it has been suggested that accurate measurements can be obtained using low-temperature scanning electron microscopy where ultra-rapid cryofixation can avoid aperture changes (van Gardingen et al., 1989).

Much of our current knowledge of stomatal physiology derives from studies using isolated epidermal strips where changes in pore size can be followed microscopically. Anatomical measurements may be used to derive estimates of the stomatal diffusion resistance (r_s) using diffusion theory (see e.g. Penman & Schofield, 1951). For the simplest case of a cylinder or other shape of constant cross-section, the resistance of the pore (s m^{-1}) is given by Eq. (3.21) as:

$$r_s = \ell/D \tag{6.1}$$

where ℓ is the length of the tube and D is the diffusion coefficient.

This resistance per unit pore area can be converted to a resistance per unit area of leaf by dividing by the ratio of average pore cross-sectional area to leaf surface area. For circular pores this ratio is $v\pi r^2$, where v is the frequency of stomata per unit leaf area and r is the pore radius. This gives the stomatal resistance on a leaf area basis as:

$$r_s = \ell/v\pi r^2 D \tag{6.2}$$

Because the pore area is much less than the leaf area there is a zone close to the pore where the lines of flux converge on the pore. Therefore this zone of the boundary layer is not effectively utilised for diffusion and there is an extra resistance or 'end effect' associated with each pore. The magnitude of this extra resistance can be derived from three-dimensional diffusion theory as being proportional to pore radius and equal to $\pi r/4D$. Therefore converting to a leaf area basis and adding this extra resistance to the pore resistance gives:

$$r_s = [\ell + (\pi r/4)]/v\pi r^2 D \tag{6.3}$$

This approach can be extended to other shaped pores and to the estimation of the small intercellular space resistance (Figure 6.3). It is also possible to correct for alterations in D that occur when the pore size is small, and for interactions with other diffusing species (Jarman, 1974), but these effects are small and usually neglected.

Equations (6.1)–(6.3) can readily be converted to obtain molar resistances if we remember that the molar resistance ($r^m = r$) can be obtained from r by multiplying by ($\mathcal{R}T/P$) (Eq. (3.23b)).

6.3.2 Infiltration

Graded solutions of differing viscosity (e.g. various mixtures of liquid paraffin and kerosene) have been widely used to estimate relative stomatal opening (Hack, 1974). The viscosity of the solution that just infiltrates the pores provides a measure of aperture. Because of differences in anatomy of pores in different species and differences in cuticular composition, this approach is only of use for studying qualitative differences within one species. A further problem is that infiltration is destructive so that the same leaves cannot be reused.

6.3.3 Viscous-flow porometers

The mass flow of air through stomata under a pressure gradient has been widely used as a measure of stomatal aperture since the first viscous-flow 'porometer' was constructed by Darwin and Pertz (1911). Air may be forced through the stomata in either of the ways illustrated in Figure 6.4 and the viscous-flow resistance derived from the measured flow rate or the rate of pressure change (see Meidner & Mansfield, 1968). The mass flow rate is inversely proportional to the viscous-flow resistance. This resistance in turn depends on the apertures of the stomata, which constitute the major resistance, and to a lesser extent on resistance to mass flow through the intercellular spaces in the leaf mesophyll (see Figure 6.4).

Unfortunately there are difficulties in absolute calibration because of the complexity of the flow paths. Furthermore the resistance obtained with trans-leaf flow porometers (Figure 6.4(a)) is the *sum* of the resistances of the upper and lower surfaces, since they are in series, and is dominated by the larger (usually that of the upper epidermis). However, for diffusive exchange of CO_2 or water vapour, the upper and lower pathways are in parallel, so that the diffusive resistance is largely dependent on the smaller of the two resistances (usually the lower epidermis). Results obtained with viscous-flow porometers must, therefore, be interpreted with care but they are useful for comparative measurements and rapid screening. Where such an instrument is operated with a constant applied pressure it is useful for continuous recording (a typical example is illustrated in Figure 6.4(c)). An alternative mode of operation is more suitable for rapid screening: this involves introducing a fixed quantity of air at a given pressure to a reservoir and then measuring the time between two pressures as the air is released (as in the CSIRO/Thermoline porometer, Rebetzke *et al.*, 2000). The mass flow conductance (sometimes referred to as the *leaf porosity*) is defined as proportional to the inverse of the time between these two pressures, though it is worth noting that theoretical and experimental

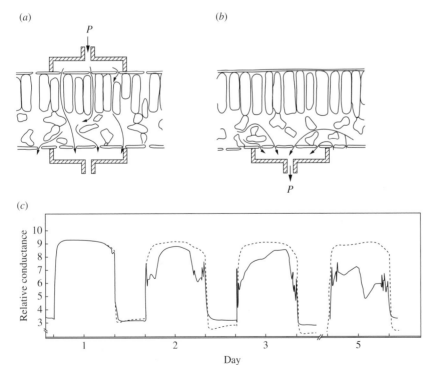

(a)   (b)   (c)

Day

Figure 6.4 Mass-flow (or viscous-flow) porometers: (a) trans-leaf type as suitable for amphistomatous leaves such as cereals, (b) one-cup type for hypostomatous leaves. The applied pressure (P) may be above or below atmospheric, and either the flow rate or the rate of change of P measured. (c) Typical viscous-flow porometer traces for a control leaf of *Helianthus annuus* (dashed line) and for a corresponding leaf on a plant whose roots had been placed in moist air near the start of day 1 (solid line) (data from Neales *et al.*, 1989).

results indicate that diffusive resistances are proportional to the square root of the viscous-flow resistance.

6.3.4 Leaf gas-exchange systems and diffusion porometers

These instruments measure diffusive transfer and are therefore the most relevant and useful for studies of leaf gas exchange. The older trans-leaf diffusion instruments that measured diffusion of gases such as hydrogen, argon or nitrous oxide (see Meidner & Mansfield, 1968) were subject to many of the criticisms of the trans-leaf viscous-flow porometers and are little used nowadays. Most current instruments measure the total diffusive water loss from plant leaves so rather than measuring the stomatal resistance they actually measure the *leaf resistance* to water vapour (including any cuticular component) together with any boundary layer resistance in the porometer chamber.

The leaf resistance may be obtained for any sample by subtracting the chamber boundary layer resistance (a chamber constant often determined by using water-saturated blotting paper in place of the leaf as described in Appendix 8) from the total resistance observed. The magnitude of this chamber boundary layer resistance may be minimised by stirring the air in the chamber with a small fan.

Transit-time instruments

The principle of this type of instrument is that when a leaf is enclosed in a sealed chamber, evaporation will tend to increase the humidity in the chamber at a rate dependent, among other things, on the stomatal diffusion resistance. The time taken for the humidity to increase over a fixed interval can be converted to resistance by the use of a previously obtained calibration curve. Calibration involves replacing the leaf by a wet surface (e.g. wet blotting paper) covered with a calibration plate perforated by a number of precisely drilled holes to generate a known diffusion resistance. A range of resistances can be obtained by varying the number and size of

holes, with the resistance for any plate being obtained from theory (Eq.(6.3)).

There are several designs of transit-time porometer, some of which are available commercially (e.g. AP-4 porometer, Delta-T Devices, Burwell, Cambridge). Different instruments may include stirring in the chamber to minimise the chamber boundary layer resistance and may be used on irregular leaves or even conifer needles, while others include automatic timing, automatic purging of the chamber with dry air between measurements, or the facility for changing the humidity range over which they operate so as to mimic natural conditions as closely as possible, together with microprocessors to calculate and store data as necessary. In practice the main source of error occurs when leaf temperature (T_ℓ) is not the same as the cup temperature, but this error is difficult to avoid in the field.

Steady-state gas exchange and porometers

The principle of this equipment has long been used in laboratory gas-exchange equipment and in recent years has been adapted to field instruments (Figure 6.5). The less sophisticated instruments that only measure water vapour loss tend to be termed porometers. When a leaf is enclosed in a chamber through which air is flowing, the evaporation rate into the cuvette (in molar units) can be calculated approximately from the product of the flow entering the chamber and the concentration differential across the chamber as:

$$\mathbf{E}^m = u_e(e_o - e_e)/(PA) = u_e(x_{Wo} - x_{We})/A \qquad (6.4)$$

where $\mathbf{E}^m$ is the evaporation rate (mol m^{-2} s^{-1}), u is the molar flow rate (mol s^{-1}), x_W is the mole fraction of water vapour (mol mol^{-1}), A is the leaf surface area (m^{-2}) exposed in the chamber, and the subscripts 'o' and 'e' refer to the outlet and inlet flows respectively. This equation is only approximate because the water transpired by the leaf changes the flow rate across the cuvette, so it is better to use the full mass balance:

$$\mathbf{E}^m = (u_o x_{Wo} - u_e x_{We})/A \qquad (6.5)$$

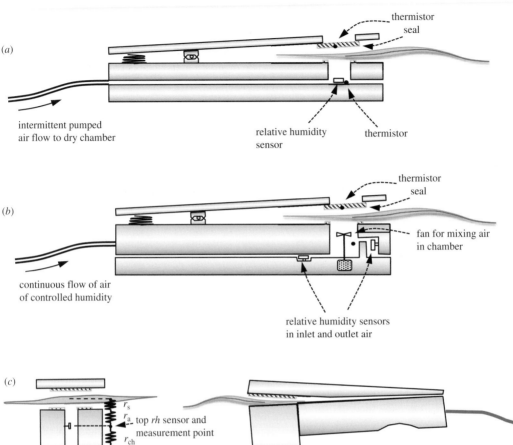

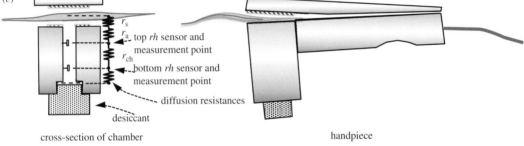

Figure 6.5 (a) Diagram of a simple transit-time porometer (e.g. Delta-T AP4 porometer, from Delta-T Instruments, Burwell, Cambridge): in this a pump is used at intervals to dry out the leaf chamber, and then the time is recorded for water vapour diffusion from the leaf to raise the humidity in the chamber between two set points. (b) Flow diagram of a typical open system continuous-flow steady-state porometer or gas-exchange system, where the leaf diffusion resistance is calculated from the flow rate and the humidities of incoming and outgoing air. (c) Illustration of the principle of the Decagon SC1-leaf porometer, where the stomatal resistance (r_s) is estimated from the differences in humidity between the leaf, two intermediate sensors and a desiccant, where the boundary layer ($r_a = r_1$) and chamber ($r_{ch} = r_2$) resistances are known, assuming that resistances for each component are proportional to the humidity differences across each component. In each case temperature sensors are used to correct the readings for temperature gradients.

The difference between u_o and u_e equals the evaporation rate (the CO_2 exchange by the leaf may be ignored because it is comparatively small and is largely balanced by O_2 exchange) so that one can write:

$$u_o = u_e + (u_o\,x_{Wo} - u_e\,x_{We}) \tag{6.6a}$$

which rearranges to:

$$u_o = u_e \left(\frac{1 - x_{We}}{1 - x_{Wo}} \right) \tag{6.6b}$$

which on substituting back into Eq. (6.5) gives:

$$\mathbf{E}^m = \frac{u_e(x_{Wo} - x_{We})}{A(1 - x_{Wo})} \tag{6.7}$$

The term $(1 - x_{Wo})$ in the denominator of Eq. (6.7) adjusts the flow rate for the amount added by the transpiration from the leaf, though for typical conditions this leads to a correction of only 2 to 4%. Equivalent equations can be written for $\mathbf{E}$ (expressed as a mass flux density) using volume flow rates and concentrations (see Chapter 3).

The evaporation rate is related to the total resistance to water vapour loss ($r_{\ell W} + r_{aW}$), where r_{aW} is the chamber boundary layer resistance, by:

$$\mathbf{E}^m = \left(\frac{x_{Ws} - x_{Wo}}{r_{\ell W} - r_{aW}}\right) \tag{6.8}$$

where the water vapour pressure at the evaporating sites within the leaf is assumed equal to the saturation vapour pressure at leaf temperature ($e_{s(T\ell)}$) and x_{Ws} is the corresponding mole fraction, and x_{Wo} is assumed to be representative of the mole fraction of water vapour in the chamber air, which will be true for a well-mixed chamber.

Eliminating $\mathbf{E}^m$ from Eqs. (6.7) and (6.8) leads to:

$$(r_{\ell W} + r_{aW}) = g_W^{-1} = \frac{A(x_{Ws} - x_{Wo})(1 - x_{Wo})}{u_e(x_{Wo} - x_{We})} \tag{6.9}$$

Application of this equation can be greatly simplified if air temperature equals leaf temperature (i.e. the system is isothermal) and if dry air is input to the chamber (i.e. $x_{We} = 0$). Making the approximation that the term $(1 - x_{Wo}) \simeq 1$, this equation simplifies to:

$$r_{\ell W} = 1/g_{\ell W} = \left(\left[(1/h) - 1\right]A/u_e\right) - r_{aW} \tag{6.10}$$

where h is the relative humidity of the outlet air ($= x_{Wo}/x_{Ws}$).

Unfortunately Eqs. (6.9) and (6.10), though adequate for many purposes, do not take account of the fact that the total transpiration from the leaf is made up of a diffusive component given by Eq. (6.8) and an additional mass flow, equal to the mean water vapour mole fraction along the diffusion pathway from the intercellular spaces times the evaporation rate. Adding this correction to Eq. (6.8) gives (von Caemmerer & Farquhar, 1981):

$$\mathbf{E}^m = \left(\frac{x_{Ws} - x_{Wo}}{r_{\ell W} + r_{aW}}\right) + \mathbf{E}^m\left(\frac{x_{Ws} + x_{Wo}}{2}\right) \tag{6.11}$$

Rearranging Eq. (6.11), and substituting from Eq. (6.7) gives the following complete expression for the total resistance or conductance to water vapour:

$$\begin{aligned} g_W &= (r_{\ell W} + r_{aW})^{-1} \\ &= \frac{u_e(x_{Wo} - x_{We})[1 - (x_{Ws} + x_{Wo})/2]}{A(1 - x_{Wo})(x_{Ws} - x_{Wo})} \end{aligned} \tag{6.12}$$

The correction involved is normally of the order of 2 to 4%, so is only important for accurate work, especially where one is attempting to calculate an intercellular space CO_2 concentration (see Chapter 7). Of greater practical importance in most cases are the errors caused by imprecise estimates of T_ℓ, and hence x_{Ws} (Mott & Peak, 2011).

Figure 6.5 shows a typical flow diagram of a continuous-flow gas-exchange system or porometer. The most usual current designs use a constant flow rate (e.g. Parkinson & Legg, 1972), in which case the relative humidity of the outlet air is uniquely related to r_ℓ (irrespective of temperature). Because of stomatal sensitivity to ambient humidity (see below), it is preferable to operate with the chamber humidity close to ambient humidity, whichever mode of operation is used. Continuous-flow porometers provide the best method currently available for rapid but accurate studies of stomatal conductance in the field; although sophisticated field photosynthesis systems are widely used for this purpose, their comparatively long equilibration times (often two or more minutes) limits the number of readings that can be obtained. One advantage over transit-time instruments is that calibration only involves calibration of the humidity sensor and flow meter and one does not need to take account of actual temperature (though see Parkinson & Day, 1980) so long as a relative humidity sensor is used. Another advantage is the fact that continuous-flow porometers have similar sensitivity over a wide range of g_ℓ; this is in contrast to transit-time instruments that are relatively insensitive at the physiologically important high values of g_ℓ.

An interesting alternative steady-state approach to estimation of leaf conductance is used in the 'SC-1 Leaf Porometer' (Decagon Devices Inc., Pullman, WA, USA). This operates on the basis of setting up a steady-state diffusion of water vapour out of the leaf through an enclosed chamber of known diffusion resistance (calculated from its dimensions using Eq. (3.21)) to a low humidity sink (Figure 6.5(b)). In its simplest form we can assume that in an isothermal steady state the diffusive flow will be constant from the leaf through the stomata into the chamber and down the chamber to the low humidity end so that the catenary flow equation (Eq. (4.26)) holds. The molar flux of water ($\mathbf{E}^m$) will then be given by (Table 3.1):

$$\mathbf{E}^m = \left(\frac{P}{\mathscr{R}T}\right)\left(\frac{x_{Ws} - x_{W1}}{r_\ell + r_1}\right) = \left(\frac{P}{\mathscr{R}T}\right)\left(\frac{x_{W1} - x_{W2}}{r_2}\right) \tag{6.13}$$

This can be simplified by dividing through by x_{Ws}, cancelling out terms and substituting from Eq. (3.21) and rearranging to give:

$$r_\ell = \frac{\ell_2}{D}\left(\frac{1 - h_1}{h_1 - h_2}\right) - \frac{\ell_1}{D} \tag{6.14a}$$

$$g_\ell = \frac{D(h_1 - h_2)}{\ell_2(1 - h_1) - \ell_1(h_1 - h_2)} \tag{6.14b}$$

where h_1 and h_2 are the relative humidities at the two humidity sensors. The leaf conductance (g_ℓ; m s^{-1}) can be converted to the molar conductance (g_ℓ) by multiplying by $P/\mathscr{R}T$. In practice it is necessary to take particular care to allow temperature equilibration and to correct for residual temperature gradients in the system (see www.decagon.com).

Estimation of leaf temperature in porometers

Accurate estimation of g_ℓ is dependent on precise estimation of T_ℓ. Although it is common practice to estimate T_ℓ in leaf chambers and porometers by

means of thermocouples appressed to the leaf surface, these tend to lead to an estimate of T_ℓ intermediate between the true value and T_a, with consequent errors in g_ℓ. An alternative approach is to estimate the leaf–air temperature difference by means of the leaf energy balance (Parkinson & Day, 1980).

6.3.5 From leaf temperature measurement

Full energy balance method

Recent improvements in the availability and accuracy of infrared thermometers and cameras have led to the development of non-contact approaches to the estimation of stomatal conductance from leaf temperature (Guilioni et al., 2008; Leinonen et al., 2006). Using similar arguments to those used in the derivation of the Penman–Monteith equation (Eq. (5.26)) we can derive the relationship between leaf surface temperature (T_ℓ) and the full range of environmental and plant factors that affect it and hence the relationship between leaf conductance and T_ℓ.

For a leaf (here using the convention of expressing fluxes per unit projected area) we can write the net radiation balance as the sum of the net shortwave absorbed and the net longwave as:

$$\mathbf{R}_n = (1 - \rho - \tau)\mathbf{R}_S + \varepsilon(\mathbf{R}_L - 2\sigma T_\ell^4) \tag{6.15}$$

where $\mathbf{R}_S$ and $\mathbf{R}_L$ are the total incoming shortwave and longwave (from above and below) and the factor 2 accounts for radiative losses from both sides of the leaf. Using the usual linearisation for the radiative losses (Section 5.1.2), noting that r_R is half the value given by Eq. (5.9) because the leaf is two sided, this gives:

$$\mathbf{R}_n = \mathbf{R}_{ni} - \rho_a c_p (T_\ell - T_a)/r_R \tag{6.16}$$

From Eqs. (5.23) and (5.22) and noting that $\gamma = P c_p/0.622 \lambda$, we can write expressions for sensible heat and latent heat losses:

$$\mathbf{C} = (\rho_a c_p)(T_\ell - T_a)/r_H \tag{6.17}$$

$$\lambda\mathbf{E} = (\rho_a c_p/\gamma)(D + s(T_\ell - T_a))/r_W \tag{6.18}$$

For a leaf one can equate the sum of these two to the net radiation (remembering that storage is zero for a leaf at the steady state, and that $\gamma = P\,c_p/0.622\,\lambda$). On rearrangement this gives an expression for the total resistance to water vapour:

$$r_W = \frac{\rho_a c_p r_{HR}}{\gamma}\left(\frac{s\,(T_\ell - T_a) + D}{r_{HR}\,\mathbf{R}_{ni} - \rho_a c_p(T_\ell - T_a)}\right) \qquad (6.19)$$

The leaf resistance can then be obtained from $r_\ell = r_W - r_{aW}$. It is important, however, to distinguish between leaves with stomata on both sides (*amphistomatous*) and those with stomata on one side only (usually *hypostomatous*) as the distribution of stomata will affect the appropriate value of r_{aW} to use, and to remember that all resistances for leaves are expressed per unit projected area. Using Eq. (6.19) one can therefore estimate the leaf resistance where data are available on T_ℓ, T_a, $\mathbf{R}_n$, D and r_a, but it can be difficult to get accurate estimates of all these terms.

Reduced equations using reference surfaces

The requirement for ancillary measurements can be reduced by the measurement of the temperatures of wet and/or dry reference surfaces (Jones, 1999). Assuming that the reference surfaces have similar optical and aerodynamic properties to real leaves, one can eliminate the need for a radiation measure by using the temperature of a dry reference surface (T_d) because in this case the energy balance reduces to:

$$\mathbf{R}_{ni} = \rho_a c_p (T_d - T_a)/r_{HR} \qquad (6.20)$$

Substituting this back into Eq. (6.19) gives after some rearrangement:

$$r_W = r_{HR}\left(\frac{s(T_\ell - T_a) + D}{\gamma(T_d - T_\ell)}\right) \qquad (6.21)$$

Further elimination of the humidity term, D, can be achieved by using both wet and dry reference surfaces. For example for hypostomatous leaves with wet reference leaves wetted on only one side, Guilioni et al. (2008) showed that:

$$r_\ell = (r_{aW} + (s/\gamma)r_{HR})\left(\frac{T_\ell - T_w}{T_d - T_\ell}\right) \qquad (6.22)$$

This equation shows that relative values of leaf ($\simeq$ stomatal) resistance for different surfaces at a given temperature can be obtained from the temperatures of the leaf, wet and dry references only. Full details of the formulae for calculation of r_ℓ for all combinations of leaf types and references are given elsewhere (Guilioni *et al.*, 2008). A great advantage of using the reduced methods with reference surfaces is that instrumental errors in temperature measurement largely cancel out.

Thermal methods for temperature measurement are particularly well suited for phenotyping for stomatal behaviour because they are readily adapted to the study of large numbers of leaves or plants (Jones *et al.*, 2009; Reynolds *et al.*, 1998). Similarly, the sensitive stomatal closure in response to water deficit stress has led to the development and widespread use of a thermal index, the so-called *Crop Water Stress Index* (CWSI) and its derivatives as a useful proxy indicator of crop 'stress'; this will be discussed further in Section 10.4.2.

6.4 Stomatal response to environment

The effects of individual factors such as radiation, temperature, humidity or leaf water status on stomatal conductance can be studied best in controlled environments or leaf chambers where each factor may be varied independently. However, the use of this information to predict the stomatal conductance in a natural environment is complicated by various factors including (i) interactions between the responses (that is, any response depends on the level of other factors); (ii) variability of the natural environment; (iii) the fact that the stomatal time response is frequently of the same order or longer than that of changes in the environment (therefore stomata rarely reach the appropriate steady-state apertures); (iv) in species with amphistomatous leaves the stomata on the upper surface tend to be more responsive than those on the lower surface; (v) endogenous rhythms tend to affect stomatal aperture independently of the current environment (e.g. night-time closure tends to occur even in continuous light); and (vi) acclimation to altered environmental conditions can occur.

6.4.1 Maximum conductance

The great variability in stomatal frequency and size that exists between different species, leaf position or growth condition (Table 6.1) leads us to expect corresponding differences in stomatal conductance. Figure 6.6 summarises a large number of reported measurements of maximum leaf conductance that have been reported for different groups of

plants. Although the range of values found within any group is very wide, there are some clear-cut differences with maximum conductances ($g_{\ell w}$ on a total leaf surface area basis) averaging less than 80 mmol m^{-1} (2 mm s^{-1}) for succulents and evergreen conifers and four times that value for plants from wet habitats.

In addition to genotypic differences, maximum leaf conductance is strongly affected by growth conditions and changes with leaf age. Characteristically, maximum conductance does not attain a peak value until several days after leaf emergence, it may then stay close to this value for a time that is characteristic of the species, finally declining to a very low value as the leaf senesces (Figure 6.7).

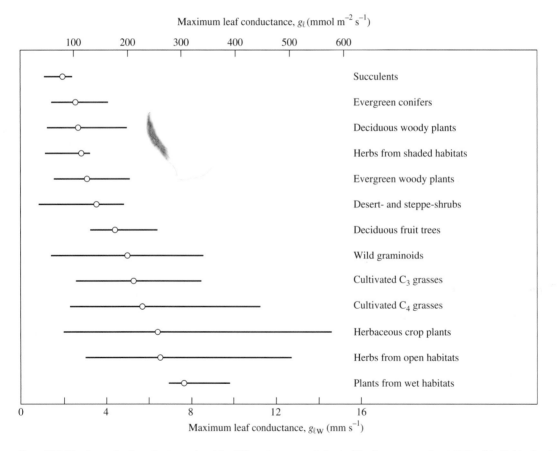

Figure 6.6 Maximum leaf conductance ($g_{\ell w}$) in different groups of plants. The lines cover about 90% of individual values reported. The open circles represent group average conductances (adapted from Körner et al., 1979).

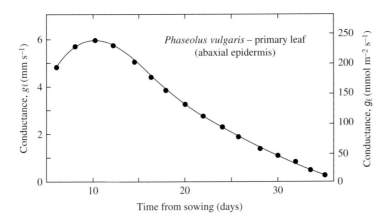

Figure 6.7 Typical developmental trends of leaf conductance of adaxial surfaces of *Phaseolus vulgaris* L. primary leaves (at 1200 µmol m^{-2} s^{-1} PAR) for a single leaf from early expansion to final senescence. (After Solárová, 1980.)

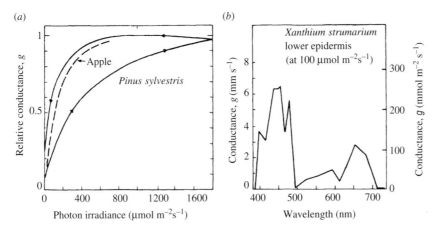

Figure 6.8 (*a*) Examples of stomatal light-response curves. The curve for apple (data from Warrit *et al.*, 1980) is typical for many species, approaching a maximum at about a quarter of full sunlight. The curves for *Pinus sylvestris* (after Ng & Jarvis, 1980) illustrate the hysteresis that can occur in certain conditions. (*b*) Action spectrum of stomatal opening in *Xanthium strumarium* (calculated from Sharkey & Raschke, 1981). Action is represented as the conductance achieved at each wavelength for a photon irradiance of 100 µmol m^{-2} s^{-1}.

6.4.2 Response to light

Perhaps the most consistent and well-documented stomatal response is the opening that occurs in most species as irradiance increases (Figure 6.8(*a*); Shimazaki *et al.*, 2007). Maximum aperture is usually achieved with irradiances greater than about a quarter of full summer sun (i.e. about 200 W m^{-2} (total shortwave) or 400 µmol m^{-2} s^{-1} (PAR)), though this value depends on species and on the natural radiation environment: stomata on shade-grown leaves open at lower light levels than do those on sun-adapted

leaves. The conductance–irradiance relationship often shows hysteresis (Figure 6.8(*a*)), particularly if time is not allowed for complete equilibrium when the light is altered. Although some of the stomatal response to light may be indirect and attributable to the decrease in intercellular CO_2 concentration that occurs on illumination (see below), there is strong evidence for a direct light response involving two independent photoreceptors.

Stomata are particularly sensitive to blue light (Figure 6.8(*b*)), a response that leads to activation of an outwardly directed plasma membrane H^+-ATPase generating the hyperpolarisation that drives K^+ accumulation in the guard cells and increased turgor pressure. Activation of the H^+-ATPase involves signal reception by blue light-absorbing flavoproteins called phototropins (Kinoshita *et al.*, 2001); these are involved in binding of 14–3–3 proteins and phosphorylation of the H^+-ATPase through a complex signalling cascade involving Ca^{2+} (see Shimazaki *et al.*, 2007). Even very low irradiances (5 $\mu mol\ m^{-2}\ s^{-1}$) can elicit substantial rapid opening in many species if given in the presence of background red light. The blue light acts as a signal with the energy required to drive the ion transport being provided by photosynthesis in red light. Blue light also stimulates the breakdown of starch to sucrose and malate.

The opening in response to red light (though note that the action spectrum actually has peaks in the blue as well as the red) requires much higher irradiances than for blue light and may act by driving photosynthesis in mesophyll and guard cell chloroplasts and therefore decreasing the intercellular space CO_2 concentration (c_i). There is, however, also some evidence that red-light responses can be independent of current photosynthesis both in guard cells and mesophyll cells (Baroli *et al.*, 2008; Messinger *et al.*, 2006). The blue-light and red-light

responses appear to act synergistically with, for example, the blue-light stimulation of malate synthesis requiring red light. Although other work has suggested that stomata actually respond to the total absorbed radiant energy, it seems likely that these results are an artefact arising from the difficulty of estimating T_ℓ accurately (Mott & Peak, 2011).

The rate of stomatal response to changing light is variable, though closing responses tend to be more rapid than opening. Half-times are generally of the order of 2 to 5 min; these are of the same order as many environmental changes. There is some evidence that stomatal closure in response to decreasing light can be potentiated by water stress. An example is shown for sorghum leaves in Figure 6.9 where mild water deficits decreased the half-time for closure to less than 1 min. The kinetics of response are likely to be particularly important when light is changing rapidly, as in sunflecks.

Although in most plants stomata open in response to light and close in the dark, the reverse is true in plants having the crassulacean acid metabolism (CAM) pathway of photosynthesis (see Chapter 7). In these plants maximal opening is in the dark, particularly in the early part of the night period. It is also worth noting that many plants maintain significant night-time leaf conductances in the dark

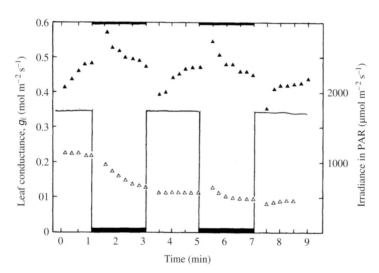

Figure 6.9 The time course of stomatal conductance for leaves of field-grown *Sorghum bicolor* in response to sudden darkening for well-watered plants (▲) or for plants subject to mild drought (△); irradiance in the PAR is indicated by the continuous line (H. G. Jones, D. O. Hall & J. E. Corlett, unpublished data). The transients obtained on changing irradiance are artefacts related to the time for the porometer system to reach equilibrium.

with a modal value among 59 species of between 20 and 60 mmol m^{-2} s^{-1}, with much of this conductance relating to stomatal opening rather than to a residual cuticular transpiration (Caird *et al.*, 2007). These rates equate to between 5 and 30% of total daily water lost. Indeed stomatal responses to temperature and to humidity deficit occur as much in the dark as in the light (Mott & Peak, 2010).

6.4.3 Response to carbon dioxide

Although the CO_2 concentration in the natural environment is relatively constant (see Chapter 11), stomata are sensitive to CO_2, responding to the CO_2 mole fraction in the intercellular spaces (x_i'). In general stomata tend to open as x_i' decreases, with the sensitivity to CO_2 being strongly species and environment dependent, being greatest in C_4 species, and at concentrations below about 300 vpm (Morison, 1987). Stomata respond to CO_2 in both light and dark, so the response cannot depend only on photosynthesis. The value of x_i' is maintained surprisingly constant (at approximately 130 vpm in C_4 species and 230 vpm in C_3 species) over a wide range of conditions and rates of photosynthesis (Wong *et al.*, 1979). This would occur if stomatal conductance (g_s) varied in proportion to assimilation rate, and has led to the suggestion that a signal from the mesophyll controls stomatal aperture, though mechanistic evidence for such a hypothesis is still lacking and it seems more likely that this results more from a close matching of g_s and assimilation rate. This close correlation between photosynthesis and stomatal conductance has provided the basis for an important class of stomatal model, the Ball–Woodrow–Berry or BWB model (Ball *et al.*, 1987); this has been found to be widely applicable at least empirically. In its original form stomatal conductance across a wide range of environments was fitted by:

$$g_s = a\mathbf{P}h/c_s' \qquad (6.23)$$

where *a* is an empirical constant, *h* is the relative humidity, c_s' is the CO_2 concentration at the leaf surface and **P** is the net photosynthesis rate. A revised

version of this model (Leuning, 1995) that better fits stomatal responses is:

$$g_s = g_o + a\mathbf{P}h/[(c_s' - \Gamma)(1 + D_s/D_o)] \qquad (6.24)$$

where Γ is the CO_2 compensation concentration, D_s is the humidity deficit at the leaf surface and D_o is a constant. Although the precise site and mechanism of CO_2 sensing is controversial, there is increasing evidence for a direct effect where CO_2 causes stomatal closure by activating anion channels and K^+-efflux channels with the necessary involvement of Ca^{2+}. Further discussion of recent work on CO_2 sensing may be found elsewhere (e.g. Kim *et al.*, 2010).

It is worth noting that the stomata can partially adapt to longer term exposure to high CO_2 concentrations (as in free-air carbon dioxide enrichment – FACE – experiments; Wullschleger *et al.*, 2002) with decreasing sensitivity to changing CO_2 concentration for plants grown at higher than normal CO_2 concentrations, though this may be at least partly a result of anatomical adaptation. The stomatal responses to CO_2 can be modified by a range of other environmental factors, especially light and humidity, with, for example, high vapour pressure deficits greatly enhancing sensitivity to CO_2 (Bunce, 2006).

6.4.4 Response to humidity

Until the early 1970s it had been thought that stomata were insensitive to ambient humidity. For example Meidner and Mansfield (1968) in their text on stomatal physiology state 'stomatal behaviour. . .is comparatively unaffected by changes in relative humidity of the ambient air'. It is now clear, however, that the stomata in many species close in response to increased leaf-to-air vapour pressure difference (D_ℓ), as shown in Figure 6.10. The sensitivity of stomata to a change in leaf-to-air vapour pressure difference tends to be greatest for temperate deciduous species, decreasing through rainforest trees and ferns to herbaceous crop plants, which are least sensitive (Franks & Farquhar, 1999). When the humidity around a leaf is reduced, the initial response is

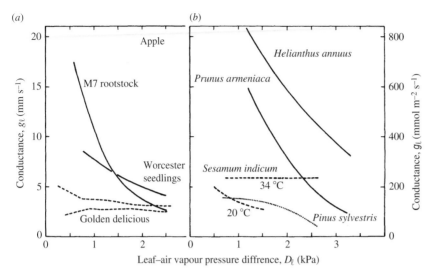

(*a*)　(*b*)

Figure 6.10 Examples of stomatal humidity responses. Data are for apple (Fanjul & Jones, 1982; Warrit *et al.*, 1980), *Helianthus* and *Sesamum* (Hall & Kaufmann, 1975), *Prunus* (Schulze *et al.*, 1972) and *Pinus* (Jarvis & Morison, 1981).

for g_s to increase transiently for about ten minutes or so but then eventually to equilibrate at a value lower than the initial. The magnitude of the steady-state response is dependent on species, growing conditions and particularly plant water status, the response being smaller at high temperature (Figure 6.10) or in stressed plants. In practice, observed responses to humidity are probably responses to changes in transpiration rate driven by changes in vapour pressure gradients, rather than to humidity per se, as stomatal closure still occurred on transfer of leaves to helox (79% helium and 21% oxygen in which diffusion of water vapour is several times faster than in air) even when maintaining e_a constant (Mott & Parkhurst, 1991).

Although the humidity response frequently takes minutes to reach full expression, it has been reported, at least for apple leaves, that as much as 90% of the total stomatal response to an increased D_ℓ can occur within 20 s of the humidity change (Fanjul & Jones, 1982). The relative rate of closing in response to increased humidity is consistently faster in droughted than in well-watered plants (Aasamaa & Sõber, 2011) but opening responses tend to be rather slower.

6.4.5 Response to water status

Stomata are sensitive to plant water status, tending to respond in such a way as to minimise imposed changes in the balance between water supply and evaporative demand. In general there is a tendency for stomata to close with decreasing leaf (Figure 6.11 (*a*)) or soil water potential. Closure occurs over a wide range of ψ_ℓ and this relationship can be modified by exposure to previous stress, or by the rate of desiccation. The effects of growth conditions on the ψ_ℓ at which g_ℓ tends to zero are summarised for several species in Table 6.2. The degree of adjustment ranged from zero for *Hibiscus* to 3.6 MPa for *Heteropogon*. At the leaf water potentials that occur normally for well-watered plants during the course of a day, however, stomatal conductance is often relatively insensitive to ψ_ℓ and may even *increase* with decreasing leaf water potential (see Figure 6.11(*b*) and also Figure 6.15(*d*); this response is what one would expect if stomatal conductance were controlling leaf water potential (through an altered transpiration rate), rather than the reverse and is discussed further below (see also Jones, 1998).

Table 6.2 Some examples of stomatal adaptation to water stress in different species.

| Species | Conditions | ψ_ℓ at which g_ℓ tends to zero (MPa) | | | |
		Maximum	Minimum	Adjustment	Reference
Apple	Seasonal change (field)	−2.7	−5.2	2.5	1
Heteropogon contortus	CE vs. field	−1.4	−5.3	3.6	2
Eucalyptus socialis	Outdoor hardening	−2.5	−3.8	1.3	3
Cotton	CE vs. field	−1.6	−2.7	1.1	4
Cotton	Drought cycles in CE	−2.8	−4.0	1.2	5
Sorghum	Field	−1.9	−2.3	0.4	6
Wheat	CE, solution culture	−1.4	−1.9	0.5	7
Hibiscus cannabinus	CE vs. field	−2.1	−2.0	0	2

CE refers to controlled environment.
References: 1. Lakso (1979); 2. Ludlow (1980); 3. Collatz *et al.* (1976); 4. Jordan and Ritchie (1971); 5. Brown *et al.* (1976); 6. Turner *et al.* (1978); 7. Simmelsgaard (1976).

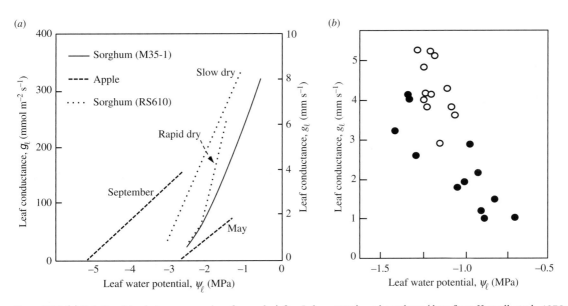

Figure 6.11 (*a*) Relationships between g_ℓ and ψ_ℓ for apple (after Lakso, 1979), and sorghum (data from Henzell *et al.*, 1976; Jones & Rawson, 1979). Slow drying was at 0.15 MPa d^{-1}, fast drying at 1.2 MPa d^{-1}. (*b*) An example for apple where a negative relationship between g_ℓ and ψ_ℓ has been observed (open symbols refer to well-watered controls and closed symbols are for droughted plants; data from Jones, 1985b).

Those species or cultivars that tend to maintain leaf water potential almost constant as soil dries are termed *isohydric* plants, while those where ψ_ℓ declines as soil water potential declines are termed *anisohydric* plants (Tardieu & Simonneau, 1998). Both herbaceous plants and trees show each type of response, with maize and poplar tending to be isohydric while sunflower and almonds are anisohydric. Even different cultivars within a species show variation in the degree to which they are isohydric (Schultz, 2003). There are indications that more drought tolerant species tend towards the more isohydric extreme as isohydric behaviour with sensitive stomatal closure tends to protect against xylem cavitation though it limits photosynthesis under moderate stresses (West *et al.*, 2008). Most species fall between the extreme behaviours illustrated in Figure 6.12

It is probable that the commonly observed close relationship between stomatal conductance and leaf water potential is largely coincidental, arising because leaf water potential tends to decline as the soil dries, while the key signal controlling stomatal closure in response to drought comes from the roots (see Section 4.3.3). Although it is clear from studies with detached leaves that stomata can respond to leaf water status through locally mediated effects on active solute accumulation in the guard cells, there are several lines of evidence indicating that stomatal closure in response to soil drying may often be controlled by factors other than through direct hydraulic control of leaf water status.

An example is shown in Figure 6.11(*b*), where it is clear that as g_ℓ decreased as ψ_ℓ increased, ψ_ℓ could not have been controlling stomata so one must invoke a long-distance signal from the roots (Jones, 1998). Other evidence includes the observation that when the leaf water status of *Helianthus annuus* and *Nerium oleander* plants was manipulated by modifying the whole-plant transpiration by changes in atmospheric humidity, while maintaining fixed environmental conditions for an individual leaf whose conductance was monitored, it was found that leaf conductance of the monitored leaf was much more closely related to soil water status than to leaf water

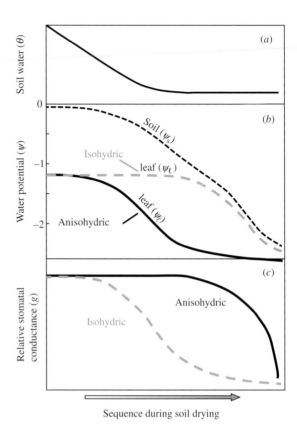

Figure 6.12 Relationships between g_ℓ and ψ_ℓ and ψ_s as soil dries in isohydric and anisohydric plants. (*a*) Progress of soil water content (θ) as it dries; (*b*) corresponding progress of soil water potential (ψ_s – short dashes) and of leaf water potential (ψ_ℓ) in extreme isohydric plants (dashed line) with early stomatal closure and in anisohydric plants (solid line); (*c*) changes in stomatal conductance to achieve the changes in ψ_ℓ shown in (*b*).

status (see Figure 6.13). The hypothesis that leaf conductance could respond directly to soil water status was tested further by pressurising the root system to maintain pressure in the xylem at zero (thus maintaining the leaf cells fully turgid), as soil was allowed to dry (Gollan *et al.*, 1986). Again leaf conductance appeared to respond to soil water content, rather than to leaf turgor (Figure 6.14). The many split-root and partial root-zone drying experiments that have demonstrated stomatal closure when part of the rooting zone has been allowed to dry out while maintaining shoot water

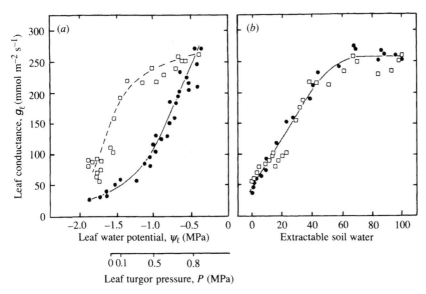

Figure 6.13 (*a*) Relationship between leaf conductance and ψ_ℓ in *Nerium oleander* as soil water content decreased and the leaf–air vapour pressure concentration difference ($Dx_{W\ell}$) was maintained at 10 Pa kPa^{-1} while the whole plant was maintained either in an atmosphere with a constant vapour concentration difference of 10 Pa kPa^{-1} (●) or where the concentration difference was as great as 30 Pa kPa^{-1} (□). (*b*) The data from (*a*) replotted as a function of extractable soil water (Gollan *et al.*, 1985; Schultze, 1986).

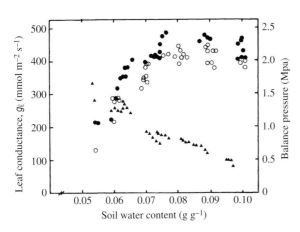

Figure 6.14 Relationship between leaf conductance (g_ℓ) and soil water content for wheat leaves, when the soil was dried either without maintaining leaf turgor by pressurising the system (○) or with maintenance of leaf turgor by applying a balancing pressure (●). The corresponding values of the balancing pressure (*P*) required to maintain turgor at each soil water content are indicated by (▲) (data from Gollan *et al.*, 1986).

status by keeping some of the roots well watered also provide very strong evidence for root–shoot signalling in the control of stomatal behaviour particularly in response to natural soil drying (Davies & Zhang, 1991).

The above observations, however, do not rule out the possibility that local signalling may also be involved in some stomatal control, especially short-term responses such as the transient 'wrong way' initial opening response to leaf excision (Iwanoff, 1928). It is commonly observed that stomatal conductance is better related to ψ_p than to ψ_ℓ with at least some of the change in response to growth conditions (Table 6.2) being related to osmotic adjustment where a given turgor pressure is achieved at a lower water potential as a result of solute accumulation in the cell sap (see Chapter 10). It is likely that a close relationship to bulk leaf ψ_p, where it is observed, is indirect, acting through an effect on ion pumping at the guard cells.

As was pointed out above (Section 6.2) it has been known for some time that the plant growth regulator abscisic acid (ABA) is an important regulator of the osmotic changes in guard cells that control stomatal opening. Water deficits promote ABA synthesis in all tissues so the increased concentration in the leaves could at least partly result from increased local synthesis in response to leaf water deficits. It is now accepted, however, that chemical signalling from roots to shoots can be critical and that ABA transport in the xylem from the roots to the shoots is central to the detection of soil water deficits. The supply of ABA from the roots depends both on the rate of xylem flow (which decreases with increasing stress and stomatal closure) and on the rate of synthesis and release of ABA into the vessels in the root (which depends on ψ_p in the root tissues). Plants have a substantial capacity to sequester ABA in leaf mesophyll cells and elsewhere, so regulation of ABA reaching the guard cells does not depend solely on the supply rate in the xylem but can also be controlled by apoplastic pH. A more acid apoplast (around pH 6.0) favours partitioning of ABA into mesophyll cells so that little reaches the stomata in the transpiration stream, while drought-induced alkalinisation of apoplastic sap favours retention of the ABA anion in the sap, thus facilitating delivery to the guard cells and amplifying the signal (Wilkinson & Davies, 2002).

When a plant is re-watered after a period of drought, the stomata may take some days to recover (depending on the severity and duration of the stress), even though leaf water potential may recover rapidly.

6.4.6 Response to temperature and other factors

Many studies of stomatal temperature responses have yielded contradictory results. Unfortunately, in many earlier studies temperature was confounded with variation in leaf–air vapour pressure difference. It is necessary to conduct temperature response studies at constant values of D_ℓ (under which conditions absolute and relative humidity increase with temperature). In general, stomata tend to open as temperature increases over the normally encountered range, though an optimum is sometimes reached (see Hall et al., 1976). The magnitude of the temperature response does, however, depend on the vapour pressure.

Stomatal aperture is also affected by many gaseous pollutants (see also Section 11.4) such as O_3, SO_2 and nitrogen oxides (Robinson et al., 1998), though the responses are rather complex according to conditions, with the same pollutant often causing either opening or closing according to the particular situation and species. At least in the case of ozone, the stomatal closure in response to pollutant increase is thought to be primarily a response to the increased internal CO_2 concentration as a result of inhibition of photosynthesis (Paoletti & Grulke, 2005), though ozone treatment also appears to reduce the rate of stomatal responses generally. Ozone treatment especially can lead to the development of heterogeneous patterns of stomatal opening ('patchiness') on leaves; the significance of patchiness for photosynthesis will be considered in Chapter 7. Many of the responses to the gaseous pollutants such as SO_2 are probably related to the toxic effects of these substances on membrane integrity and the consequent damage to epidermal cells. Stomatal aperture is also dependent on many other factors such as leaf age, nutrition and disease, while night-time opening can also enhance pollutant damage.

6.4.7 Stomatal behaviour in natural environments

Empirical models

A number of empirical models have been proposed to describe stomatal behaviour in the field. As pointed out earlier, it is difficult to determine stomatal responses from measurements in the field. Typical results for leaf conductance of extension leaves of apple trees during the course of one day are presented in Figure 6.15, together with the irradiance on a horizontal surface. These results are also plotted against irradiance and ψ_ℓ in Figure 6.15(b) and (d). The variability in g_ℓ is a result partly of the fluctuations in irradiance and

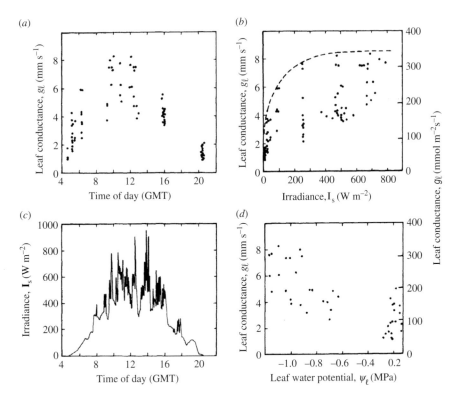

Figure 6.15 Typical variation during one day (26 July 1980) for g_ℓ of mid-extension leaves of Bramley apple leaves (a), together with irradiance above the canopy (c), and g_ℓ as a function of I (b) or ψ_ℓ (d). For details see text. (H. G. Jones unpublished data.)

partly of leaf-to-leaf variability and differences in orientation. Much of the scatter in the plot of g_ℓ against I (Figure 6.15(b)) results from the stomatal time constant being longer than that of the fluctuations of irradiance.

Several empirical approaches can be used to determine the stomatal response to individual factors or to predict g_ℓ for given conditions, using results such as those in Figure 6.15. One approach is to use boundary-line analysis. In Figure 6.15(b) a hypothetical ideal boundary-line relationship between g_ℓ and I is shown as the dashed line, which may approximate the response when no other factors are limiting. It is assumed that many points fall below this line as a result of factors such as lowered ψ_ℓ. Unfortunately this approach is difficult to quantify statistically as the upper points that define the boundary line are measured with some degree of error.

Another widely used approach is that of multiple regression, where g_ℓ is regressed against various independent variables to give an equation of the form:

$$g_\ell = a + bI + cD + d\psi_\ell + \ldots \qquad (6.25)$$

where a, b, c, d, etc. are regression coefficients. Non-significant terms can be omitted and non-linear relations fitted by the use of higher order polynomials (e.g. quadratic). This technique can give useful summaries of large amounts of data, but it is again entirely empirical and thus has rather little predictive value for untried combinations of environmental conditions. The same is true for principal components analysis, where results tend not to be repeatable across data sets. It is frequently necessary to include time of day as an independent variable.

It is worth noting that the 'important' variates found in multiple regression models are often not the same as the individual variates most closely related to g_ℓ when regressed singly. For example, in an analysis of several data sets for apple in England, Jones and Higgs (1989) showed that the most useful three variates in a multiple regression were D, $\mathbf{I}_{50}$ (a derived variate $= \mathbf{I}/(50 + \mathbf{I})$), and T_a, even though the best individual variates were either vapour pressure (e) or relative humidity (h). This latter observation is of particular interest in relation to the suggestion that g_ℓ tends to vary directly in proportion to assimilation rate scaled by the relative humidity at the leaf surface as in the BWB model (Eqs. (6.23) and (6.24)).

In order to get a robust model for predicting g_ℓ, Jones and Higgs (1989) proposed the replacement of the environmental variates by differences (Δ) from typical values so that Eq. (6.25) becomes:

$$g_\ell = g_o + \beta_1 \Delta \mathbf{I} + \beta_2 \Delta D + \beta_3 \Delta \psi_\ell + \ldots \qquad (6.26)$$

where g_o is a reference value (at typical environmental conditions) and the β_i are multiple regression coefficients. For prediction of g_ℓ for new data sets, the sizes of the coefficients can be normalised by dividing through by g_o and rearranging to give:

$$g_\ell = g_o(1 + b_1\Delta\mathbf{I} + b_2\Delta D + b_3\Delta\psi_\ell + \ldots) \qquad (6.27)$$

where the $b_i = \beta_i/g_o$. The model is scaled to different data sets by using an appropriate g_o; absolute errors in predicted g_ℓ that result from incorrect regression coefficients are minimised by this technique. Using this approach it was shown that a model derived for one data set could fit data sets obtained for other orchards in different years as successfully as a freely fitted model. The model using D, T_a and $\mathbf{I}_{50}$ explained between 32 and 62% of the variance in g_ℓ for different sets of apple data (Jones & Higgs, 1989).

In many cases a multiplicative model (rather than the additive model in Eqs. (6.25)–(6.27)) provides the best method for analysing stomatal conductance (Jarvis, 1976). In this case one can write:

$$g_\ell = g_o\, \mathrm{f}(\mathbf{I})\, \mathrm{g}(D)\, \mathrm{h}(\psi_\ell) \ldots \qquad (6.28)$$

where the forms of the individual functions ($\mathrm{f}(\mathbf{I})$, $\mathrm{g}(D)$, $\mathrm{h}(\psi_\ell)$, etc.) are obtained from controlled environment studies. Some particularly useful functions are summarised in Figure 6.16. It is worth noting that for well-watered plants the stomata do not usually close in response to the normal diurnal fall in ψ_ℓ.

In fact an opposite effect is sometimes observed (Figure 6.15(d)). This somewhat paradoxical result (compare with Figure 6.10) arises because here ψ_ℓ is falling as a result of the increased evaporation rate as stomata open, rather than control operating the other way around.

Mechanistic and semi-empirical models

Although the basic mechanisms involved in the control of stomatal aperture are increasingly well understood, it has proved surprisingly difficult to simulate stomatal behaviour in the field. There appear to be a number of reasons including the fact that stomatal control involves both long- and short-distance chemical and hydraulic signalling and because responses to environmental changes occur with differing lags according to age and previous history of the plant.

Perhaps the most successful semi-empirical models are those based on the BWB model (Eqs. (6.23) and (6.24)); when combined with effective submodels of photosynthesis they can provide quite good predictions of stomatal behaviour. A wide range of other models have aimed to incorporate our understanding of chemical signalling (e.g. by ABA) and hydraulic signalling (Tardieu & Simonneau, 1998) and these have been combined with the BWB approach (Dewar, 2002).

6.5 Stomatal resistance in relation to other resistances

Typical values for the components of leaf and boundary layer resistances to water vapour loss from single leaves (see Figure 6.3) are summarised in Table 6.3. Although the cuticular resistance tends to be by far the largest, the total resistance is largely determined by the stomatal component (which is much smaller), because the resistance of two parallel resistors is determined primarily by the smaller.

The high cuticular resistance results from the low water permeability (liquid and vapour) of the

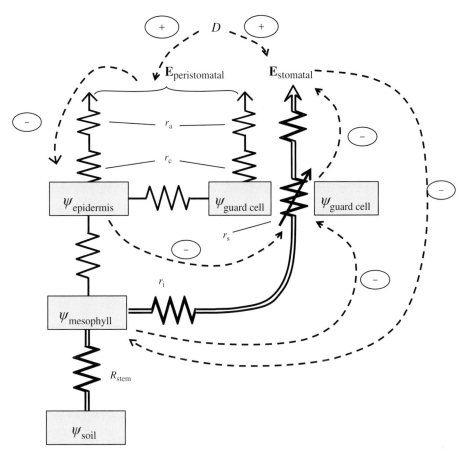

Figure 6.16 A simplified diagram of the various feedback and feedforward control loops involved in humidity effects on stomatal aperture. The solid lines and resistors represent the flows of water with the majority of the flow occurring through (and controlled by) the stomatal pores (double lines) and with only a small proportion of peristomatal evaporation directly through the epidermis or outer guard cell walls (single lines). The dotted lines indicate control processes with the signs indicating the effect of an increase in the first factor on the second (e.g. the negative sign relating $E_{stomatal}$ to $\psi_{mesophyll}$ indicates that an increase in $E_{stomatal}$ results in a decrease of $\psi_{mesophyll}$). The resistances shown include the intercellular space resistance (r_i), the stomatal resistance (r_s), the boundary layer resistance (r_a) and the cuticular resistance (r_c) and the liquid phase stem resistance (R_{stem}).

hydrophobic cuticle and the overlying wax layer. The thickness, composition and morphology of the cuticle and wax layers is very dependent on species and growth conditions (Goodwin & Jenks, 2005; Nawrath, 2006), being most developed (with highest r_c) in plants from arid sites where water conservation is most crucial. All cuticles consist of two highly lipophilic materials: cutin, a tough and rigid complex cross-linked polymer of hydroxy- and epoxy- C_{16} and C_{18} fatty acids, and wax, a mixture

of straight chain monomeric C_{20} to C_{60} cuticular waxes. The waxes may vary in form from smooth layers or platelets to rods or filaments several micrometres long. Some epidermal wax structures are shown in Figure 6.2(e).

Mathematical and physical models of gas diffusion in the intercellular spaces suggest that most transpired water originates from cell walls close to the stomatal pore, so that r_{iW} is small (probably less than 0.5 m^2 s mol^{-1}) (Jones, 1972;

Table 6.3 Relative values of the different water vapour resistances (see Figure 6.3) and the corresponding conductances for single leaves. The leaf resistance (r_ℓ) is dominated by the stomatal component.

Component		Resistance		Conductance	
		r^m	r	g^m	g
		$(m^2 \text{ s mol}^{-1})$	$(s \text{ cm}^{-1})$	$(mmol \text{ m}^{-2} \text{ s}^{-1})$	$(mm \text{ s}^{-1})$
Intercellular space and wall	$(r_a + r_w)$	<1	<0.4	>1000	>25
Cuticular	(r_c)	50->250	20->100	4–20	<0.1–0.5
Stomatal	(r_s)				
at minimum r for many succulents, xerophytes and conifers		5–25	2–10	40–200	1–5
at minimum r for mesophytes		2–6	0.8–2.4	170–500	4–13
at maximum r when closed		>125	>50	8	<0.2
Boundary layer	(r_a)	0.25–2.5	0.1–1	400–4000	10–100

Tyree & Yianoulis, 1980). Because the sites of CO_2 assimilation are more evenly distributed throughout the leaf, the intercellular space conductance to CO_2 (r_{iC}) may be significant and perhaps as much as 2.5 m^2 s mol^{-1}.

There has been controversy about the magnitude of the 'wall' resistance, r_w, since Livingston and Brown (1912) reported evidence for non-stomatal control of water loss. They envisaged that the evaporation sites retreated into the cell walls thus increasing the gas-phase diffusion path, a process they called 'incipient drying'. An alternative possibility is that the effective water vapour concentration at the liquid–air interface is significantly below saturation at leaf temperature. From Eq. (5.14), the relative lowering of the partial pressure of water vapour (e) compared with that at saturation (e_s) is related to water potential by:

$$e/e_s = \exp(\overline{V}_w/\mathscr{R}T) \qquad (6.29)$$

Application of this equation predicts that at 20°C, e/e_s would be 0.99 at –1.36 MPa, and only falls to 0.95 at –6.92 MPa, a figure found only in the most severely stressed leaves. For a relative humidity of 0.50 in the ambient air, even this latter figure would introduce less than a 10% error in the driving force for evaporation (or equivalently r_w would be less than 10% of the total resistance).

In addition to effects of bulk leaf water potential on ψ at the evaporating surface, ψ could also be lowered by accumulation of solutes, or because of the presence of a large internal hydraulic resistance. Experimental results for several species and theoretical calculations (see Jones & Higgs, 1980) all suggest that r_w is small (<<1 m^2 s mol^{-1}) at physiological water contents.

Table 6.3 shows that, for single leaves, r_ℓ is normally at least an order of magnitude greater than r_a. In canopies, however, the relative importance of the boundary layer resistance increases. This is because all the individual leaf resistances are in parallel and therefore the total leaf resistance decreases as leaf area index increases. In addition, for a crop, the boundary layers for the individual

Table 6.4 Canopy leaf (or physiological) resistance to water vapour (r_L) and canopy boundary layer resistance (r_A) for different types of vegetation. Values in mass units (s m^{-1}) and approximate equivalents in molar units (m^2 s mol^{-1}) are given.

	r_L per unit ground area		r_L per unit ground area		r_A per unit ground area	
	s m^{-1}	m^2 s mol^{-1}	s m^{-1}	m^2 s mol^{-1}	s m^{-1}	m^2 s mol^{-1}
Grassland/heathland	100	2.5	50	1.25	50–200	1.25–5
Agricultural crops	50	1.25	20	0.5	20–50	0.5–1.25
Plantation forest	167	4.2	50	1.25	3–10	0.08–0.25

leaves must be added to an overall crop boundary layer (see Figure 5.4). Some typical ranges for the values of the canopy boundary layer resistance (r_A) and the canopy leaf resistance (r_L) are given in mass units (because these are more commonly used for canopy studies) together with the equivalents in molar units in Table 6.4. The values shown indicate that the ratio between r_A and r_L can vary widely between different plant communities. Some implications of differences in this ratio between grassland and aerodynamically rough canopies such as tall forests, were discussed in Section 5.3.4.

6.6 Stomatal function and the control loops

The main control systems involved in the regulation of stomatal aperture are related to the fluxes of water vapour and CO_2 that need to be controlled (Figure 6.1).

6.6.1 Water control loop

The main function of stomata appears to be the control of water loss. In general stomata respond to factors that lower ψ_ℓ in such a way as to minimise further increases in stress. The responses involve either *feedback* via leaf water status itself, or else direct *feedforward* control (Figure 6.1). Feedback

occurs where the output of a controller feeds back to affect the controller (either positively or negatively), while feedforward occurs where the control signal is independent of the output. The distinction between feedback and feedforward can be illustrated by the possible effects of altered D_ℓ (see Figure 6.16).

Negative feedback is illustrated by the control pathway in the right hand side of Figure 6.16. Here a decrease in air humidity (increased D) increases **E**, since:

$$\mathbf{E} = (2.17/T)g_\ell D_\ell \tag{6.30}$$

This in turn tends to lower bulk ψ_ℓ because, from Eq. (4.26) (representing the sum of all the soil and plant resistances by R_{sp}):

$$\psi_\ell = \psi_s - R_{sp}\mathbf{E} \tag{6.31}$$

This lowered ψ_ℓ then leads to stomatal closure, which finally has a negative feedback effect on **E**. Assuming a linear relationship (Figure 6.17(d)):

$$g_\ell = a + b\psi_\ell \tag{6.32}$$

and, combining this with Eq. (6.31) gives:

$$g_\ell = a + b(\psi_s - R_{sp}\mathbf{E}) = c - d\mathbf{E} \tag{6.33}$$

where a, b, c and d are constants. From Eqs. (6.30) and (6.31) the relationships describing this negative feedback are:

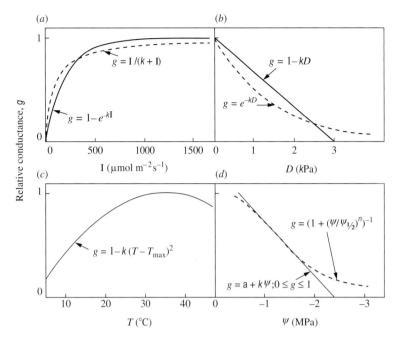

Figure 6.17 (*a*) to (*d*) The most useful functions for describing stomatal response to environment (solid lines), and other useful functions (dashed lines). *a, k, n* are constants, as are T_{max} (*T* at which *g* is maximal) and $\psi_{1/2}$ (the value of ψ at when *g* is half maximum). All functions give *g* as a fraction of a reference value. Details of the ψ functions are discussed by Fisher *et al.* (1981).

$$\mathbf{E} = cD_{\ell}/(1 + dD_{\ell}) \tag{6.34}$$

$$g_{\ell} = c/(1 + dD_{\ell}) \tag{6.35}$$

and are illustrated by the solid lines in Figure 6.18. Note that Eq. (6.34) is a saturation-type curve and that negative feedback cannot cause a steady-state reduction in transpiration (dotted curve in Figure 6.18) as D_{ℓ}, increases. (Any reduction in **E** implies increases in ψ_{ℓ} and hence g_{ℓ}, which would immediately restore **E**.) A feedback loop with a high *gain* would tend to maintain **E** relatively constant. However, a high gain can lead to instability and regular stomatal oscillations where there are time delays in the responses (Cowan, 1977).

Feedforward is illustrated in the left side of Figure 6.16. In this case the environment affects the controller (the stomata) directly *without* depending on changes in the flux that is being controlled (i.e. evaporation through the stomata). An increased D_{ℓ} increases the rate of peristomatal transpiration (water loss not passing through the stomatal pore). This would potentially lead to

lowered guard-cell turgor and consequent stomatal closure, as long as the hydraulic flow resistance in the pathway to the evaporating sites on the outer surface of the guard cells is great enough (see Buckley, 2005). Using an argument similar to that used to derive Eq. (6.35), it is possible to show that in this case (cf. Figure 6.17):

$$g_{\ell} = e - fD_{\ell} \tag{6.36}$$

where *e* and *f* are constants. Combining with Eq. (6.30) and rearranging gives:

$$\mathbf{E} = e'D_{\ell} - f'(D_{\ell})^2 \tag{6.37}$$

Depending on the relative values of the new constants e' and f', **E** may even fall with increased evaporative demand (dotted curve in Figure 6.18). Such responses have been observed in several species. In practice, both feedback and feedforward responses usually occur together. Further discussion of feedback and feedforward in this context may be found in articles by Farquhar (1978) and Buckley (2005).

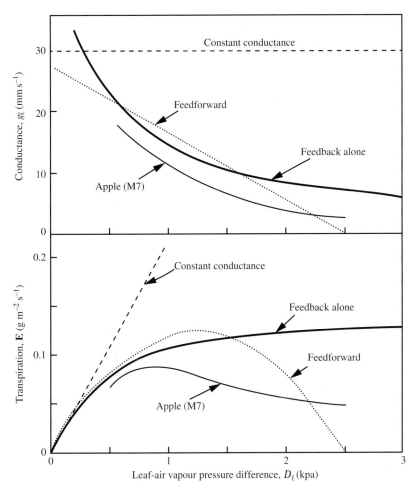

Figure 6.18 Types of stomatal humidity response and the consequent relation between **E** and D_ℓ. The thick solid lines show negative feedback (Eqs. (6.34) and (6.35)), the dotted lines show feedforward (Eqs. (6.36) and (6.37)), while the dashed line indicates a constant g_ℓ. Data for M7 apple from Figure 6.10 are shown for comparison.

Figure 6.18 also shows the expected linear response of **E** to D_ℓ when there is no stomatal response (dashed curve).

The general processes involved in stomatal response to guard cell or mesophyll water status were outlined above, and it was pointed out that although hydropassive movements can occur, and that they may be responsible for the very rapid movements observed on detaching leaves, active changes in ion pumping and osmoregulation are much the more important and probably underlie even the local hydraulic feedbacks discussed here.

6.6.2 Carbon dioxide control loop

The central role of the stomata in the control of photosynthesis depends on sensing the photosynthetic rate. Most evidence indicates that the photosynthetic feedback control (Figure 6.1) depends on sensing the intercellular space CO_2 concentration. For example, as light increases, the rate of CO_2 fixation increases, thus lowering the intercellular CO_2 concentration, and the stomata open to compensate. Other changes that reduce the intercellular space CO_2 generally have a similar effect. Such a mechanism appears to underlie the stomatal responses to photosynthesis described by the BWB model (Section 6.4.3).

There is a strong feedback control that tends to maintain intercellular CO_2 remarkably constant (at a mole fraction of around 130×10^{-6} in C_4 species and 230×10^{-6} in C_3 species – see Chapter 7). The night-time stomatal opening in CAM plants can also be explained on the same basis: dark CO_2 fixation at night lowers intercellular CO_2 and stomata open.

6.7 Sample problems

6.1 A leaf detached from a plant and suspended in a moving airstream ($h = 0.2$) initially loses mass at a rate of 80 mg m^{-2} s^{-1}, falling to a constant rate of 2 mg m^{-2} s^{-1} after about 20 min. A piece of wet blotting paper suspended alongside the leaf loses water at a rate of 230 mg m^{-2} s^{-1}. Assuming that leaf, air and blotting paper are all at 20°C, calculate (i) the boundary layer resistance, r_{aW}; (ii) the cuticular resistance, r_{cW}; and (iii) the initial value of the stomatal resistance, r_{sW}, if the leaf has equal numbers of stomata on each surface.

6.2 Calculate (i) the stomatal diffusion resistance (r_{sW}) and (ii) the corresponding conductance (g_{sW}) for a leaf with 200 stomata mm^{-2} on each surface, each pore being 10 μm deep and circular in cross-section ($d = 5$ μm).

6.3 A continuous-flow porometer with a chamber area of 1.5 cm^2 is attached to the lower surface of a leaf. If the flow rate is 2 cm^3 s^{-1}, and the relative humidity of the outlet air is 35% (inlet air is dry), calculate $g_{\ell W}$ assuming (i) that $T_\ell = T_a$, or (ii) that $T_\ell = 25°C$ and $T_a = 27°C$.

6.4 Assuming that g_ℓ for a particular species decreases linearly from 10 mm s^{-1} at $D = 0$ kPa to 0 at $D = 3$ kPa: (i) plot a graph of the relationship between E and D; what is E when $D = 1$ kPa? (ii) If g_ℓ is also sensitive to ψ_ℓ falling by 50% per MPa below zero, plot the dependence of E on D if ψ_ℓ falls linearly at a rate of 1 MPa for each 0.1 g m^{-2} s^{-1} increase in evaporation; what is E when $D = 1$ kPa?

7 Photosynthesis and respiration

Contents

The most important characteristic of plants is their ability to harness energy from the sun to 'fix' atmospheric CO_2 into a range of more complex organic molecules. This process of photosynthesis provides the major input of free energy into the biosphere: some of the free energy stored in these photosynthetic assimilates is then transferred in the process of respiration to high-energy compounds that can be used for synthetic and maintenance processes. The net rate of photosynthetic CO_2 fixation by a photosynthesising plant (net photosynthesis, $\mathbf{P_n}$) is the instantaneous difference between the gross rate of CO_2 fixation by the photosynthetic enzymes ($\mathbf{P_g}$) and the rate of respiratory CO_2 loss ($\mathbf{R}$). Because of the central nature of these two processes to all aspects of plant growth and adaptation, they will be discussed in some detail in this chapter.

7.1 Photosynthesis

The overall reaction of photosynthesis can be represented by:

$$CO_2 + 2H_2O + light \rightarrow CO_2 + 4H + O_2$$
$$\rightarrow (CH_2O) + H_2O + O_2 \tag{7.1}$$

The net effect is the removal of one mole of water and the production of one mole of O_2 for every mole of CO_2 that is reduced to the level of sugar ($(CH_2O)_6$). Many of the individual reactions take place in specialised organelles called chloroplasts within the leaf mesophyll cells (Figure 7.1). These are bounded by a double membrane and contain a network of vesicles called thylakoids arranged either as single lamellae or stacked up to form characteristic granal stacks. The ground material of the chloroplast is called the

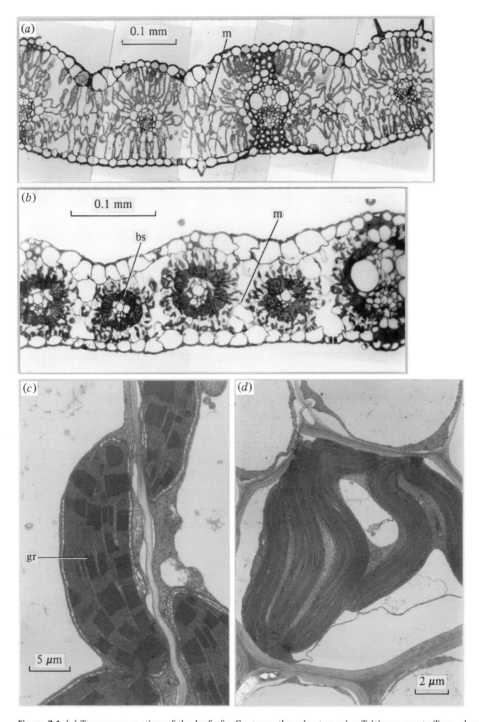

Figure 7.1 (a) Transverse section of the leaf of a C$_3$ grass, the wheat species *Triticum urartu* Tum. showing the photosynthetic mesophyll cells (m); (b) transverse section of a leaf of the C$_4$ grass millet (*Pennisetum americanum*) showing 'Krantz' anatomy with the distinct mesophyll (m) and bundle-sheath (bs) cells; (c) electron micrograph of mesophyll chloroplast from (b) showing the photosynthetic lamellae and granal stacks (gr); (d) an agranal chloroplast from the bundle-sheath of (b). Courtesy of Dr M. L. Parker, Plant Breeding Institute, Cambridge.

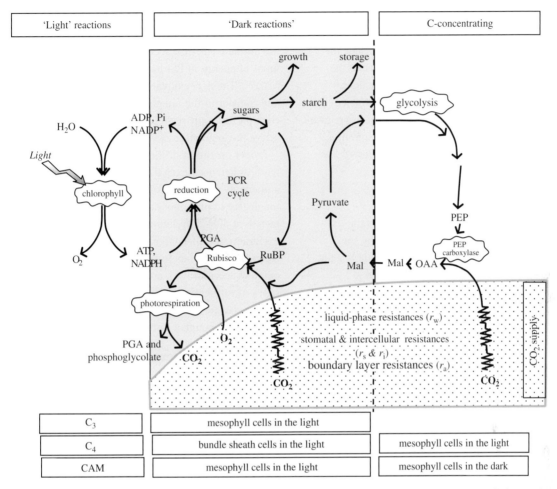

Figure 7.2 A schematic illustration of the photosynthetic process in plants. The left-hand panel shows the 'light reactions' where sunlight is absorbed by chlorophylls and accessory pigments in the chloroplast lamella membranes causing the generation of reducing power and ATP. These high-energy compounds are then used to drive the 'dark reactions' (centre panel) where CO_2 is fixed by the enzyme Rubisco and converted to sugars. In C_3 plants this occurs in the mesophyll cells in the light. The supply of CO_2 from the atmosphere through the stomata and into the mesophyll cell chloroplasts is illustrated in the lower part of the main diagram, which shows the CO_2 uptake resistances in the boundary layer, the stomata and the intercellular spaces, together with the photorespiratory pathway where oxygenation of RuBP leads to loss of CO_2 with some PGA returned to the Calvin cycle. The right-hand panel shows the carbon concentrating mechanisms used to enhance the availability of CO_2 to the mesophyll chloroplasts either by spatial separation as in C_4 plants or by temporal separation as in CAM plants (see text). The temporal and physical separation of processes in C_3, C_4 and CAM photosynthesis is summarised at the extreme bottom.

stroma while the lumen is the compartment within the lamellae. Details of the biochemistry and physiology of photosynthesis can be found in appropriate texts (Falkowski & Raven, 2007; Hall & Rao, 1999; Lawlor, 2001; Taiz & Zeiger, 2010) and in useful web resources such as http://bioenergy.asu.edu/photosyn/ photoweb/index.html and http://www.biologie. uni-hamburg.de/b-online/e00/index.htm.

As is illustrated in Figure 7.2, photosynthesis can be conveniently treated as three separate components: (i) light reactions, in which radiant energy is absorbed and used to generate the high-energy compounds adenosine triphosphate (ATP) and reduced nicotinamide adenine dinucleotide phosphate (NADPH); (ii) light-independent (or dark) reactions, which include the biochemical reduction of CO_2 to sugars using the high-energy compounds generated in the light reactions; and (iii) supply of CO_2 from the ambient air to the site of reduction in the chloroplast. It is also relevant to consider the linked process of photorespiration, which will be discussed in Section 7.2 below.

7.1.1 Light reactions

The primary process in photosynthesis is the absorption of incoming solar radiation by the pigments located on the grana and stroma lamella membranes of the chloroplasts. The main pigments, the chlorophylls, are most effective at absorbing in the red and blue (Figure 2.4) while carotenoids and other accessory pigments permit the absorption of other wavelengths in the PAR (400–700 nm). In green plants the chlorophylls are embedded in three chlorophyll–protein complexes: the light harvesting protein complexes (LHC), the photosystem I antenna complex (PSI) and the photosystem II antenna complex (PSII) (details of these structures have been reviewed by Nelson and Yocum, 2006). The complete electron transport pathway is represented schematically in Figure 7.3.

Radiation absorption in the antenna pigments causes excitation of electrons in the pigment molecules, with excitation being funnelled via the light harvesting protein complexes to one of the two specialised 'reaction centres' (P_{680} in PSII and P_{700} in PSI) by resonance transfer, a process that can occur where the energy available corresponds to that required for excitation in the receptor. This energy transfer results in the excitation of the chlorophyll in the reaction centre and charge separation with transfer of an electron to a nearby acceptor molecule and subsequent energetically 'downhill' transfers along an electron transfer chain located in the thylakoid membranes. In PSII the oxidised P_{680} is returned to a ground state by accepting an electron from the splitting of water, thus releasing protons (H^+) into the thylakoid and molecular oxygen, while the ultimate electron acceptor is CO_2 after a second input of energy in PSI has been used to reduce $NADP^+$. The net removal of protons from the stroma and transfer to the lumen during the whole-chain electron transport sets up a proton gradient that drives the generation of ATP.

Proper operation of the photosynthetic system requires balanced excitation of the two photosystems and energy transfer between the two is regulated by the degree of phosphorylation of LHC protein, with increased phosphorylation increasing the transfer of excitation to PSI. Charge separation at the PSII reaction centre leads to the reduction of the primary quinone acceptor (Q_A), and hence transport through a series of further electron acceptors to PSI where a further charge separation occurs with the electrons transferred through a further series of electron carriers (Q_B, the cytochrome b_6f complex, plastocyanin, etc.) to reduce $NADPH^+$ to NADPH.

It is thought that the major pathway of electron flow is the non-cyclic pathway from H_2O to $NADP^+$ as shown in the classic Z-scheme in Figure 7.3. In this pathway one molecule of O_2 and two molecules of $NADP^+$ are reduced to NADPH for each four electrons that flow, thus requiring eight photons (four for each photosystem). The coupling of this electron flow to the generation of ATP involves Mitchell's chemi-osmotic mechanism where the reaction centres and redox carriers in the electron transport pathway are asymmetrically arranged in the membranes such that electron transport gives rise to transport of H^+ from the stroma to the lumen inside the thylakoid thus decreasing the pH of the lumen. The return flow of H^+ then drives the production of ATP by its linkage to a membrane-bound ATPase. It is not certain how many ATP molecules can be generated by non-cyclic electron flow, but it is thought to be somewhere between 1 and 2 ATP per 2 e^-. The electron pathway is somewhat flexible, with cyclic and pseudocyclic pathways that lead to ATP production without reduction of $NADP^+$ being possible, as also occurs in the direct reduction of oxygen (the Mehler reaction).

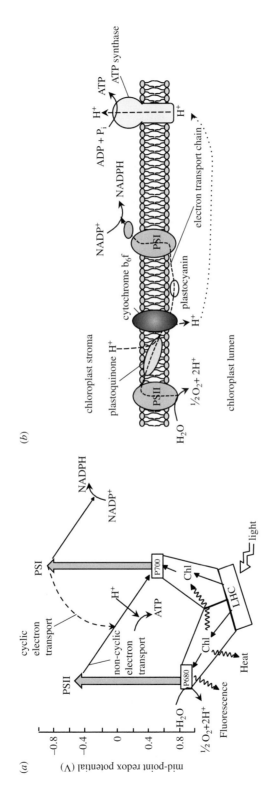

Figure 7.3 (*a*) Schematic representation of the redox potentials at different stages in the photosynthetic electron transport chain (the 'Z'-scheme) showing light trapping by the major chlorophyll protein complexes and energy transfer to the reaction centres of the two photosystems with fluorescence from PSII, energy transfer to PSI and dissipation as heat competing with photochemistry at PSII. (*b*) Schematic representation of the electron transfer chain in the thylakoid membrane from water splitting at PSII to the reduction of NADP$^+$ to NADPH, together with the resulting phosphorylation as H$^+$ ions transferred to the lumen are transported out from the lumen through the ATPase. (LHC = light harvesting complex, Chl = chlorophyll, P680 and P700 are the reaction centres of photosystems II and I, respectively.)

Blue light is about 1.5 times as energetic as red (i.e. 4.62×10^{-19} J quantum^{-1} at 430 nm compared with 3.00×10^{-19} J quantum^{-1} at 662 nm), but because the upper singlet state of chlorophyll (excited by blue light) is very unstable, decaying rapidly to the lower excited state with the generation of heat, blue quanta tend to be no more effective than red (Figure 2.4). In addition to driving electron transport with the consequent generation of ATP and NADPH, the energy released when excited electrons return to the ground state can also be lost as heat or else it may be lost by re-radiation (a process called *fluorescence*). At room temperature in vivo the majority of fluorescence arises from chlorophyll *a* in PSII and this fluorescence can be used as a powerful probe of photosynthetic functioning (Baker, 2008). Of the PAR incident on the leaf (**I**), only a fraction (α_{leaf}, usually assumed $\simeq 0.84$) is absorbed by the leaf; of this only a fraction (f_{PSII}, often assumed $\simeq 0.5$) is transferred to PSII with the remainder going to PSI. Considering only the absorbed PAR that is received by PSII ($= \mathbf{I}\,\alpha_{leaf}\,f_{PSII}$) we need to recognise that fluorescence of PSII is only one of a number of competing processes for de-excitation of these excited chlorophyll molecules. The probability that an excited chlorophyll molecule will de-excite by fluorescence (equal to the quantum yield for PSII fluorescence, ϕF = number of quanta emitted as fluorescence (F)/number of quanta absorbed by PSII ($\mathbf{I}\,\alpha_{leaf}\,f_{PSII}$)) is given by the ratio of the rate constant for fluorescence to that of all competing processes:

$$\phi F = F/(\mathbf{I}\,\alpha_{leaf}f_{PSII}) = k_F/(k_F + k_H + k_P f_{open}) \quad (7.2)$$

where k_F, k_H and k_P, respectively, are the rate constants for de-excitation through fluorescence, thermal dissipation as heat and PSII photochemistry with all reaction centres open, and f_{open} is the fraction of PSII reaction centres that are open (see Section 7.4.1). Normally ϕF is of the order 0.01 to 0.02.

7.1.2 Dark reactions

Plants can be classified into at least three major groups (C_3, C_4 and CAM) on the basis of differences in the biochemical pathway by which they fix CO_2. The characteristic anatomical differences between C_3 and C_4 species are illustrated in Figure 7.1(*a,b*), while the characteristics of each pathway are listed in Table 7.1 and will be discussed in detail in following sections. The essential features of the three pathways are summarised in Figure 7.2.

C_3 pathway

C_3 plants use the enzyme ribulose bisphosphate carboxylase-oxygenase (*Rubisco*) for the primary fixation of CO_2 in the chloroplast to form the 3-carbon compound 3-phospho-glyceric acid (PGA), which is then converted to triose phosphate using ATP and NADPH. Most of the triose phosphate then takes part in a complex reaction sequence (the photosynthetic carbon reduction (PCR) or Calvin cycle) that requires further ATP to regenerate the substrate (ribulose-1,5-bisphosphate, RuBP) for the initial carboxylation reaction. Some of the triose is siphoned off as the net product of photosynthesis to form sugar phosphate (fructose-1,6-bisphosphate) and sugars. This cycle requires 3 ATP and 2 NADPH per CO_2 converted to sugar phosphate. The C_3 pathway is the dominant photosynthetic pathway of species from cool, temperate or moist habitats and is the only pathway found in trees (with very few exceptions), or lower plants. The majority of crop plants use the C_3 pathway, including all the temperate cereals (wheat, barley, etc.), root crops (e.g. potato and sugar beet) and leguminous species (beans, etc.).

An important feature of Rubisco is that it can catalyse either the carboxylation of RuBP as in photosynthesis, or its oxidation to phosphoglycolate in the process of photorespiration. At high concentrations of CO_2 carboxylation dominates, but as the O_2 concentration increases or CO_2 decreases the competition between oxidation and carboxylation of RuBP leads to increased photorespiration. As a consequence of photorespiration, C_3 plants tend to have a CO_2 compensation concentration (Γ, the CO_2 concentration at which net CO_2 exchange is zero) between about 30 and 80 vpm.

Table 7.1 Characteristics of the main photosynthetic pathways (data collated from a variety of sources).

	C_3	C_4	CAM Day	CAM Night
Anatomy				
'Kranz' anatomy (distinct bundle sheath)	No	Yes	No (Succulent)	
Frequency of leaf bundles	Low	High	Low	
Leaf air space volume (%)				
monocots	10–35%	<10%	Low	
dicots	20–55%	<30%	Low	
Biochemistry				
Early products of ^{14}C fixation	PGA	C_4 acids	PGA	C_4 acids
Primary carboxylase	Rubisco	PEPCase	Rubisco	PEPCase
Discrimination against ^{13}C ($\delta^{13}C$, ‰)	–22 to –40	–9 to –19	C_3-like	C_4-like
Absolute sodium requirement	No	Yes	No	Yes
Physiology				
CO_2 compensation point (Γ, vpm)	30–80	<10	c. 50	<5
Post-illumination burst of CO_2	Yes	Slight	Yes	–
Enhancement of $\mathbf{P}_n$ in low O_2	Yes	No	Yes	No
Quantum requirement	15–22	19	–	–
Internal (liquid-phase) resistance				
r_i' (m^2 s mol^{-1})	7–15	1.2–5	c. 20	?
r_i' (s cm^{-1})	3–6	0.5–20	c. 8	?
Relative stomatal sensitivity to environment	Insensitive	Sensitive	Reversed cycle	
Intercellular space CO_2 partial pressure	~0.7 p_a'	~0.4 p_a'	~0.5 p_a'	?
Maximum photosynthetic rate (μmol m^{-2} s^{-1})	14–40	18–55	6	8
Maximum photosynthetic rate (mg CO_2 m^{-2})	0.6–1.7	0.8–2.4	0.25	0.3
Optimum day temperature (°C)	c. 15–30 (Wide acclimation)	25–40	c. 35 (Needs low night temperature)	
Light response saturating well below full sunlight	Usually	Rarely	Usually	–

Table 7.1 (*cont.*)

	C_3	C_4	CAM Day	CAM Night
Ecology				
Regions where commonest	Temperate	Tropical, Arid	Arid	
Transpiration ratio	High	Low	Medium	Very low
(g H_2O lost/g CO_2 fixed)	450–950	250–350	50–600	<50
Max. growth rate (g m^{-2} day^{-1})	34–39	51–54	7	
Average productivity (tonne ha^{-1} yr^{-1})	*c.* 40	60–80	Low	

Despite its importance for photosynthesis, Rubisco is rather slow and inefficient and catalyses only 2 to 12 carboxylations per catalytic site per second. This results in the need for a large proportion of leaf protein to be allocated to Rubisco to maintain photosynthesis; indeed Rubisco comprises around 50% of soluble leaf protein in C_3 plants and includes about 20 to 30% of leaf nitrogen (Evans, 1989). The enzyme itself is composed of two types of subunit (large and small) and it needs to bind Mg^{2+} for its activity. Full activity in plants also requires an enzyme, Rubisco activase, which maintains the enzyme in an active (carbamylated) form and protects the enzyme from a range of different inactivating events including binding by the inhibitor 2-carboxy-D-arabitinol 1-phosphate (Ca1P) which can accumulate in the dark.

C_4 pathway

In this pathway, the initial carboxylation reaction occurs in the mesophyll cells and involves phospho*enol*pyruvate carboxylase (PEP carboxylase) fixing in this case bicarbonate (HCO_3^-) rather than CO_2 and producing the four-carbon compound oxaloacetate (OAA) as the first product of fixation, with other 4-carbon compounds (especially malate and aspartate) also being formed very rapidly. The next stage in these plants is the transfer of these 4-carbon compounds to specialised 'bundle sheath

cells' (see Figure 7.1(*b*)) where they are decarboxylated. The CO_2 released is then refixed in the bundle sheath cells using the normal C_3 enzymes of the PCR cycle. The initial fixation by PEP carboxylase in the mesophyll cells acts as a CO_2 'concentrating' mechanism because PEP carboxylase has a higher effective affinity for CO_2 (calculated for CO_2 in equilibrium with the actual substrate, HCO_3^-) than does Rubisco, partly because O_2 is not a competitive substrate and because at the pH of the cytosol the equilibrium between CO_2 and HCO_3^- (catalysed by carbonic anhydrase) favours HCO_3^-.

C_4 plants have been subdivided into groups depending on the C_4 acid translocated to the bundle sheath (aspartate or malate) and on the basis of the decarboxylation enzyme employed (NADP-malic enzyme, NAD-malic enzyme or PEP carboxykinase), but as these three types are physiologically and ecologically rather similar, they will be treated together. In addition there is evidence that there is some flexibility in the balance between the translocation and decarboxylation mechanisms according to the environment (Furbank, 2011). As will be discussed in later sections, the C_4 pathway has particular adaptive value in hot, dry environments and is common in species from tropical and semi-arid habitats including the cereals maize, millet and sorghum. An important feature of C_4 plants is that their

CO_2 compensation concentration is usually close to zero as a consequence of their lack of photorespiration.

The C_4 pathway is a striking example of convergent evolution as it appears to have evolved over 60 times in 19 families of vascular plants (including both monocots and dicots). In addition there are at least 21 lineages (mostly in the dicots) that have a C_3–C_4 intermediate type of photosynthesis (including well-known examples from *Moricandia*, *Flaveria* and *Panicum*) (Sage *et al.*, 2011). These intermediates usually have a CO_2 compensation concentration that tends to be intermediate between C_3 and C_4 species, though only some have advanced to the stage of having enhanced PEP carboxylase activity or even an active C_4 cycle. Although the capacity for photorespiration appears to be similar in C_3 and C_3–C_4 intermediate species, there is a difference in the distribution of the enzyme (glycine decarboxylase) that catalyses the release of CO_2 in photorespiration. In the C_3–C_4 intermediates (e.g. *Moricandia arvensis*) that have been studied it is only present in the mitochondria of bundle sheath cells, while in C_3 species it is present in mitochondria of all chloroplast-containing cells (Hylton *et al.*, 1988). This localisation of the photorespiratory CO_2 release at the inner wall of the bundle sheath cells probably explains the very effective light-dependent recapture of photo-respiratory CO_2 in these species because any CO_2 released must diffuse out through the overlying chloroplasts. This may have been an early step in the evolution of the C_4 pathway.

Crassulacean acid metabolism (CAM pathway)

In many ways the CAM pathway (Figure 7.2) resembles the C_4, in that CO_2 is initially fixed into C_4 compounds using PEP carboxylase and is subsequently decarboxylated and refixed by Rubisco (Ting, 1985). In this case, however, the initial carboxylation occurs during the dark night period when large quantities of C_4 acids accumulate in the mesophyll cell vacuoles. During the day, malic enzyme acts to decarboxylate the stored malate thus providing CO_2 as a substrate for the normal C_3 enzymes. In CAM plants, the two carboxylation

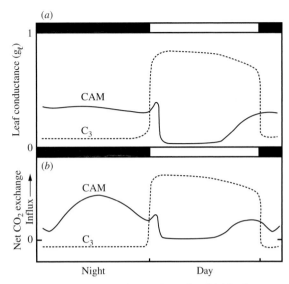

Figure 7.4 Characteristic day–night cycles of (*a*) leaf conductance and (*b*) net CO_2 exchange in C_3 and CAM plants. The early-morning and late-afternoon opening and net photosynthesis in the CAM plant disappear when water stressed.

systems occur in the same cell but are separated temporally, while in C_4 plants the two carboxylation systems operate at the same time, but are spatially separated in different cells. A consequence of the night-time activity of the primary carboxylase is that in CAM plants the stomata tend to be open during the night and closed during the day (Figure 7.4). This is clearly advantageous in terms of water conservation. The CAM pathway is usually found in succulent plants, such as the cacti, that occur in arid areas. The few economically important CAM plants include *Ananas comosus* (pineapple), *Agave sisalana* (sisal) and *Agave tequilana* (tequila).

Although some plants such as *Opuntia basilaris* always use the CAM pathway, a number of species (e.g. *Agave deserti*) are facultative CAM plants in that they can operate either as C_3 plants when water supply is adequate or may have a varying expression of CAM activity depending on the environment. This may range from full CAM expression to an intermediate form called 'idling', where there is no net night fixation of CO_2 but some vacuolar acidification occurs. The switch between C_3 and CAM activity appears to be controlled by water stress (von Willert et al., 1985).

It is notable that CO_2 concentrating mechanisms are not confined to higher plants with a number of different systems being found also in the cyanobacteria and the algae (Falkowski & Raven, 2007).

7.1.3 Carbon dioxide supply

The CO_2 that is fixed in photosynthesis must diffuse from the ambient air (where the partial pressure averages nearly 39 Pa – equivalent to a volume fraction of approximately 390 ppm) through a series of resistances to the carboxylation site (Figure 7.2). The first part of the pathway through the leaf boundary layer, the stomata and the intercellular spaces to the mesophyll cell walls is in the gas phase and is very similar to the pathway for water vapour loss in transpiration. In order to derive the gas-phase resistance to CO_2 uptake from measurements of H_2O loss it is necessary to correct for the differences between the diffusion coefficients for CO_2 and water vapour as discussed in Chapter 6, while it should be noted that the greater dispersion of the sinks for CO_2 than the sources of H_2O (the latter generally closer to the stomatal pore) also tends to increase the intercellular space resistance to CO_2 (Parkhurst, 1994). The remainder of the transport pathway through the cell wall to the carboxylation site in the chloroplasts is in the liquid phase. This will be discussed in detail in Section 7.5.1 below.

7.2 Respiration

Of the CO_2 fixed each day by photosynthesis, a very large proportion (30 to 50%) may be released back to the atmosphere in respiration, with at least half of this quantity attributable to respiration in the leaves.

7.2.1 Dark respiration

The oxidation of carbohydrate to CO_2 and H_2O in living cells is generally termed respiration. Details of respiratory metabolism in plants may be found in

many reviews (e.g. Atkin & Macherel, 2009; Foyer et al., 2009; Millar et al., 2011; Moore & Beechey, 1987; Taiz & Zeiger, 2010) while a more general review of respiration in relation to productivity may be found in Amthor (1989). In photosynthetic organisms there are two main types of respiration. The first is often called 'dark' respiration ($\mathbf{R}_d$; though confusingly dark respiration can also occur at the same time as photosynthesis during the light, though generally at a lower rate (e.g. Krömer, 1995; Wang et al., 2001)) and includes various pathways of substrate oxidation such as glycolysis, the oxidative pentose phosphate pathway and the tricarboxylic acid (TCA or Krebs) cycle (Figure 7.5) that conserve some of the free energy in carbohydrate in the high-energy bonds of ATP, reduced pyridine nucleotide (NADH) and $FADH_2$. The term dark respiration also covers the further oxidation of NADH and $FADH_2$ by transfer of electrons through the various electron transfer complexes including cytochrome oxidase (COX) of the mitochondrial electron transport pathway in the mitochondrial membranes to O_2 as the final electron acceptor. There are a number of routes for this electron transfer, all of which feed through ubiquinone. As only three of the six main complexes transport protons across mitochondrial membranes and hence allow the energy to be harnessed in ATP synthesis (by a mechanism similar to that in chloroplasts), up to 4.5 protons may be transported across the membrane per electron depending on the exact pathway used (Wikström & Hummer, 2012). The overall yield of ATP/ mole of glucose oxidised can be up to 29 moles (Amthor, 2000), though it will frequently be lower.

7.2.2 Alternative oxidase

Normally efficient mitochondrial respiration leads to phosphorylation of somewhat less than three ADP per oxygen atom (a P/O ratio of up to three), but this can be greatly reduced if significant electron flow occurs through the so-called alternate oxidase system (AOX), which bypasses cytochrome c and does not generate a trans-membrane potential so is therefore non-phosphorylating (see Figure 7.5 and van Dongen

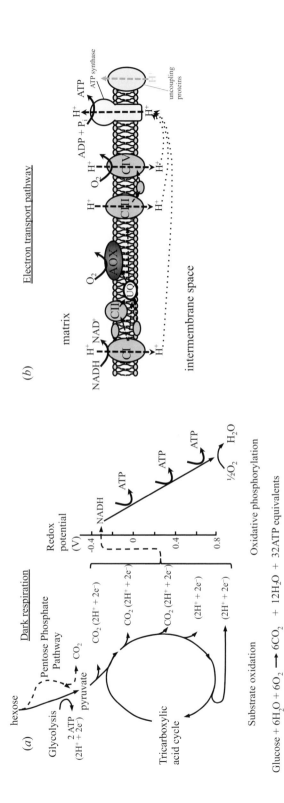

Figure 7.5 (*a*) Summary of the main pathways involved in dark respiration in plants. The $2H^+ + 2e^-$ represents reductant that ultimately goes to reduce oxygen. (*b*) A simplified diagram of the main components of the electron transport pathway (for further details see Millar *et al*., 2011). (CI, CII, CII and CIV are the respiratory complexes; UQ is ubiquinone; AOX is the alternative oxidase.)

et al., 2011; Vanlerberghe & McIntosh, 1997). The alternate oxidase is not inhibited by cyanide and is especially important in thermogenesis (e.g. in the *Arum* spadix and in other specialised situations). This thermogenesis arises not because there is a substantially greater enthalpy change for this pathway, but rather because it allows a high respiratory rate (and hence high heat output) as it is not linked to the requirement of generating ATP. Engagement of this alternative oxidase pathway occurs when the ATP needs of the plant are adequately met, so the main function of this pathway appears to be to balance carbon metabolism and electron transport, and partitioning of electrons to this pathway depends both on the amount of AOX present and on fine biochemical regulation. In leaves, AOX also helps to dissipate excess reducing equivalents derived from the chloroplast.

The engagement of the AOX pathway may be estimated by comparison of the O_2 consumption with and without specific inhibitors of the different pathways (e.g. cyanide for COX and salicylhydroxamic acid (SHAM) for AOX). Because the two pathways discriminate differentially against the heavier isotope of oxygen (^{18}O), it can, however, be more reliable to estimate the engagement of the AOX pathway from measurements of the oxygen isotopic discrimination using mass spectrometry. The fractionation factor (Δ) can be determined as the slope of a line of regression of $\ln(R/R_0)$ against $-\ln(f)$, where R is the $^{18}O/^{16}O$ ratio in a given sample, R_0 is the isotopic ratio in an initial reference sample and f is the fraction of O_2 remaining unconsumed (Guy *et al.*, 1989). Depending on the species, the AOX has a discrimination (Δ_a) of between 25‰ and 31‰ while the COX discrimination (Δ_c) is 16 to 21‰ (Florez-Sarasa *et al.*, 2007; Guy *et al.*, 1989; Nagel *et al.*, 2001). Where Δ is the observed fractionation for a particular plant, the proportion of flow through the AOX pathway (τ_a), therefore, is given by:

$$\tau_a = (\Delta - \Delta_c)/(\Delta_a - \Delta_c) \qquad (7.3)$$

An alternative mechanism for the dissipation of excess energy is through activation of uncoupling proteins (UCPs) in the mitochondria, which dissipate the proton gradient developed by operation of the normal mitochondrial electron transport chain, thus uncoupling ATP generation from electron flow. Uncoupling proteins appear to be primarily involved in longer term modulation of energy equilibrium while AOX may regulate shorter term changes (van Dongen *et al.*, 2011).

7.2.3 Photorespiration

The second type of respiration in plants is called photorespiration (see reviews by Foyer *et al.*, 2009; Krömer, 1995). This is the pathway of CO_2 production via the photorespiratory carbon oxidation (PCO) cycle in the peroxisomes (also known as the glycolate pathway) (Figure 7.6) and an associated cycle of ammonia recycling. The same enzyme (Rubisco) that catalyses the carboxylation of RuBP as the first step of the PCR cycle can also catalyse the oxygenation of RuBP to phosphoglycolate as the first step of the PCO cycle. The relative rates of carboxylation and oxygenation depend on the relative concentrations of CO_2 and O_2 at the active site of Rubisco and on its preference for CO_2 over O_2 known as the *specificity factor*. Under the low planetary O_2 concentrations that pertained when photosynthesis evolved this pathway would have been insignificant, but with present day O_2 concentrations (21%) photorespiratory losses of CO_2 can account for a large fraction of photosynthetic fixation and can represent a major limitation to carbon gain by C_3 plants.

The rate of photorespiration in C_3 plants is between about 10 and 30% of photosynthesis (often determined as the difference in $\mathbf{P}_n$ between that measured in normal air and at 2% O_2), but it can increase substantially under stress conditions. It is unclear, however, how much of the CO_2 released is actually re-assimilated before it escapes the leaf. Loreto *et al.* (1999) have estimated that as much as 80% of photorespiratory CO_2 release may be re-assimilated, on the basis of experiments using rapid changes between $^{13}CO_2$ and $^{12}CO_2$ as the substrate for photosynthesis, and making use of the fact that

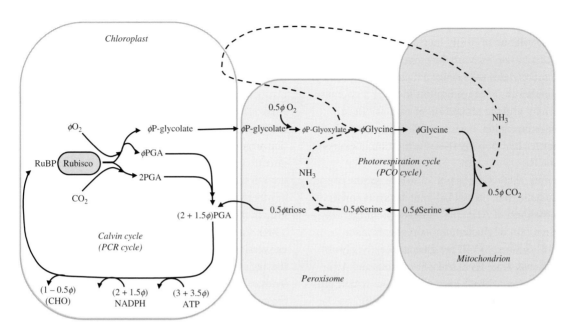

Figure 7.6 Simplified diagram of the photorespiration (PCO) and Calvin (PCR) cycles, expanded from Figure 7.2, showing the stoichiometry (Farquhar *et al.*, 1980b) where ϕ molecules of RuBP are oxygenated to every one carboxylated and the associated N cycling. For further details see Foyer *et al.* (2009).

some infrared gas analysers can be insensitive to $^{13}CO_2$ because the heavier isotope shifts the infrared absorption peak of CO_2 outside the instrument's detection range.

Many of the physiological characters of C_4 plants listed in Table 7.1 indicate a lack of external symptoms of photorespiration in these plants. These include a CO_2 compensation point near zero (indicating little or no respiratory release of CO_2), a low mesophyll resistance (see below), a lack of photosynthetic response to O_2 partial pressure, high rates of photosynthesis and even a high temperature optimum (the oxygenase to carboxylase ratio tends to increase with temperature). Many of these can be explained in terms of efficient refixation of any respired CO_2, but this cannot explain the lack of any O_2 effect on the efficiency of radiation use (quantum efficiency), as this would be expected to change with photorespiration rate, but such effects have not yet been detected. In spite of this, there is good evidence that the appropriate enzymes are present in bundle sheath cells (albeit at fairly low activities), and are

indeed essential for survival (Zelitch *et al.*, 2009), so that the lack of photorespiration probably results from high internal CO_2 concentrations that shift the path of RuBP metabolism almost entirely to carboxylation.

There are a number of physiologically important differences between photorespiration and dark respiration:

1. True photorespiration is obligatorily coupled to operation of the PCR cycle for generation of its substrate (RuBP), so it only occurs in photosynthetic cells in the light. Dark respiration, on the other hand, can occur in all cells in the dark or light, and usually continues in photosynthetic cells in the light, albeit at rates between about 30 and 80% of the rate in darkness (Krömer, 1995; Raven, 1972; Vilar *et al.*, 1995; Wang *et al.*, 2001). The total respiration in the light is the sum of CO_2 production of leaves by photorespiration ($\mathbf{R_p}$) and dark (mitochondrial) respiration ($\mathbf{R_d}$) pathways in light.
2. Photorespiration is affected by the concentrations of both O_2 and CO_2. This is because of the

competitive nature of oxygenation versus carboxylation of RuBP. Increases in CO_2 concentration increase the proportion of RuBP that is carboxylated (and hence net photosynthesis), but increases in O_2 concentration increase oxygenase activity and the proportion of carbon lost in photorespiration. In contrast dark respiration is not affected by CO_2 or by O_2 concentration in excess of 2 to 3%.

3. In dark respiration, about 35 to 40% of the energy available from the oxidation of sugars is conserved in the form of ATP. The energetic yield for respiration of glucose, for example, is made up as follows: 6 ATP per glucose for glycolysis (2 via substrate level phosphorylation and 4 via mitochondrial electron transport from NADH (up to a P/O ratio of 2 via this route)), 2 ATP via the TCA cycle substrate level phosphorylation, 24 via mitochondrial phosphorylation with a P/O of 3 and 2 ATP for each of the $FADH_2$ from the TCA cycle. This gives a total of 36 ATP per glucose (though this is likely to be an overestimate). From these figures about 1.10 MJ mol^{-1} of the 2.87 MJ mol^{-1} released by complete oxidation of glucose is potentially trapped in ATP, though this is only an approximation as it assumes standard conditions, which do not pertain in plant cells. In contrast to dark respiration, however, the PCO cycle requires a net input of energy to drive it (28 ATP equivalents per CO_2 released – see Figure 7.6 and Lorimer & Andrews, 1981).

7.2.4 Respiratory quotients and respiratory efficiencies

The ratio of the numbers of moles of CO_2 released to the number of moles of O_2 absorbed in respiration is known as the respiratory quotient. Respiration of glucose and other hexoses gives rise to a respiratory quotient of unity, but oxidation of reduced compounds such as fats or proteins yields a respiratory quotient of less than 1 (0.7 for many lipids, 0.8 for some proteins), while oxidised compounds such as organic acids yield respiratory quotients

greater than unity (about 1.33 for citric acid). Overall a value of about 1 is usually applicable.

There is evidence for genetic variation in respiratory efficiency with large differences between plants in the ratio of seasonal R to seasonal P_n varying from about 0.35 to 0.8 (Amthor, 2000). The lower extreme appears to be close to the minimum that would support all the processes required for growth, while anything above 0.8 implies a very slow growth rate. A low activity of the alternative pathway or other wasteful processes would tend to favour high growth rates (some results are summarised by Amthor, 1989). Calculations of the theoretical efficiency of conversion of substrate to biomass can be made on the basis of known biochemical pathways (Penning de Vries et al., 1983). The amount of CO_2 released during the synthesis of different compounds (mg CO_2 per g of compound formed) ranges from −11 for organic acids, through 170 for carbohydrates, 544 for proteins to as much as 1720 for lipids. Overall the average for leaf tissue is 333, while a peanut seed could be over 1000.

The assimilatory quotient (the ratio of net CO_2 fixation rate to the net O_2 evolution rate) is also of interest as this ratio indicates the degree of coupling between linear photosynthetic electron transport and CO_2 fixation into sugars.

7.2.5 Function of respiration

Dark respiration acts as a source of the reduced NADH and ATP needed for biosynthetic and maintenance processes and also provides the essential carbon skeletons. The oxidative pentose phosphate pathway may also be an important source of NADPH for reactions in the cytoplasm. Dark respiration is often separated into growth (R_g) and maintenance (R_m) components, on the basis that R_g provides energy for growth and synthesis of new cell constituents and R_m is used for maintenance of existing cell structure. Although there may in reality be no biochemical distinction between these components, R_m is assumed to be proportional to dry mass and strongly temperature sensitive, while R_g is assumed to be directly dependent on photosynthesis and insensitive to temperature.

The function of photorespiration is more controversial since its operation only seems to lower plant productivity, which hardly seems evolutionarily advantageous! It has been suggested that photorespiration may be an evolutionary 'hangover' serving no useful purpose. However, it does have substantial impacts on cell energetics: for example it promotes export of reducing equivalents from the chloroplasts and facilitates energy dissipation that might protect plants from photo-oxidation under conditions of high irradiance but low potential CO_2 fixation (e.g. water stress when the stomata are closed), and so avoid damage to the photosynthetic apparatus (termed *photo-inhibition* – see below). One suggestion is that oxygenation is an inevitable consequence of the chemical mechanism of carboxylation of RuBP, so that the PCO cycle acts to recycle any phosphoglycolate formed. It is notable that the photorespiratory pathway also plays an important part in recycling of carbon and nitrogen and in redox exchanges between subcellular compartments, while the generation of hydrogen peroxide (H_2O_2) in the peroxisome appears to have an important regulatory signalling role (Foyer *et al.*, 2009). The idea that photorespiration is unnecessary or at least excessive has led to attempts by plant physiologists and breeders to increase net photosynthesis by decreasing photorespiration and/or enhancing photosynthesis by increasing the specificity factor of Rubisco and its rate constant or affinity for CO_2 (discussed below in Section 7.9.4).

7.3 Measurement and analysis of carbon dioxide exchange

The main techniques for measuring the carbon balance and gas exchange of plants were well reviewed by Šesták *et al.* (1971), but as this text is not widely available the main approaches that include growth analysis, the use of radioactive or stable isotope tracers, and net CO_2 or O_2 exchange are

surveyed here. Other useful sources include Pearcy *et al.* (1991) and websites such as *PrometheusWiki* (http://prometheuswiki.publish.csiro.au/tiki-custom_home.php).

7.3.1 Growth analysis

Growth analysis (Evans, 1972; Hunt *et al.*, 2002; see also the related software tool to facilitate calculations http://aob.oupjournals.org/cgi/content/full/90/4/485/DC1) is a powerful and widely applicable method for estimating long-term net photosynthetic production (photosynthesis minus respiration). It is based on readily obtainable primary measurements of plant dry weight and leaf dimensions made at intervals on growing plants or plant stands. It is also useful for analysing physiological adaptations of different species in terms of their partitioning of carbohydrate into leaves and other organs such as roots or seeds. This partitioning is at least as important as photosynthetic activity per unit area in determining productivity of different plant stands. The growth rate or rate of change of total plant dry mass (dW/dt) is obtained from a series of destructive harvests. It can be calculated for single plants or for plant stands and either expressed per unit total dry mass as a *relative growth rate* (RGR):

$$\text{RGR} = (1/W)(dW/dt) \tag{7.4}$$

or else it is expressed per unit ground area as a *crop growth rate* (CGR). It is possible to derive the net photosynthetic rate per unit leaf area (A), called unit leaf rate or *net assimilation rate* (NAR):

$$\text{NAR} = (1/A)(dW/dt) \tag{7.5}$$

from either of the following relations:

$$\text{NAR} = \text{RGR/LAR} \tag{7.6}$$

or:

$$\text{NAR} = \text{CGR}/L \tag{7.7}$$

where LAR is the leaf area ratio (leaf area divided by total dry mass) and L is the leaf area index (leaf area per unit ground area). Note that NAR allows for respiratory losses at night and from non-photosynthesising tissues so is not quite equivalent to an instantaneous $\mathbf{P}_n$ measured on single leaves. Particularly in natural ecosystems the CGR is frequently now referred to as the net primary production (NPP; see Section 7.9.1).

7.3.2 Use of isotopic tracers

Photosynthesis can be measured using radioactive tracers for carbon or oxygen. One approach is to monitor the rate of loss of activity in the air in a closed chamber containing a photosynthesising leaf and initially supplied with a known specific activity of $^{14}CO_2$. More usually the leaf is allowed to fix $^{14}CO_2$ for a short period and then it is killed and the amount of ^{14}C incorporated into the leaf is determined in a scintillation counter. For short exposures (less than about 20 s), ^{14}C incorporation is a measure of *gross* photosynthesis (though it is an underestimate because the ^{12}C released by respiration may dilute the ^{14}C supplied). As the exposure time increases, the proportion of ^{14}C initially fixed that is re-released during the exposure increases so that with long exposures the technique may tend to estimate *net* photosynthesis. Another method that is useful for studying photosynthesis with stomatal control eliminated is to use thin slices of leaf tissue (see e.g. Jones & Osmond 1973), or isolated protoplasts or chloroplasts incubated with $H^{14}CO_3^-$ in solution.

Stable isotopes (e.g. ^{13}C rather than the usual ^{12}C) can also be used as tracers, though it is more usual to use variation in their natural abundance for the study of photosynthesis as discussed below (Section 7.6). Their use as tracers in the study of respiration has already been noted (Sections 7.2.1 and 7.2.3).

7.3.3 Net gas exchange

Measurement of CO_2 concentration: infrared gas analysers

The most usual method for CO_2 detection is by means of infrared gas analysers (IRGAs) that make use of the strong CO_2 absorption in the IR (especially that at

4.26 μm: see Figure 2.6). In the standard differential Luft-type of sensor, radiation from an IR source passes to a detector through either a fixed-volume analysis cell containing the sample gas or through a corresponding reference cell containing a reference gas; these two light beams are then passed through a pair of sealed detector cells where further absorption heats the gas and raises its pressure. A high concentration of CO_2 in the analysis cell leads to more absorption there and hence less heating of the corresponding detector cell; the differential pressure sensor between the detector cells is then used to determine the CO_2 concentration in the analysis cell. Cross sensitivity to other atmospheric gases (especially water vapour, which has a common absorption band with CO_2 at 2.7 μm) can be eliminated by use of interference filters to restrict sensitivity to the 4.26 μm absorption band.

There is increasing use, especially for eddy covariance systems, of open path absolute IRGAs that can measure rapid changes of H_2O or CO_2 concentration in free air. In these detectors IR radiation is passed through a fixed path between the source and the detector; by alternating the wavelength of the radiation between one where the gas being measured is a strong absorber and one where it does not absorb, it is possible to calculate the gas concentration from the ratio between the transmitted power at the two wavelengths after appropriate calibration.

Infrared gas analysers are usually calibrated in terms of the volume fraction of CO_2 (obtained for example by mixing known volumes of CO_2 and CO_2-free air or nitrogen with precision mixing pumps). In the atmosphere, the mole fraction (equal to the volume fraction) of CO_2 is now about 395×10^{-6} or 395 volume parts per million (vpm). Use of Eq. (3.6) allows us to calculate the equivalent concentration as 776 mg m^{-3} at 100 kPa and 20°C. Other measures of CO_2 concentration can be obtained using the conversions described in Chapter 3 (i.e. $p' = x'P$; and $c' = x'PM_C/\mathscr{R}T = p'M_C/\mathscr{R}T$). Although the IRGA actually measures the molar concentration ($= c'M_C$) in the analysis cell it is more convenient to use the more conservative quantities of mole fraction or partial pressure. For example, as the analysis cell is usually maintained at a constant temperature, the IRGA reading is constant for a given mole fraction in the cuvette whatever the cuvette temperature, though it is necessary to correct for any pressure difference between measurement and calibration, i.e.:

$$x'_{true} = x'_{observed} P^o / P \tag{7.8}$$

where P^o is the calibration pressure.

Oxygen exchange

As an alternative to the measurement of CO_2 exchange for studies of photosynthesis and respiration one can measure O_2 exchange but most O_2 sensors tend not to be sensitive enough for accurate photosynthesis measurements in open systems, partly because of the need to measure small changes in a large background. The simplest sensors are the ambient temperature electrochemical sensors that depend on diffusion of O_2 from the air through a membrane or small pore to a galvanic cell where the reduction of O_2 at the cathode generates a current proportional to O_2 concentration. The polarographic (Clark electrode) sensors measure the current when a fixed polarising voltage is applied across the cell. Alternative types of gas-phase analyser include several types of paramagnetic analysers and tuned diode laser sensors operating at 760 nm. As the best sensors currently have a maximum sensitivity of only about 10 ppm they are not very useful for open system measurements so are most usually used in closed system studies of respiration. Sensitive polarographic electrodes have been widely used for the study of photosynthetic O_2 evolution in the liquid phase for algae or for cells or leaf slices in solution (Jones & Osmond, 1973) while gas-phase systems (Delieu & Walker, 1983) are particularly useful for the measurement of maximum O_2 evolution rates. Although stomatal closure limits the value of measurements at normal CO_2 concentrations, it is possible to make measurements at higher CO_2 concentrations than are possible using IRGAs, so that it is possible to estimate the photosynthetic capacity. For stressed tissue with stomatal closure it may be

necessary to make measurements with as much as 15% CO_2 to obtain a true $\mathbf{P}_{max}$ (though beware possible inhibition of photosynthesis in C_4 plants).

Micrometeorological measurements

Net CO_2 exchange of large areas of vegetation or complex plant communities is best estimated from micrometeorological measurements, as they interfere little with the environment and provide effective averages over both area and time. In general, net CO_2 flux density per unit area of ground ($\mathbf{J}_C$) is negative during the day (representing absorption by the surface, $\mathbf{P}_n$) and positive at night (representing net respiratory loss) and may be obtained using the familiar transport equation (see Eq. (3.34)):

$$\mathbf{J}_C = -\mathbf{P}_n = -\mathbf{K}_C (M_C / \mathcal{R} T)(\mathrm{d}p'/\mathrm{d}z) \simeq -\mathbf{K}_C (\mathrm{d}c'/\mathrm{d}z) \tag{7.9}$$

from measurements of the gradient of CO_2 partial pressure ($\mathrm{d}p'/\mathrm{d}z$) within the crop boundary layer. Note that a prime (′) will be used throughout this chapter to distinguish various measures of CO_2 concentration (c', m', p' and x') and resistances (r') and conductances (g') to CO_2 diffusion.

The main approaches for estimation of the transfer coefficient ($\mathbf{K}_C$) from aerodynamic or energy balance measurements as well as direct estimation of $\mathbf{P}_n$ by eddy covariance have been introduced in Section 3.4.1, with the most generally useful approach being the use of eddy covariance.

Cuvette measurements

The commonest technique involves measurements of net gas exchange in cuvettes that can range in size from small (c. 1 cm^2) single-leaf chambers for portable porometers (cf. Chapter 6) to large (>10 m^3) chambers that can enclose whole plants or areas of canopy (see e.g. Barton et al., 2010; Pearcy et al., 1991). The degree of environmental control in these cuvettes can vary in sophistication up to completely independent control of temperature, light, humidity and of CO_2 and O_2 concentrations in the best laboratory systems. The largest chambers are often used for studies of the impact of elevated CO_2 concentrations on plant functioning, but it should be noted that very substantial and expensive cooling capacity is required to maintain internal conditions close to ambient, especially in high-irradiance environments.

Although closed systems (where the flux is calculated from the rate of change of CO_2 concentration in the cuvette) are used for some specific purposes, for example for studies of soil respiration, most measurements are based on open or semi-open gas-exchange systems. In these the flux of CO_2 is obtained from the difference between the flow of CO_2 into the cuvette ($u_e r_e'$) and the flow of CO_2 out ($u_o r_o'$) as:

$$\mathbf{P}_n^m = (u_e r_e' - u_o r_o')/A \simeq u_e (r_e' - r_o')/A \qquad (7.10)$$

where $\mathbf{P}_n^m$ is the net assimilation rate (in molar units – mol m^{-2} s^{-1}), A is the reference area (usually the leaf area in the chamber, m^2, though it would be ground area for larger canopy chambers) and u is the molar flow rate (mol s^{-1}). This corresponds to Eq.(6.4) for the estimation of evaporation. Accurate estimates of stomatal and other resistances depend on effective stirring of the air in the chamber so that all leaves are exposed to similar conditions. For precise work it is necessary to allow for the small difference between the volume flow entering the cuvette (u_e) and the outflow (u_o) that results from water vapour efflux from leaves (any CO_2 flux can be neglected as it is balanced by an O_2 flux). For a typical ($e_o - e_e$) of 1 kPa the relative increase in u is only about 1% (i.e. ($e_o - e_e$)/P). In larger chambers it is common to control the CO_2 concentration by injection of pure CO_2; this flux needs also to be considered in the calculation of fluxes.

Estimation of respiration in the light

Although many methods have been proposed for estimating photorespiration (Sharkey, 1988), none is altogether reliable because of interactions between the various biochemical processes and our incomplete understanding of the gas-exchange processes within leaves. Most methods suffer from the difficulty of separating photorespiration via the phosphoglycolate pathway from any dark respiration continuing in the light. The most obvious estimate of photorespiration is the magnitude of the enhancement of $\mathbf{P}_n$ in 2% O_2 as compared with 21%, but this is an overestimate because it includes some stimulation of carboxylase resulting from relief of competitive inhibition by the oxygenase and, at the low O_2 concentrations necessary, dark respiration may also be inhibited. Other approaches include those based on the size of the transient burst of the CO_2 release observed on darkening a leaf (because photorespiration continues longer than assimilation on darkening), a comparison of the rate of $^{14}CO_2$ fixation after short-term exposure to $^{14}CO_2$ (as a measure of gross photosynthesis) with the net fixation by gas exchange, and studies using simultaneous measurements of exchange of oxygen isotopes.

In contrast to direct measurement of photo-respiration, more accurate estimates can be obtained from estimates of the CO_2 concentration at which CO_2 loss resulting from oxygenation at Rubisco equals CO_2 fixation by carboxylation (termed Γ_*); this can be considered as the compensation point in the absence of dark respiration. At this point the velocity of carboxylation (V_c) equals half the velocity of oxygenation (V_o) (because half a mole of CO_2 is evolved for every oxygenation). Brooks and Farquhar (1985) showed that Γ_* can be estimated as the point where $\mathbf{P}_n/p_i'$ curves intersect when irradiance is varied (Figure 7.7); at that point $\mathbf{P}_n$ equals $\mathbf{R}_d$, and the specificity S of Rubisco equals $0.5\,[O_2]/\Gamma_*$. The rate of photorespiration as a fraction of carboxylation is given by $\Gamma_*/[CO_2]$, where $[CO_2]$ is the CO_2 concentration.

Although $\mathbf{R}_d$ in the light is often assumed to equal the rate in the dark many studies have indicated that $\mathbf{R}_d$ in the light is only between 30 and 80% of that in the dark. One approach to the estimation of $\mathbf{R}_d$ was originally proposed by Laisk and is based on Γ_* as outlined above, while the other main method is based on the discontinuity of the photosynthetic light-response curves at low irradiances (the Kok effect) where $\mathbf{R}_d$ in the light ($\mathbf{R}_{d-L}$) is estimated as the extrapolated value of the photosynthetic light response at low irradiances. These methods are illustrated in Figure 7.7.

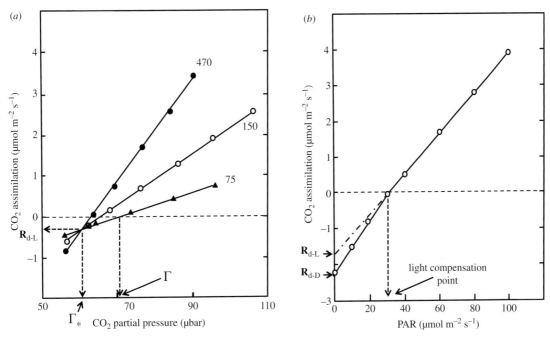

Figure 7.7 Estimation of respiration in the light. (*a*) Illustration of Laisk's method for spinach leaves measured at irradiances of 470, 150 and 75 μmol m^{-2} s^{-1}, and at 20°C and 380 mbar O$_2$, where Γ_* is the CO$_2$ concentration at which photorespiration balances assimilation and Γ is the CO$_2$ concentration at which the total respiration balances assimilation (for the lowest irradiance). The net gas exchange at Γ_* represents dark respiration in the light. (Data from Brooks & Farquhar, 1985.) (*b*) A representative light-response curve near the light compensation point for *Xanthium strumarium* showing the Kok effect, where R$_{d-D}$ is the mitochondrial respiration in darkness and R$_{d-L}$ is the mitochondrial respiration in the light. (Data from Wang *et al.*, 2001.)

7.4 Photosynthetic models

7.4.1 Photosynthetic biochemistry

The most widely used approach to modelling of photosynthetic biochemistry is the Farquhar–von Caemmerer–Berry (FvCB) model (Farquhar *et al.*, 1980b) and its more recent variants and extensions (see e.g. Dreyer *et al.*, 2001; Müller & Diepenbrock, 2006). In this model it is assumed that photosynthesis is either limited by supply of CO$_2$ and its rate (**P$_c$**) determined by the properties of Rubisco (Rubisco limited), or else CO$_2$ fixation is limited by the regeneration of the substrate (**P$_j$**; RuBP-limited), a process primarily dependent on photosynthetic electron flow. A third type of limitation occurs where use of the triose phosphate products of photosynthesis

determines the rate of assimilation (**P$_t$**; triose phosphate utilisation or TPU-limited); this largely sets the maximum rate (**P$_{max}$**). The basis of the model is outlined in Box 7.1.

Because the model splits the photosynthetic response curve into three separate segments it is challenging to estimate the required parameters accurately from the usual **P$_n$**:c_i curves (see below). Approaches to the estimation of the necessary parameters have been discussed in detail by Sharkey *et al.* (2007) and by Gu *et al.* (2010). Extensions of the approach to describe the biochemistry of C$_4$ photosynthesis have also been developed (von Caemmerer & Furbank, 1999). An important parameter is the degree of leakage of CO$_2$ back out of the bundle sheath.

7.4.2 Empirical models

Although mechanistic models such as the FvCB model give the best fits to experimental data, simpler equations are useful for many predictive purposes as the FvCB model implies a discontinuous function for the CO_2 response. Some of the more useful empirical models that have been used to fit photosynthetic responses to CO_2 and light have been outlined by Thornley and France (2007). Many semi-mechanistic models are based on early applications (Maskell, 1928; Rabinowitch, 1951) of Michaelis–Menten (rectangular hyperbolic) kinetics giving rise to non-rectangular hyperbolae. These latter curves saturate more sharply and more realistically than do simple exponential or rectangular hyperbolic functions and more readily approximate the 'Blackman type' responses with two straight lines (a constant initial slope where photosynthesis is entirely CO_2-limited, switching suddenly to a horizontal light-limited portion) that are often observed. Non-rectangular hyperbolae are also found to give the best fit to light-response data, though simple rectangular hyperbolae or exponential

---

Box 7.1   An outline of the Farquhar–von Caemmerer–Berry model for photosynthetic biochemistry (Farquhar et al., 1980b).

The model can be expressed in the following form:

$$\mathbf{P} = \min[\mathbf{P}_c, \mathbf{P}_j, \mathbf{P}_t] \qquad (B7.1.1)$$

where

$$\mathbf{P}_c = V_{c,max}[p_c{'} - \Gamma_*]/[p_c{'} + K_C(1 + p_c^O/K_O)] - \mathbf{R}_d \qquad (B7.1.1a)$$

$$\mathbf{P}_j = ETR[p_c{'} - \Gamma_*]/[4p_c{'} + 8\Gamma_*] - \mathbf{R}_d \qquad (B7.1.1b)$$

$$\mathbf{P}_t = 3\,TPU - \mathbf{R}_d \qquad (B7.1.1c)$$

where $V_{c,max}$ is the maximum velocity of Rubisco for carboxylation, $p_c{'}$ is the partial pressure of CO_2 at Rubisco, p_c^O is the partial pressure of O_2 at Rubisco, K_C is the Michaelis constant of Rubisco for CO_2, K_O is the corresponding constant for O_2, Γ_* is the p' for CO_2 at which photorespiration balances assimilation (i.e. oxygenation rate is twice that of carboxylation), $\mathbf{R}_d$ is the dark respiration in the light, ETR is the rate of photosynthetic linear electron transport (see Section 7.5), and TPU is the rate of use of triose phosphates, which can be approximated for most purposes as a constant.

It is worth noting that the above set of equations uses the partial pressure of CO_2 at Rubisco, while conventional gas-exchange analysis (Section 7.6.1) estimates the intercellular space concentration ($p_i{'}$). The value of $p_c{'}$ can be estimated as:

$$p_c{'} = p_i{'} - \mathbf{P}_n^m/g_m{'} \qquad (B7.1.2)$$

where $g_m{'}$ is the mesophyll conductance as discussed in Section 7.6.1 (below). As shown in Figure B7.1 we see that at low CO_2 concentrations assimilation is Rubisco limited, as CO_2 supply increases we get to the electron-transport (RuBP-regeneration) limited phase, and then ultimately

Box 7.1 | (continued)

the TPU phase (this can be modelled simply as a fixed limit or as a decreasing function with increasing CO_2 (von Caemmerer, 2000)). Details of estimation of the various parameters from gas-exchange measurements are outlined by Gu *et al.*, (2010).

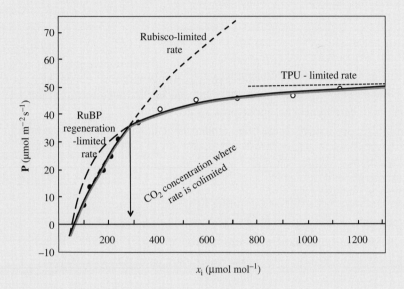

Figure B7.1 Illustration of the three functions comprising the Farquhar–von Caemmerer–Berry model of photosynthesis (data for *Glycine max* from Dubois *et al.*, 2007).

functions are often adequate. Some of the equations for empirical modelling of photosynthetic responses are presented in Box 7.2.

7.5 Chlorophyll fluorescence

As indicated above, the analysis of chlorophyll *a* fluorescence provides a powerful probe of the functioning of the intact photosynthetic system. Two main techniques are available for fluorescence analysis: (i) analysis of the rapid (ms timescale) and slow (s to min timescale) components of the changes in chlorophyll fluorescence (the Kautsky effect) that are observed on illuminating a dark-adapted leaf

(Figure 7.8); and (ii) use of modulated fluorescence to detect the fluorescence from a normally photosynthesising leaf in white light. When combined with the use of high intensity (saturating) pulses of light, this latter technique permits analysis of the processes that quench (i.e. decrease) fluorescence and provides a non-destructive measure of the relationship between the light and dark reactions of photosynthesis. Other techniques are discussed in Section 7.5.4 below.

In a traditional Kautsky system, the leaf would be illuminated with an exciting light with wavelengths less than *c.* 620 nm, and fluorescence from PSII (peak emission at 695 nm) measured with a sensitive photodetector having a narrow band

Box 7.2 | Empirical modelling of leaf photosynthesis.

Because the FvCB biochemical model of photosynthesis is a non-continuous function of CO_2 mole fraction at the carboxylase (x_c') or in the intercellular spaces (x_i') it is difficult to fit to real data, so semi-mechanistic models based on Michaelis–Menten kinetics are commonly used. The most useful is a non-rectangular hyperbola, which can be written in the form (Jones & Slatyer, 1972):

$$\mathbf{P}_n = \mathbf{P}_{max}^C (x_i' - \Gamma - b\mathbf{P}_n)/(a + x_i' - \Gamma - b\mathbf{P}_n) \tag{B7.2.1}$$

where a and b are constants and $\mathbf{P}_{max}^C$ is the photosynthetic rate at saturating CO_2. This is a quadratic equation where the parameter a determines the sharpness of the transition from the linear phase to saturation and b determines the initial slope of the response.

Responses of photosynthesis to irradiance ($\mathbf{I}$) are often modelled using a rectangular hyperbola, as in:

$$\mathbf{P}_n = (\mathbf{I} - \mathbf{I}_c).\mathbf{P}_{max}^I/(a + (\mathbf{I} - \mathbf{I}_c)) \tag{B7.2.2}$$

where $\mathbf{P}_{max}^I$ is a maximum photosynthetic rate at saturating light (and is an appropriate function of temperature) and $\mathbf{I}_c$ is the light compensation point. Combination of Eqs. (B7.2.1) and (B7.2.2) gives a useful general expression for photosynthesis, whose behaviour is illustrated in Figure B7.2. This figure shows a close fit to observed responses (e.g. Figure 7.16) with more rapid saturation than can be achieved with a rectangular hyperbola.

Alternatively simple exponential functions, such as the Mitscherlich equation are often used:

$$\mathbf{P}_n = \mathbf{P}_{max}^I[1 - \exp(-\varepsilon_p(\mathbf{I} - \mathbf{I}_c)/\mathbf{P}_n^{max})] \tag{B7.2.3}$$

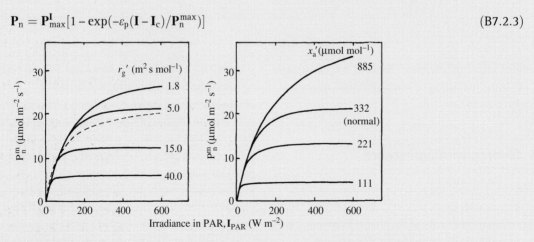

Figure B7.2 Photosynthesis–irradiance curves for typical C_3 leaves calculated using Eqs. (B7.2.1) and (B7.2.2) for a range of values of r_g' and x_a' (parameter values unless stated otherwise: $x_a' = 332$ µmol m^{-2} s^{-1}, $\Gamma = 55$ µmol m^{-2} s^{-1}, $K_m^C = 11.1$ µmol m^{-2} s^{-1}, $K_m^I = 200$ W m^{-2}, $P_{max}^{max} = 45.5$ µmol m^{-2} s^{-1}, $r_i' = 7.5$ m^2 s mol^{-1}, $r_g' = 5$ m^2 s mol^{-1}). The dashed line indicates a rectangular hyperbola.

Box 7.2 | **(continued)**

where ε_p is the quantum yield or photosynthetic efficiency at the light compensation point. Again, however, improved fits can usually be obtained with a non-rectangular hyperbola, which can be written in the form (Johnson *et al.*, 2010):

$$\theta(\mathbf{P}_n)^2 - (\varepsilon_p\mathbf{I} + \mathbf{P}_n^{max})\mathbf{P}_n + \varepsilon_p\mathbf{I}\mathbf{P}_n^{max} = 0 \tag{B7.2.4}$$

where $\mathbf{P}_n^{max}$ is a maximum photosynthetic rate at saturating light (and is an appropriate function of temperature) and θ is a dimensionless curvature parameter ($0 \le \theta \le 1$). An improvement can be achieved by replacing $\mathbf{I}$ by $(\mathbf{I} - \mathbf{I}_c)$.

Appropriate temperature functions have been discussed by Thornley and France (2007) among others.

---

filter (centred on 695 nm). This arrangement ensures that the small fluorescence signal can be separated from the much larger amount of reflected exciting light.

On first illuminating a dark-adapted leaf, where all components of the electron transport chain would be fully oxidised, fluorescence (F) immediately rises to a level (F_o) that is characteristic of open PSII reaction centres and fully oxidised primary electron acceptor (Q_A). The quantum yield of fluorescence in this situation is given by Eq. (7.2) with $f_{open} = 1$. As light is absorbed and charge separation occurs, thus closing reaction centres and reducing Q_A, fluorescence rises in a complex manner (Figure 7.8) reaching a peak (F_m) when all of Q_A is fully reduced. At this point, because all the reaction centres are closed ($f_{open} = 0$), chlorophyll excitation can no longer decay through photochemistry (i.e. k_p f_{open} falls to zero), so the quantum yield of fluorescence increases to:

$$\phi F = k_F/(k_F + k_H + k_P f_{open}) = k_F/(k_F + k_H) \tag{7.11}$$

As electron transport restarts, and photosynthesis increases, fluorescence slowly declines (is *quenched*) through a number of transients to a steady state.

Unfortunately the Kautsky system cannot be used to study photosynthesis in white light in the field because of the need to separate the actinic and fluorescent wavelengths. The development of modulated fluorescence systems overcomes this limitation and enables one to distinguish the various quenching processes that cause the fluorescence in *actinic light* (light that drives photosynthesis) at any time (F') to decline from its peak value after equilibration in the dark (F_m). The principle of modulated systems is that a very weak (~1 to 5 μmol m^{-2} s^{-1}) light source is switched on and off rapidly (modulated), and the measuring system is set to detect only the fluorescence signal that corresponds to the modulated exciting light. By the use of sensitive electronics it is possible to distinguish this signal from the very much larger signal arising from the illuminating white light. As long as the modulated light is so weak that it cannot drive electron transport, the fluorescence obtained with illumination by the modulated source alone is equivalent to F_o. It is common to normalise measurements using F_o so that differences in leaf chlorophyll content or sensor geometry that give rise to different absolute levels of fluorescence can be neglected. It is also important to ensure that the detector used does not respond to fluorescence at wavelengths longer than 700 nm, otherwise results will be confounded by some PSI fluorescence.

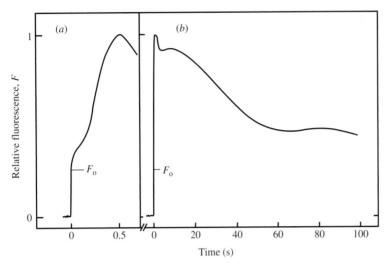

Figure 7.8 Characteristic fluorescence induction kinetics (Kautsky curve) on illuminating a dark-adapted leaf, showing the very rapid rise to F_o followed by a fairly fast rise to a maximum, followed by a slower decay to a steady state as photosynthesis starts up: (a) fast kinetics, (b) slow kinetics.

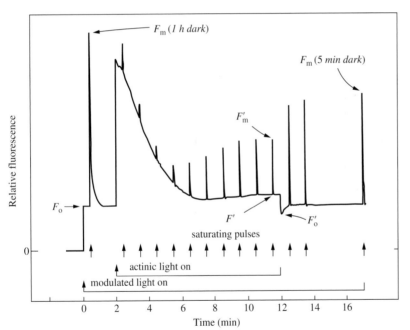

Figure 7.9 A fluorescence trace for a wheat leaf obtained with a modulated fluorescence system, showing quenching analysis with the terms defined in Box 7.2. The modulated light has an irradiance of 1 µmol m^{-2} s^{-1} while the actinic irradiance was 560 µmol m^{-2} s^{-1} and the saturating pulse was 4600 µmol m^{-2} s^{-1}.

7.5.1 Quenching analysis

As outlined by Maxwell and Johnson (2000) and Baker (2008) a lot of information about the functioning of the photosynthetic system can be obtained from an analysis of fluorescence quenching. The various fluorescence parameters that can be obtained using a modulated system and the method of calculation of the different quenching coefficients is illustrated in Figure 7.9 and summarised in Table 7.2. Note that we use a prime to distinguish fluorescence at any time in the light from the dark-adapted values. (As a word of warning it is worth noting that although there has been some attempt at standardisation of terminology, a wide range of different symbols exist in the literature so great care needs to be taken in their use.)

Table 7.2 Commonly used chlorophyll fluorescence parameters (see Figure 7.9).

Symbol	Definition
F, F'	Fluorescence emission from dark-adapted or light-adapted leaf, respectively
F_o, F_o'	Basal fluorescence emission (when Q_A is maximally oxidised and PSII centres open) from dark-adapted or light-adapted leaf, respectively. The latter is achieved by the use of weak far-red light
F_m, F_m'	Maximal fluorescence emission (when Q_A is maximally reduced and PSII centres closed) from dark-adapted or light-adapted leaf, respectively
F_v, F_v'	Variable fluorescence from dark-adapted (i.e. $F_m - F_o$) or light-adapted leaf (i.e. $F_m' - F_o'$), respectively
F_q' ($= F_m' - F'$)	Photochemical quenching of fluorescence by open PSII reaction centres
F_v/F_m	Maximal quantum efficiency of PSII photochemistry
F_v'/F_m'	Maximal quantum efficiency of PSII photochemistry at a given irradiance
F_q'/F_m'	PSII operating efficiency in the light: this measures quantum yield of linear electron transport (often termed $\Delta F/F_m'$ or ϕ_{PSII} in the literature)
F_q'/F_v'	q_P or PSII efficiency factor at a given irradiance; this is non-linearly related to the proportion of PSII centres that are oxidised (open)
$(F_m/F_m') - 1$	NPQ or non-photochemical quenching as an indicator of heat loss from PSII
NPQ	Non-photochemical quenching
q_E, q_I, q_T	Quenching associated with energy-dependent processes, photo-inhibition and state transitions (dependent on PSII light harvesting complexes), respectively
q_P	PSII efficiency factor or coefficient of photochemical quenching (F_q'/F_v')
q_L	Estimates fraction of open reaction centres (f_{open}) ($= q_P F_o'/F'$)
ϕF	Quantum yield of fluorescence (number of fluorescence events per photon absorbed)

The main types of fluorescence quenching are photochemical quenching and non-photochemical quenching.

Photochemical quenching

Photochemical quenching (q_P) results from the use of excitation energy within PSII to drive electron flow from P680 to Q_A and is therefore an estimate of the redox state of Q_A, and hence of the fraction of 'open'

PSII reaction centres (f_{open}). It ranges from zero when Q_A is fully reduced (i.e. at F_m where the maximum amount of incident energy is re-emitted as fluorescence because all reaction centres are closed) to unity when Q_A is oxidised and all PSII reaction centres are open to accept excitation energy (i.e. a minimum amount of excitation is re-emitted as fluorescence). The value of q_P at any instant is obtained from the fluorescence at that time (F') according to:

$$q_P = (F_m' - F')/(F_m' - F_o') = F_q'/F_v' \qquad (7.12)$$

where F_m' is the fluorescence obtained at the same time with a flash of light strong enough to saturate photochemistry and reduce all of Q_A, and F_o' is the basal fluorescence (with all reaction centres open) at that time as obtained by using a weak pulse of far-red light (>680 nm) to allow full oxidation of Q_A. Other symbols are explained below with definitions given in Table 7.2. Note that the saturating pulse required to oxidise all of Q_A can be 4000 to 6000 μmol m^{-2} s^{-1} in high-light adapted plants. It follows that q_P is therefore a measure of the fraction of the maximal PSII efficiency ($= F_v'/F_m'$) that is actually realised under the measurement conditions used and is therefore now often termed the *PSII efficiency factor*. Although q_P is often used as a measure of the redox state of Q_A, this relationship is not necessarily linear and a more accurate estimate of f_{open} (q_L) is given by (Baker, 2008):

$$q_L = q_P F_o'/F' \qquad (7.13)$$

Non-photochemical quenching (*NPQ*)

This term is used to describe a wide range of mechanisms that result in an increase in the rate at which excitation energy in PSII is lost as heat. Its value at any instant can be estimated from the amount that the maximal fluorescence as obtained with a saturating flash, F_m', falls below the maximum value obtained after a 'long' (often overnight) period in the dark (F_m). These are normally lumped together in *NPQ*, which is calculated from:

$$NPQ = (F_m - F_m')/F_m' \qquad (7.14)$$

NPQ is linearly related to heat dissipation and can take any value from 0 upwards. (The older term to describe non-photochemical quenching, q_N, is somewhat less sensitive as it lies in the range 0 to 1 and is not now recommended). Note that *NPQ* is calculated relative to the dark-adapted state. The main components of *NPQ* are:

1. *Energy-dependent quenching* (q_E). This quenching in the photosynthetic antennae is associated with the development of a thylakoid pH gradient (ΔpH, with acidification of the thylakoid lumen) as a result of electron transport. Acidification activates violaxanthin de-epoxidase, which catalyses the conversion of the carotenoid violaxanthin to zeaxanthin, which binds to PSII and is an efficient quencher of excitation (Demmig-Adams & Adams III, 1992). This quenching in the PSII antennae leads to a greater proportion of energy being dissipated as heat (i.e. k_H in Eqs. (7.2) and (7.11) increases) and less fluorescence. This heat loss helps to dissipate the excess energy that is available for causing damage to the photosynthetic machinery when the energy cannot be utilised in photosynthesis (e.g. at low CO_2 concentrations or under water stress).

2. *State transitions* (q_T). The maximum fluorescence from PSII can also be decreased by a transfer of some of the excitation energy from PSII to PSI (this transfer is under control of the phosphorylation state of the LHC protein). Although the kinetics of recovery of q_E and q_T are similar, the latter is generally rather small and only makes a contribution at low light.

3. *Photo-inhibition* (q_I). A particularly important type of quenching is termed photo-inhibition (q_I) and is generally taken to include irreversible, or only slowly reversible, effects on the photosynthetic system that lower F_m(*1 hr dark*) below the value found in a healthy plant. Photo-inhibition defined in this way includes both protective processes and damage to PSII reaction centres.

The distinction between the different forms of non-photochemical quenching is largely based on their different rates of relaxation in the dark. Although q_E is normally thought to disappear within a few seconds of darkening in isolated chloroplasts, it may take a minute or more to disappear in an intact leaf. For leaves, this rapidly decaying component of *NPQ* is usually ascribed to q_E, while more slowly decaying quenching can be attributable to state transitions. Any *NPQ* that takes longer than an hour or so to relax is often operationally defined as photo-inhibition.

All types of non-photochemical quenching act to dissipate excess energy absorbed by the chloroplasts in high light conditions when it cannot be used in photosynthesis (e.g. when photosynthesis is inhibited by cold, water stress or low CO_2 concentrations). In the absence of these dissipation mechanisms, photo-inhibitory and photobleaching damage would be much greater. In some situations, especially in stressed plants, it is also possible for F_o to be quenched (Figure 7.8). This must be allowed for in calculating the various quenching coefficients (see Table 7.2).

7.5.2 Efficiency of photochemistry

Using Eq. (7.2), we see that after a period in the dark when all reaction centres are open (i.e. $f_{open} = 1$), $F = F_o = (\mathbf{I}\,\alpha_{leaf}f_{PSII})\,\phi F$, so we can write:

$$F_o = (\mathbf{I}\,\alpha_{leaf}f_{PSII})k_F/(k_F + k_H + k_Pf_{open})$$
$$= (\mathbf{I}\,\alpha_{leaf}f_{PSII})k_F/(k_F + k_H + k_P) \qquad (7.15)$$

When the reaction centres are closed by a flash of saturating light ($f_{open} = 0$), we get the maximum fluorescence (F_m), given by:

$$F_m = (\mathbf{I}\,\alpha_{leaf}f_{PSII})k_F/(k_F + k_H) \qquad (7.16)$$

Defining a variable fluorescence (F_v) as the difference between F_m and F_o and substituting from Eqs. (7.15) and (7.16) we get, after cancelling terms and some rearrangement:

$$F_v/F_m = (F_m - F_o)/F_m = k_P/(k_F + k_H + k_P) \qquad (7.17)$$

which is the maximum quantum yield for photochemistry.

The ratio F_v/F_m for dark-adapted healthy plants is normally close to 0.83 (Björkman & Demmig, 1987), showing that with open reaction centres $k_P >> (k_F + k_H)$, though slightly different values are obtained with different measuring instruments. Not only does photo-inhibition lower F_v/F_m from this optimal level, but so do all other non-photochemical quenching mechanisms. Exposure of plants to both biotic and abiotic stresses in the light can substantially reduce F_v/F_m, so that measurement of F_v/F_m has been widely proposed as a tool for monitoring stress, but the causes of such reductions are often difficult to identify. Reductions in F_v/F_m may relate to increases in F_o for example as a result of loss of PSII reaction centres or to reductions in F_m resulting from non-photochemical quenching. Corresponding measurements in the light are of F_v'/F_m'; which is the maximum efficiency of PSII at the given irradiance (i.e. assuming that all the reaction centres are open). These values tend to be less than for the dark-adapted F_v/F_m because of quenching of F_m'.

Because q_P measures the proportion of PSII reaction centres that are oxidised or 'open', and F_v'/F_m' measures the efficiency of these open centres for electron transport, it has been argued by Genty et al. (1989) that the product of these two variables gives the quantum yield of non-cyclic electron transport through PSII (ϕ_{PSII}). Substituting from Eq. (7.12) and cancelling terms gives:

$$\phi_{PSII} = (F_v'/F_m')q_P$$
$$= ((F_m' - F_o')/F_m')((F_m' - F'))/(F_m' - F_o'))$$
$$= (F_m' - F')/F_m' \qquad (7.18)$$

where $(F_m' - F') = F_q'$ is the difference between steady state and maximal fluorescence at any time under the same conditions of non-photochemical quenching. It is therefore very simple to estimate ϕ_{PSII} using measurements of only the steady-state fluorescence (F') and the F_m' obtained with a saturating flash. The non-cyclic electron transport rate through PSII (ETR) can be estimated as:

$$\mathrm{ETR} = \mathbf{I}\cdot\alpha_{leaf}\cdot f_{PSII}\cdot\phi PSII \qquad (7.19)$$

Because ETR is often closely correlated with the rate of CO_2 fixation, chlorophyll fluorescence estimates of ETR have been widely used as estimators of photosynthesis. ETR can be estimated on any photosynthesising leaf in situ simply from a measure of fluorescence followed by a measurement of the corresponding value after a saturating flash. Unfortunately the relationship between ETR and assimilation can break down as a result of variation in rates of competing processes such as photorespiration, nitrogen metabolism and even electron donation to oxygen (the Mehler reaction), so ϕ_{PSII} is more usually just a measure of relative quantum yield. A further problem is that the equation itself relies on some major assumptions: although α_{leaf} is often approximated as 0.84, improvements can be made by use of measurements of the leaf absorptance using an integrating sphere; more seriously, however, the partitioning of energy between PSI and PSII (f_{PSII}) can vary substantially though this is particularly difficult to estimate.

Another point to consider in using fluorescence as a measure of assimilation, is that fluorescence arises from only the top few layers of photosynthesising cells while CO_2 exchange depends on the whole leaf thickness. Simultaneous measurements of F and $\mathbf{P}_n$ can be used to estimate the ETR per CO_2 fixed under non-photorespiratory conditions (e.g. 2% O_2). If one assumes that this holds in normal air the rate of photorespiration can then be derived.

7.5.3 Fluorescence imaging

At a laboratory scale it is now possible to image the variation in fluorescence across leaves with several instruments now commercially available (e.g. Oxborough & Baker, 1997); this provides a useful tool for the study of leaf heterogeneity, especially that arising in response to biotic and abiotic stresses. It is relatively easy to obtain all the key fluorescence parameters (F_o, F_m, F_o', F_m' and F') and hence to calculate the variation over the leaf surface in parameters such as F_v/F_m, NPQ and ϕ_{PSII} or ETR from the ratios of successive images.

7.5.4 Remote sensing of fluorescence

A limitation of fluorescence measurement is the need to provide a saturating pulse strong enough to oxidise Q_A (normally this needs to be several thousand μmol m^{-2} s^{-1} at the position of the leaf); this normally restricts measurements to the laboratory, or to very close-range measurements in the field.

A problem in attempting to measure fluorescence in naturally illuminated plants is that F is generally only a few % of the radiation reflected by leaves, and therefore difficult to distinguish from reflected solar radiation. One approach to the remote detection of the steady-state chlorophyll fluorescence (F') is the use of *solar induced fluorescence* (SIF); this is a passive technique that makes use of infilling of the very sharp atmospheric absorption bands for O_2 and H_2 (the Fraunhofer lines). Within these atmospheric absorption bands (e.g. the O-bands at 687 nm or 760.6 nm, which overlap significantly with the emission spectrum for chlorophyll fluorescence) atmospheric absorption substantially reduces the amount of reflected radiation reaching a remote detector. The method, which has been well reviewed by Meroni *et al.* (2009), requires sensors with very high spectral resolution and narrow half-bandwidths (preferably of 1 nm or less). For ground-based instruments it is feasible to estimate F' from measurements of the reflected radiance at the middle of a Fraunhofer line (L_{in}) and at a nearby reference outside the absorption line, together with measurements of incoming irradiance at these wavelengths (or of reflected radiance from a white reflecting reference surface) (Figure 7.10). Where we have measurements of incoming irradiance (W m^{-2}) as shown in Figure 7.10(a), the radiance (W m^{-2} steradian^{-1}) at any wavelength that reaches the detector is the sum of reflected radiance and emitted fluorescence. Using the subscripts $_{in}$ and $_{out}$ to distinguish values inside the absorption line and outside the absorption line, we can write (using π to correct the reflected irradiance to radiance):

$$L_{in} = \rho_{in}E_{in}/\pi + F_{in}' \tag{7.20}$$

$$L_{out} = \rho_{out}E_{out}/\pi + F_{out}' \tag{7.21}$$

If we assume that the wavelengths λ_{in} and λ_{out} are sufficiently close that the fluorescences (F_{in}' and F_{out}') and the reflectances (ρ_{in} and ρ_{out}) can each be assumed to be identical, these equations can be rearranged to give:

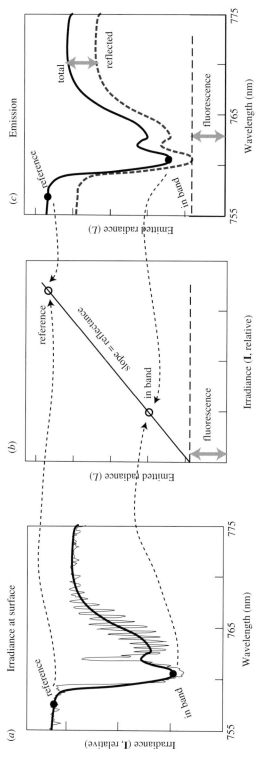

Figure 7.10 The Fraunhofer line depth method for measurement of chlorophyll fluorescence illustrated for the O₂–A atmospheric absorption band at 760.6 nm, showing the 'in band' and 'out of band' measurement wavelengths. (*a*) Irradiance at the surface around 760 nm, as measured using a sensor with a half band-width of 0.13 nm (fine line) or 1 nm (heavy line). (*b*) Plot showing the relationship between emitted radiance and irradiance inside and outside the absorption band showing that the fluorescence can be estimated from the intercept on the x-axis of the line joining these points. (*c*) Interpretation of the measured emittance spectrum (solid line) as the sum of fluorescence and reflectance.

$$F' = (E_{out}L_{in} - L_{out}E_{in})/(E_{out} - E_{in}) \qquad (7.22)$$

The diagnostic power of SIF is limited by the fact that it is not possible to obtain any value for F_m, so we only have access by this means to the steady-state fluorescence (F'), which is affected by many aspects of leaf physiology in a rather complex manner. Further work is needed to identify useful correlations with physiological processes and stress responses.

An alternative approach is the active measurement of fluorescence in a way that is more analogous to the conventional pulse-modulation scheme used in laboratory fluorimeters. Although lasers can be used to deliver the necessary saturating pulses over metre-scale distances, a greater range can be achieved by the laser induced fluorescent transient (LIFT) technique as this does not require saturation and estimates the true F_m' (and other fluorescence parameters such as F', F_o') by curve fitting to the response to bursts of very rapid (as short as 0.1 µs) lower energy pulses (Kolber et al., 1998, 2005). The general approach is illustrated in Figure 7.11.

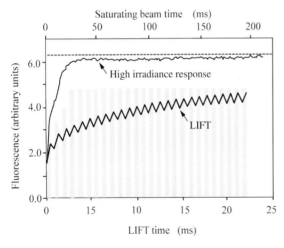

Figure 7.11 Comparison of the effects of a high-irradiance saturating pulse (4600 µmol m^{-2} s^{-1}) on F' for *Malva parviflora* L. leaves in the dark (saturating at $F_m' \simeq 6.15$ after about 30 ms) and bursts of rapid laser pulses at 342 µmol m^{-2} s^{-1} showing the LIFT transients with the fitted $F_m' \simeq 6.32$ (dashed line). (Data from Kolber et al., 2005.)

7.5.5 Photochemical reflectance index (PRI)

A very different approach to the remote study of photosynthetic performance is to make use of the fact that the degree of xanthophyll epoxidation is an effective indicator of the degree of q_E quenching and hence of *NPQ* and stress (Gamon et al., 1992) with increasing de-epoxidation of violaxanthin to zeaxanthin at high light. The state of epoxidation can be estimated remotely from changes in absorptance or reflectance at 531 nm. The use of narrow-band measurements of reflectance at 531 nm and at 570 nm (to act as a control wavelength where absorptances of violaxanthin and zeaxanthin are similar) allows the calculation of what has been termed a *photochemical reflectance index* (PRI) as:

$$PRI = (\rho_{570} - \rho_{531})/(\rho_{570} + \rho_{531}) \qquad (7.23)$$

This is a normalised vegetation index that indicates the proportion of absorbed radiation that is dissipated as heat and not used for photosynthesis. This is of similar form to *NDVI* (Section 2.7.1) with lower values indicating a higher light-use efficiency (*LUE*). As *PRI* is a useful indicator of light-use efficiency, it becomes possible to obtain estimates of canopy assimilation by remote sensing as the product $I_S \cdot PRI \cdot fAPAR$ (where *fAPAR* is the fraction of absorbed PAR – see Chapter 3), while for any image where incident radiation can be assumed constant, *PRI·fAPAR* is a useful relative measure of net primary productivity. Although *PRI* can be reasonably reliable when used on single leaves, when applied to canopies results are less consistent with differing proportions of the leaves in full sun and shade. Therefore it is found that the relationship between *PRI* and *LUE* can be rather variable (see e.g. Jones & Vaughan, 2010).

7.6 Control of photosynthesis and photosynthetic 'limitations'

It is of general interest to be able to determine the relative importance of different processes, such as the light reactions, the dark reactions, diffusion through the boundary layer or the stomata, or transfer in the liquid phase to Rubisco, in controlling the rate of net assimilation. For example, workers concerned to improve crop photosynthesis need to be able to identify those processes that are most important in restricting the rate of assimilation in order to best target their efforts. Early studies were guided by Blackman's classic (1905) paper enunciating the *Principle of limiting factors*: 'When a process is conditioned as to its rapidity by a number of separate factors, the rate of the process is limited by the pace of the "slowest" factor'. Although the concept that the rate is limited by the slowest factor may be useful to a first approximation, it has led to a widespread failure to recognise that in well-adapted plants several factors (e.g. gas-phase and liquid-phase processes) may be evenly balanced and all contribute to the overall limitation; this implies a tendency towards optimal allocation of limiting resources. Before discussing methods for estimating the relative limitation imposed by stomatal and mesophyll processes it is useful to introduce a method that has increasingly been used for the analysis of control in biochemical pathways.

7.6.1 Control analysis

A widely applicable quantitative approach that can be applied to the analysis of any complex metabolic pathway, where changes in one component affect the performance of other steps (for example, by altering substrate concentration or by feedback regulation), as occurs in photosynthesis, has been introduced by Kacser and Burns (1973). These authors outlined a general theory of 'control' in relation to biochemical pathways that can be used to predict system response to *small* changes in one of the components, from a knowledge of the properties of the components in isolation. Central to the approach is the calculation of *control coefficients* or *sensitivity coefficients* (C) that reflect the response of steady-state fluxes (**J**) through the whole system to small changes in certain parameters (b_i) such as individual enzyme activities. In order to make the control coefficients independent of the units chosen

for **J** and b_i, they are defined in terms of fractional changes in these two quantities:

$$C_{bi} = (\partial \mathbf{J}/\mathbf{J})/(\partial b_i/b_i) \qquad (7.24)$$

Defined in this way the control coefficient varies from 0 (when the component being studied has no effect on the overall process) to 1 (when that step exerts complete control). The relative size of the different control coefficients in a pathway is a measure of their relative importance in controlling flux. An important property of the C_{bi} is that all the control coefficients in one pathway sum to 1.

A particular advantage of control analysis is that it provides a technique for estimating system behaviour (that is, control coefficients) from a knowledge of the *elasticities* (ε_i) of all the individual reactions when isolated from the whole pathway, where the individual ε_i are defined as the fractional changes in the *local* reaction rate for a fractional change in substrate concentration at that step (with all other components of the overall sequence being held constant). Further details of the calculation of elasticities and control coefficients have been described elsewhere (Fell, 1997; Kacser & Burns, 1973) while methods for the analysis of more complex metabolic networks are now being developed (Libourel & Sachar-Hill, 2008). There have been a number of attempts to apply control analysis to the study of photosynthesis using transgenic plants with altered expression of component enzymes such as Rubisco or different Calvin cycle enzymes (see Stitt & Sonnewald, 1995). Although control analysis is strictly only appropriate for small changes in any component the transgenic studies have successfully determined control coefficients for Rubisco that can vary from around 0.1 to as much as 0.8 (implying an 8% change in **P**$_n$ for a 10% change in Rubisco) at high light.

7.6.2 Resistance analysis

Traditionally, partitioning limitations has been achieved by the use of the electrical analogues that were introduced in Chapter 3, though, as we shall see below, these are not entirely appropriate for the description of biochemical processes. (Note that because of the series nature of the CO_2 uptake pathway it is most convenient to use resistances (as these sum linearly) rather than conductances, for this purpose. Though when considering one component alone it is common to refer to conductances because these are proportional to the rate.) In the resistance approach the whole photosynthetic system is treated as a linear process to which the transport equation applies, with the overall resistance to CO_2 uptake (r') being partitioned into, for example, gas-phase (r_g') and liquid-phase (otherwise known as 'internal' or 'mesophyll') (r_i') components, according to:

$$\mathbf{P}_n^m = (x_a' - x_x')/\Sigma r' = (x_a' - x_i')/r_g' = (x_i' - x_x')/r_i'$$

(7.25)

where x_a', x_i', and x_x', respectively, are the mole fractions of CO_2 in the ambient air, in the air in the intercellular spaces at the surface of the mesophyll cells, and 'internal' to the carboxylation site.

Unfortunately, as discussed below, there is some controversy about the meaning of a CO_2 concentration internal to the carboxylase and, though it is often assumed equal to the CO_2 compensation concentration (Γ), it results in r_i' being variable. This result means that resistance terminology is not really appropriate to describe limitations that include a biochemical component.

7.6.3 Calculation of intercellular carbon dioxide concentration (x_i')

An essential step in all approaches for partitioning photosynthetic limitations is the estimation of CO_2 mole fraction in the intercellular spaces (x_i'). The usual approach is first to determine the gas-phase resistance to water loss ($r_{gW} = r_{aW} + r_{\ell W}$, where $r_{\ell W}$ is the parallel sum of cuticular and stomatal resistances) using the methods outlined in Chapter 6; this is then converted to the corresponding gas-phase CO_2 transfer resistance (r_g') assuming that the paths for CO_2 and water vapour are similar (see Table 3.2):

$$r_g' = r_a' + r_\ell' = (D_W/D_C)^{2/3} r_{aW} + (D_W/D_C) r_{\ell W}$$
$$= 1.39 r_{aW} + 1.64 r_{\ell W} \simeq 1.6 r_{gW}$$

(7.26)

The approximate conversion factor of 1.6 is normally used where r_{aW} and $r_{\ell W}$ are not measured separately. Significant cuticular water loss can cause Eq. (7.26) to underestimate r_g' because the large liquid-phase component of the cuticular pathway would lead to a large extra resistance to CO_2 transport. Another slight complication in the conversion from a water vapour resistance to a CO_2 resistance arises from the fact that the distribution of sources for H_2O is primarily close to stomatal pores while the distribution of CO_2 sinks is more evenly distributed through the mesophyll (Parkhurst, 1994).

The final step is calculation of x_i' from a rearranged Eq. (7.25) as:

$$x_i' = x_a' - \mathbf{P}_n^m r_g' = x_a' - \mathbf{P}_n^m/g_g'$$

(7.27)

For accurate work, instead of using the very convenient mole fraction, it is slightly better to express CO_2 concentrations in terms of the partial pressure of CO_2 (p_C or p') replacing x' by p'/P. The use of partial pressure has an advantage over mole fraction for the study of CO_2 fluxes in the liquid phase because photosynthesis at Rubisco depends on the chemical potential of dissolved gas, which itself depends on the effective pressure, or fugacity, of the gas. This is proportional to the actual partial pressure with constant of proportionality being known as the fugacity coefficient, which is a measure of departure from the behaviour of an ideal gas. Von Caemmerer and Farquhar (1981) derived the following improved approximation:

$$x_i' = [(g_a' - (E/2)) x_a' - \mathbf{P}_n^m]/[g_g' + (E/2)]$$

(7.28)

Although the approximation represented by Eq. (7.27) is adequate for many purposes, in precise studies it is better to take account of the ternary interactions between air, H_2O and CO_2 (Farquhar & Cernusak, 2012).

7.6.4 Liquid-phase (internal) resistance

The liquid-phase internal (or intracellular) resistance (r_i') or its corresponding conductance is a complex term that depends on (i) the diffusion of CO_2 from the intercellular spaces through a series of resistances to the sites of carboxylation in the chloroplasts (r_m') and

(ii) an 'enzyme' component that depends on the activities of photochemical and biochemical processes including respiration (r_x'). Therefore the internal resistance in Eq. (7.25) can be further expanded into its two main components:

$$\mathbf{P}_n^m = (x_i' - x_x')/r_i' = (x_i' - x_c')/r_m' = (x_c' - x_x')/r_x'$$
(7.29)

where x_c' is the CO_2 mole fraction at the carboxylation site.

The diffusive component: mesophyll resistance (r_m')

The component relating to liquid-phase transfer from the intercellular spaces to Rubisco is usually now termed the *mesophyll resistance* (or equivalently the *mesophyll conductance* for its reciprocal), though it is worth noting that in some publications this term is applied to the whole internal pathway including the enzyme component. In its more restricted sense, the main components of r_m' include diffusion barriers due to transfer through the cell wall, the plasmalemma and chloroplast membranes, and within the cytosol and the chloroplast (Evans *et al.*, 2009; Pons *et al.*, 2009). As it is difficult to determine all these components accurately it is common to refer to the overall mesophyll conductance, g_m'. The cell wall thickness probably contributes c. 50% of g_m' and variation in the surface area of chloroplasts appressed to the intercellular spaces is another important contributor (Scafaro *et al.*, 2011). Application of the FvCB model of photosynthesis requires a knowledge of the mesophyll resistance to allow estimation of the CO_2 concentration at Rubisco.

Methods for estimation of the mesophyll resistance are based either on simple gas exchange, on gas exchange plus fluorescence (to estimate non-cyclic $ETR = \mathbf{I} \cdot \alpha_{leaf} \cdot f_{PSII} \cdot \phi_{PSII}$), or on gas exchange plus carbon isotope discrimination (for details of the calculations see Pons *et al.*, 2009; Tazoe *et al.*, 2011).

Biochemical resistance (r_x')

The major component of r_x' is usually related to carboxylation. The definitions of r_i' in Eq. (7.25) and of r_x' using Eq. (7.29) involve an important extension of the transport equation to biochemical processes. There are two main difficulties with Eqs. (7.25) and (7.29) for defining these 'biochemical' resistances. The first difficulty arises because the calculated r_x' increases as CO_2 saturation approaches, so that unlike the situation for the gas-phase resistance (a true transport resistance), the concentration drop across the mesophyll does not vary in direct proportion to the flux. The second problem is the need to have an estimate of r_x'; although it was originally assumed equal to 0 (Gaastra, 1959), probably with some validity for C_4 plants, in other cases this assumption leads to the calculated r_x' being variable for all x_i'. Both these problems limit the general predictive value of Eqs. (7.25) and (7.29). A reasonable approach is to limit application of the resistances (r_x' and r_i') to the linear part of the CO_2 response curve and to assume that x_x' equals the compensation concentration (Γ), that is, the CO_2 concentration at which $\mathbf{P}_n^m$ is zero. Therefore the overall resistance (r_i') can be obtained from either:

$$r_x' = 1/g_x' = (x_i' - \Gamma)/\mathbf{P}_n^m$$
(7.30)

or:

$$r_x' = 1/g_x' = dx_i'/d\mathbf{P}_n^m$$
(7.31)

That is, r_i' is given by the inverse of the *initial* slope of the $\mathbf{P}_n^m - x_i'$ curve. This restricted definition of r_i' is the most useful as it is always valid (behaving as a true resistance). In all other cases (e.g. at non-limiting CO_2 or when x_i' is not equal to Γ) it is more appropriate to call the calculated internal resistance a 'residual' resistance.

Some typical values for $\mathbf{P}_n$ and $\mathbf{R}_d$, and for r_m' and r_ℓ' are given in Table 7.3. This shows that the ratio r_ℓ'/r_m', tends to be greater in C_3 than in C_4 plants. Further discussion of the relative contribution of different processes to the overall internal resistance (r_i') or conductance (g_i') may be found in Tholen and Zhu (2011).

7.6.5 Photosynthetic limitations

Relative limitation due to stomata and mesophyll

A very widely used graphical device for investigating the relative contribution of gas- and liquid-phase resistances was introduced by Jones (1973b) and is

Table 7.3 Some representative values for various carbon dioxide exchange parameters for different species. P_n at saturating light and normal ambient CO_2 (μmol m^{-2} s^{-1}), R_d (μmol m^{-2} s^{-1}), Γ (μmol mol^{-1}), minimum r_m' (sum of mesophyll and biochemical components; m^2 s mol^{-1}), and corresponding values of r_ℓ' (m^2 s mol^{-1}) and r_ℓ'/r_m'. Values converted from original units assuming a temperature of 20°C.

	P_n	R_d	Γ	r_m'	r_ℓ'	r_ℓ'/r_m'	Reference
C_3 plants							
Atriplex hastata (young plants)	25	2.5	50	6.8	1.8	0.26	1
Wheat (9 species)	24	–	41	7.5	2.5[a]	0.33	2
Sitka spruce (field)	11	0.7	50	12.5	10.8	0.86	3
4 tropical legumes (30°C)	18	1.8	35	8.3	1.8	0.22	4
Larrea divaricata	23–27	3–4	–	6.5[b]	2.5	0.38	6
C_4 plants							
Atriplex spongiosa	40	2.3	0	1.5	4.3	2.8	1
6 pasture grasses (30°C)	36	2.5	0	2.5	2.1	0.85	4
Zea mais	14–55	1.4	0	2.0	6.0	3.0	7
CAM plants							
Kalanchoe diagremontiana (light)	6	–	51	20	21	1.06	5
Kalanchoe diagremontiana (dark)	5	–	–	–	25	–	5

[a] $= r_g'$
[b] $=$ estimated

References: 1. Slatyer (1970); 2. Dunstone *et al.* (1973); 3. Ludlow and Jarvis (1971); 4. Ludlow and Wilson (1971); 5. Allaway *et al.* (1974); 6. Mooney *et al.* (1978); 7. Gifford and Musgrave (1973) and El-Sharkawy and Hesketh (1965).

illustrated in Figure 7.12. Figure 7.12(*a*) shows a response curve relating P_n^m to x_i', obtained by varying ambient CO_2 concentration and plotting P_n^m obtained against the calculated x_i'; this curve represents the photosynthetic 'demand function'. Figure 7.12(*b*) shows the relationship between P_n^m and x_i' calculated according to Eq. (7.25) for two different values of r_g'; these straight lines represent photosynthetic 'supply' functions, and indicate the drop in mole fraction across the gas phase, showing how x_i' would fall below x_a' as P_n^m increases. The actual value of x_i' and P_n^m in any particular conditions is obtained from the

intersection of the supply and demand functions (Figure 7.12(*c*)).

A discussion of the rationales of different methods proposed for partitioning limitations may be found in a number of papers (Farquhar & Sharkey, 1982; Grassi & Magnani, 2005; Jones, 1985a; Lawlor, 2002). As we have seen above, implicit in the use of resistance analogues is the assumption that they provide a means for quantifying the relative contributions of different processes to the control of P_n. In principle, the ratio of the resistances of different components (e.g. r_g' and r_i'), or the concentration drops across

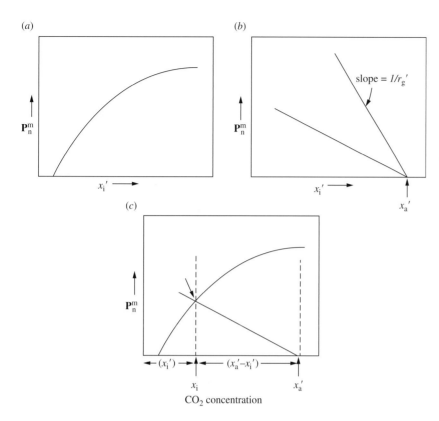

Figure 7.12 (a) Response curve or 'demand' function relating P_n^m to x_i', (b) two 'supply' functions relating P_n^m to x_i' for different gas-phase resistances, with the slopes of the lines equal to $-1/r_g'$ ($= -g_g'$), and (c) simultaneous solution of supply and demand functions to obtain x_i' at the operating point indicated by the arrow.

them, is a measure of their relative control of $\mathbf{P}_n$ (see Figure 7.13). On this basis the gas-phase limitation (ℓ_g') can be defined as the relative contribution of gas-phase diffusion to the overall limitation as in:

$$\ell_g' = r_g'/(r_g' + r_i') = r_g'/(\Sigma r') \tag{7.32}$$

Unfortunately this definition of limitation is only useful under CO_2-limiting conditions and fails as CO_2 saturation is approached. The reason may be illustrated by taking an extreme case: consider a plant where the photosynthetic system is operating at its maximum rate ($\mathbf{P}_{max}$) so that diffusion is non-limiting. In this case the calculated value of ℓ_g' may be finite because r_g' and r_i' remain finite, but clearly it is not reasonable to say that there is a finite gas-phase limitation as a small change in stomatal conductance would have no effect on $\mathbf{P}_n$.

The second alternative approach suggested by Farquhar & Sharkey (1982) involves calculation of the gas-phase limitation as the relative reduction in

actual $\mathbf{P}_n$ below the potential rate ($\mathbf{P}_n^o$) at infinite gas-phase conductance (see Figure 7.13). Unfortunately the assumption of infinite gas-phase conductance as used in this method is unrealistic and little better (and no more likely to be achieved in practice) than the logical alternative of deriving it at infinite mesophyll conductance.

Although this approach (Farquhar & Sharkey, 1982) is often used, where possible it seems more appropriate to use a method that is more generally valid and does not involve unrealistic extrapolation. Returning to control analysis, it is clear that one such option, by analogy with the definition of a control coefficient, would be to define the gas-phase limitation (ℓ_g') as the relative sensitivity of $\mathbf{P}_n$ to a small change in r_g':

$$\ell_g' = (\partial\mathbf{P}_n/\mathbf{P}_n)/(\partial r_g'/r_g') = (\partial\mathbf{P}_n/\partial r_g')/(r_g'/\mathbf{P}_n) \tag{7.33}$$

It can be shown (Jones, 1973b) that this is equivalent to:

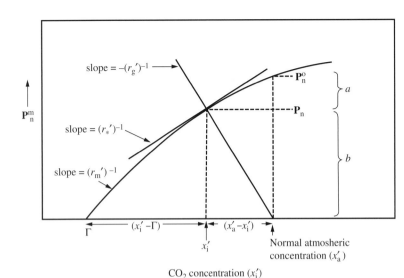

Figure 7.13 Calculation of photosynthetic limitations (see text for details). The heavy line represents the demand function relating P_n^m to x_i' for a hypothetical leaf. The CO_2 concentration drop across all the internal limitations in the mesophyll is represented by $(x_i' - \Gamma)$, and that across the gas phase by $(x_a' - x_i')$. Farquhar and Sharkey's (1982) definition of ℓ_g' is given by $a/(a + b)$; the recommended definition (Eq. (7.34)) is $r_g'/(r_g' + r_*')$.

$$\ell_g' = r_g'/(r_g' + r_*') \qquad (7.34)$$

where r_*' is the slope dx_i'/dP_n at the operating point (Figure 7.13). Advantages of this approach include: (i) it does not involve extrapolation, and (ii) it is simple to apply. It is particularly useful in plant breeding or in studies of plant response to environment where one is concerned with small changes in stomatal resistance. This approach can be extended to partition the non-stomatal component of limitation further into the mesophyll and biochemical components (Grassi & Magnani, 2005).

Thus far only the relative limitation by gas-phase and mesophyll processes under one set of conditions has been considered. In practice, it is useful to determine the relative contributions of different components to a *change* in P_n in response to factors such as water stress (see Jones, 1985a for a detailed discussion of possible methods). Unfortunately there is again no absolute method for this as the result strictly depends on the exact sequence of changes. Nevertheless if one considers only the initial and final states one can define the change in stomatal limitation ($\sigma_{1,2}$) between two states P_{n1} and P_{n2} as:

$$\sigma_{1,2} = \left(\ell_g' P_{n1} - \ell_g' P_{n2}\right)/\left(P_{n1} - P_{n2}\right) \qquad (7.35)$$

Effect of stomatal heterogeneity

A critical assumption in gas-exchange studies and especially in the calculation of the CO_2 concentration (expressed in terms of mole fraction) at the cell wall and the mesophyll resistance is that the stomata are equally open over the whole area of leaf whose gas exchange is being measured. If that assumption does not hold, as for example where the stomata close in 'patches', P_n^m, r_g' and the calculated values of x_i' and r_i' only represent average values; this can give very misleading indications of physiological responses to environment. The reason for this is illustrated in Figure 7.14. All leaves are to some extent heterogeneous and evidence is accumulating that there can be variation in stomatal aperture over the leaf surface, especially when abscisic acid is applied exogenously (Terashima et al., 1988), but also in many other situations (Mott & Buckley, 1998). The implications of heterogeneity are most important where its degree changes as a result of an environmental change. It is therefore essential to confirm whether stomatal heterogeneity exists for any material being studied before applying resistance analysis. Recent advances in imaging technology, such as the use of thermal imaging to study differences in leaf temperature resulting from patchy stomatal closure or the use of chlorophyll

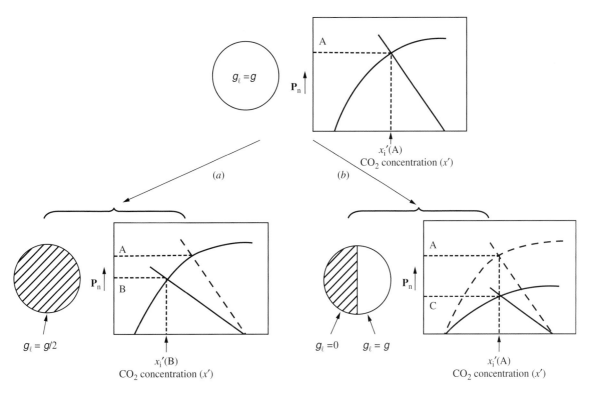

Figure 7.14 An illustration of the effect of patchy stomatal closure on calculated x_i' and on the estimated mesophyll properties. The top figure represents a healthy leaf with the photosynthesis demand function given by the solid line, resulting in a photosynthetic rate A. Assuming no change in the photosynthetic properties of the leaf we can compare what happens if (a) all the stomata close by 50%, or (b) if half the stomata (in discrete patches unconnected via the intercellular spaces) close completely. In (a) stomatal conductance per unit area falls to $g/2$, assimilation decreases according to the response curve, falling to B (>A/2), and the calculated x_i' decreases significantly (implying increased stomatal control of photosynthesis). In (b) on the other hand, although the average stomatal conductance over the large area also falls to $g/2$, assimilation would now decrease to C (= A/2) as only half the area is now photosynthesising, but because both A and g_ℓ have changed similarly it follows from Eq. (7.26) that x_i' is apparently unchanged. In addition there is a large apparent shift in the photosynthetic demand function with the photosynthetic capacity apparently reduced by 50% even though there is no real change in the mesophyll properties.

fluorescence imaging to follow changes in photosynthetic electron transport, provide powerful tools for the study of the dynamics of stomatal patches (West *et al.*, 2005).

7.7 Carbon isotope discrimination

The two stable isotopes of carbon (^{13}C and ^{12}C) occur in the molar ratio of 1:89 in the atmosphere, but the relative abundance of ^{13}C in plant material is commonly less than its abundance in atmospheric CO_2.

The value of the molar abundance ratio, ^{13}C/^{12}C, in plant material provides information on the physical and chemical processes involved in its synthesis because processes such as diffusion and carboxylation discriminate against the heavier isotope by different amounts (see Ehleringer *et al.*, 1993; Farquhar *et al.*, 1989 for useful reviews). These 'isotope effects' result in an amount of discrimination (Δ), usually defined by:

$$\Delta = \frac{(^{13}C/^{12}C)_{reactants}}{(^{13}C/^{12}C)_{products}} - 1 \qquad (7.36)$$

Table 7.4 Typical values for the molar abundance ratio ($^{13}C/^{12}C$), the deviation of isotopic composition from that of the PDB standard ($\delta^{13}C$ or δ) and the isotopic discrimination (Δ) (from data quoted by Farquhar et al., 1989).

	$^{13}C/^{12}C$	$\delta^{13}C$ (‰)	Δ (‰)
PDB standard	0.011237	0	–
Free atmospheric CO_2 (in 1988)	0.01115	−7.7	–
C_4 plant material	0.01085 to 0.01102	−20 to −35	13 to 28
C_3 plant material	0.01107 to 0.01116	−7 to −15	−1 to 7
CAM plant material	0.01099 to 0.01112	−10 to −22	2 to 15
Coal	0.01087	−32.5	–

In practice Δ is not measured directly, and isotopic abundances (determined using a mass spectrometer) are usually expressed as deviations $\delta^{13}C$ or δ from the abundance ratio in a fossil belemnite formation (the PDB standard, where $^{13}C/^{12}C = 0.01124$), as:

$$\delta = \frac{(^{13}C/^{12}C)_{sample}}{(^{13}C/^{12}C)_{standard}} - 1 \qquad (7.37)$$

By combining Eqs. (7.36) and (7.37) it follows that:

$$\Delta = (\delta_a - \delta_p)/(1 + \delta_p) \qquad (7.38)$$

where the subscripts 'a' and 'p' refer to the air and plant material respectively. Some typical values of Δ and δ are presented in Table 7.4, which shows that although δ tends to be negative when using the PDB standard, the value of the discrimination, Δ, is positive. Because both Δ and δ are small numbers they are conventionally presented as parts per thousand (‰) so that using ‰ as equivalent to 10^{-3} we can write the typical Δ for C_3 plants ($0.02 = 20 \times 10^{-3}$) as 20‰.

The value of δ_a has been declining as a result of burning fossil fuels (from −6.7‰ in 1956 at 314 ppm to −7.9‰ in 1982 at 342 ppm), and can also vary seasonally (by about 0.2‰) and diurnally (Farquhar et al., 1989). In large conurbations these fluctuations may be as large as 2‰ as a result of human activities. This variation in the isotopic composition of the source air provides another reason for the general use

in physiological studies, where possible, of Δ rather than δ. The value of δ is expected to decline as the atmospheric CO_2 concentration increases. For example, at the bottom of a dense tropical forest where respiratory processes release fixed C (with a smaller proportion of ^{13}C than in air) and raise the concentration of CO_2, to 400 vpm or more, δ_a can be as low as −11.4‰.

Discrimination can result from both *equilibrium effects*, such as the different relative concentrations that occur in the gas phase and liquid phase when in equilibrium, and *kinetic effects* such as those involved in enzyme reactions or transport processes. The main factors determining Δ in C_3 plants are diffusion in the air (including in the boundary layer and through the stomata) where Δ is about 4.4‰, and carboxylation at Rubisco where Δ is about 30‰. Farquhar et al. (1982) showed that the net effect is approximated by:

$$\Delta = 0.0044(x_a' - x_i')/x_a' + 0.030 x_i'/x_a'$$
$$= 0.0044 + 0.0256 x_i'/x_a' \qquad (7.39)$$

(The true situation is slightly complicated by the fact that the actual substrate (HCO_3^-) has a higher $^{13}C{:}^{12}C$ ratio than the CO_2 with which it is in equilibrium.) This equation shows that the value of Δ depends on x_i'/x_a', which itself depends on the stomatal aperture: when stomata are closed x_i' declines and Δ also falls. Other fractionations that are relevant to photosynthesis include solubility of CO_2 in water (1.1‰), CO_2 diffusion in water (0.7‰), hydration of CO_2 (−9.0‰) and fractionation as a result of photorespiration, but the net effect of these extra effects can usually be disregarded.

In C_4 plants, Δ is relatively insensitive to x_i'/x_a' as the large discrimination by Rubisco (c. 30‰) is replaced by a much smaller or negative effective discrimination by PEP carboxylase (c. −5.7‰). This net discrimination by PEP carboxylase is the result of a 2‰ Δ for fixation of HCO_3^-, combined with factors dependent on the 'leakiness' of the bundle sheath and the equilibrium between gaseous CO_2 and HCO_3^- in solution.

In practice, environmental effects on photosynthesis can substantially affect the value of $\delta^{13}C$. The major effect relates to factors such as stomatal closure or variation in irradiance that affect x_i'/x_a'. In the case of species that can switch between CAM and C_4 metabolism, their observed discrimination also changes as expected. The use of Δ in studies of water use efficiency is discussed in Chapter 10.

7.8 Response to environment

Maximum values of $\mathbf{P}_n^m$ at normal CO_2 concentrations and saturating light are usually 14 to 40 $\mu mol\ m^{-2}\ s^{-1}$ for C_3 leaves, 18 to 55 $\mu mol\ m^{-2}\ s^{-1}$ for C_4 and up to about 10 $\mu mol\ m^{-2}\ s^{-1}$ for CAM plants (see Table 7.5). Leaf respiration rates in the dark are often in the range 0.5 to 3 $\mu mol\ m^{-2}\ s^{-1}$. There are large variations in respiration rates of other tissues: for example, fruits tend to have high respiration rates during early cell division and also to have another peak called the 'climacteric rise' near maturity. The rate of photorespiratory carbon loss in photosynthesising C_3 leaves is probably between 20 and 25% of $\mathbf{P}_n$.

7.8.1 Carbon dioxide and oxygen concentration

The main reason for determining the photosynthetic CO_2 response is to provide mechanistic information for interpreting photosynthetic responses to

Table 7.5 Typical maximum rates of photosynthesis at saturating light and normal carbon dioxide for leaves and canopies, together with short-term crop growth rates.

	C_3	C_4	CAM
$\mathbf{P}_n$ ($\mu mol\ m^{-2}\ s^{-1}$)			
Single leaves	14–60	18–55	8
Crops	14–64	64	–
Crop growth rate (g m^{-2} day^{-1})			
Typical non-stressed	15–30	15–50	3–5
Maximum	34–39	51–54	7

Data from Allaway et al., 1974; Cooper, 1970; Kluge and Ting, 1978; Milthorpe and Moorby, 1979; Monteith, 1976, 1978; Nobel, 2009; Slatyer, 1970.

environment. It is only in controlled environments and glasshouses where x_a' can be artificially manipulated and, in relation to long-term climatic changes in atmospheric CO_2 (see Chapter 11), that the CO_2 response curves are of direct interest.

The main physiological differences between photosynthetic pathways are illustrated by the CO_2 responses of C_3 and C_4 leaves in normal air (21% O_2) or low oxygen (Figure 7.15). Reduction of O_2 concentration only enhances $\mathbf{P}_n$ in C_3 plants (and in CAM plants during the light period). The value of Γ is normally close to zero for C_4 (or night-time CAM) where the primary fixation is by PEP carboxylase, as it is for C_3 plants when photorespiration is inhibited by low O_2. Plotting these results as a function of x_i' shows the smaller r_i' in C_4 leaves (averaging 2 to 5 $m^2\ s\ mol^{-1}$) compared with 5 to 12.5 $m^2\ s\ mol^{-1}$ in C_3 leaves.

In C_3 plants x_i' is often maintained at between 0.6 x_a' and 0.7 x_a' (200 to 230 vpm at normal ambient concentrations) compared with 0.3 x_a' to 0.4 x_a' (about 100 to 130 vpm) in C_4 plants (Wong et al., 1979). However, because of the different time constants of intracellular and stomatal responses to factors such as

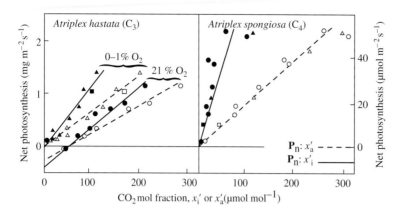

Figure 7.15 The initial linear portions of CO_2 response curves for typical C_3 and C_4 species, showing a response to O_2 concentration in C_3 but not C_4 leaves. Zero O_2 – triangles, 1% O_2 – squares, 21% O_2 – circles; solid lines and closed symbols give P_n: x_i' response curves, dashed lines and open symbols refer to P_n: x_a' response. (After Slatyer, 1970.)

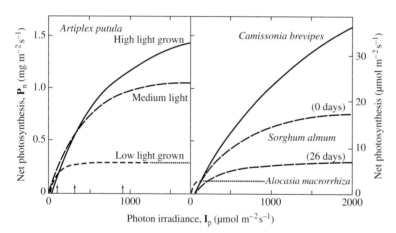

Figure 7.16 Photosynthetic light-response curves for (a) *Atriplex patula* grown at three different irradiances as shown by the arrows (after Björkman et al., 1972a) and (b) *Alocasia macrorrhiza* from extreme shade (after Björkman et al., 1972b), newly unfolded or 26-day old *Sorghum almum* leaves (after Ludlow & Wilson, 1971) and *Camissonia brevipes* (after Armond & Mooney, 1978).

irradiance, x_i' is likely to be rather more variable in the fluctuating environment in the field.

The potential impacts on photosynthesis of increasing atmospheric CO_2 concentrations will be discussed further in Section 11.3.3 where the differences between short-term and long-term responses and the differences between leaf-scale and canopy-scale responses will be addressed.

7.8.2 Light

Typical light-response curves for several different species, and for one species grown at a range of irradiances, are shown in Figure 7.16. Although there is a slight tendency for P_n in C_4 species to continue increasing at higher irradiances than in C_3 species, there are greater differences between sun and shade

species, or between leaves of one species grown at different irradiances (Figure 7.16). In shade species or in shade-grown leaves, P_n may be saturated at less than 100 μmol m^{-2} s^{-1} (PAR), which is approximately 5% full sunlight. Sun leaves, on the other hand, often continue to respond up to typical values for full sunlight. The light compensation point also varies from as low as 0.5 to 2 μmol m^{-2} s^{-1} in extreme shade species such as *Alocasia macrorrhiza* growing in the Queensland rainforest to as much as 40 μmol m^{-2} s^{-1} in sun leaves.

As there are now many hundreds of publications on photosynthetic light-response curves of different species there is increasing effort on meta-analyses that aim to generalise results from a range of disparate independent experiments (Wright et al., 2004). One such meta-analysis has been the derivation of models that could predict the actual light

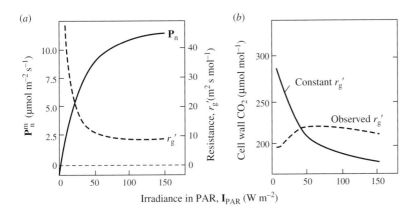

Figure 7.17 (a) Light-response curves for P_n and r_g' in Sitka spruce (data from Ludlow & Jarvis, 1971). (b) The same data plotted to show calculated variation of x_i' with constant r_g' or with the observed r_g'.

responses for different species on the basis of traits such as nitrogen (N) content or leaf mass per unit area (Marino *et al.*, 2010). A difficulty with interpreting light-response curves is that the CO_2 concentration at the cell wall or photosynthetic site is not constant with changing light. This is because the internal CO_2 concentration is a function of P_n and r_g', and (see Eq. (7.26)), as shown in Figure 7.17, though the effect of the stomatal response to light is to minimise these changes in x_i'.

Several factors contribute to the differences in the photosynthetic behaviour of sun and shade leaves. There is good evidence that all components of the photosynthetic system adapt together: for example, high-light leaves tend to be thicker with a greater internal surface area than shade leaves and have more chlorophyll and very much more carboxylase per unit area. In addition, although there is evidence that r_i' per unit cell surface area is approximately constant, internal resistances tend to decrease with increasing irradiance as a result of changes in A_{mes}/A (the ratio of mesophyll cell surface area to leaf surface area: see e.g. Nobel *et al.*, 1975), as do stomatal resistances. The light reactions are also affected by growth irradiance; there is more extensive granal development in shade leaves but the capacity for electron transport is markedly reduced. For example, electron transport through both photosystems (when expressed on a chlorophyll basis) may be as much as 14 times higher in chloroplasts extracted from sun plants than in those from shade plants (Boardman *et al.*, 1975).

Part of the effect may result from a slightly smaller photosynthetic unit size (the ratio of collector chlorophyll to reaction centres) but at least for photosystem I, the ratio of chlorophyll to the reaction centre (P_{700}) is often quite constant. It is the cytochrome f and cytochrome b components of the transport chain that are particularly reduced in low-light plants. The various adaptations in response to altered irradiance can occur within days of the change.

In spite of large differences in light-saturated P_n, only small differences have been observed in the initial slope of the photosynthetic light-response curves in healthy leaves. The reciprocal of this slope, the quantum requirement (quanta per CO_2 fixed), is a measure of the efficiency of photosynthesis and is relatively constant at a value of about 19 for C_4 plants, but is strongly dependent on temperature and oxygen concentration in C_3 plants (Figure 7.18).

Very high irradiances can damage the photosynthetic system, particularly in shade-adapted leaves, or in leaves where photosynthetic metabolism has been inhibited by other stresses such as extreme temperature or water stress. The damage can be a result of photo-oxidation, where bleaching of the chlorophyll occurs. Where no bleaching is observed the damage is usually termed photo-inhibition, though this term usually also includes reductions in photosynthetic electron transport that result from protective mechanisms. Avoidance of high-light damage involves a range of mechanisms for sensing and adapting to the high light (Li *et al.*, 2009). For

(a)

(b)

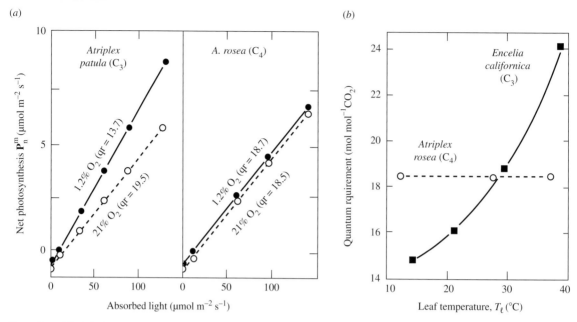

Figure 7.18 (a) Light-limited photosynthetic rates in *Atriplex patula* and *A. rosea* leaves at 1.2% and 21% O_2, showing effects on the quantum requirement (qr) (after Björkman *et al.*, 1970). (b) Temperature dependence of quantum requirement in C_3 and C_4 plants (after Ehleringer & Björkman, 1977).

example movement of the chloroplasts so that they become oriented parallel to the radiation, thus reducing radiation absorbed, is mediated by the detection of excess blue light by phototropins, while cryptochromes mediate a range of changes in gene expression, including the enhanced synthesis of protective anthocyanins. In addition the direct biochemical or biophysical effects of excessive electron transport including altered thylakoid lumen pH, altered redox states (especially over-reduction of the plastoquinone pool) and production of reactive oxygen species can directly lead to adaptive processes such as enhanced q_E and *NPQ* and reductions in photosynthetic antenna size.

We have already noted that mitochondrial respiration in the light is slower than in the dark. In addition, dark respiration rates are also influenced by growth irradiance, being as low as 0.1 µmol m^{-2} s^{-1} in shade plants compared with 1.0 to 3.4 µg m^{-2} s^{-1} in sun leaves. This difference can contribute to the net photosynthetic advantage at low light that is often exhibited by shade leaves.

7.8.3 Water status and salinity stress

The effects of water deficits and salinity on assimilation have also been the subject of many hundreds of papers over the years (see reviews such as Chaves *et al.*, 2009; Lawlor & Tezara, 2009). Although both stresses generally lead to reductions in assimilation there is still debate concerning the relative importance of stomatal closure and the complex changes in mesophyll photosynthetic capacity in causing the observed decreases in assimilation. In natural situations control is usually distributed between the different components of the photosynthetic pathway, even though many earlier rapid-stress or pot studies suggested that changes in stomatal conductance were the main cause of the decreased photosynthesis. Most field studies and those using more realistic longer term drying cycles have indicated that there is coordinated regulation of most photosynthetic components such that their relative limitations remain nearly constant (e.g. Figure 7.19 and Table 7.6). Because of this shared control it is not

Table 7.6 Effect of medium-term water stress on respiration and photosynthesis of potted apple 'mini-trees' (unpublished data of L. Fanjul and H. G. Jones). A useful review of the many experiments on the effects of water stress on carbon dioxide exchange in different species may be found in Lawlor and Tezara (2009).

	ψ_ℓ	P_n	$R_p{}^a$	R_d	$\Gamma_{21\% \, O_2}$	$\Gamma_{2\% \, O_2}$	$(R_d + R_p)/P^n$
	(MPa)	(μmol m^{-2} s^{-1})			(μmol m^{-2} s^{-1})		
cv. James Grieve (24 days stress)							
Control	−1.3	27.3	14.1	1.6	49	7[b]	0.47
Moderate stress	−2.0	17.3	7.0	2.3	32	1[b]	0.53
Severe stress	−2.8	9.5	3.9	1.6	71	28[b]	0.56
cv. Egremont Russet (14 days stress)							
Control	−1.0	14.1	5.7	2.0	74	32	0.55
Moderate stress	−1.8	10.0	3.2	2.7	78	37	0.59
Severe stress	−3.6	2.0	1.4	1.8	136	83	1.56

[a] R_p was estimated from the difference in P_n at 2% and 21% O_2, so is an overestimate of true photorespiration.
[b] 1% O_2.

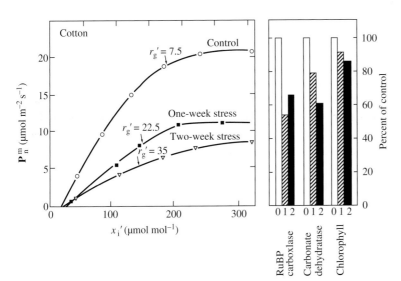

Figure 7.19 Effect of moderate-term water stress (zero, one- or two-week stress) on various photosynthetic parameters in cotton (resistances, m^2 s^1 mol^{-1}). (Data from Jones, 1973c.)

generally useful to attempt to identify any single factor as all-important, though the balance between components does vary with the type of stress, even though one often finds that the stomatal component may dominate in short-term mild stresses. Over the development of longer term stresses much of the parallel variation in activities of different photosynthetic components involves coordinated changes in transcription and protein synthesis (Chaves *et al.*, 2009). Evidence from chlorophyll fluorescence, however, suggests that electron transport is relatively little affected by low water potential, though high irradiance during the water stress can lead to photo-inhibitory damage at PSII. The possibility exists that non-uniform stomatal closure has been the cause of some reports of lowered mesophyll photosynthetic capacity and constant x_i', but this is unlikely to be the case for data such as those in Figure 7.19 where enzymes were extracted from the whole leaf.

The use of O_2 evolution measurements at saturating CO_2 can provide a useful independent check of whether the potential maximum photosynthesis has been affected by stress, though by itself this does not indicate whether such reductions contribute to decreases in photosynthesis. In general C_4 plants seem to be at least as sensitive, if not more so, to water deficits than are C_3 plants (Ghannoum, 2009). Estimates of changes in C_4 photosynthetic *capacity* using the O_2 electrode are made difficult by the strong sensitivity of C_4 stomata to the elevated CO_2 needed to saturate photosynthesis to obtain potential rates, with as much as 20% CO_2 being necessary in some cases, even though 2% is usually adequate in C_3 plants. Because of concerns that high CO_2 may interfere with biochemistry, some workers have proposed using leaf discs with the epidermis removed or leaf slices (see Jones, 1973d; Kaiser, 1987; Tang *et al.*, 2002). Results from such experiments confirm that photosynthetic capacity is affected by water stress as much in C_4 as in C_3 plants, but neither such an effect nor the changes in activity of the photosynthetic enzymes that are often observed with water stress (Ghannoum, 2009)

provide any evidence that these changes actually determine the reduction in photosynthesis (rather than just being an example of co-adaptation). The topic of water stress effects on CO_2 exchange will be discussed further in Chapter 10, where the importance to crop growth of changes in leaf area will be emphasised.

Although respiration and photosynthesis are independent processes, there is some tendency towards homeostasis in the ratio R/P_n; however, this does depend on the proportion of young tissues (which have relatively high respiration rates), on the temperature and on the tissue water status (Atkin *et al.*, 2007). Respiration tends to be substantially less sensitive to drought than is photosynthesis, so the ratio R/P_n usually increases, thus explaining some of the decline in P_n and the tendency for Γ to increase with stress (Table 7.6). In general water deficits inhibit respiration, certainly in roots and in whole plants, though in some cases slight increases or no effect have been reported for mature leaves (Atkin & Macherel, 2009). The cases where there is little effect of drought on leaf respiration (and cases where respiration increases) may relate to an increased need to oxidise the excess redox equivalents generated as CO_2 fixation is inhibited; this can be achieved by a shift from the COX to the AOX pathway (Ribas-Carbo *et al.*, 2005). Although most studies have concentrated on the effects of water deficits on respiration in the dark, it appears that mitochondrial respiration in the light is similarly inhibited. The observed changes in respiration are modulated by changes in the proton gradient in the mitochondria, by changes in partitioning of electrons between the ATP-generating COX and the AOX pathway and, on occasions, by changes in substrate availability.

7.8.4 Temperature and other factors

Over the normal physiological temperature range, P_n shows a broad optimum response as will be discussed in more detail in Chapter 9. A useful discussion of modelling of the temperature response of photosynthesis on the basis of appropriate extensions of the FvCB photosynthesis model may be found in Dreyer *et al.* (2001). The decline at high temperatures results partly from the more rapid increase of respiration with temperature and partly from a time-dependent photosynthetic inactivation at high temperatures. The fact that Γ increases markedly with temperature (Bykov *et al.*, 1981) is a manifestation of this changing balance between CO_2 fixation and release including enhanced photorespiration at high temperatures (see Figure 7.18).

On the basis that dark respiration can be partitioned into a temperature-sensitive maintenance component ($\mathbf{R}_m$) and a temperature-insensitive growth component ($\mathbf{R}_g$), McCree (1970) proposed the following expression for total $\mathbf{R}_d$ over a 24 h period:

$$\mathbf{R}_d = \mathbf{R}_g + \mathbf{R}_m = a\mathbf{P}_g + bW \tag{7.40}$$

where $\mathbf{P}_g$ (g m^{-2} day^{-1}) is total photosynthesis (excluding that lost in photorespiration), W (g m^{-2}) is the leaf mass in CO_2 equivalents ($44/12 \times$ carbon content of the leaf $\simeq 44/30 \times$ dry mass), and a (dimensionless) and b (day^{-1}) are constants. The value of a is usually 0.25 to 0.34 and is relatively independent of species or temperature though it depends on the length of the light period. The value of b ranges from 0.007 day^{-1} to 0.015 day^{-1} at 20°C, approximately doubling for every 10°C rise in temperature, so that:

$$b_{(T)} = b_{(20)} 2^{0.1(T-20)} \tag{7.41}$$

where $b_{(20)}$ is the value at 20°C and T is in °C. Temperature responses are discussed in more detail in Chapter 9.

Many other factors such as ambient humidity (Chapter 6) and wind speed (Chapter 11), which both primarily affect the gas-phase resistance, affect $\mathbf{P}_n$. In addition, factors such as leaf age (responses parallel the stomatal response shown in Figure 6.7; see also Figure 7.13), altered nutrition, disease, etc. may affect any of the component processes (Grassi & Magnani, 2005). An important factor is the internal control of $\mathbf{P}_n$ exercised by the demand for photosynthate. $\mathbf{P}_n$ can be inhibited as much as 50% by removal of sinks such as developing grain in cereals (e.g. King *et al.*, 1967), and there are many reports that $\mathbf{P}_n$ is higher in fruiting than in non-fruiting plants (e.g. Hansen, 1970), with at least part of the difference being attributable to differences in stomatal aperture.

As well as leaves and leaf-like structures, several other plant organs can make important photosynthetic contributions. For example, ears of cereals, especially the awns, and the pods and stems of many species can have high rates of photosynthesis. Many green stems and fruits have relatively impermeable 'skins' with low conductances. In such cases their photosynthetic capacity acts to refix the respired CO_2 rather than to perform net photosynthesis (see Jones, 1981b).

7.9 Photosynthetic efficiency and productivity

7.9.1 Photosynthetic efficiency of single leaves

The photosynthetic efficiency (ε_p) may be defined thermodynamically as the available energy stored in plant dry matter expressed as a fraction of incoming radiant energy. The free energy content of the immediate product of photosynthesis (sucrose) is about 480 kJ (mol C)$^{-1}$ or 16 kJ g^{-1}. However, plant material also contains a range of other compounds (proteins, fats, etc.) that lead to an average energy content for dry matter of about 17.5 kJ g^{-1} (Monteith, 1972); this is approximately equivalent to 525 kJ (mol C)$^{-1}$.

The efficiencies of various steps involved in conversion of solar energy to carbohydrate in photosynthesis are illustrated in Table 7.7. If we consider only the radiation actually intercepted by green leaves, we first note that photosynthetically active radiation (PAR) comprises only around half of the energy in solar radiation (depending on solar altitude and atmospheric conditions). Next, a proportion (around 10 to 15%) of the intercepted PAR

Table 7.7 The maximal efficiencies (%) of different steps in energy conversion of intercepted solar radiation in photosynthesis (following Zhu *et al.*, 2010) showing the corresponding percentage of the original energy remaining after each step (net efficiency).

	Efficiency of step (%)		% remaining	
	C_3	C_4	C_3	C_4
Proportion in PAR	48.7		48.7	
Proportion absorbed by chlorophylls	89.9		43.8	
Photochemical efficiency	74.9		37.2	
Thermodynamic limit	63.0		26.0	
Carbohydrate biosynthesis	48.5	32.7	12.6	8.5
Photorespiration	51.6	100.0	6.5	8.5
Mitochondrial respiration	70.0	70.0	4.6	6.0

is lost by reflection or transmission by the leaves. Further losses include the photochemical inefficiency that relates to the fact that all photochemistry is driven by the energy of red photons (with the extra energy available in the more energetic blue photons being lost as heat) and the thermodynamic limits set by charge separation.

There are also inefficiencies associated with the fixation of CO_2 to sugars. C_3 photosynthesis requires a minimum of three ATP and two NADPH to fix one molecule of CO_2; under optimal conditions these can be provided by eight photons driving whole chain electron transport. C_4 photosynthesis, at least in the most efficient (NADP-malic enzyme) subtype, requires an extra two ATP, which could be generated by cyclic phosphorylation at PSI by three extra photons absorbed by PSI (making a total of 11).

Assuming an average energy content in the PAR for solar radiation of 220 kJ mol^{-1} photons, this makes the overall efficiency of gross conversion of absorbed PAR to free energy in sugars as approximately 27% for C_3 plants (i.e. $100 \times 480/(8 \times 220)$) and 18% for C_4 photosynthesis ($100 \times 480/(12 \times 220)$). Further losses to consider are the losses due to photorespiration,

which may be as much as 50% of fixed carbon for C_3 plants, and mitochondrial respiration, which probably accounts for at least 30% of the remaining fixed carbon and fuels the conversion of sugars to average plant dry matter. Overall, therefore, as seen in Table 7.7 only a maximum of 4.6% of the energy in solar radiation is fixed in C_3 plants with somewhat higher efficiencies in C_4 plants. These extra losses bring the typical minimum quantum requirements up to between about 15 and 22 (depending on temperature) for C_3 leaves and 19 for C_4 leaves (Figure 7.18). The advantage of C_4 arises because of the lack of photorespiratory losses, which more than compensates for the less efficient fixation process.

For a bright day ($I_S = 900$ W m^{-2}) even a high photosynthetic rate of 2.4 mg CO_2 m^{-2} s^{-1} (which is equivalent to a sucrose accumulation rate of 1.8 mg m^{-2} s^{-1} ($2.4 \times 30/44$)) represents an efficiency of only 3.2%. This low efficiency at high irradiance results from light saturation. In practice most single leaves do not achieve even this high efficiency because of non-optimal nutrition, water status or temperature, or because of internal factors such as senescence or sink limitation.

7.9.2 Scaling up from leaf to canopy photosynthesis

An important application of photosynthesis models is in the prediction of productivity in different environments of canopies with different structure (e.g. erect or horizontal leaves) or photosynthetic characteristics (e.g. C_3 or C_4). Models of canopy carbon (and water) exchanges are also critical components of land-surface models used in weather and climate modelling. The approaches used in the different models range from complex multilevel simulation models that take account of detailed leaf distributions and radiation penetration through the canopy (Whisler *et al.*, 1986) to simpler 'big leaf' models (Lloyd *et al.*, 1995; Sellers *et al.*, 1997). Details of particular models may be found elsewhere (de Pury & Farquhar, 1997; Harley & Baldocchi, 1995; Johnson *et al.*, 2010; Thornley & France, 2007; Whisler *et al.*, 1986).

The multilayer simulation models combine leaf photosynthesis models with a number of submodels for predicting (i) leaf angle distribution in the canopy; (ii) light distribution within the canopy (see Chapter 2 for details); (iii) stomatal resistance as a function of water status and irradiance at different levels in the canopy (see Chapter 6); and (iv) respiratory CO_2 loss (including that from non-photosynthesising tissues

and soil organisms). Overall canopy assimilation is then calculated by summing the contributions of each stratum (Figure 7.20). Unfortunately it is necessary to estimate a large number of parameters for this approach and calculations can become very tedious. For many purposes, especially for application to global carbon models, simpler models can be adequate, especially where daily totals are adequate.

For example, since leaves only reflect or transmit between 10 and 20% of incident PAR, in photosynthesis models it is only necessary to take account of the first or second interception of any ray of light. Various limiting cases can be considered.

1. *All leaves are at an acute angle to any direct radiation.* In this case direct irradiance at the surface of each leaf ($\mathbf{I_o} \cos \theta$) is low, so that each leaf is light limited and $\mathbf{P_n} \propto \mathbf{I}$ or $\mathbf{P_n} = \varepsilon_p \mathbf{I}$. Assuming that any diffuse irradiance is also within the light-limited range, total canopy photosynthesis is therefore proportional to light interception with a proportionality constant ε_p. At a high leaf area index, therefore, where all radiation is intercepted, canopy photosynthesis is approximated as $\varepsilon_p \mathbf{I_o}$.

2. *Very low leaf area index canopies.* For homogeneous canopies where $L < 1$, one only needs to take account of the first interception of

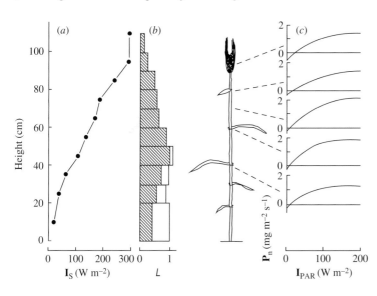

Figure 7.20 A method for determining canopy photosynthesis from (*a*) irradiance at different levels in the canopy, (*b*) corresponding leaf area index (*L*) (green leaf area shown hatched), and (*c*) photosynthesis-light response curves for each organ. Measurements are for a barley crop in central England on 28 June. (After Biscoe *et al.*, 1975a).

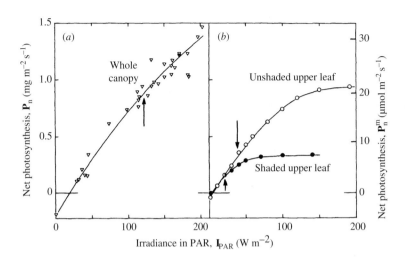

Figure 7.21 Photosynthetic light-response curves for tomato plants in daylit cabinets: (*a*) whole canopy and (*b*) individual leaves. The irradiance giving the highest photosynthetic efficiency is the point on each curve where a tangent to the curve passes through the origin; these points are indicated by arrows. (Data from Acock *et al.*, 1978.)

any radiation. For a horizontal-leaved canopy with no mutual shading, $\mathbf{I} = \mathbf{I}_o$ for all leaves. Therefore single-leaf photosynthetic models, taking account of the fraction of radiation intercepted by leaves, can be applied successfully.

In practice most canopies are between these two extremes, with their light response showing less tendency to light saturation than single leaves (Figure 7.21). As a result, canopy photosynthesis is greater than for single leaves, up to about 2.8 mg m^{-2} s^{-1} (64 µmol m^{-2} s^{-1}: Table 7.5). Another important consequence is that the irradiance giving maximal photosynthetic efficiency (ε_P) is higher for a crop canopy than for single leaves (Figure 7.21). In this example the optimal irradiance (giving maximal $\mathbf{P}_n/\mathbf{I}$) was between 15 and 25 W m^{-2} for single leaves and over 100 W m^{-2} for the whole canopy.

One application of canopy photosynthesis models has been the prediction of effects of leaf angle on productivity. A simple model to study these effects can be obtained by combining (i) the single-leaf photosynthesis model (assuming $K_m^l = 55$ µmol mol^{-1} and $r_g' = 5$ m^2 s mol^{-1}, and other parameters as for Figure B7.2) with (ii) a model for predicting shortwave irradiance on a horizontal surface (Eq. (2.11)) with β obtained according to Appendix 7) and (iii) the simple Beer's law model for predicting sunlit leaf area and

irradiance per unit sunlit area (Eq. (2.19) where $k = 1$ for a horizontal-leaved canopy and $k = 2/\pi \tan\beta$ for a vertical-leaved canopy). The results of some simulations for low, medium and high leaf area indices, for clear days at tropical and mid-temperate latitudes are shown in Figure 7.22.

These curves show that, as expected, erect leaves are only advantageous at high solar elevations and where L is moderately high. At the low solar elevations reached in temperate latitudes in the spring, the difference between total daily photosynthesis of erect- and horizontal-leaved canopies can be very small, even at high leaf area index. The midday depression of photosynthesis predicted for vertical leaves in the tropics arises when the sun is immediately overhead. Alteration of the model to include diffuse radiation would tend to reduce the effect of leaf orientation on daily photosynthesis.

'Big leaf' simplifications tend to be based on the assumption that canopy photosynthetic capacity varies with depth in the canopy according to the long-term average irradiance at that depth and also to a parallel change in leaf nitrogen (i.e. amount of Rubisco). Unfortunately these cannot adequately account for the changing penetration of radiation through the canopy during the day. Another approach to simplification of canopy models is to use canopy radiation transfer models to determine the fraction of sunlit and shaded

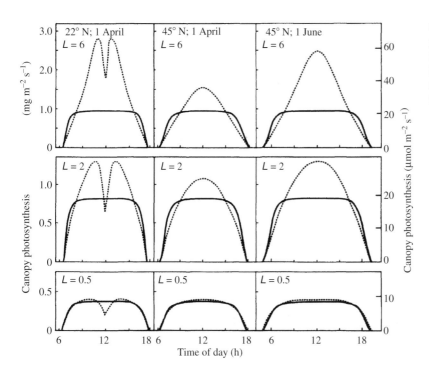

Figure 7.22 Simulations of photosynthesis by erect leaved (----------) and horizontal leaves (————) canopies of high (6), medium (2) and low (0.5) leaf area index (see text).

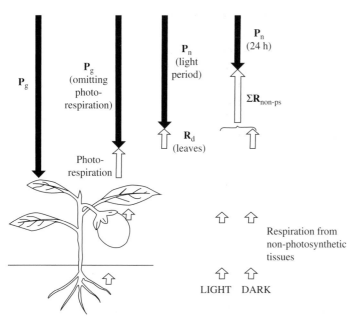

Figure 7.23 Diagrammatic representation of the CO_2 exchanges showing how net photosynthesis over a 24 h period is derived.

leaves and to assume that the shaded leaves are light limited and respond linearly to irradiance, while the leaves in direct sunlight are light saturated and hence independent of irradiance. Thus simple averages of irradiance in each of these classes gives a reasonable estimate of canopy assimilation, though these can be improved by incorporating a full biochemical model into the treatment of each class (de Pury & Farquhar, 1997).

Table **7.8** Components of carbon dioxide balance (see Figure 7.23) of a barley crop (per unit ground A) at two stages of development. Dark respiration by leaves assumed to be half the total R_d.

	Mid-May (vegetative)	Mid-July (grain filling)
Standing dry matter (g m^{-2})		
a Standing dry matter in CO_2 equivalents	526	1605
Respiration (g CO_2 m^{-2} week^{-1})		
b Soil microbial respiration	23	18
c Measured night-time R_d losses	43	64
d Day-time R_d, estimated from c by adjusting for temperature differences	78	138
e ΣR_{non-ps} ($= c + 0.5 \times d$)	82	133
f Maintenance respiration (R_m as % of total)	30	64
g Growth respiration (R_g as % of total)	70	36
h Root respiration as % of total	14	5
Photosynthesis (g CO_2 m^{-2} week^{-1})		
i Net photosynthesis for 24 h	167	102
j Net photosynthesis in light ($\Sigma P_n = e + i$)	249	235
k Gross photosynthesis in light (omitting photorespiratory losses) ($= c + d + i$)	288	304

Data from Biscoe *et al.*, 1975b.

7.9.3 Crop growth, primary production and net ecosystem exchange

When scaling up from instantaneous to longer term measurements and from single leaves to plants, canopies and ecosystems it is useful to define some new terms to describe CO_2 fluxes. *Net primary production* (NPP) is the rate at which CO_2 (usually expressed in terms of biomass, but sometimes in terms of CO_2 or energy) is accumulated by vegetation over daily or longer time periods and integrates the difference between gross photosynthesis and all respiration by the plant (during the day and night) per unit ground area over at least a daily cycle. *Net*

ecosystem exchange (NEE) is the net rate of carbon accumulation by an ecosystem and equates to gross photosynthesis minus all respiratory losses including those from the non-photosynthetic parts of the plant and from other organisms in the soil.

The various components of the CO_2 balance of a crop or ecosystem are illustrated in Figure 7.23 and summarised for a barley crop in Table 7.8. Although gross photosynthesis changed relatively little over the season, respiratory losses (especially the maintenance component) increased markedly. The respiratory loss factor represents the loss by non-photosynthesising tissues (ΣR_{non-ps}), that is, the sum of the loss from non-photosynthetic tissues in the light and all tissues in the dark. On this basis the respiratory loss factor can be defined as $(\Sigma P_n - \Sigma R_{non-ps})/\Sigma P_n$, where ΣP_n is the total net photosynthesis during the light period. If one assumes for the barley data that half the respiratory losses in Table 7.8 come from photosynthetic tissues, in May, $\Sigma R_{non-ps} = 43 + (78/2) = 82$ mg m^{-2} week^{-1} and $\Sigma P_n = 167 + 82 = 249$ mg m^{-2} week^{-1}. Therefore the respiratory loss factor is 0.68 in May and is 0.43 in July. These results are comparable to other results (Monteith, 1972) and imply that between about 40 and 70% of net photosynthesis is conserved as dry matter growth. Assuming a typical respiratory loss factor of 0.6 would give an expected maximum efficiency of use of solar radiation of $0.6 \times 5.4 = 3.2\%$.

The maximum short-term crop growth rates that have been observed for C$_3$ crops of approximately

36 g m^{-2} day^{-1} and for C$_4$ crops of 52 g m^{-2} day^{-1} (Table 7.5) correspond to efficiencies of 3.1% for rice (20 MJ m^{-2} day^{-1} insolation) and between 3.1 and 4.5% for maize (Monteith, 1978). The difference between C$_3$ and C$_4$ species is maintained in spite of their similar quantum requirement at low light, probably as a result of the smaller tendency for light saturation in C$_4$ plants. A useful approximation is that dry matter production, particularly during the vegetative phase of plant growth, is a linear function of the amount of radiation *intercepted* (as expected for limiting case in Section 7.9.2(1) above). Figure 7.24 shows that total dry matter production for a wide range of different crops was approximately 1.4 g dry matter per MJ intercepted solar radiation (an efficiency of 2.5%, which is not far from the theoretical value).

Nutrition and water stress apparently exert a large proportion of their effects on yield by altering leaf area index and consequent light interception, though in one water-stressed barley crop ε_p was reduced by 20% (Legg et al. 1979). Similar results have been obtained in many subsequent experiments. Much of the limitation to the yield of annual crops arises from poor light interception in the early stages of crop growth when leaf area index is small. Low temperatures during winter also inhibit growth in temperate climates and contribute much towards the different productivity achieved in temperate and tropical environments. Actual productivities of different plant communities and estimated efficiencies of utilisation of incident solar radiation are presented in Table 7.9. This shows that in extreme environments photosynthetic efficiency can be several orders of magnitude less than potential.

7.9.4 Enhancing crop yield

Although there has been much interest, and success, in improving crop yields in the past 50 to 100 years, rather little of this advance has related to improvements in photosynthesis (see Chapter 12 for further discussion). Nevertheless, as many of the more obvious agronomic advances have been made, and

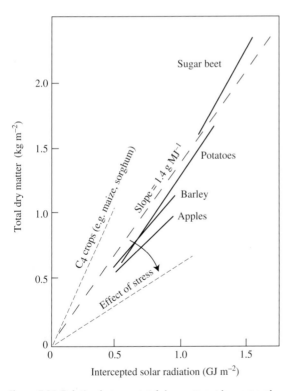

Figure 7.24 Relation between total dry matter at harvest and total radiation intercepted by the foliage for C$_3$ and C$_4$ crops, modified from Monteith (1977) and Jones & Vaughan (2010).

Table 7.9 Typical values for net primary production and efficiency of conversion of incident solar radiation for different ecosystems.

	Net primary production	Incident solar radiation (I_S)	Efficiency (ε_p)
Over whole year	(g m^{-2} year^{-1})	(GJ m^{-2} year^{-1})	(%)
Dry desert	3–(250)	7	0.0007
Arctic tundra	100–600	3	0.06–0.36
Temperate grassland	500–2000	3.3	0.25–1.1
Sugar cane	2500–7500	6	0.7–2.2
120-day growing season	(g m^{-2})	(GJ m^{-2})	(%)
Potatoes	1500	2	1.3
Temperate cereals	2000	2	1.8
Rice	3000	2	2.6

Data from Cooper, 1970, 1975; Milthorpe and Moorby, 1979; Trewartha, 1968.

genetic advances in harvest index are reaching a plateau, future improvements in yield potential are likely to depend heavily on improving photosynthesis (Flood *et al.*, 2011; Zhu *et al.*, 2010). Approaches might include improvement of canopy structure to optimise light absorption by canopies (for example by avoiding saturation by vertical leaf orientation), potential engineering of carboxylases with higher specificity for CO_2 and even by manipulation of PS pathways (e.g. current attempts to introduce the C_4 pathway to crops such as rice or to reduce photorespiratory losses). For example there is good evidence for substantial genetic variation in Rubisco specificity factors and catalytic rates (Parry *et al.*, 2011) that might provide the basis for higher photosynthetic rates.

One early unsuccessful attempt to improve photosynthesis was to screen for C_4 photosynthesis in mutagenised populations of C_3 plants by enclosing the plants in a sealed illuminated chamber together with a C_4 plant (Menz *et al.*, 1969). Because C_4 plants can deplete the CO_2 in the chamber below Γ for C_3

plants, any C_3 plants lose CO_2 until they eventually die. No C_4-like mutants have been detected among the several hundreds of thousands of lines tested by this means. In other studies, crosses have been made between C_3 and C_4 species from the same genus. For example, although the cross between *Atriplex rosea* and *Atriplex triangularis* has been taken beyond the first generation, no fully C_4 recombinants have been detected even though individuals with Kranz anatomy or high PEP carboxylase activity have been found (Nobs, 1976), There has been substantial recent effort aimed at transferring the C_4 pathway into C_3 crops, with the introduction to rice being the subject of a major research initiative (von Caemmerer *et al.*, 2012), even though such a project is highly speculative. Further details of opportunities for exploiting C_4 photosynthesis more widely have been discussed in a special issue of the *Journal of Experimental Botany* (Roberts, 2011).

A detailed analysis of crop improvement strategies and the role of genetic and agronomic improvements in crop yield will be presented in Chapter 12.

7.10 Evolutionary and ecological aspects

C_3 photosynthesis evolved early in the history of life, while the more recent evolution of C_4 photosynthesis seems to have occurred as a result of a need to minimise the photorespiratory losses of fixed carbon, which may be as much as 40% of photosynthesis. Despite the complexity of the C_4 pathway and its requirement for many coordinated changes in anatomy, stomatal behaviour, transport and enzymology, it appears to be strongly polyphyletic having evolved on more than 60 separate occasions (Sage *et al.*, 2011). The evolution of C_4 and other CO_2 concentrating mechanisms was initially stimulated by the rapid reduction in global atmospheric CO_2 concentration (and hence increased photorespiration) in the Oligocene (about 30 million years ago) with the centres of origin of C_4 being concentrated in semi-arid regions of the world, as an adaptation especially to high-temperature arid environments. Since then new occurrences of C_4 have continued to appear in appropriate climatic regions, with most appearances in the dicots probably having occurred in the last 5 million years (Sage, 2004). It is also notable that many of the C_4 lineages also contain one or more types of C_3–C_4 intermediate species; these usually appear to be ancestral to the C_4 form. Although C_3–C_4 inter-mediates all appear to have a localisation of glycine decarboxylase in bundle sheath cells, at least three groups of C_3–C_4 intermediates have been recognised, each with different expression of C_4 characteristics ranging from those with refixation of photorespired CO_2 in enlarged bundle sheath cells to those that have an operating C_4 cycle but still have Rubisco in the mesophyll (Sage, 2004).

Ecological analysis of the distribution of C_4 species shows that the C_4 pathway is most advantageous in high-light, high-temperature environments where water is limited (Ehleringer *et al.*, 1997). In these conditions leaf photosynthetic rates of C_4 species generally exceed those of C_3 species, though there are C_3 species that have photosynthesis rates equal to the highest known in C_4 species (Mooney *et al.*, 1976) and also C_3 species that can achieve high rates of net photosynthesis at temperatures above 45°C (Mooney *et al.*, 1978). The higher water use efficiency of C_4 species (see Chapter 10) that arises from their CO_2 concentration mechanism, and their consequent ability to sustain any given photosynthetic rate with more closed stomata than C_3 species, is probably the most important factor determining their common occurrence in many deserts and other subtropical regions. The dominance of C_3 species in cool or shady habitats probably results from their lower quantum requirement (and hence greater potential productivity) at temperatures below 30°C (Ehleringer *et al.*, 1997). In spite of this, some C_4 *Atriplex* and *Spartina* species can operate effectively at low temperatures (Long *et al.*, 1975), while some C_4 trees (genus *Euphorbia*) occur in the Hawaiian rainforest understory (Pearcy & Troughton, 1975).

The very high water use efficiency and ability to survive long periods without rainfall of CAM plants enables them to survive in extremely arid regions, in spite of their low maximum productivity. In a multiple regression study of species distribution in North America in relation to climatic variables, Teeri *et al.* (1978) showed that CAM abundance was greatest in areas of low soil moisture (the dry Arizona deserts) while C_4 species occurrence was associated with high minimum temperature or high evaporative demand (Teeri & Stowe, 1976). In any one region it is common to find a balance between species having different photosynthetic pathways. For example in the East Thar Desert in northwest India, both C_3 and C_4 shrubs and annual species are common, although the grasses are almost exclusively C_4. The C_3 species are probably active mainly in the cooler winter months when they might be expected to be superior, and the C_4 species during the hotter months. Probably because of the high night temperatures in the summer, CAM species are rare, with only *Euphorbia caducifolia* being widespread in this environment.

Many of the species that are known as CAM plants are in fact facultative CAM plants, using the C_3 pathway as seedlings, or when night temperatures are above 18°C, or in conditions of adequate water supply, but switching to the CAM pathway under conditions of salinity or water stress (Kluge & Ting, 1978). In these plants, the pattern of stomatal opening changes to increased night opening as the CAM pathway develops. The facultative CAM plants are found in the Aizoaceae and Portulacaceae while other families (e.g. Cactaceae) tend towards being obligate CAM plants.

In an interesting analysis, Körner *et al.* (1979) have shown a correlation between the maximum stomatal conductance and photosynthetic capacity of different species from different ecological groups. All C_3 species fell on one line, from plants such as conifers with low photosynthetic capacity to the more photosynthetically active herbs from open habitats and plants from aquatic habitats and swamps. C_4 plants fell on a different line, having a higher photosynthetic capacity for any stomatal conductance than C_3 plants.

7.11 Sample problems

7.1 For a leaf where the steady-state level of fluorescence (all in arbitrary units) is 1.2, the maximum fluorescence achieved with a saturating flash (applied during steady-state photosynthesis) is 3.2, F_m (1 h dark) is 3.7, F_o is 1.0 and F_o' is 0.9 calculate: (i) F_v/F_m, (ii) F_q', (iii) q_P, (iv) NPQ and (v) the quantum yield of electron transport through photosystem II.

7.2 Using a well-stirred open gas-exchange system where the volume rate into the cuvette is 5 cm^3 s^{-1}, leaf area in the chamber is 10 cm^2, $T_\ell = 23°C$, $P = 100$ kPa, $e_e = 0.5$ kPa, $e_o = 1.5$ kPa, $c_e' = 600$ mg m^{-3} and $c_o' = 450$ mg m^{-3} calculate: (i) u_e, (ii) u_o, (iii) x_e', (iv) P_m, (v) the molar gas-phase conductance to CO_2.

7.3 For the control photosynthesis response curve in Figure 7.19, calculate the stomatal limitation to photosynthesis according to (i) the resistance analogue method (Eq.(7.32)), (ii) Farquhar and Sharkey's method, or (iii) the sensitivity approach (Eq. (7.34)).

7.4 Assuming that incident global radiation is 14.5 MJ m^{-2} day^{-1} in May and 17 MJ m^{-2} day^{-1} in July, calculate for the data in Table 7.8 the efficiency of (i) gross photosynthesis, (ii) net productivity in terms of incident radiation or incident PAR.

8 Light and plant development

Contents

8.1 Introduction

The ability of plants to modify their patterns of development appropriately in response to changes in the aerial environment is a major factor in their adaptation to specific habitats. These morphogenetic responses are usually taken to include quantitative changes in growth (both cell division and cell expansion), as well as differentiation of cells and organs and even changes in metabolic pathways. Important examples include: the tendency for stem elongation to be greater in certain classes of shade plants, thus enabling them to outgrow competitors; the development of characteristic 'sun' and 'shade' leaves with appropriate biochemical and physiological characteristics (see Chapter 7); the induction of flowering or other reproductive growth at an appropriate season and the induction of dormancy. For many of these developmental responses some feature of the light environment provides the main external signal, though other important signals can include temperature (see Chapter 9) and water availability.

In addition to affecting development through effects on photosynthesis (for example, rapid growth depends in part on high rates of photosynthesis) and on cellular damage (e.g. DNA damage caused by high irradiance UV-B), light can influence growth and development in a number of ways. These photomorphogenic responses, which are summarised in Table 8.1 include:

1. *Phototropism* – those directional alterations in growth that occur in response to directional light stimuli, as shown by the growth of shoots towards the light.
2. *Photonasty* – reversible light movements and related phenomena that occur in response to directional and non-directional light stimuli.
3. *Photoperiodism* – non-directional developmental responses to non-directional but periodic light stimuli.
4. *Photomorphogenesis* – other non-directional developmental responses to non-directional and non-periodic light stimuli.

Table 8.1 Summary of the main light-sensitive plant developmental processes indicating the main involvement of different photosensing systems, where phyA is primarily involved in the VLFR and FR-HIR responses and phyB in the reversible FLD responses. (See text for details.)

Class of response	Light sensors	Modifiers	Comments
Photoperiodic responses			Classical LFR
Flowering			
Tuberisation	phyB	cry	Depend on circadian clock, itself
Bud break, leaf fall	phyA	ZTL	entrained by light cycles
Dormancy, cold hardiness			
Tropic responses			
Leaf movement			
Hypocotyl curvature	phot1		
Root curvature	(phot2 at	phyA, cry	Primarily blue-light responses, with growth determined by auxins and GA
Chloroplast movement	high **I**)		
Stomatal opening			
Nastic responses			
Leaflet closing	phyA phyB	(phyE), cry?	Involves circadian clock
Photomorphogenesis			
Stomatal development	phyB		Depends on light quantity
Seed germination	phyA phyB	(phyE)	Promotes in VLFR Inhibits in FR-HIR Canopy space detection FR inhibits germination
De-etiolation	phyA	(phyB) (cry1)	VLFR response
Hypocotyl extension	phyA	cry2 (phot)	
Shade responses	phyB		Primary sensor of R/FR
	cry1		Primary sensor of quantity
Leaf expansion	phyA		phyA antagonist of shade response
Chalcone synthesis			
UV-B stimulated morphogenesis	UVR8	(cry1)	Low irradiance UV-B
Tissue damage/necrosis	DNA		Direct damage by high irradiance UV-B
Plant defence	phyB UVR8		Defence – often involves jasmonate signalling

5. *Other altered gene expression* – other signal transduction leading to changes in metabolism and physiological functioning such as altered flavonoid synthesis in response to UV-B radiation.

This chapter provides an introduction to the role of light in the control of plant morphogenesis and plant functioning, though a detailed treatment of mechanisms involved in the tight control and coordination of developmental responses is outside the scope of this book. General information on photomorphogenesis may be found in introductory plant physiology texts (Salisbury & Ross, 1995; Taiz & Zeiger, 2010) and in valuable earlier reviews (Kendrick & Kronenberg, 1994; Smith, 1995; Vince-Prue, 1975; Vince-Prue *et al.*, 1984) while more detail on photoreceptors and their modes of action may be found in several more recent reviews (Chen & Chory, 2011; Christie, 2007; Jackson, 2009; Jenkins, 2009; Li & Yang, 2007; Lin & Shalitin, 2003; Möglich *et al.*, 2010).

8.2 Detection of the signal

In addition to the molecules that absorb light in photosynthetic energy transduction (e.g. chlorophyll in light harvesting centres), there are many different photoreceptors that are involved in signalling and the control of plant development. These photoreceptors are proteins bound covalently or non-covalently to an associated chromophore; most are cytoplasmic and water soluble, though phytochromes for example need to be translocated to the nucleus for their action. The best known photoreceptors in plants include phytochrome (phy; a family of red/far-red reversible chromo-proteins) and the blue-light or ultraviolet absorbing receptors using flavin adenine dinucleotide (FAD; utilised in phototropins and cryptochrome). In addition there is increasing evidence for the existence of a specific UV-B receptor, and one that responds to green light, and a number of other photoreceptors in algae. Different photomorphogenic responses involve at least one, and often more, of these pigments; these sensors then interact with the full range of phytohormones (including gibberellins, auxins, brassinosteroids, cytokinin and ethylene) to determine the final responses.

8.2.1 Cryptochromes

Cryptochromes (cry) are a class of photoreceptors that respond to UV-A and blue light in the range 320 to 500 nm (Lin & Shalitin, 2003). They are flavoproteins involved in a range of photomorphogenic responses including entrainment of the circadian clock, regulation of gene expression, de-etiolation and stimulation of leaf expansion, and the control of photoperiodic processes such as flower induction (Li & Yang, 2007) as well as a range of other morphogenetic processes, generally operating in conjunction with phytochromes.

8.2.2 Phototropins and F-box proteins

A second class of blue-light receptors was identified in 1998. These are the phototropins (phot), which are based on '*Light-oxygen-voltage*' (LOV) domains that use flavin cofactors to detect blue light. Phototropins have been implicated in a number of rapid light-induced photomovement responses such as phototropism, leaf expansion, chloroplast and leaf movements, and in stomatal opening (Christie, 2007). They are responsible for recognising blue-light directional information. Proteins of the ZEITLUPE (ZTL) family, which are sometimes treated as a separate class of photoreceptor, though they also utilise LOV domains, are involved in slower responses such as the control of flowering, regulation of phototropism and entrainment of the circadian clock (Demarsy & Frankhauser, 2009). Some lower plants also have a photoreceptor (neochrome) that includes both phototropin and phytochrome domains.

8.2.3 Phytochrome

Phytochrome has been known since the 1950s but it is only recently that the complexity of the phytochrome signalling system in plants has become clear. Not only are there at least five distinct phytochrome

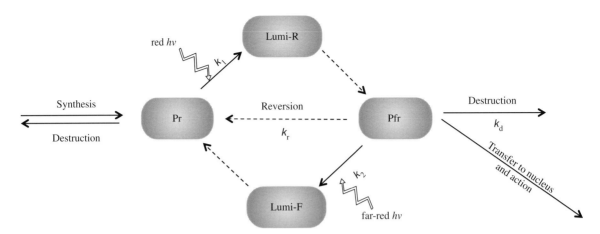

Figure 8.1 A simplified schematic of phytochrome interconversions showing the light-driven interconversions via the Lumi-R and Lumi-F intermediates, the dark reversion, and synthesis and destruction (Rockwell *et al.*, 2006). The rate constants of each reaction are shown as k_1, k_2, k_r, k_d. Not shown are further interconversions that include the interconversions between Pr and Pfr in the nucleus and in nuclear bodies (Rausenberger *et al.*, 2010).

molecules that participate in regulation of different photomorphogenic events, but there is also a series of complex downstream signalling pathways, which is only now beginning to be understood (Chen & Chory, 2011). Phytochromes use a linear tetrapyrrole bilin chromophore to sense the ratio of red light (R) to far-red light (FR) and are apparently unique among higher plant photoreceptors in that they can exist in two photo-interconvertible forms: a red-light absorbing form (Pr) that has an absorption maximum in the red (660 nm) and a far-red absorbing form (Pfr) that has an absorption maximum in the far-red (730 nm). It is in this region of the spectrum that natural radiation shows the greatest variation so phytochrome is perhaps well fitted as a detector of the spectral quality of incident radiation and is therefore involved in morphogenetic responses such as shade avoidance or the initiation of growth on exposure to light. Absorption spectra of Pr and Pfr are illustrated in Figure 2.4 (along with the absorption spectra of riboflavin and chlorophyll).

Phytochromes are synthesised as the biologically inactive (at least in plants) Pr form. Biological action is inititated by photoconversion to the far-red absorbing Pfr form; this triggers rapid (within minutes) translocation of phytochrome into the nucleus where the activated photoreceptor interacts with a family of transcription factors (the phytochrome interacting factors, or PIFs) and initiates a cascade of changes in gene expression that themselves are detectable within minutes. In general Pfr depletion leads to increased stability of the growth-promoting PIFs and enhanced production of auxins and gibberellins. Interactions of phytochromes with PIFs also control turnover of the phytochrome molecule. Further details of the roles of the different PIFs in phytochrome signalling may be found in appropriate reviews (see Bae & Choi, 2008; Franklin, 2008; Franklin & Quail, 2010). The main interconversions of phytochrome are illustrated in Figure 8.1: absorption of light (especially in the red) by the Pr form converts it to Pfr, while light absorption by Pfr converts it back to Pr. Red light therefore tends to convert most of the phytochrome present to the Pfr form. These transformations, which each proceed through a number of intermediate structures, are driven by the light energy absorbed with quantum yields (ε_q) of between about 0.07 and 0.17 (Jordan *et al.*, 1986).

The Pr form of phytochrome accumulates rapidly in the dark. This indicates a relatively slow rate of Pr destruction. Photoconversion to Pfr, a much more labile form (at least in dark-grown plants), results in rapid loss of phytochrome by destruction. The biological responses controlled by phytochrome are primarily effected by the Pfr form and reflect the ratio between the Pr and Pfr forms as determined by the light environment and the forward and reverse rates of photoconversion and by the rates of thermal interconversion. The amount of Pfr present at any time depends both on processes that determine the relative amount of Pr and Pfr, and on processes that change the total amount of phytochrome (synthesis and destruction). The relative amounts of Pr and Pfr in a population of phytochrome molecules are determined both by phototransformation and, at least in dicotyledons, by an additional dark reaction involving the thermal reversion of Pfr to Pr. This dark reversion appears to be absent in monocotyledons and in the Centrospermae, and though it is a moderately rapid process ($t_{1/2} \simeq 8$ min), this is so much slower than phototransformation that it is only likely to affect the ratio of Pfr to total phytochrome at very low irradiances (less than about 3 μmol m^{-2} s^{-1}). The significance of dark reversion is not clear.

The plant phytochrome apoproteins (PHY) fall into two classes: the 'Type I phytochromes', which are the light labile PHY-As, and four to five 'Type II phytochromes', which are relatively light stable phytochrome apoproteins (PHY-B, PHY-C, PHY-D and PHY-E), though note that not all are found in all plants. The various phytochromes differ in their spectral sensitivities and the processes they are involved in controlling. For example, PHY-A is the predominant phytochrome in etiolated seedlings but rapidly degrades on exposure to light. The phytochrome signalling response is dependent on relocalisation from cytoplasm to the nucleus. A further complication is that phytochrome normally exists as a dimer, with the possibility that the heterodimer (Pfr:Pr) is more unstable than either of the homodimers.

8.2.4 UV-B receptors

Perhaps the best known responses to high irradiance ultraviolet (UV) light are direct non-specific damage to DNA and the generation of damaging reactive oxygen species. Such non-specific damage is even greater with highly energetic UV-C radiation (<280 nm) but fortunately this is largely screened out by the atmosphere so is not usually a problem in natural environments. The damage processes can lead to many changes in gene expression and a range of developmental responses. There is, however, evidence that UV-B can modulate many physiological and morphogenetic processes not only through such high irradiance-caused damage (often acting through the generation of reactive oxygen species), but also through a range of separate processes operating at irradiances as low as 0.1 μmol m^{-2} s^{-1} (less than 5% of the UV-B in sunlight). These low irradiances or very rapid (low energy requiring) responses include suppression of hypocotyl extension and modification of chalcone synthase expression (Jenkins, 2009). The action spectrum of UV-B responses in different plants generally has a peak between about 280 and 300 nm. The fact that aromatic amino acids, especially tryptophan, absorb at around 280 nm suggests that direct UV-B absorption by proteins may be involved in low irradiance UV-B responses (see e.g. Jenkins, 2009), with separate receptors accounting for responses that peak between 300 and 320 nm and in the UV-A region. Perception of low irradiance UV-B does not appear to depend on the cryptochrome or phototropin receptors or on DNA damage; rather there is a specific protein, UVR8 that is now thought to be involved. Useful reviews of UV-B reception and plant responses have been provided by Jordan (2011) and Jenkins (2009) with the main mechanisms summarised in Figure 8.2.

8.3 Phytochrome control of development

Many developmental responses use phytochrome as the primary light receptor but these responses are complex in higher plants, with the different phytochromes having overlapping but often distinct roles. Phytochrome responses are often divided into *inductive* (or photoreversible) responses and *high*

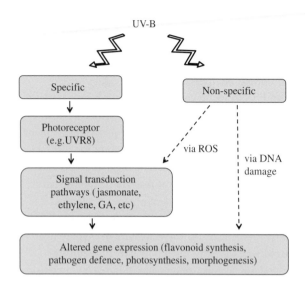

Figure 8.2 The main pathways of UV-B signal detection and transduction in plants. (Adapted from Jordan, 2011).

irradiance responses (HIR). In addition, the highly sensitive, non-reversible function of PHY-A in response to low quantities of light, such as in the de-etiolation response on first illuminating seedlings, are termed *very low fluence rate* (VLFR) responses.

Although PHY-A is generally responsible for the VLFR response and the far-red HIR (FR-HIR), other phytochromes usually control the red/far-red reversible low fluence rate (LFR) responses. There are, however, exceptions with, for example, the LFR being mediated by both PHY-A and PHY-B while PHY-E is involved in the FR-HIR for seed germination (Bae & Choi, 2008).

Inductive responses, which include rapid (15 to 30 s) effects of light on leaflet closing, to slow (days to weeks) photoperiodic effects on flowering, use the so-called low-energy phytochrome system and are typically saturated by radiant exposures of less than 1000 J m^{-2} (often 1 to 60 J m^{-2}), and are usually proportional to the logarithm of the incident energy up to this saturation point. This implies that a high irradiance for a short time or a low irradiance for a longer time are interchangeable in their effect. In addition these responses are usually red/far-red reversible, though in some cases reversibility is

incomplete, either because the response is initiated extremely rapidly or because the amount of Pfr required to saturate the response is in excess of that established by the far-red (reversing) treatment.

High irradiance responses classically refer to phenomena such as stem extension and anthocyanin formation in dark-grown seedlings, and they characteristically show irradiance dependence and a requirement for continuous exposure to relatively high irradiances. They often involve an action spectrum with a peak in the blue (implying some involvement of cryptochrome) in addition to the peak in the far-red that would be expected for a phytochrome response. These responses have been most extensively studied in etiolated systems where it has been concluded that they depend on the total amount of Pfr present, while there are indications that the corresponding responses in green plants may be somewhat different.

An important factor in all phytochrome responses to continuous irradiation is that in any constant radiation environment, a dynamic equilibrium is set up, such that the rate of conversion of Pr to Pfr exactly balances the rate of the reverse transformation of Pfr to Pr. Such an equilibrium is termed a photoequilibrium, which can be expressed either in terms of the ratio of the concentrations Pfr and Pr (Pfr/Pr $= f$) or more commonly as ϕ, the steady-state ratio of the concentration of Pfr to total phytochrome, i.e.:

$$\phi = \mathrm{Pfr}/(\mathrm{Pr} + \mathrm{Pfr}) = \mathrm{Pfr}/\mathrm{P_{total}} = f/(1 + f) \qquad (8.1)$$

Using the rate constants for each of the transformations involving Pfr indicated in Figure 8.1 we see that the rate of synthesis of Pfr ($= \mathrm{Pr} \times k_1$) equals the sum of the back reaction, reversion and destruction ($= \mathrm{Pfr} \times (k_2 + k_r + k_d)$) where Pr and Pfr are the concentrations of the two phytochromes. Equating and rearranging gives:

$$\phi = k_1/(k_1 + k_2 + k_r + k_d) \qquad (8.2)$$

For monochromatic light at any wavelength, λ, the rate constant for the conversion Pr $\rightarrow$ Pfr ($k_{1\lambda}$) is given by the product of the incident irradiance at λ ($\mathbf{I}_{p\lambda}$), the absorptance of Pr at λ ($\alpha_{\lambda(\mathrm{Pr})}$) and the quantum yield for the conversion ($\varepsilon_{q(Pr)}$). A similar expression can be written for the reverse reaction, $k_{2\lambda}$. Substituting these into Eq. (8.2) gives:

$$\phi = \frac{\mathbf{I}_{p\lambda}a_{\lambda(Pr)}\varepsilon_{q(Pr)}}{\mathbf{I}_{p\lambda}(a_{\lambda(Pr)}\varepsilon_{q(Pr)} + a_{\lambda(Pfr)}\varepsilon_{q(Pfr)}) + k_r + k_d} \quad (8.3)$$

The value of ϕ at equilibrium depends on the spectral distribution of the incident radiation, with ϕ decreasing as the proportion of far-red increases. Even monochromatic radiation gives rise to a photoequilibrium because both the Pr and Pfr forms each absorb some radiation over a wide range of wavelengths (Figure 2.4). Substituting appropriate values has led to estimates for red light at 660 nm of ϕ_{660} ranging between about 0.84 and 0.87 (Jordan et al., 1986).

Unfortunately the amount of Pfr present at any time depends not only on the photoequilibrium, but also on the total amount of phytochrome present. For example, it is possible that continuous far-red light may result in a higher total concentration of Pfr than does red light, even though it leads to a smaller ϕ. It has been proposed that such an effect could result from the greater susceptibility of Pfr than Pr to destruction, thus leading to a smaller amount of total phytochrome in red light. Other phenomena that may be involved in phytochrome responses in continuous light include 'photoprotection' (where it is postulated that phytochrome is increasingly protected from destruction as irradiance increases because high light causes rapid cycling between Pr and Pfr, so that a smaller proportion of time is spent in forms that are susceptible to destruction), and an irradiance dependent inactivation.

Another problem is that it is difficult to measure ϕ in an intact green plant because, with the usual photometric technique, absorption due to the small amount of phytochrome is swamped by that of the much larger amount of chlorophyll present. However, it is possible to determine ϕ spectrophotometrically in dark-grown etiolated tissues (which have no chlorophyll). This value of ϕ can be related to the ratio of the photon flux densities in the red (655–665 nm) and far-red

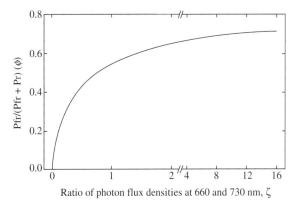

Figure 8.3 Relation between phytochrome photoequilibrium (ϕ) and photo flux density ratio (ζ). (Data of Smith & Holmes, 1977.)

(725–735 nm) portions of the spectrum (ζ), so that ζ is given by:

$$\zeta = \mathbf{I}_{p(660)}/\mathbf{I}_{p(730)} = 0.904\,\mathbf{I}_{e(660)}/\mathbf{I}_{e(730)} \quad (8.4)$$

The factor 0.904 is required to convert radiant (energy) flux densities to photon flux densities as required by the definition of ζ. The relationship between ϕ and ζ is shown in Figure 8.3, so that ζ can be used to estimate ϕ. Although not identical to the value of ϕ in green tissues in a corresponding radiation climate, the value obtained for etiolated tissue is often assumed to be a good estimate. Figure 8.3 shows that ϕ is most sensitive to ζ over the natural range of variation ζ. The maximum value of ζ in unfiltered sunlight is approximately 1.2, while the main factor that affects ζ is the filtering of light through leaves, though there is also a tendency for ζ to decrease with decreasing solar angle (Table 8.2 and Figure 8.4). As was pointed out by Hughes et al. (1984), however, the variation in ζ as solar angle decreases is probably too variable and unreliable a signal to provide an effective photomorphogenic signal. The values of ϕ and ζ corresponding to the spectra shown in Figure 8.4 are summarised in Table 8.2. As described in Chapter 2, the relative enrichment in longer wavelengths when the sun is low in the sky, results from Rayleigh scattering removing most of the shorter wavelengths from

Table 8.2 Approximate values of ζ and calculated phytochrome photoequilibrium (ϕ) for the spectral distributions under different numbers of shading leaves as shown in Figure 8.4 (data from Smith, 1975).

Time of day	No. of shading leaves	ζ	ϕ
Midday	0	1.00	0.50
	1	0.12	0.20
	2	<0.01	0.06
Sunset or sunrise	0	0.63	0.35
	1	0.08	0.09
	2	<0.01	0.02

Certain artificial light sources such as fluorescent tubes that are used in controlled environment chambers may give values of ζ from 2 to more than 9.5, with correspondingly high values of ϕ. This can be an important factor in the somewhat unnatural growth often experienced in growth chambers.

A more precise estimate of ϕ than can be obtained using measured photon irradiances at only two wavelengths can be achieved by taking into account the distribution of photon irradiance over the whole shortwave spectrum by integrating Eq. (8.3) over all wavelengths, but any improvement in precision is likely to be small in most situations.

Further details of the involvement of phytochrome and other photoreceptors in different photomorphogenic responses are outlined below.

8.4 Physiological responses

8.4.1 Phototropism

Phototropic responses are widely distributed in the plant kingdom, occurring in fern and moss tissues and the sporangiophores of some fungi, as well as in the growing regions of leaves, roots and stems of higher plants. Directional responses in angiosperms respond only to blue light, while in cryptograms red light may also be involved. In a natural environment, phototropic alterations in the direction of growth (usually towards the light source for shoots and away for roots) are of great importance in optimising the interception of available solar energy and ensuring correct location of roots. Unfortunately most research has concentrated on the short-term phototropic responses of etiolated grass coleoptiles, but these are probably rather unrepresentative of normal green plant tissues. As with other tropic movements, phototropic curvature can occur only in tissues that still have some capacity for growth; fully differentiated stems do not bend.

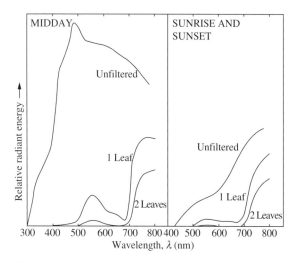

Figure 8.4 Spectral distribution of radiant energy in direct sunlight at midday and at sunrise or sunset, together with the effects of filtering through one or two layers of sugar beet leaves. (After Smith, 1973.)

the direct solar beam. Even further enrichment in the far-red component occurs as radiation passes through a canopy of leaves. This results from the much greater transparency of leaves to infrared than to PAR (see Figure 2.14), so that ζ can fall to below 0.1 in the deep shade below several layers of leaves.

Phototropic curvature of etiolated coleoptiles has been found to be related to the radiant exposure, that is the magnitude of the response is a function of the total radiation dose ($\int Idt$) rather than either flux density or time. This is often known as the *Bunsen–Roscoe reciprocity law* and has been taken as evidence that only one photoreceptor system is operative. In fact the dose–response curve is rather complex with positive curvature (towards the light) increasing up to a maximum radiant exposure of about 0.1 J m^{-2}, but as the dosage increases further the response falls off and a slight negative curvature can occur. With still greater doses the curvature again becomes positive (the second positive curvature), and in this region the reciprocity law also breaks down. Although the directional responses in angiosperms are primarily sensitive to blue (<500 nm) light detected by phototropin receptors (and hence do not involve phytochrome) the actual sensitivity observed is known to be modulated by the phytochrome system and cryptochromes (Kami *et al.*, 2012; Tsuchida-Mayama *et al.*, 2010).

Although it was shown as long ago as 1880 by Charles Darwin that in coleoptiles the tip region appears to be the most sensitive to the phototropic stimulus, details of the mechanisms underlying the differential growth in response to directional radiation are only now becoming clear. In roots the blue-light receptor, phototropin 1 (phot1), is located in the growing region where curvature occurs, while phytochrome is in the root tip and although not directly involved in the directional response appears to modulate the response to blue light (Kutschera & Briggs, 2012). In hypocotyls also, phot1 is the primary photoreceptor for the blue-light response, though phototropin 2 (phot2) is also involved in the response to high irradiances. These receptors are located on the plasma membrane where they influence auxin transporters that control the gradients of auxin that lead to differential growth and hence curvature. One difficulty in interpreting many experimental results is that plant tissues can transmit light to shaded regions perhaps by a light-piping mechanism similar to that occurring in optical fibres. Although research has concentrated on the first two responses of coleoptiles, it is much more likely that something corresponding to the second-positive curvature is most relevant for green shoots growing in natural environments where they will rapidly receive a great deal more than 0.1 J m^{-2}!

The movements of leaves in relation to the sun (especially common in the Fabaceae), though sometimes termed helionastic, are probably better termed heliotropic in view of their directional nature. Some examples of heliotropic movements have been reviewed by Ehleringer and Forseth (1980). When well watered, the leaves of appropriate species tend to remain perpendicular to the solar beam throughout the day thus maximising light interception and photosynthesis (Figure 2.21): that is, they are diaheliotropic. When water stressed, however, the leaves tend to become paraheliotropic and orientate parallel to the incident radiation, thus reducing irradiance at the leaf surface, conserving water and preventing overheating. Other organs, such as the flower heads in sunflower, also move in relation to the sun. The effects of heliotropic movements on light interception were discussed in Chapter 2, and ecological implications for temperature regulation and drought tolerance are discussed in Chapters 9 and 10.

Leaf angle also changes as part of the shade-avoidance response outlined below, with the leaf angle varying as a function both of the red to far-red ratio, and the absolute irradiance (see e.g. Mullen *et al.*, 2006), though the response is not directional so should perhaps be treated as a nastic response.

8.4.2 Photonasty

Examples of photonastic phenomena include the opening or closing of flowers with a change in the general level of irradiance, and the nyctinastic 'sleep movements' of leaves that fold up at night. Although some nastic movements involve differential growth, the reversible photonastic movements of leaves of many Fabaceae (such as the lupin, *Lupinus albus*, and the sensitive plant, *Mimosa pudica*) are brought about by reversible turgor changes in the specialised hinge cells or pulvini at the base of the leaflets. These changes in turgor apparently involve a similar mechanism to that of stomatal movement, including H^+ extrusion from the hinge cells with consequent uptake of K^+ as a major osmoticum. A similar mechanism is involved in many heliotropic movements.

The light sensor for the 'sleep movements' is not known, though there appears to be an involvement of both blue and red-light sensors and they have a strong endogenous circadian rhythm with movements able to continue for several days even in continuous light.

8.4.3 Photoperiodism

The daylength at any time and place depends on season and latitude (Figure 8.5). Daylength is constant throughout the year at the equator and the seasonal variation increases with latitude, changing by as much as 5.5 h between winter and summer at 40° (north or south) and by 24 h at the Arctic and Antarctic circles. These seasonal changes are utilised by many species as a reliable signal for phasing the conversion of the growing apices from a vegetative to a floral state as well as in the regulation of various other developmental processes including tuberisation, bud break and dormancy. Using the photoperiod as the signal for floral induction ensures that flowering

can occur at the optimum time for any particular species in relation to the climate at that location; for example, long enough before the average onset of frosts for seeds to ripen or before the onset of the dry season in more tropical climates.

The shortwave irradiance is sufficient to influence the photoperiodic detection system during the period of twilight when the sun is just below the horizon, so it is necessary to allow for this extension of daylight in calculation of the daylength. Although the exact threshold irradiance for photoperiodic effects varies with species and climate, an appropriate, though arbitrary, definition of daylength for photoperiodic studies is the period of daylight between sunrise and sunset together with the period of civil twilight (i.e. the period when the sun is less than 6° below the horizon). This period may be calculated by suitable rearrangement of Eq. (A7.1) (see Appendix 7), giving the curves in Figure 8.5.

Since Garner and Allard (1920), it has been known that angiosperms can be classified into at least three main groups on the basis of their flowering response to daylength. Some plants, such as tobacco and the weed *Chenopodium rubrum*, have been found to require a certain minimum number of days where the uninterrupted dark period exceeds a certain minimum duration for flower induction to occur. These are called short-day (SD) plants. In long-day (LD) plants, such as ryegrass, on the other hand, flowering only

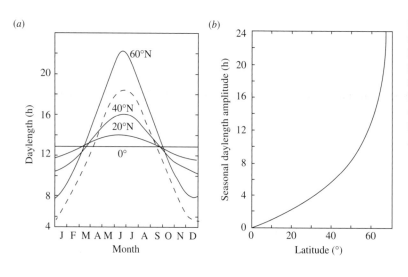

(*a*) (*b*)

Figure 8.5 (*a*) Seasonal variation of daylength at different latitudes. Solid lines include civil twilight, dashed line shows the daylight period when sun is above the horizon (at 60° N). (*b*) Amplitude of seasonal daylength changes at different latitudes (calculated according to Appendix 7).

occurs when the photoperiod exceeds a certain critical daylength for a minimum number of cycles; this varies greatly between species. Other species, particularly those of tropical origin, are day-neutral in their flowering response, being unaffected by daylength. This class includes, among others, many desert ephemerals. The critical daylength varies greatly between species.

In many species either the daylength requirement changes with age, or else they have a requirement for a particular sequence of environmental conditions such as a period of long days followed by a period of short days (LSD plants), or short days followed by long days (SLD plants) in order to flower. Similarly some plants have an absolute or obligate photoperiodic requirement that must be satisfied before flowering can occur, while others may show hastened flowering in appropriate photoperiods, but will ultimately flower even under unfavourable daylength (quantitative photoperiodic plants). Some examples of the different classes of photoperiodic plants are listed in Table 8.3. As the flowering response is often also very dependent on temperature, this table also includes a sample of those plants where a period of low-temperature vernalisation (see Chapter 9) is either required or can accelerate flowering when combined with the appropriate daylength treatment.

The SD plant group includes many plants that originated in low latitude regions, where daylength never exceeds about 14 h, including important crop species such as maize, millet, rice and sugar cane. There are, however, some SD plants in temperate regions including plants such as chrysanthemum that only flower late in the summer as the days shorten. Long-day plants are typically temperate region plants that flower during the long days of summer, and include many of the common temperate crop species.

Although a very precise photoperiodic response may be advantageous for a plant growing in a specific environment, it limits its adaptability to other regions. Wild species and crop plants that occur over a wide latitudinal range tend to be differentiated into ecotypes or local races with different daylength responses, thus restricting the area over which they can grow successfully. Many modern varieties of soybean, wheat and rice, however, are less rigorously controlled by daylength than are related wild species. In fact, breeding for independence of daylength was a major objective in the development of the new 'green revolution' varieties of wheat in Mexico and rice in the Philippines, to aid trans-world adaptation. This selection for photoperiod insensitivity, however, can have some disadvantages. For example, some daylength sensitivity is useful in wheat grown at high latitudes to delay ear development until the risk of frost is past. Another example is in 'floating rice', where photoperiod sensitivity is used to delay flowering until the monsoon floods recede, or else the grain cannot be harvested.

Various other plant processes are known to be controlled or influenced by photoperiod. These include onset and breaking of bud dormancy in woody perennials, leaf abscission, rooting of cuttings, seed germination, bulb and tuber formation in herbaceous plants, and the development of frost resistance. For example, the increasing length of the dark period in late summer is used as a signal by many woody plants for the initiation of dormant buds and for enhancing winter hardiness in anticipation of winter, even though climatic conditions may still be favourable for growth. This could provide an important ecological adaptation where the relatively slow process of developing winter hardiness is started before the first frosts occur, thus avoiding damage to the plant. Conversely the long-day induced formation of resting buds in the desert liverwort *Lunularia cruciata* provides an example of an adaptation reducing the effects of summer water stress.

8.4.4 The circadian clock and photoperiodic mechanism

Photoperiodic responses in higher plants involve complex interactions of photoreceptors with an endogenous timekeeping system (the *circadian clock*) and with the signalling of the output physiological

Table 8.3 Some examples of photoperiodic plants, with the right column also showing examples of those plants where a low temperature (vernalisation) treatment is either required or accelerates flowering (data from Vince-Prue, 1975). Note that different genotypes of one species may fall into more than one class.

No vernalisation requirement	Require or accelerated by vernalisation
Short-day plants (SD)	
(a) Obligate SD plants	
Kalanchoe blossfeldiana	*Chrysanthemum morifolium* (chrysanthemum)
Coffea arabica (coffee)	*Xanthium strumarium* (cocklebur)
Fragaria × *ananassa* (summer strawberry)	
(b) Quantitative SD plants	
Cannabis sativa (hemp)	*Allium cepa*
Gossypium hirsutum (cotton)	*Chrysanthemum morifolium* (chrysanthemum)
Long-short-day plants (LSD)	
Bryophyllum crenatum	
Long-day plants (LD)	
(a) Obligate LD plants	
Avena sativa (oats – spring cvs)	*Triticum aestivum* (wheat – winter cvs)
Spinacia oleracea (spinach)	*Lolium perenne* (rye grass)
Lolium temulentum (ryegrass)	*Arabidopsis thaliana*
(b) Quantitative LD plants	
Brassica rapa (rape)	*Beta vulgaris* (beet)
Hordeum vulgare (barley – spring cvs)	*Triticum aestivum* (wheat – winter cvs)
Short-long-day plants (SLD)	
Campanula medium (Canterbury bell)	*Dactylis glomerata* (cocksfoot)
Trifolium repens (creeping clover)	*Poa pratensis* (meadow grass)
Day-neutral plants (DN)	
Cucumis sativus (cucumber)	*Daucus carota* (carrot)
Fragaria vesca (everbearer strawberry cvs)	*Vicia faba* (broad bean)

response (Jackson, 2009). These interactions have been studied by using a wide range of types of experiment including those that involve short periods of illumination during the dark period (night-breaks). These studies have shown that whether or not short-day plants flower in light/dark cycles is dependent primarily on the length of the dark period rather than on the duration of light. The fact that plants (such as some rice varieties) are sensitive to changes in daylength of as little as 15 to 20 min implies a precise endogenous time-measuring mechanism that must be combined with accurate detection of the light 'on' and light 'off' signals. Photoperiodic responses rely on the interplay between the photoperiod sensor, an endogenous circadian clock and the response pathways such as the control of flowering or other physiological processes. Rapid advances have been made in recent years, largely through the application of genetic approaches, in our understanding of the circadian clock in plants, how it is entrained by photoperiodic signals, and how it interacts with phytochrome and other light sensors in controlling photoperiodic responses such as flowering (see e.g. Gardner *et al.*, 2006; Hotta *et al.*, 2007).

Much of our understanding of the circadian clock in plants is based on experiments with the facultative LD plant *Arabidopsis*, though similar mechanisms are widespread across different classes of organism from cyanobacteria to plants and animals (Harmer, 2009; Sung *et al.*, 2010). The circadian clock is based on a number of negative-feedback loops that can continue to cycle in constant light or dark for several days, though as their natural rhythm tends to be more than 24 hours the central circadian oscillator needs to be entrained by diurnal environmental signals (diurnal fluctuations in light or temperature). There is evidence that light regulation of the clock involves ZTL, cryptochrome and phytochrome sensors. In *Arabidopsis* the circadian clock regulates the timing of expression of the *CONSTANS* (*CO*) locus, with daytime expression of *CO* only occurring in LD. Once *CO* expression increases in light at the end of the light period, the CO acts as a transcriptional activator of the flowering gene *FLOWERING LOCUS T*

(*FT*); the protein from expression of *FT* is translocated to the shoot meristem where it initiates expression of the flowering genes. *FT* expression is inhibited in those plants that are in a juvenile phase, or, in those that require low-temperature vernalisation, until adequate chill has been received. This mechanism prevents biennial plants flowering in their first year. Homologues of *CO* such as the *Heading date 1* (*Hd1*) gene in the SD plant rice appear to fill the same role in other plants. Related mechanisms appear to be involved in the induction of and release from dormancy by SD in perennials (Rohde & Bhalerao, 2007).

8.4.5 Photomorphogenesis

The term photomorphogenesis is given to the wide range of light-controlled developmental responses that were not conveniently discussed under any of the above headings (Smith, 1995). These include effects on seed germination, stem elongation, leaf expansion, the development of chloroplasts, and the synthesis of chlorophyll and many secondary products. Phytochrome responses have also been implicated in the regulation of plant defence against insect attack, with far-red radiation (as in dense canopies) tending to downregulate responses (Ballaré, 2009).

Germination
Although many seeds are not affected by light during germination, there are some species that are strongly light dependent. In some plants (e.g. varieties of lettuce, *Lactuca sativa*, and beech, *Fagus sylvatica*) germination is stimulated by white light, while in others (e.g. varieties of *Cucumis sativus*) white light is inhibitory. Interestingly, seeds of relatively few cultivated species are light sensitive, perhaps as a result of artificial selection. Some weed species, however, are polymorphic for their light response. The potential adaptive significance of contrasting light adaptation by seeds is clear. Light inhibition would ensure germination only when the seeds were buried. On the other hand, light-stimulated seeds could

remain dormant for long periods until exposed, even if only briefly, by soil disturbance. This behaviour could be advantageous in spreading out seed germination over many years while its sensitivity to very low irradiances suggests that it is mediated by PHY-A. Phytochrome also appears to be involved in the stimulation of germination that is involved in the colonisation of gaps in forest canopies. In this case, however, sustained periods of high R:FR characteristic of large canopy gaps are required for stimulating germination of the forest seed bank (Vásquez-Yanes & Smith, 1982). This is characteristic of the PHY-B system.

The light required for germination varies with species; some such as lettuce need as little as one minute's irradiation by low light, while others may require repeated exposure for several hours per day. Studies by Bliss and Smith (1985) have shown that seeds can respond to the very low irradiances occurring within the surface layers of the soil. For example, *Digitalis pupurea* required light for germination, yet germinated well when covered by 10 mm of soil. Transmission through soil attenuates the shorter wavelengths preferentially, thus decreasing the red/far-red ratio; this effect is greatest for dry sand.

Plant morphology

Both quality and quantity of light have complex effects on plant morphology. Seedlings grown in complete darkness become etiolated: that is, they grow very long and pale. This provides a clear adaptation enabling a plant to extend above the soil surface into the light before expanding its leaves. The process of de-etiolation is known to require the action of all three main photoreceptor systems.

The further development of plants once they have been exposed to light is also controlled by the phytochrome and blue-light systems. For example there is extensive evidence that stem and leaf expansion are sensitive to light quality acting through the high-energy phytochrome system. Different

Table 8.4 Some developmental changes in *Tripleurospermum maritimum* and *Chenopodium album* grown under fluorescent (F, $\phi = 0.71$) or incandescent (I, $\phi = 0.38$) lights (equal photon flux densities in the PAR) (data from Holmes & Smith, 1975).

	T. maritimum		*C. album*	
	F	I	F	I
Height (cm)	29.7	59.4	15.0	28.4
Internode length (cm)	0.8	3.5	-	-
Leaf dry mass (g)	0.48	0.46	0.34	0.31
Stem dry mass (g)	0.33	0.58	0.10	0.20
Leaf area (cm^2)	-	-	107	78
Chlorophyll $a + b$ (mg/g fresh mass)	-	-	112	94

light sources give rise to different photoequilibria, with different proportions of the phytochrome in the Pfr form. Much of the information on morphological responses has derived from research to determine appropriate lighting systems for controlled environments and glasshouses for producing 'normal' plants or for obtaining particularly short or tall specimens, or for maximising productivity.

For example, illumination under fluorescent or incandescent lamps, arranged to provide equal photon flux densities in the PAR resulted in significant differences in growth (Table 8.4). In particular, the incandescent lamps gave greater total dry matter production and a much greater stem extension rate. These results are typical of what happens when the proportion of far-red light is increased: the incandescent lamps have a high proportion of radiation in the red and far-red ($\phi = 0.38$), while the fluorescent radiation is mainly in the blue ($\phi = 0.71$).

In general, it is found that the logarithmic rate constant for stem extension tends to be inversely proportional to ϕ, but exponentially related to ζ (see Figure 8.6).

Shade avoidance and proximity sensing responses

As shown in Figure 8.4, a characteristic feature of shade light is a relative enrichment in the longer wavelengths with consequent lowering of the phytochrome photoequilibrium (ϕ). The decreased R:FR ratio characteristic of shade light, and even

brief exposure to supplemental far-red at the end of a light period, can both have large morphogenetic effects (Table 8.5). In natural environments this sensitivity to light quality is an important factor in adaptation to shade and leads to increased internode elongation, petiole extension and leaf expansion together with other responses such as an upward reorientation of leaves (hyponasty); these responses are often lumped together as the 'shade avoidance syndrome'. Many species, particularly arable weeds, are shade avoiders that show dramatic stem extension when shaded by other plants, thus enabling them to outgrow competitors. Shade-adapted woodland herbs, on the other hand, are shade tolerators and show much smaller responses to shade light adopting morphological and biochemical changes such as reduced leaf thickness and lowered chlorophyll a:b ratios and enhanced PSII:PSI ratios to optimise use of low irradiances. Recent knowledge on the complex sensing mechanisms involving phytochromes and cryptochromes and the various downstream signalling pathways involving growth regulators such as gibberellin, ethylene, brassinosteroids and auxin and their various interacting factors (e.g. DELLAs and PIFs) has been summarised by Gommers *et al.* (2013). Although much of the response to shade light is a response to its *quality*, the characteristically low total flux density or *quantity* is also an important factor in

Table 8.5 Summary of the effect on growth of tobacco seedlings (*Nicotiana tabacum*) of 0.5 h illumination with fluorescent (F) or incandescent (I) lamps at the end of an 8.5 h photoperiod with F and I at 670 μmol m^{-2} s^{-1}. Average for five cultivars (calculated from Downs & Hellmers, 1975).

	F	I
Stem length (cm)	7.1	12.9
Leaf fresh mass (g)	11.4	14.5
Stem fresh mass (g)	2.0	4.1
Width of fifth leaf (cm)	16.4	19.0
Length of fifth leaf (cm)	9.3	9.7

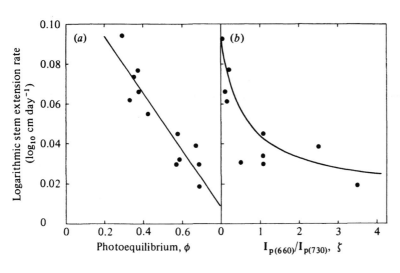

Figure 8.6 Relation between stem growth rate and (*a*) photoequilibrium ϕ and (*b*) photon flux density ratio ζ for *Chenopodium album*. (After Smith, 1981.)

many shade responses (Young, 1975). Low irradiances tend to result in maximum leaf development at the expense of stem extension. This contrasts with the effect of low values of ζ alone, which result in maximum stem extension at the expense of leaf expansion. Low irradiance also tends to increase the length of stem per unit mass and give larger but thinner leaves. In fact the morphological and biochemical differences between sun and shade leaves (see Chapter 7) involve interactions between quality and quantity perception.

Even in sparse canopies the selective scattering of far-red radiation from neighbouring plants may be enough to signal their proximity, without the need to consider spectral modification of radiation by passage through leaves (Smith, 1995). This proximity detection can be important in controlling morphological responses to plant density, where the red/far-red ratio appears to be the key regulator at lower plant densities, but at higher leaf area index the total amount of radiation (largely sensed by blue-light sensors) appears to be at least as important. As a result, both PHY-B sensing of the R:FR ratio and cry1 sensing of blue light may operate in parallel, though both pathways appear to involve the PIF4 and PIF5 in their downstream action (Keller *et al.*, 2011). The precise mechanism, however, whereby differences in plant spacing are distinguished from, for example, values of ζ that occur at different times of day is still unclear. There is probably involvement of the circadian clock and the response may be weighted for the irradiance received. Shade avoidance stimulated by low R:FR, e.g. by reflected FR from other plants or from filtering by transmission through leaves, leads to lowering of Pfr of PHY-B, which in turn leads to accumulation of growth-promoting transcription factors and elongation and other shade-avoidance responses (Franklin, 2008).

Stomatal development

Irradiance also affects the differentiation of stomata, with higher absolute irradiance tending to enhance the stomatal index (the fraction of epidermal cells that are stomata). Casson *et al.* (2009) have demonstrated the importance of PHY-B in this response.

8.4.6 UV-B responses

Exposure of plants to UV-B gives rise to many characteristic developmental and physiological responses. In general UV-B inhibits extension growth and leaf expansion and promotes branching while it also stimulates the synthesis of a range of secondary metabolites, especially the UV-protective flavonoids, which can accumulate in the epidermis where they reduce penetration of UV into the leaf. High irradiance UV, especially in the shorter wavelengths, gives rise to DNA damage and can disrupt many cellular processes, often through the generation of reactive oxygen species, and commonly resulting in the rapid appearance of tissue necrosis. UV-B, however, is not only directly damaging but it also is an important environmental signal that operates at low irradiances to modulate many developmental and biochemical responses and even plays an important role in protection against pests and pathogens. The tolerance of plants to UV-B can vary substantially between species and even between genotypes of the same species, while prior exposure to UV-B can greatly enhance tolerance of later UV-B exposure through an acclimation process. The genes stimulated by exposure to high irradiance UV-B are those normally associated with defence, wounding or general stress responses, while those stimulated at lower irradiances or by brief exposures are primarily those associated with UV protection or damage repair.

The downstream developmental responses to low irradiance UV-B involving UVR8 depend on, in *Arabidopsis* at least, a close interaction with the *COP1* genes, which are closely involved as positive regulators of many photomorphogenic signals (Jenkins, 2009).

Methods for study of UV-B responses

It is necessary to comment here on the methodology used for the investigation of UV responses in plants. In many cases results obtained may not be representative of the responses of plants growing in natural environments as a result of flaws in the experiments. For example, it is often forgotten that the UV-B response observed in any situation depends on the background PAR irradiance, so supplementation by even natural irradiances of UV-B in growth chambers where the background PAR is less than one tenth of natural field irradiances will give misleading results. Similarly UV-B treatments often apply unnaturally high UV-B irradiances that may substantially exceed the UV-B content of typical summer daylight (summer daylight in the UK typically has a maximal UV-B irradiance of only about 1.5% of I_S (or 15 W m^{-2}) with the daily maximum spectral irradiance at 300 nm being around 1 to 10 mW m^{-2} nm^{-1}).

One approach to field investigation of the effects of UV-B radiation is to filter out differing amounts of the incoming UV-B using polyester-film filters: unfortunately this approach cannot give the enhanced UV-B irradiances that are expected in higher latitudes with thinning of the ozone layer. Therefore it is usual to supplement incoming UV-B with additional radiation from UV-B fluorescent lights with the most realistic approach being to modulate the supplemental UV-B so that it continually represents a given percentage enhancement over the natural radiation at that time; this mimics effectively the effect of thinning of the ozone layer. Since the fluorescent lights used in such experiments produce substantial quantities of UV-C and shorter wavelength UV-B it is usual to filter out excess short-wave radiation using cellulose di-acetate films, though these need frequent replacement as their filtering ability rapidly degrades with age (Mepsted et al., 1996). Unfortunately, even the small amount of UV-A enhancement using this system can lead to some morphogenetic effects, so for studies of UV-B effects it is also necessary to use as a control, a treatment where all the UV-B is filtered out using a polyester film, just leaving the supplementary UV-A (Newsham et al., 1996). An advantage of the modulated supplementation approach is that it avoids the unnatural aspects of a 'square wave' UV-B supplementation.

8.5 The role of plant growth regulators

In most cases control of the altered development discussed in this chapter involves participation of one or more plant growth regulators. Some examples include the close interaction between phytochrome signalling and the downstream PIFs and the DELLA families of transcriptional regulators, which are intimately involved in the gibberellin (GA) signalling pathway, which itself is central to many of the final growth responses that are required for the morphogenetic responses such as leaf expansion and hypocotyl bending (see e.g. discussion in Franklin, 2008), while auxin redistribution is a key component of phototropic curvature responses. Ethylene is another plant growth regulator that appears to be involved in shade avoidance responses (see Ballaré, 2009). Similarly, control of germination by light acting through Pfr involves a complex interaction between both the abscisic acid (ABA) and GA pathways. The details of the involvement of plant growth regulators in these signalling and control pathways in response to light are outside the scope of this book, so the reader is directed to appropriate texts and reviews (for references see e.g. the *Annual Review* compilation by Kamiya, 2009).

8.6 Sample problem

8.1 Assuming that the incident energy in sunlight is 1.1 times greater at 660 nm than at 730 nm, (i) what is (a) ζ in unattenuated sunlight, (b) ζ in sunlight filtered through one layer of leaves ($\tau = 0.08$ at 660 nm, $\tau = 0.35$ at 730 nm)? (ii) Estimate the corresponding values of the phytochrome photoequilibrium (ϕ).

Temperature

Contents

Plants can survive the whole range of atmospheric temperatures from –89°C (recorded at Vostok in Antarctica www.ncdc.noaa.gov/oa/climate/globalextremes.html#sites) to 56.7°C (recorded in Death Valley, California; El Fadli *et al.*, 2012) that occur on the surface of the Earth, as well as the associated higher temperatures (up to about 70°C) that occur in the surface of desert soils and in the surface tissues of slowly transpiring massive desert plants such as cacti (Nobel, 1988). The even higher surface temperatures of up to 300°C that occur in bushfires can be survived by fire-tolerant plants. Seeds are particularly hardy, though other tissues of some species can also survive an extremely wide temperature range. Most plants can only grow, however, over a much more limited range of temperatures from somewhat above freezing to around 40°C, while growth approaches the maximum over an even more restricted temperature range that depends on species, growth stage and previous environment. Useful information on plants and temperature may be found in Larcher (1995) and Long and Woodward (1988).

In this chapter the physical principles underlying the control of plant temperatures are described and the physiological effects of high and low temperatures outlined. The final section considers the more ecological aspects of plant adaptation and acclimation to the thermal environment.

9.1 Physical basis of the control of tissue temperature

As outlined in Chapter 5, the temperature of plant tissue at any instant is determined by its energy balance. Neglecting any metabolic storage, the energy balance equation (Eq.(5.1)) reduces to:

$$\mathbf{R}_n - \mathbf{C} - \lambda\mathbf{E} = \mathbf{S} \qquad (9.1)$$

where, as we have seen, $\mathbf{R}_n$ is the net radiation, $\mathbf{C}$ is the sensible heat transfer, λ is the latent heat of evaporation of water, $\mathbf{E}$ is the evaporative flux and $\mathbf{S}$ is the amount of energy going into physical storage. Any imbalance in the energy fluxes goes into physical storage, thus altering tissue temperature. Although what follows refers mainly to leaf temperatures, the same principles apply to all above-ground tissues. Further discussion of biophysical aspects of control of plant temperature may be found in appropriate texts (Campbell & Norman, 1998; Gates, 1980; Monteith & Unsworth, 2008).

9.1.1 Steady state

In the steady state, when leaf temperature is constant, Eq. (9.1) reduces to:

$$\mathbf{R}_n - \mathbf{C} - \lambda\mathbf{E} = 0 \tag{9.2}$$

This may be expanded by substituting the following versions of equations for the sensible heat (Eq.(3.29)) and latent heat (Eq.(5.20)) losses:

$$\mathbf{C} = \rho_a c_p (T_\ell - T_a)/r_{aH} \tag{9.3}$$

$$\begin{aligned}\lambda\mathbf{E} &= (0.622\rho_a\lambda/P)(e_{s(T\ell)} - e_a)/(r_{aW} + r_{\ell W}) \\ &= (\rho_a c_p/\gamma)(e_{s(T\ell)} - e_a)/(r_{aW} + r_{\ell W})\end{aligned} \tag{9.4}$$

Using the resulting expanded equation, it is possible to determine leaf temperature when values for absorbed radiation, air temperature, humidity, and leaf and boundary layer resistances are known by means of an iterative computing procedure (Campbell & Norman, 1998; Gates, 1980; Jones & Vaughan, 2010; Monteith & Unsworth, 2008). A rather more convenient analytical expression for leaf temperature can be obtained by using the Penman linearisation as used for the combination equation for evaporation (Eq. (5.26)). The procedure is to use Eq. (5.21) to replace the leaf–air vapour pressure difference in Eq. (9.4) by the humidity deficit of the ambient air (D) and the leaf–air temperature difference. Subsequent combination of Eqs. (9.2) to (9.4) yields the following:

$$\begin{aligned}T_\ell - T_a &= \frac{r_{aH}(r_{aW} + r_{\ell W})\gamma\mathbf{R}_n}{\rho_a c_p[\gamma(r_{aW} + r_{\ell W}) + s\,r_{aH}]} \\ &\quad - \frac{r_{aH}D}{[\gamma(r_{aW} + r_{\ell W}) + s\,r_{aH}]}\end{aligned} \tag{9.5}$$

This equation shows that the leaf temperature excess is given by the sum of two terms, one depending on net radiation and the other on the vapour pressure deficit of the air.

There are two important approximations involved in the derivation of Eq. (9.5). The first is the Penman linearisation, which assumes that the rate of change of saturation vapour pressure with temperature (s) is constant between T_a and T_ℓ. This introduces negligible error for normal temperature differences. The other approximation is that net radiation is an environmental factor unaffected by leaf conditions, but $\mathbf{R}_n$ is actually a function of leaf temperature itself. It is possible to allow for this effect by using the concept of isothermal net radiation that was introduced in Section 5.1.2. Replacing $\mathbf{R}_n$ in Eq. (9.5) by $\mathbf{R}_{ni}$ and replacing r_{aH} by r_{HR} gives:

$$\begin{aligned}T_\ell - T_a &= \frac{r_{HR}(r_{aW} + r_{\ell W})\gamma\mathbf{R}_{ni}}{\rho_a c_p[\gamma(r_{aW} + r_{\ell W}) + s\,r_{HR}]} \\ &\quad - \frac{r_{HR}D}{[\gamma(r_{aW} + r_{\ell W}) + s\,r_{HR}]}\end{aligned} \tag{9.6}$$

Using Eq. (9.6) we can now investigate how the leaf–air temperature difference depends on environmental and plant factors. Although in practice there are complex feedback effects on r_ℓ (see Chapter 6), a useful summary of how the various factors in Eq. (9.6) affect leaf temperature can be obtained by varying each independently. The results of such an approach are shown in Figure 9.1 and summarised below.

Leaf resistance

Where the surface is dry so that there is no latent heat term in the energy balance (this is equivalent to $r_\ell = \infty$), Eq. (9.6) reduces to:

$$T_\ell - T_a = \mathbf{R}_{ni} r_{HR}/\rho_a c_p \tag{9.7}$$

In this case, $T_\ell - T_a$ is proportional to $\mathbf{R}_{ni}$, with the leaf being warmer than air when $\mathbf{R}_{ni}$ is positive (as it usually is during the day), and cooler than air when $\mathbf{R}_{ni}$ is negative. Because r_{HR} includes both a radiative and a convective component, $T_\ell - T_a$ is not linearly related to r_{aH}.

When the surface is perfectly wet, as might occur when it is covered in dew, $r_\ell = 0$. In this case the latent heat cooling is maximal for any boundary layer resistance. As r_a tends to 0, the value of $T_a - T_\ell$ tends to $D/(\gamma + s)$, the theoretical wet bulb depression.

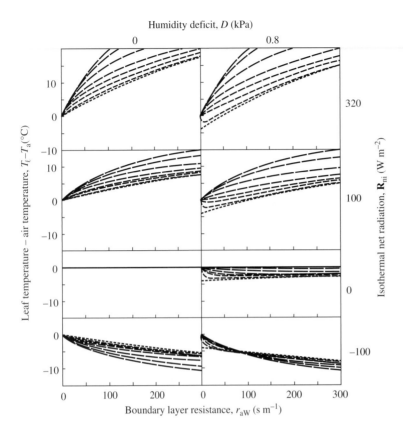

Humidity deficit, D (kPa)

Boundary layer resistance, r_{aW} (s m^{-1})

Leaf temperature – air temperature, $T_\ell - T_a$ (°C)

Isothermal net radiation, $\mathbf{R}_{ni}$ (W m^{-2})

Figure 9.1 Calculated dependence of leaf–air temperature difference on D, $\mathbf{R}_{ni}$, r_a and r_ℓ for $T_a = 20°$C. The different lines represent r_ℓ ranging from 0 s m^{-1} with short dashes, through 10, 50, 100, 200, 500, 2000 to ∞ with the longest dashes.

When r_ℓ is finite, leaf temperature tends towards air temperature as the boundary layer resistance approaches zero. With normal values for the boundary layer resistance, the amount of transpirational cooling increases as r_ℓ decreases. Whether this transpirational cooling is adequate to cool the leaf below air temperature depends on other factors, particularly $\mathbf{R}_{ni}$ and D. Leaf temperature tends to rise as r_ℓ increases (see Figure 9.1); the anomalous behaviour in Figure 9.1, when net radiation is negative, arises when condensation (i.e. dewfall) is occurring. In this case the leaf resistance is zero so the calculated curves for higher values of r_ℓ are physically unrealistic.

Vapour pressure deficit

The effect of humidity deficit on T_ℓ depends on the total resistance to water vapour loss. Where the surface is dry (or where $r_\ell = \infty$) so no latent heat loss can occur, D is irrelevant to leaf temperature (Eq. (9.7)). In all other cases, any increase in D lowers T_ℓ especially when r_ℓ is low.

Net radiation

Increasing the radiative heat load on a leaf while maintaining other factors constant always tends to increase T_ℓ (Figure 9.1). When $\mathbf{R}_{ni}$ is negative (it is commonly as low as -100 W m^{-2} on a clear night) T_ℓ must be below T_a. The net radiation absorbed by a leaf is very dependent on the value of the reflection coefficient for solar radiation (ρ_S, see Chapter 2).

Boundary layer resistance

The effect of r_a on leaf temperature is complex, especially with low boundary layer resistances where evaporative cooling can lower leaf temperature below air temperature. Increasing r_a can increase or decrease T_ℓ depending on the environmental conditions and on r_ℓ. When T_ℓ is above T_a, increases in r_a always tend to increase T_ℓ. The value of r_a itself is dependent on wind speed and leaf size and shape as outlined in Chapter 3.

Air temperature (T_a)

The effect of ambient air temperature on leaf temperature is two-fold. First, it provides the reference temperature to which T_ℓ tends. Second, there are two major effects of T_a on the value of $T_\ell - T_a$: the value of s increases with temperature so that any leaf temperature excess decreases with increasing temperature (Eq. (9.6)), and for any given value of relative or absolute humidity, D increases with increasing temperature, therefore increasing latent heat loss and lowering T_ℓ with respect to T_a as shown in Figure 9.2. These two latter effects lead to large positive values of $T_\ell - T_a$ at low temperatures.

Water deficit

Because the leaf–air temperature differential is related to the leaf conductance, it has been suggested that T_ℓ or the leaf–air temperature differential can be used as a measure of the degree of water stress to which a plant is subject. Both empirical and theoretical approaches have been used to calculate a 'stress index', based on this property. The theoretical and practical basis of such a stress index is discussed in detail in Chapter 10.

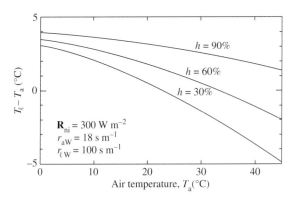

Figure 9.2 Effect of air temperature on leaf–air temperature difference at various constant relative humidities (h, %) according to Eq. (9.6).

9.1.2 Non-steady state

In a natural environment, irradiance and wind speed, particularly, are continually varying so that steady plant temperatures are rarely attained. When any component of the energy balance changes so that **S** is no longer zero, leaf temperature alters in the direction needed to return the net energy exchange to zero. For example, if $\mathbf{R}_n$ increases, leaf temperature also increases until the increased sensible and latent heat losses again balance the new value of $\mathbf{R}_n$.

The rate of change of leaf temperature depends on the heat capacity per unit area of the tissue $(\rho^* c_p^* \ell^*)$:

$$\frac{dT_\ell}{dt} = \mathbf{S}/(\rho^* c_p^* \ell^*) = (\mathbf{R}_n - \mathbf{C} - \lambda\mathbf{E})/(\rho^* c_p^* \ell^*) \quad (9.8)$$

where ρ^* and c_p^* are the density and specific heat capacity, respectively, of leaf tissue and ℓ^* is a volume to area ratio that equals the thickness for a flat leaf, $d/4$ for a cylinder or $d/6$ for a sphere (where d is the diameter). If the equilibrium temperature (T_e) for any environment is defined as that value of T_ℓ attained in the steady state, Eq. (9.8) (see Appendix 9) can be rewritten as:

$$\frac{dT_\ell}{dt} = \frac{\rho_a c_p (T_e - T_\ell)}{\rho^* c_p^* \ell^*}((1/r_{HR}) + [s/\gamma(r_{aW} + r_{\ell W})]) \quad (9.9)$$

Equation (9.9) is in the form of *Newton's law of cooling*, which states that 'the rate of cooling of a body under given conditions is proportional to the temperature difference between the body and the surroundings'. This is a first-order differential equation (like Eq. (4.31)) that after substitution of appropriate boundary conditions (i.e. the leaf is initially at equilibrium at T_{e1} and the environment is altered instantaneously at time zero to give a new equilibrium T_{e2}) can be solved by standard techniques to give:

$$T_\ell = T_{e2} - (T_{e2} - T_{e1})\exp(-t/\tau) \quad (9.10)$$

where the time constant τ is given by:

$$\tau = \frac{\rho^* c_p^* \ell^*}{\rho_a c_p((1/r_{HR}) + [s/\gamma(r_{aW} + r_{\ell W})])} \quad (9.11)$$

The time constant may readily be estimated in any situation by fitting Eq. (9.10) to any sequence of temperature measurements made following a step change in equilibrium 'environmental' temperature (e.g. caused by a change in absorbed radiation resulting from leaf shading). Any suitable curve-fitting routine can be used to estimate the three parameters (T_{e1}, T_{e2} and τ) in this equation, though I have found the use of the Solver Add-in in Microsoft Excel (Microsoft Corporation) to be particularly convenient for this purpose.

9.1.3 Thermal time constants for plant organs

The thermal time constant (τ) provides a measure of how closely tissue temperatures track T_e in a changing environment. The value of τ depends on the wind speed and on the size and shape of the organ (thickness affects heat capacity per unit area, while size, shape and wind speed affect r_a), on stomatal resistance and on air temperature (which affects the values of the constants, particularly s). Thermal properties of various materials are summarised in Appendix 5. If one assumes an average specific heat capacity of 3800 J kg^{-1} K^{-1} for leaves and fruits (this is close to the value for pure water (4180 J kg^{-1} K^{-1} at 20°C) because usually between 80 and 90% of tissue fresh mass is water), and an average leaf density of 700 kg m^{-3}, this gives $\rho^* c_p^*$ as $\simeq 2.7$ MJ m^{-3}.

Using this value for all plant tissues enables us to calculate approximate time constants for plant organs of different size and shape and for two wind speeds (Table 9.1). This table shows that τ is likely to be significantly less than a minute for all but the largest leaves. Stems and fruits have longer time constants than leaves because they have a larger mass per unit

Table 9.1 Thermal time constants for leaves, stems and fruits treated as simple geometric shapes at 20°C calculated using Eq. (9.11), where d is the breadth of a leaf (or diameter for a cylinder or sphere) and ℓ^* is the volume to area ratio. Values for τ are for non-transpiring organs except those in parentheses, which assume an r_ℓ of 50 s m^{-1} (Monteith, 1981).

| | Dimensions | | Calculated time constant (τ) | |
	d (cm)	ℓ^* (cm)	$u = 1$ m s^{-1}	$u = 4$ m s^{-1}
Leaves				
Grass	0.6	0.05	0.18 (0.13)	0.09 (0.08)
Beech	6	0.10	0.94 (0.52)	0.55 (0.36)
Alocasia	60	0.15	2.90 (1.34)	2.00 (1.01)
Stems				
Small	0.6	0.15	1.4	0.68
Medium	6	1.5	31	16
Large	60	15	540	330
Fruits				
Rowan	0.6	0.1	0.71	0.33
Crab apple	6	1	16	7.7
Jack fruit	60	10	300	170

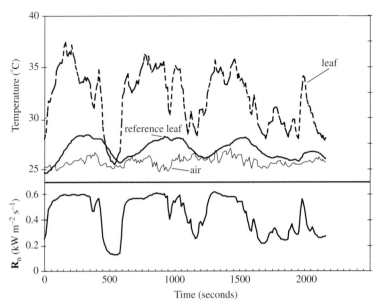

Figure 9.3 (*a*) Temperature fluctuations of a *Phaseolus* leaf (approximately 30 cm^2) in the field, together with simultaneous fluctuations of air temperature and the temperature of a relatively massive artificial wetted reference surface (aluminium covered with filter paper) and (*b*) the simultaneous fluctuations of net radiation. Temperatures measured with 42-gauge copper constantan thermocouples (H. G. Jones, unpublished data).

area, so that for the trunks of mature trees τ can be of the order of one day. The value of τ for leaves is significantly increased by stomatal closure. Figure 9.3 illustrates just how rapidly the temperature of medium-sized (*c.* 30 cm^2) transpiring leaves can fluctuate in the field, while the temperature changes of thicker and more massive artificial 'reference' leaves are much more heavily damped. Note that the air temperature itself also shows rapid changes that are not detected when using a measuring instrument with a long time constant such as a mercury-in-glass thermometer. Observed values of τ for leaves of different species are close to the values predicted in Table 9.1, ranging from about 0.15 to 0.45 min for species as diverse as vine, cotton, *Salix arctica* and *Pinus taeda* (Paw U, 1992) though reaching 7 min for the very thick leaves of *Graptopetalum* in a low wind (Ansari & Loomis, 1959; Linacre, 1972; Thames, 1961; Warren Wilson, 1957) and values in excess of 2 hours for a cactus stem (Ansari & Loomis, 1959).

One approximation in the derivation of Eq. (9.9) is the assumption of a uniform surface temperature, a condition that is not usually satisfied in the field, though the error involved is usually small. A second problem is that with bulky tissue, the rate of heat conduction to the surface is important, because the thermal conductivity of plant tissues is quite low

(being of the same order as water: Appendix 5). The lateral thermal conductivity of leaves, for example, ranges from about 0.24 to 0.50 W m^{-1} K^{-1} (Nobel, 2009) while the thermal conductivity of apple fruits is less than 0.9 W m^{-1} K^{-1} (Thorpe, 1974). This means that the temperature at the centre of large organs lags behind that at the surface, and that the time constant at the centre of stems may be longer that that given by Eq. (9.11), which may in turn be longer than that at the surface. The low thermal conductivity of plant tissue also means that unequal radiation absorption on different sides of large organs can lead to large temperature gradients. For example, temperature differences as large as 10°C between the two sides of an apple (Vogel, 1984) or a cactus (Thorpe, 1974) have been observed with high irradiance. Nobel (1988) has developed a model for calculating stem surface temperatures in different *Ferocactus* species that includes effects of plant size, apical pubescence and shading by spines. The results of the model were in close agreement with field observations and could be related to the natural distribution of the different species.

The actual temperature dynamics in any situation can be derived by solution of Fourier's equation for heat flow (see Section 3.2.2), which in one dimension is:

$$\partial T/\partial t = D_H \partial^2 T/\partial z^2 = (k/C_v)\partial^2 T/\partial z^2 \qquad (9.12)$$

where D_H is the thermal diffusivity of the material ($m^2\ s^{-1}$), k is the thermal conductivity ($W\ m^{-1}\ K^{-1}$), and C_v is the volumetric heat capacity of the material ($J\ m^{-3}\ K^{-1}$). Note that C_v is the product of the density (ρ; $kg\ m^{-3}$) and the specific heat capacity (c_p; $J\ kg^{-1}\ K^{-1}$) of the material; these vary as a function of composition (the fractions of mineral content (for soils), organic matter, water and air) as shown in Appendix 5. Further details may be found in Campbell and Norman (1998) and Hillel (2004).

9.1.4 Particular cases of the time course of temperature changes

Step change
Where the equilibrium environmental temperature changes instantaneously from one value to another, the time course of leaf temperature, for example, is given by Eq. (9.10) and is illustrated in Figure 9.4(a).

Ramp change
Where the environment is changing at a steady rate, tissue temperature lags behind the equilibrium temperature, but when the time of the steady change exceeds about $3 \times \tau$, the rate of increase of tissue temperature equals the rate of increase of equilibrium temperature (Figure 9.4(b)).

Harmonic change
Particularly important situations in environmental studies occur where the environmental temperature oscillates or where the energy input (e.g. solar radiation) oscillates. Both the diurnal and the seasonal changes in temperature, for example, can be approximated by sine waves so that we can write the notional variation of equilibrium temperature with time t ($T_{e,t}$) for a single cycle as:

$$T_{e,t} = T_{ave} + \Delta T_e \sin(\omega t) \qquad (9.13)$$

where T_{ave} is the average temperature over the cycle, ΔT_e is the amplitude or half the peak to peak range

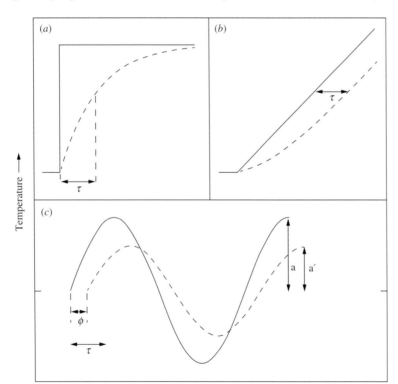

Figure 9.4 The change of surface temperature (broken line) in response to changing environmental temperature (solid line) (modified from Monteith & Unsworth, 2008). (a) Response to a step change, where τ is the t for a 63% change (see also Figure 4.14); (b) response to a ramp change, where τ is the constant time lag eventually established; (c) response to a sinusoidally varying environmental temperature (Eq. (9.13)).

of the cycle of the equilibrium temperature and ω is the angular frequency of the driving energy input ($= 2\pi/P$; where P is the period of oscillation, so that ω equals 7.27×10^{-5} s^{-1} for a diurnal cycle and 2×10^{-7} s^{-1} for an annual cycle). Using this harmonic input, Eq. (9.12) can be solved (see e.g. Monteith & Unsworth, 2008) to give the temperature at any time (t) and depth (z) as:

$$T_{s,t} = T_{ave} + \Delta T_z \sin (\omega t - \phi) \qquad (9.14)$$

where ΔT_s is the amplitude of the surface temperature and ϕ is the *phase lag* of the surface temperature behind the driving input (radian). The general effect is illustrated in Figure 9.4(c) for the same value of the time constant as in Figures 9.4(a) and (b). The effect of increasing τ is two-fold: first, it causes a damping of the amplitude of the oscillation and, second, it increases the magnitude of the phase lag between the driving temperature and the sample temperature. The phase lag, is related to the time constant by, $\phi = \tan^{-1}(\omega\tau)$, while the damping of amplitude ($\Delta T_s/\Delta T_e$) equals cos (ϕ).

Temperature profile in massive tissues and soils

The analysis can be extended to allow prediction of variation in temperature with depth in soil or in massive tissues such as tree trunks (Campbell & Norman, 1998; Hillel, 2004). The finite thermal conductivity of soil or large plant organs leads to slow transfer of heat from the surface to the interior and an increase both of the lag and in the amount of damping that occurs. Solution of Eq. (9.12), assuming a constant temperature at infinite depth equal to T_{ave}, gives:

$$T_{z,t} = T_{ave} + \Delta T_s \exp (-z/Z) \sin (\omega t - \phi - z/Z) \quad (9.15)$$

where $T_{z,t}$ is the variation of temperature at depth, z, with time t, and Z is known as the *damping depth* and is given by:

$$Z = \sqrt{2D_H/\omega} \qquad (9.16)$$

Results from application of Eq. (9.15) to a typical soil are summarised in Figure 9.5. Although this adequately describes variation for many purposes it should be recognised that this represents a substantial simplification as both annual and diurnal cycles are superimposed, the rate of heat loss to the atmosphere varies with weather conditions and the thermal diffusivity depends on soil water content.

9.2 Physiological effects of temperature

9.2.1 Effect of temperature on metabolic processes

Although most metabolic reactions are strongly influenced by temperature, some physical processes such as light absorption are relatively insensitive, while the rate of diffusion is generally intermediate in sensitivity.

Temperature dependence arises where the process requires that the molecules involved have a certain minimum energy (usually in the form of kinetic energy). In general, a high minimum energy requirement leads to greater temperature sensitivity. The reason for this temperature effect may be discussed in relation to a simple chemical reaction where, before the reaction can take place, the molecule or molecules involved must be raised to a state of higher potential energy (Figure 9.6). The energy involved in this reaction 'barrier' is called the activation energy (E_a). How the value of the activation energy affects the temperature response can be seen if one considers the distribution of energies between different molecules in a population of similar molecules at a given temperature. Although the mean kinetic energy increases with temperature, the number in the 'high-energy tail' of the frequency distribution increases more rapidly. The number ($n(E)$) that have an energy equal to or greater than E_a is given by the Boltzmann energy distribution, which can be expressed on a molar basis in the following form:

$$n(E) = n \exp(-E/\mathcal{R}T) \qquad (9.17)$$

where n is the total number, $\mathcal{R}$ is the gas constant and T is the absolute temperature. The rate of a reaction that has a particular activation energy would be expected to be proportional to the number of

(a)

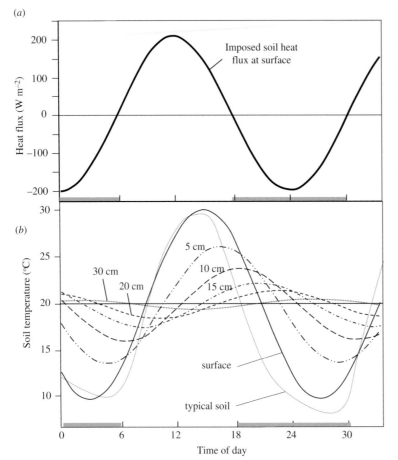

(b)

Figure 9.5 Diurnal trends in (a) imposed soil heat flux at the soil surface resulting from solar energy input on a clear day (with positive values corresponding to a positive flux, **G**, into the soil) and (b) the corresponding fluctuations in soil temperature at various depths in the soil. A typical actual diurnal trend of soil surface temperature for a representative soil is also shown (after Jones & Vaughan, 2010).

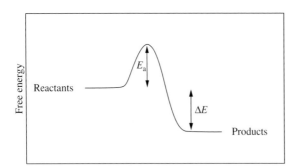

Figure 9.6 The energy threshold for a chemical reaction showing the activation energy (E_a) of the reaction and the net free energy change (ΔE).

molecules with the appropriate energy so that the rate constant (k) is given by:

$$k = A\exp(-E_a/\mathscr{R}T) \tag{9.18}$$

where A is approximately constant and depends on the type of process. If one takes logarithms of this equation, which is known as the *Arrhenius equation*, one obtains:

$$\ln k = \ln A - E_a/\mathscr{R}T \tag{9.19}$$

which predicts that the natural logarithm of the rate constant should be linearly related to $1/T$ with a slope of $-E_a/\mathscr{R}$.

Another way of describing the temperature sensitivity of biochemical processes is in terms of the temperature coefficient, Q_{10}, which is the ratio of the rate at one temperature to that at a temperature ten degrees lower. This coefficient is somewhat arbitrary and potentially misleading, especially when applied outside the rather limited range of conditions where the temperature response is exponential, but as long as its limitations are remembered it can be useful, and is widely used. The equations that follow can readily be reformulated for coefficients over temperature ranges other than 10°C. Therefore from Eq. (9.18):

$$Q_{10} = \frac{A \exp[-E_a/\mathcal{R}(T+10)]}{A \exp(-E_a/\mathcal{R}T)} \quad (9.20)$$

$$= \exp[10 E_a/\mathcal{R}T(T+10)]$$

From this it can be shown that, at 20°C, a Q_{10} of 2 arises where the activation energy is 51 kJ mol^{-1} (i.e. $2 = \exp [(10 \times 51\,000)/(8.3 \times 293 \times 303)]$). The rate of respiration, for example, often has a Q_{10} of 2 at normal temperatures (Figure 9.7), though it varies with the state of the tissue and also decreases at high temperatures that damage tissues. Where the activation energy is lower, as for example for the diffusion of mannitol in water where $E_a = 21$ kJ mol^{-1}, the Q_{10} is lower: in this case 1.3.

In practice the Q_{10} may be obtained from the reaction rates k_1 and k_2 at any two temperatures T_1 and T_2, by using the approximation:

$$Q_{10} \simeq \left(\frac{k_1}{k_2}\right)^{[10/(T_2 - T_1)]} \quad (9.21)$$

Although the rates of simple chemical reactions increase exponentially with increasing temperature, most biological reactions show a clear optimum temperature, with reaction rates declining with any temperature increase above the optimum. There are several reasons why the rates of biological reactions do not continue to increase indefinitely with increasing temperature. One factor is that the rate limiting reactions for any process may change, as rates increase, from highly temperature-sensitive ones to those such as diffusion, which have lower temperature coefficients. Another factor is that many processes are the net result of two opposing reactions with different temperature responses. The most important reason, however, is that most biological reactions are enzyme catalysed. Enzymes act to lower the activation energy, and hence to decrease temperature sensitivity and to increase the rate at any given temperature. However, as temperatures rise the catalytic properties of most enzymes are harmed and the total amount of enzyme present may fall as a result of increased rates of denaturation. These factors are discussed below in relation to net photosynthesis.

It is particularly notable that the rates of many plant processes, such as development, which integrate many individual components, are frequently approximately *linearly* related to temperature over a wide range of normal temperatures, though an optimum with a subsequent decline is reached at high temperatures.

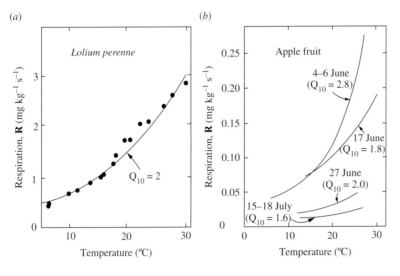

Figure 9.7 Temperature response of respiration for (a) simulated sward of ryegrass (*Lolium perenne*) together with the curve for $Q_{10} = 2$ (data from Robson, 1981); (b) fitted exponential curves for respiration from apple fruits on different dates (data from Jones, 1981a).

A convenient empirical equation to simulate many temperature responses, such as photosynthesis (see below) is (for $0 \leq k \leq 1$):

$$k = \frac{2(T + B)^2 (T_{max} + B)^2 - (T + B)^4}{(T_{max} + B)^4} \qquad (9.22)$$

where T_{max} is the temperature at which the coefficient k reaches a maximum of 1.0 and B is a constant (Figure 9.8).

9.2.2 Temperature response of net photosynthesis

Photosynthesis is one of the most temperature-sensitive aspects of growth. Some photosynthetic temperature responses for species from different thermal environments are shown in Figure 9.9, illustrating the tendency for net photosynthesis of temperate zone plants to be maximal between about 20 and 30°C, with species from hotter habitats having higher temperature optima. In addition many species show marked temperature acclimation when grown in different temperature regimes, as illustrated for *Eucalyptus* and *Larrea*.

At higher temperatures the shapes of the temperature-response curves depend on the duration

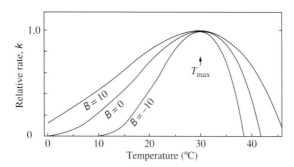

Figure 9.8 Simulation of temperature responses using Eq. (9.22), for $T_{max} = 30°C$, and for different values of the constant B.

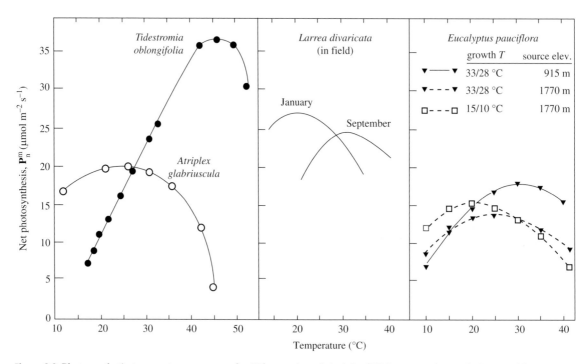

Figure 9.9 Photosynthetic temperature responses for *Tidestromia* and *Atriplex* (Björkman *et al.*, 1975), *Larrea tridentata* (Mooney *et al.*, 1978) and for *Eucalyptus pauciflora* (Slatyer, 1977), illustrating temperature acclimation in *Eucalyptus* and *Larrea*.

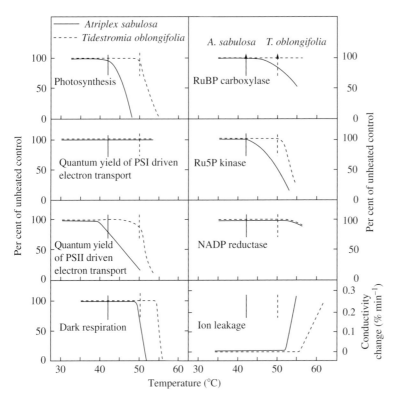

Figure 9.10 Time-dependent temperature inactivation of various photosynthetic components in *Atriplex sabulosa* and *Tidestromia oblongifolia* (see Björkman *et al.*, 1980 for data sources). Rates were measured at 30°C after 10 or 15 min pretreatment at the indicated temperature. Vertical lines indicate the temperatures at which time-dependent inactivation of photosynthesis sets in for the two species. PSI and PSII represent photosystems I and II, respectively.

of exposure to these temperatures, because thermal instability leads to time-dependent inactivation of the photosynthetic system. There are species differences in the temperature at which this inactivation occurs. For example, thermal inactivation occurs when leaf temperatures exceed about 42°C in *Atriplex sabulosa* (a C$_4$ species from a cool coastal environment), but does not occur in the desert species *Tidestromia oblongifolia* until about 50°C (Figure 9.10).

Some results from an extensive series of experiments investigating differences between *A. sabulosa* and *T. oblongifolia* in photosynthetic stability at high temperatures are presented in Figure 9.10. Some processes, such as photosystem I driven electron transport and activity of enzymes such as NADP reductase, showed little sensitivity to short exposures to high temperatures. Other features, such as membrane permeability (measured by ion leakage), dark respiration and carboxylase activities, were sensitive to high temperatures, with species differences being apparent. However, the inhibitory

temperatures for these processes were significantly higher than those damaging photosynthesis. The temperature sensitivity of photosynthesis was most closely related to that of the quantum yield of photosystem II driven electron transport and some photosynthetic enzymes (such as ribulose 5-phosphate kinase). Other evidence indicated that species differences did not result from differences in stomatal behaviour. Overall, the available evidence supports the conclusion that much of the difference in thermal stability between species results from differences in thermal stability of the chloroplast membranes, and in particular the integrity of photosystem II. The involvement of primary damage at the photosystem II system is particularly clearly demonstrated by the sharp increase in basal level of chlorophyll fluorescence (F_o) above a threshold temperature that corresponds to the onset of leaf necrosis (Bilger *et al.*, 1984).

In contrast, there is evidence that differences between species in their photosynthetic rate at low

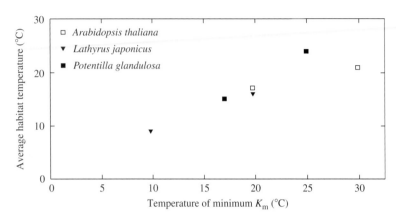

Figure 9.11 The relationship between the assay temperature giving the minimum K_m and the average habitat temperatures for two populations of each of three species. The enzymes studied were glucose 6-phosphate dehydrogenase for *Arabidopsis* and malate dehydrogenase for the other species. (Data from Teeri, 1980.)

temperatures is strongly correlated with the capacity of specific rate-limiting enzymes such as RuBP carboxylase and fructose bis-phosphate phosphatase. There is also evidence, at least for some enzymes, that their kinetic properties may be adapted to the normal environmental temperature of that ecotype. The assay temperature giving the minimum Michaelis constant (K_m) can be closely related to the habitat temperature (Figure 9.11). Small changes in K_m of about two-fold over the normal temperature range may provide a mechanism for maintaining the rate of the catalysed reaction relatively insensitive to temperature fluctuation. In some cases at least, thermal acclimation is based on changes in the profile of the isoenzymes synthesised (Scandalios *et al.*, 2000) with much of this protein diversity resulting from alternative splicing (Syed *et al.*, 2012).

9.3 Effects of temperature on plant development

9.3.1 Thermal time

The rates of many plant developmental processes, and hence the timing of phenological stages, are strongly temperature dependent. If under a particular set of conditions a particular developmental stage takes t days, the corresponding rate of development (k_d) is $1/t$; this implies that the time taken to complete the developmental stage is inversely proportional to k_d.

The rate of development is usually a strong function of temperature, so that in a constant environment:

$$1/t = k_d = f(T) \tag{9.23}$$

We can denote the state of plant development at any time t (measured from a suitable starting date such as sowing) as $S(t)$, where S is a fractional state of development on a zero to one scale. In a fluctuating environment where T is a function of time (written as $T(t)$), $S(t)$ is given by:

$$S(t) = \int_{t=0}^{t} k_d dt = \int_{t=0}^{t} T(t) dt \tag{9.24}$$

In practice it is often found that, between a threshold temperature for development to occur known as the base temperature (T_{base}) and an optimum temperature (T_o), the rate of development is approximately linearly related to temperature so that between these temperatures:

$$k_d = a(T - T_{base}); \text{ for } T_{base} \leq T \leq T_o \tag{9.25}$$

where a is a constant. Temperatures below T_{base} therefore do not contribute to development. Thus if temperature remains constant over time one can substitute into Eq. (9.24) to get:

$$S(t) = a(T - T_{base}) \int_{t=0}^{t} dt = a(T - T_{base})t \tag{9.26}$$

It follows that the value of the temperature integral, $(T - T_{base})t$, that is required to complete the developmental stage being considered (i.e. $S(t) = 1$)

is equal to $1/a$. For convenience this temperature integral is given the symbol D and measured in day-degrees. Under fluctuating temperature conditions we can write:

$$S(t) = a \int_{t=0}^{t} (T(t) - T_{base})dt$$

$$= (1/D) \int_{t=0}^{t} (T(t) - T_{base})dt \qquad (9.27)$$

Calculation of this integral requires a knowledge of the relationship between temperature and time. For convenience the temperature sum (D) is often obtained by summing for each day the excess of the daily mean temperatures (T_m) above the threshold according to:

$$D = \sum_{d=1}^{n} (T_m - T_{base}); \text{ for } T_{base} \leq T_m \leq T_o \qquad (9.28)$$

Completion of the developmental stage ($S = 1$) requires that this temperature sum, D, often incorrectly referred to as a 'heat sum', equals or exceeds $1/a$. D is known as the *thermal time* or the accumulated temperature required for completion of the developmental stage being considered. Other terms such as growing degree days (GDD) or even, unfortunately, 'heat' units (HU) are frequently used as synonyms for thermal time. It is

worth noting in passing that this recommended method of calculation uses the daily average temperature ($T_m = (T_{max} + T_{min})/2$, irrespective of whether or not T_{min} falls below T_{base}).

Estimates of the appropriate threshold or base temperature can be obtained in two main ways. Where studies can be made in controlled environments at constant temperature, it is possible to plot rate of development against temperature and T_{base} is given by the intercept on the x-axis. Where temperatures are fluctuating, as in the field, it is necessary to calculate D with different thresholds and determine which gives the best linear fit to Eq. (9.26) across a number of different years or sites. In practice the base temperature varies with species, with calculations for some temperate crops such as wheat, barley and *Brassicas* often using 0°C as a threshold, while others such as peas and forage often use 5°C. For maize, soybeans and tomato a value of 10°C is often used while thresholds for some tropical crops such as cowpea often reach around 15°C. Many published calculations of GDD (e.g. for maize) use 10°C as a standard base temperature value, but it is important to remember that this may not be appropriate for all crops. Some typical values for thresholds and GDD are shown in Table 9.2; information on specific crop

Table 9.2 Some published values of the number of growing degree days (GDD) required for different growth stages for different plants.

	T_{base} (°C)	GDD to flowering (°C d)	GDD to maturity (°C d)	Reference
Pea	3.0		824–926	(Bourgeois *et al.*, 2000)
Cherry	4.0	243 (from 1 Mar)		(Zavalloni *et al.*, 2006)
Rice	10.0	1350–1484	1810–1915	(Islam & Sikder, 2011)
Pearl millet	10.0	667–944	1150–1220	(Cardenas, 1983)
Barley	0.0	738–936	1269–1522	(Miller *et al.*, 2001)
Brassica rapa	0.0	630–726	1152–1279	(Miller *et al.*, 2001)
Maize (OCHU)[a]	5.0 (night) 10.0 (day)		2500–3500	(Ma & Dwyer, 2001)

[a] Ontario corn heat units.

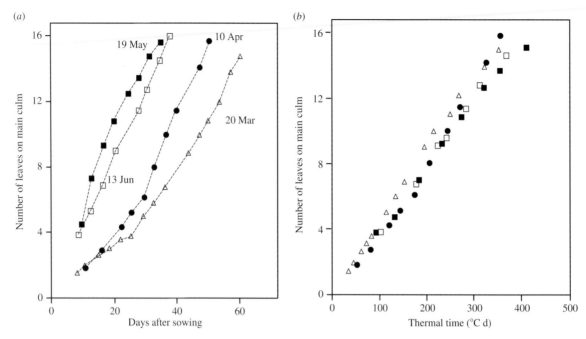

Figure 9.12 (a) Time course of appearance of leaves on the main stem of *Pennisetum typhoides* for four different sowing dates. (b) The same data plotted against thermal time calculated for soil temperature at 5 cm depth and using a threshold temperature of 12.4°C. (Data from Ong, 1983.)

requirements together with calculations of GDD for local climatic data helps farmers to determine which crops will grow successfully in any area.

An example of the effectiveness of thermal time as a tool for the study of plant development is presented in Figure 9.12; this figure shows the dependence of a plant's rate of development on the sowing date, with more rapid development occurring for later sowings because of the higher temperatures. When, however, the rate of leaf appearance along the main stem, in this case for pearl millet, is expressed as a function of thermal time rather than real time it is clear that the rate of appearance of leaves falls on one line.

Extensions of thermal time

The concept of thermal time as a replacement of chronological time has been in use for phenological studies for over 200 years (Wang, 1960). Although the simple formula is adequate for most purposes, non-linearities in the temperature responses and interaction with other environmental factors have been incorporated into a number of more sophisticated equations. For example, polynomial or other functions of T can be substituted for the linear form used in Eqs. (9.25) and (9.26), while the 'Ontario corn heat unit' accumulates maximum and minimum temperatures separately (Dwyer *et al.*, 1999).

For seed germination studies it is common to assume that germination rate increases linearly with temperature between T_{base} and T_o, and then decreases linearly above T_o (see Figure 9.13), so that above the optimum:

$$k_d = (T_{max} - T)/D_2; \text{ for } T_o \leq T \leq T_{max} \tag{9.29}$$

where T_{max} is the upper limit for germination and D_2 is an appropriate temperature sum. The behaviour at

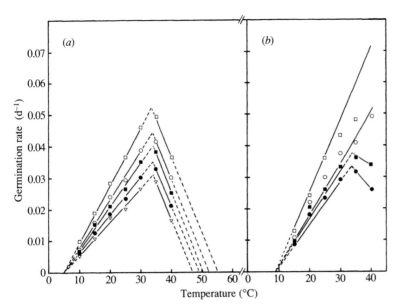

Figure 9.13 The relation between temperature and rate of progress to germination (at constant temperature) for (*a*) soybean and (*b*) cowpea. Broken lines represent extrapolation beyond experimental values, and the different symbols represent various percentage germination: 10% (□), 30% (○), 50%, (■) 70% (•) and 90% (▽). (Data from Covell *et al.*, 1986.)

supraoptimal temperatures, however, is complicated by the fact that there is a time × temperature dependency. Although high temperatures may damage the seed there can still be an underlying tendency for germination to be speeded up by higher temperatures. This is illustrated for the cowpea data in Figure 9.13 where the earliest germinating seeds emerged before high temperature damage became apparent.

Effective day-degrees

Inclusion of other environmental variables enables one to describe development where temperature is not the only environmental variable affecting the process. A particularly useful approach is the definition of what have been called 'effective day-degrees' (D_{eff}) (Scaife *et al.*, 1987) to include effects of radiation and temperature. In this approach D is replaced by D_{eff}, defined by:

$$D_{eff}^{-1} = D^{-1} + b\mathbf{I}^{-1} \qquad (9.30)$$

where b is a constant describing the relative importance of irradiance ($\mathbf{I}$) and temperature. Where b is zero, D_{eff} becomes equal to conventional day-degrees.

9.3.2 Dormancy, chilling, stratification and vernalisation

Temperature shows marked seasonal fluctuation, particularly at higher latitudes, and although there may be considerable short-term variability these seasonal changes are a major factor in the control of flowering, often interacting with photoperiodic control. Many perennial species show a well-marked period of dormancy, usually during the winter in temperate species, but in the summer in many Mediterranean species. A characteristic of dormant tissues is that their meristems are not able to initiate growth even in favourable conditions. A feature of winter dormancy is that it also involves hardening of tissues in a way that enhances cold tolerance.

Seasonal temperature fluctuation plays an important role in the induction of dormancy and leaf abscission in autumn, and in the subsequent release from dormancy after the winter, in each case often acting in conjunction with daylength effects. Similarly, biennial species usually require exposure to a period of low temperatures before they can flower. For example 'winter' varieties of wheat must be exposed to several months of low temperatures to permit flowering. If sown in the spring (rather than the autumn), they remain vegetative as a consequence of failure to satisfy their vernalisation requirement. Spring varieties, on the other hand, have only a minimal or no vernalisation requirement so they can be successfully sown in the spring. The requirement for a period of low temperatures can ensure an adequate period of growth before flowering, otherwise, in natural situations seeds that germinate in late summer might flower in the same year and fail to mature. Similarly, release from dormancy in many perennial plants, including most temperate fruit species, also requires a period of cold (chilling) to break dormancy and as a prerequisite for effective and synchronous bud break.

Vernalisation

As with photoperiodic responses (Section 8.4.3), vernalisation requirements vary from obligate to quantitative, with the rate of vernalisation being a variable function of temperature according to the species. Vernalisation is required for the meristem to change from a vegetative state (producing leaves) to a floral state (producing flowers). Unfortunately the effect of temperature on this transition confounds the effect of temperature on vernalisation with the effect of temperature on meristem development (as discussed above). It is therefore difficult to isolate the true temperature response of vernalisation: the common interpretation of the vernalisation response

for wheat is illustrated in Figure 9.14 where the rate is very slow at 0°C or below, reaches a maximum at about 2 to 4°C, and has an upper limit of about 11°C (Evans et al., 1975). It has been argued, however, on the basis of a developmental analysis (where rate of vernalisation is defined as the reciprocal of the number of days of treatment until achievement of the minimum number of leaf primordia) that the rate of vernalisation in wheat actually increases linearly from 0°C to about 11°C, declining thereafter (Brooking, 1996). Plants may be vernalised as seeds (a process known as stratification) or may achieve adequate chilling when exposed as leafy plants to low temperatures.

Although the pathway of vernalisation is complex, there appears to be substantial similarity between the genetic control of vernalisation in monocots and dicots, with the central *FLOWERING LOCUS C* (*FLC*) gene in *Arabidopsis* (or the *VRN2* homologue in wheat) acting to suppress flowering. Exposure to extended periods of low temperature, however, leads in *Arabidopsis* to a stable repression of *FLC*, and in wheat to stimulation of *VRN1*, which represses *VRN2*; these changes then set off subsequent genetic cascades leading to floral development. An interesting feature is that repression is maintained even after the temperature increases through epigenetic modification of the histones in the appropriate part of the DNA (Dennis & Peacock, 2007).

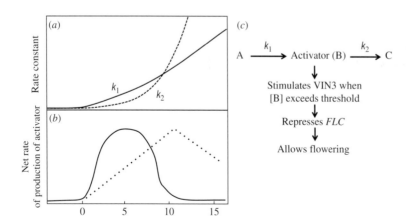

Figure 9.14 A hypothetical mechanism for vernalisation. (*a*) Shows the temperature responses of the rate of synthesis (k_1) and degradation (k_2) of an activator (B), (*b*) the resulting temperature dependence of the rate of net synthesis of B and (*c*) a schematic of the role of this in vernalisation response in *Arabidopsis* when enough B accumulates after prolonged exposure to vernalising temperatures (see text for further detail). The dotted line in (*b*) shows the temperature sensitivity of the 'rate of vernalisation' for wheat proposed by Brooking (1996).

An interesting feature of vernalisation is the very low temperature optimum, which is unusual for biological processes. Very little is known of how the plant senses the low temperature exposure, though it is possible that some component of the cold-acclimation process (e.g. modification of membrane lipids and membrane fluidity or changes in protein phosphorylation patterns in response to low temperatures) may be involved. There is evidence that cold acclimation and vernalisation involve independent pathways (Bond *et al.*, 2011). What is known is that in *Arabidopsis* the *VIN3* gene appears to be the gene most closely linked to signalling of the cold exposure. A hypothetical scheme (see e.g. Sung & Amasino, 2005 for a different variant) that could give rise to such a response would be a two-stage reaction where upregulation of *VIN3* would occur when the concentration of an activator exceeds a threshold as a result of a period of exposure to permissive temperatures, with the concentration of the activator at any time depending on the competing processes of synthesis and degradation each having different temperature responses as indicated in Figure 9.14.

Many seeds also have a requirement for a period (often extensive) of low temperature before they can germinate. These low temperatures are most effective at overcoming seed dormancy if given when the seeds are moist.

Chilling and winter dormancy

Woody perennial plants, such as temperate fruit trees, that enter a dormant period during the winter require exposure to adequate chilling for breaking the 'endodormancy' and to ensure synchronous bud break. It has been noted in some cases, however, that it is possible partly to substitute the cold requirement by drought or by defoliation; for example Jones (1987b) was able to initiate in apple trees a second phase of flowering within one year by drought treatments that were severe enough to cause defoliation on re-watering.

The amount and depth of chilling that is required to satisfy the chilling requirement varies substantially between species and even between cultivars. A number of models have been proposed to estimate the amount of physiologically significant chilling from meteorological data and have been used as an aid to the identification of genotypes appropriate for specific regions. The most common models for describing satisfaction of the chilling requirement are illustrated in Figure 9.15 and can be classified into (i) simple temperature accumulation models of which the commonest is the number of hours below an empirically derived threshold of 7.2°C (or the number of hours between 0°C and 7.2°C); (ii) more sophisticated temperature weighting models, such as the 'Utah' model for peach bud development (Richardson *et al.*, 1974), which includes a negative

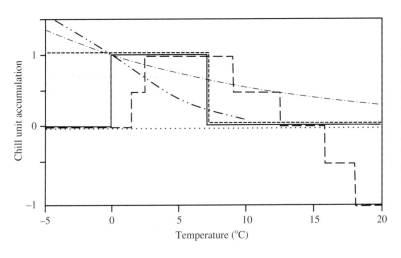

Figure 9.15 Various chilling functions for accumulation of chill units: these include those that are best suited to crops such as peach ('Utah' units, — — — —) and 0 to 7.2°C (——), and those that also respond to temperatures below 0°C including the <7.2°C units (-----), Bidabé's exponential function (– · – · – · –) and the best fit-curve for blackcurrant chilling requirement (– ·· – ·· – ·· –) (Jones *et al.*, 2012).

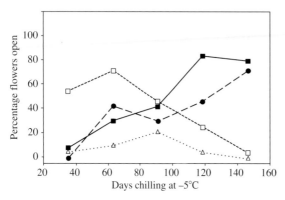

Figure 9.16 Effect of different periods of chilling at –5°C on flowering (recorded as % flowers open after 45 days in a permissive environment at 20°C) of typical blackcurrant cultivars that either require large amounts of chilling (Ben Avon, ——■——; Hedda, – – ● – –) and cultivars where flowering can be inhibited by excess chilling (Andega, ---□---; Amos Black, ···△···). (Data from Jones *et al.*, 2012.)

effect of higher temperatures, and models that assume that chilling 'effectiveness' increases exponentially (Bidabé, 1967) with decreasing temperature or according to some other function (Jones *et al.*, 2012). In deriving phenological models for predicting the dates of spring events such as bud burst and flowering it is usual to use sequential models that combine both the chill accumulation phase required to release endodormancy and a warming or 'anti-chill' phase (often referred to as the *forcing* phase) to describe the rate of release from dormancy and the actual date of flowering (see Jones *et al.*, 2012). This forcing phase is commonly calculated on the basis of a temperature summation (e.g. GDD above an appropriate base temperature). Because spring events may also be at least partially dependent on daylength, the best predictions of effects of climate change on developmental events may be achieved by incorporation of extra components such as daylength (Blümel & Chmielewski, 2012).

Although many chill models are based on hourly accumulation of chill, the necessary hourly temperature data are often not available, so Sunley *et al.* (2006) have shown that hourly temperatures may be estimated adequately from daily

maximum/minimum data by assuming sinusoidal temperature variation according to Eq. (9.13), substituting $(T_{max} - T_{min})/2$ for T_{ave} in this equation.

The most appropriate function varies substantially between species and indeed within species, with more 'arctic' genotypes such as blackcurrant (*Ribes nigrum*) gaining most effect from chilling temperatures below 0°C. For blackcurrant, the number of chilling hours <7.2°C required to satisfy the chilling requirement varies from <1300 h to >2000 h for late flowering UK cultivars such as 'Ben Lomond' or Nordic cultivars. In some cases, excess chilling can even be inhibitory (Figure 9.16), with longer exposures to temperatures below –5°C reducing flowering in some cases (Jones *et al.*, 2012).

9.4 Temperature extremes

The ability of different plants to survive extreme temperatures depends both on their innate physiology and on the degree to which they have been acclimated by a process of 'hardening'. The survivable temperatures for fully hardened plants of different plant groups from different regions is summarised in Table 9.3. Although there are major differences between plant groups and provenances in their ability to tolerate low temperatures, the upper temperature limits are remarkably similar across the different groups.

9.4.1 High temperatures

High-temperature damage to cells and tissues normally involves loss of membrane integrity with consequent ion leakage, together with deactivation and denaturation of many enzymes. Cell death can readily be assessed by means of the ability of cells to take up a vital stain such as neutral red. On the tissue scale, high-temperature damage can usually be seen as tissue necrosis.

As has already been indicated (e.g. Figure 9.9), however, many plants have a great capacity to adapt to temperature extremes. This ability to acclimate is widespread: for example, Nobel (1988) has collated results from a number of experiments on 33 species

Table 9.3 Summary of temperature thresholds for survival of extreme temperatures by leaves of plants from different climatic regions, where thresholds are defined at the temperatures giving 50% damage after 2 h exposure to the given temperature (extracted from Larcher, 1995). Seeds are generally more tolerant of extremes.

	Threshold T for cold injury (hardened) (°C)	Threshold T for heat injury in summer (°C)
Tropics		
Trees	+5 to −2	45 to 55
Herbs	+5 to −3	45 to 48
Mosses	−1 to −7	–
Subtropics		
Evergreen woody plants	−8 to −12	50 to 60
Subtropical palms	−5 to −14	55 to 60
Succulents	−5 to −10 (−15)	58 to 67
C$_4$ grasses	−1 to −5 (−8)	60 to 64
Temperate		
Evergreen woody plants	−7 to −15 (−25)	46 to 50 (55)
Deciduous woody plants	(−25 to −35)[a]	c. 50
Herbaceous plants	−10 to −20 (−30)	40 to 52
Graminoids	(−30 to −196)[a]	60 to 65
Succulents	−10 to −25	(42) 55 to 62
Homoiohydric ferns	−10 to −40	46 to 48
Boreal		
Evergreen conifers	−40 to −90	44 to 50
Boreal deciduous trees	(−30 to −196)[a]	42 to 45
Arctic-alpine dwarf shrubs	−30 to −70	48 to 54
Arctic-alpine herbs	(−30 to −196)[a]	44 to 54
Mosses	−50 to −80	–

[a] Vegetative buds.

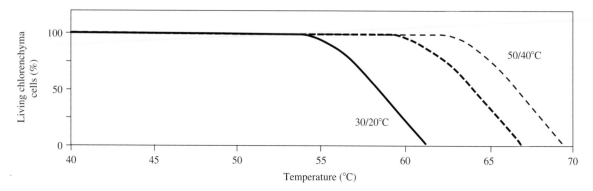

Figure 9.17 Influence of day/night growth temperatures on the high-temperature tolerance (measured as the % of chlorenchyma cells taking up vital stain) of *Opuntia ficus-indica*. The solid line is for plants grown at 30°C/20°C, the intermediate dashed line is for plants three days after shifting to a 50°C/40°C regime and the final dashed line is after three weeks at the higher temperature. (Data from Nobel, 1988.)

of agaves and cacti where it was shown that the temperature leading to 50% apparent cell death increased by between 1.6 and 15.8°C for different species when the growth temperature was raised from 30°C/20°C (day/night temperature) to 50°C/40°C. For nearly all species the temperature tolerated under the higher temperature growth regime was in excess of 60°C (Figure 9.17). Indeed a number of cacti including *Opuntia ficus-indica* (prickly pear) can tolerate temperatures in excess of 70°C for one hour, a common duration for the most extreme temperatures in the field. Such a high temperature appears to be lethal to other vascular plants that have been studied.

The underlying basis of acclimation to high temperatures is not well understood, but there is evidence that binding of the variant histone H2A.Z to nucleosomes acts as a thermosensor with occupancy by H2A.Z decreasing as temperature increases over a wide range of ambient temperature (Kumar & Wigge, 2010). These changes mediate genome-wide transcriptional changes (for a review see Iba, 2002) in response to temperature. There appear to be some parallels between thermal adaptation and light responses with the phytochrome signalling pathway and PIF4 being involved in both responses. As the temperature rises above a critical value (often around 38 to 40°C in plants) normal protein synthesis stops and is replaced by the rapid and coordinated synthesis

of a characteristic set of heat shock proteins (HSPs). Many of these HSPs show a very high degree of homology in all organisms that have been studied (e.g. HSP70), though higher plants also synthesise a unique group of small (15 to 18 kDa) proteins. The rapid synthesis of HSPs is related to increased transcription and the appropriate mRNAs can increase within 3 to 5 min of high-temperature stress. Heat shock protein levels tend to decline fairly rapidly, even if the high temperatures continue.

Although the detailed functions of the many HSPs are not well known, they tend to bind to structurally unstable proteins and act as molecular chaperones both helping the correct folding of proteins after synthesis, and preventing aggregation of denatured proteins and on occasions promoting renaturation. The synthesis of HSPs tends to be associated with the development of thermotolerance caused by exposure to even brief periods at high temperature. A common HSP, ubiqitin, is thought to be involved in tagging thermally denatured proteins for subsequent proteolysis by a special protease. Interestingly, some of the plant HSPs can be induced by other factors such as abscisic acid, heavy metals, osmotic stress, arsenite or anaerobiosis, though the physiological significance is not clear. In addition, many of these stresses and other stresses such as pathogenesis also give rise to specific stress proteins (Sachs & Ho, 1986). Other changes during acclimation to high temperatures include alterations in membrane lipids and changes in the complement of compatible solutes.

9.4.2 Low temperatures

There are two main types of low-temperature injury (Levitt, 1980). The first, which is common in plants of tropical or subtropical origins (such as beans, maize, rice and tomatoes) is called *chilling injury*. It is usually manifest as wilting or as inhibited growth, germination or reproduction, or even complete tissue death, and occurs in sensitive species when tissue temperatures are lowered below about 8 to 10°C (though this varies with species and degree of acclimation and may occur at temperatures as high as 15°C). As long as exposure to chilling temperatures is of short duration, the damage is usually reversible. The other major type of low-temperature injury is *freezing injury*, which occurs when some of the tissue water freezes. All growing tissues are sensitive in some degree to frost though sensitivity varies. In many plants that have not been given a chance to acclimate, tissues are killed by freezing to only −1 to −3°C. After acclimation the range of temperatures that can be survived is very wide, depending on species and tissue. Many seeds, for example, can withstand liquid nitrogen temperatures (−196°C), while many tissues from acclimated frost-tolerant species can survive −40°C or below (Table 9.3).

Chilling injury

An early event in the signalling cascade involved in cold sensing by plants is the transient influx of calcium into the cytosol (Knight *et al.*, 1991). It is also clear that an important early, if not primary, effect of chilling involves damage to the cell membranes with an early symptom in whole plants often being wilting, which is characteristic of inhibited water uptake. For example, within minutes of lowering the temperature of the roots of chilling-sensitive species such as cucumber to 8°C, root water uptake decreases and any root pressure decreases. Many associated changes in root-cell ultrastructure rapidly become visible; these include alterations to cell walls, nuclei,

endoplasmic reticulum, plastids and mitochondria (Lee *et al.*, 2002). There is also evidence that chilling injury can be related to the breakdown of cytoplasmic microtubules; this is supported by evidence that microtubule-disrupting agents enhance chilling injury, while abscisic acid tends to increase chill resistance and also retards disruption of microtubules (Rikin *et al.*, 1983). Chilling-damaged tissues tend to lose electrolytes rapidly because of increased permeability of the plasma membrane, while there is also particular evidence for damage to chloroplast and mitochondrial membranes.

There have been many studies that suggest that membrane properties in chilling-sensitive plants undergo a sudden change at about the temperature where chilling injury occurs, while chilling-resistant plants show no such abrupt change. It has been suggested, on the basis of changes in the slopes of *Arrhenius plots* (ln(rate) against $1/T$) at this critical temperature, that this results from a phase change in the membranes from a relatively fluid form to a more solid gel structure, so that normal physiological activity can only occur above the critical temperature. It appears that this phase change does not occur in the bulk membrane though it may be localised in small 'domains' within the membrane. There is some evidence that the temperature of any phase change is correlated with the fatty acid composition of the lipids, with a high proportion of saturated fatty acids occurring in the membranes of chilling-sensitive species, but the correlations are not always very good. Fructose-based polymers called fructans appear to play a major role in stabilising membranes in stress conditions including low-temperature stress where they can stabilise membranes by direct hydrogen-bonding to membrane lipids (Valluru & Van den Ende, 2008). Other factors, such as changes in membrane protein composition, may also be involved.

9.4.3 Mechanisms of damage: freezing

With freezing injury the damage results not from the low temperatures directly but from the formation of ice crystals within the tissue. These ice crystals disrupt the protoplasmic structures and cell membranes (Levitt, 1980). Although freezing-tolerant species can apparently withstand some extracellular ice formation, intracellular ice formation is commonly fatal to cells. A precise value for frost sensitivity in different plants is difficult to determine because the actual damage depends on the rate of thawing (though not generally on the time frozen), as well as on the lowest temperature reached. Membrane damage is a universal result of freezing damage, though it is still not certain whether it is the primary effect. As temperatures are lowered, ice starts to form in the extracellular water (e.g. in the cell walls). Because ice has a lower vapour pressure (and chemical potential) than liquid water at the same temperature, extracellular freezing causes water to be removed from within the cells to the sites of extracellular freezing. This leads to rapid dehydration of the cell (see Levitt, 1980) so that at least some of the effects of ice formation result from this dehydration.

The amount of water that must be lost from a cell before equilibrium is reached between intracellular water and extracellular ice depends on the temperature and on the osmotic properties of the cell. Water continues to be lost until the reduction in cell volume causes the cell water potential to balance the extracellular water potential.

The cell water potential that is in equilibrium with pure extracellular ice at the same temperature can be obtained, using Eq. (5.14), from:

$$\psi = \frac{\mathcal{R}T}{\overline{V}_W} \ln(e_{ice}/e_{s(T)}) \qquad (9.31)$$

where e_{ice} is the vapour pressure over pure ice and $e_{s(T)}$ is the vapour pressure over pure water at the temperature T (see Appendix 4). The value of ψ given by Eq. (9.30) decreases by approximately 1.2 MPa $°C^{-1}$ below 0°C. The equivalent increase in solute concentration is given by the van't Hoff relation (Eq. (4.8)) as approximately 530 osmol m^{-3} $°C^{-1}$ lowering of temperature (i.e. $-\psi/\mathcal{R}T = 1.2 \times 10^6/2270$).

Since solute concentration, c_s, is proportional to $1/V$ it follows that relatively large absolute changes in cell volume are required to maintain equilibrium for

a 1°C drop in temperature near the freezing point compared with lower temperatures. That is, the relationship between cell volume and temperature at equilibrium should be hyperbolic. As expected from this it is found that the amount of liquid water present in tissue at any temperature (minus the 'bound' water), expressed as a fraction (f) of the liquid water in unfrozen tissue (minus the bound water), decreases hyperbolically as temperature falls below that at which freezing first occurs (adjusted from Gusta et al., 1975) according to:

$$f = (\Delta T_f/T) \qquad (9.32)$$

where ΔT_f is the temperature at which freezing first occurs, known as the freezing point depression, and T is the temperature (°C).

A difficulty with simple dehydration hypotheses of freezing damage is that most water freezes at relatively high temperatures (above about −10°C), so that differences in tolerance of lower temperatures would require sensitivity to relatively small changes in hydration. For example, for a typical cell sap concentration, ΔT_f is −1.5°C so it follows from Eq. (9.32) that less than 20% of the original water would remain in the cell at −10°C.

9.4.4 Hardening and mechanisms of frost tolerance

Many plants show some degree of acclimation to chilling or freezing temperatures such that the damage temperature is lowered after exposure to a period of low temperatures. Figure 9.18 shows the changes in minimum survival temperatures that occur in two winter cereals over the winter period. Some degree of winter hardiness can also be induced by exposure to drought or salinity, while conversely, frost hardening can induce a degree of drought or salinity tolerance. At least in insects cold hardening can be rapid, occurring within two hours, where it has been associated with glycerol accumulation (Lee et al., 1987).

Various plant factors have been associated with freezing tolerance in different species, with the main aspect of cold acclimation being the increased stabilisation of the membranes against freezing

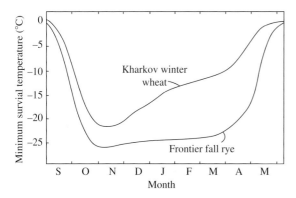

Figure 9.18 Minimum survival temperatures (LD_{50}) for a winter wheat and a winter rye, showing development of cold tolerance during the winter. (After Gusta & Fowler, 1979.)

damage. Many mechanisms are involved in this stabilisation including lowered osmotic potentials, increased concentrations of soluble carbohydrates (especially the so-called 'compatible solutes' such as glycine-betaine), changes in lipid composition, synthesis of membrane stabilising proteins and small cell size (Levitt, 1980; Thomashow, 1999). Cold hardening and cold tolerance involve the expression of a large suite of specific genes that contribute to cold tolerance, these include both those that lead to the production of proteins and other molecules that act to stabilise membranes against freeze-induced injury and the production of a series of transcription factors (especially in the DREB/CBF family) that regulate expression of such genes (for details see Thomashow, 1999). The plant growth regulator abscisic acid (ABA) has also been implicated in the development of freezing resistance. It may be that there are several alternative mechanisms that can increase freezing tolerance. It is interesting to note that although agaves and cacti also show the ability to increase their frost hardiness when grown at low temperatures, the alteration in damage temperature is much smaller than the degree of hardening observed at high temperatures.

Another mechanism that may be important in the frost tolerance of some species is supercooling. The equilibrium state is for ice to form when the temperature falls below the freezing point depression (ΔT_f) appropriate for the solute concentration. As the absolute value of ΔT_f increases approximately 1°C for every 1.2 MPa, typical cell osmotic potentials of down to –3 MPa would lower the freezing point by less than 3°C. In fact the cell contents rarely freeze. This is partly because of the absence of suitable ice nucleation sites within the cells, allowing the water to remain in the unstable supercooled condition. Similarly, the cell-wall water may remain liquid far below the theoretical freezing point, though in most plant tissues only a few degrees of supercooling can be achieved. Ice nucleation usually starts in the extracellular water, either because of the lower solute concentration there or because of the presence of ice-nucleating bacteria. Some hardwoods, including various oak, elm, maple and dogwood species, however, can supercool to the homogeneous nucleation temperature (Burke & Stushnoff, 1979). This is the temperature at which ice forms spontaneously without a requirement for nucleation sites, being between about –41 and –47°C for plant tissues.

Certain tissues and cells in these hardwood species, such as ray parenchyma cells and flower buds, show supercooling. This mechanism for avoiding freezing damage cannot work below the deep supercooling temperature of about –41°C, so that species that rely on deep supercooling cannot survive in areas where lower temperatures occur. Those species that occur in regions where lower temperatures are likely (e.g. pines, willows, etc.) survive by tolerating extracellular freezing.

A wide range of avoidance mechanisms can also contribute to the ability of plants to survive freezing temperatures. For example, the biophysical effect of dense canopies that can shield the sensitive tissues from direct radiative cooling, or bulky organs with a high heat capacity and long thermal time constants may also help avoid damaging tissue temperatures. Both these mechanisms contribute to the frost survival of the giant rosette forms of *Lobelia* and *Senecio* and other genera that are typical of tropical alpine regions with their characteristic dense covering of closely arranged hairy insulating leaves and massive stems.

9.4.5 Frost protection

Because of the economic importance of frost damage in temperate climates, much work has been done to develop techniques for protecting high-value crops from the damaging effects of frost (Snyder & de Melo-Abreu, 2005a,b). Plant and air temperatures may fall below freezing, either by advection of a cold air mass (e.g. from polar regions), or as a result of the net heat loss by longwave radiation that can occur on calm clear nights. Such a *radiation frost* causes the build up of a stable inversion layer where the air near the ground is cooler than that at higher levels.

The various methods of frost protection may be separated into (i) passive protection approaches and (ii) active methods.

Passive approaches

1. *Site and plant selection.* Perhaps the most important principle is to avoid frost-prone sites for the planting of sensitive crops. Because radiation frosts are associated with calm or very light winds, cold air drainage occurs down slopes, with the cold air accumulating in 'frost pockets' at the bottom of slopes. For an orchard situated on a slope it is feasible to divert cold air draining down the slope away from the crop by installing impermeable fences around the uphill side of the crop. It is also worth noting that other aspects of site selection can be important, with, for example, the presence of large water bodies upwind of the site also tending to reduce frost frequency. The chance of frost damage can also be reduced by the selection of cultivars that flower later, or even, somewhat counter-intuitively, to plant crops on north-facing rather than south-facing slopes because the lower daytime temperatures there can also delay bud burst. Management of the soil to maximise diurnal heat storage also acts to maintain higher minimum soil temperatures; this is achieved by maximising soil thermal conductivity and heat capacity, noting that sandy soils tend to have higher thermal conductivity than clay or peat soils, while there is usually an optimum soil moisture content.

2. *Use of covers.* Plant covers can usefully protect against frost as they may reduce radiative heat loss from the leaves, especially if they are impermeable to longwave radiation, because they increase the net downward radiation to the leaves as the temperature of the covers would normally be substantially higher than the clear sky temperature ($<-40°C$). Where condensation forms on the plastic this would also help as the latent heat released would also warm the plastic. Covers may also inhibit convective heat loss to the bulk air. Covering the soil with plastic can also usefully raise the soil temperature, though it should be noted that the use of soil mulches can actually decrease the diurnal minimum soil temperature (and hence the chance of damage) because of its effect on reducing the thermal conductivity.

Active methods

1. *Water sprinkling.* As water freezes it liberates a large amount of heat (the latent heat of fusion is $334\ J\ g^{-1}$). Water sprayed onto sensitive tissues liberates heat as it freezes, preventing tissue temperatures falling below zero, as long as enough water is supplied to maintain liquid on the surface of the tissue. The effect can potentially be enhanced by the inclusion of ice-nucleating bacteria, which ensures that nucleation and frost formation occurs above the critical damage temperature. Since sprinkler irrigation systems are commonly available in fruit orchards, such systems can be relatively cheap to operate in comparison with other active approaches, though it should be noted that the cooling effect of droplet evaporation can counter the thermal benefits from the initial water temperature and the latent heat of fusion.

2. *Direct heating.* Various types of orchard heater have been used to keep air temperatures above freezing, though their main effect is to help break up the inversion layer and to replace the cold air near the surface by warmer air from higher levels. A secondary effect of some heaters results from smoke or soot released, which may act as a screen minimising net longwave radiation losses.

3. *Air mixing.* With radiation frosts a particularly useful approach is to break up the inversion layer near the ground by using fans or propellers. These machines work, not by producing heat, but by redistributing sensible heat that is already available in the air. Clearly this approach only works in conditions where there is a strong temperature inversion.

9.5 Comments on some ecological aspects of temperature adaptation

Different plants exhibit a wide range of adaptations that enable them to live in different thermal climates (Table 9.3). These include biochemical tolerance mechanisms as well as seasonality and morphological or physiological adaptations that lead to avoidance of temperature extremes. Not only is the success of any species dependent on its tolerance or avoidance of extreme temperatures, but also it depends on its capacity to grow and compete at more normal temperatures. An ability to tolerate the short-term extreme temperatures that occur during bush fires is also an important factor in environments where fires are common.

Some species are widely adaptable, being able to grow over a wide temperature range, while others are more specialised. Examples of these two types of plant are found in the flora of Death Valley, California, USA, where the mean maximum temperature ranges from less than 20°C in January to more than 45°C in July (Figure 9.19). Some species, such as the evergreen perennials *Larrea divaricata* and *Atriplex hymenelytra* grow all the year, acclimating to the changing temperatures. Others, for example the perennial *Tidestromia oblongifolia*, are summer active and cannot grow at normal winter temperatures, while winter annuals such as *Camissonia claviformis* grow only at low temperatures. These species differences are reflected in seasonal changes of maximum photosynthetic capacity (e.g. Figure 9.9) and in their ability to grow in, and acclimate to, different temperatures in terms

Table 9.4 Total dry matter yields (final dry mass/initial dry mass) for four species over a 22-day growth period in two contrasting thermal regimes.

Species	16°C day/ 11°C night	45°C day/ 31°C night
Atriplex glabriuscula (coastal C$_3$)	24.4	0.1 (died)
Atriplex sabulosa (coastal C$_4$)	18.2	0.3 (died)
Larrea divaricata (desert C$_3$)	5.4	3.2
Tidestromia oblongifolia (desert C$_4$)	<2.5	88.6

of dry matter yields (Table 9.4). Table 9.4 also illustrates the general point that the more widely adaptable species (e.g. *L. divaricata*) tend not to have as high growth rates in any environment as the appropriate 'specialists'.

9.5.1 Experimental manipulation of temperature

As was pointed out in Chapter 1, it is difficult to manipulate temperature in a natural way to study, for example, the potential effects on plant growth of small temperature differences as may occur with climate change. The use of *common garden* experiments where plants are grown at different sites where temperatures are naturally different (e.g. at different altitudes) can be a useful approach, though results may be confounded by other climatic factors that may vary in parallel. Although it is straightforward to manipulate temperature in controlled environments and glasshouses, other aspects of the microenvironment, including aspects such as atmospheric coupling and the temperature gradients between shoots and roots tend to be very unnatural. Even the use of covers, polyethylene

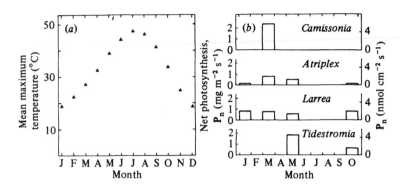

Figure 9.19 Seasonal changes of (*a*) monthly mean maximum daily temperatures for Furnace Creek, Death Valley, California and (*b*) maximum photosynthetic capacity of native species: *Camissonia claviformis*, *Atriplex hymenelytra*, *Larrea divaricata* and *Tidestromia oblongifolia* (after Mooney *et al.*, 1976).

tunnels and open-top chambers in the field (with or without supplementary heating or cooling) seriously affect the aerial environment and energy balance of the plants and can lead to misleading results, even where a paired design is used with similar covers being used for the different temperature regimes. While soil warming may conveniently be used to manipulate soil temperatures (Siebold & von Tiedemann, 2012) this approach does not substantially enhance aerial temperatures and leads to unnatural temperature gradients and differences between root and shoot temperatures. Some workers have recommended the use of infrared heating as an experimental tool (Aronson & McNulty, 2009), but again this approach does not well mimic natural temperature gradients within the system. In all cases it is important to consider implications of raised temperatures on evaporative demand, with the vapour pressure deficit and potential evaporation increasing with temperature if the temperature increase is applied only locally because the water vapour content of the air is determined by the airstream flowing over the crop.

9.5.2 Avoidance mechanisms

There are many possible avoidance mechanisms that can result in particular species not being subject to unfavourable tissue temperatures. (The terms 'avoidance' and 'strategy' used below, are used in a strictly non-teleological sense in that they do not imply that the plant can be rational or have a sense of

purpose. Rather, they provide simple descriptions of particular physical or biological responses.)

Perhaps the most important avoidance mechanism is seasonality (see e.g. Figure 9.19). Annual species may avoid periods of potentially damaging temperatures by completing their life cycle entirely within the period of favourable temperatures. Perennial species, on the other hand, often have the capability of going into a dormant state where the tissues are less sensitive to temperature extremes. In other plants the sensitive tissues may avoid being subject to the extremes of daily temperature range as a result of the damping (long thermal time constant) that results from a high thermal capacity of the tissues or of their immediate surroundings or from a high degree of insulation. This is particularly important in the grasses, where the meristematic tissues (at least in the vegetative phase) are at or below ground level and thus not subject to the temperature fluctuations found in the aerial environment. Similarly, the long time constants for large stems or fruits (Table 9.1) can lead to a significant reduction in the amplitude of daily temperature fluctuations.

Because the thermal time constants for most other aerial plant tissues are relatively short (Table 9.1), the steady-state energy balance can be used to obtain a good indication of their temperature control mechanisms. These may be discussed in relation to the different components of the steady-state energy balance (R_n, **C** and λE), with radiation absorption, and leaf and boundary layer resistance being of prime importance. The principles described above (Eq. (9.6), and Figures 9.1 and 9.2) will now be discussed in relation to some examples.

One general factor in the avoidance of extreme tissue temperatures is a consequence of the temperature dependence of s and D. At high air temperatures (above about 30 to 35°C), T_ℓ tends to be below T_a, but the converse occurs at low temperatures (Figure 9.2). The exact changeover temperature is a function of plant and environmental factors.

Hot environments

Leaf temperature may be kept low either by decreasing net radiation absorbed or by increasing latent heat loss. Whether or not sensible heat transfer needs to be minimised or maximised depends on whether the leaf is below or above air temperature.

For the case where water is freely available, the most productive strategy is to maximise latent heat loss (λE) by having a low leaf resistance, but to minimise any sensible heat gain from the air by having large leaves with a large boundary layer resistance. This combination is common in drought-evading desert annuals and in some perennials with access to ample water. As an example, *Phragmites communis*, although a temperate species, can grow in summer in a wet part of Death Valley. It survives the high temperatures there because leaf temperature is maintained as much as 10°C below air temperature by a large latent heat loss (Pearcy et al., 1972). A similar mechanism probably explains why *Caliotropis procera*, a shrub with leaves 10 cm or more wide, is common in areas of the Thar Desert of northwest India. It is worth noting that transpirational cooling to below air temperature is most effective for large leaves, as sensible heat transfer (which tends to negate the effect) is then minimal.

Where water is limited, however, as it often is in hot environments, large leaves are likely to be disadvantageous. In this case, small leaves with their low boundary resistance and efficient sensible heat exchange can avoid heating up much above air temperature, but equally they cannot cool much below air temperature even when transpiring rapidly. For example, *Tidestromia oblongifolia* has small leaves that closely track air temperature (e.g. Pearcy et al., 1971). Many summer-active desert perennials have small leaves.

Either strategy for latent and sensible heat exchange can be combined with minimising radiation absorption by means of vertical leaf orientation (as in the Eucalypts of Australia), wilting, leaf rolling, para-heliotropic movements keeping leaves edge-on to the solar beam, or high leaf reflection coefficient. In a rather different fashion, shading by other tissues (as, for example, by the spines on cacti (Nobel, 1988)) can also protect sensitive tissues from an excessive heat load. The effects of leaf angle and reflectivity are additive and can be evaluated by means of Eq. (9.6). As an example, it has been calculated (Mooney et al., 1977) that an alteration of leaf angle from horizontal to 70° could be expected to lower leaf temperature of *Atriplex hymenelytra* in Death Valley by 2 to 3°C, while halving the total absorptance for shortwave radiation can lower leaf temperature by a further 4 to 5°C. A factor in determining the low leaf temperatures in *Phragmites* is that the leaves are normally fairly erect.

Leaf reflectance depends on a variety of characters (see Chapter 2) including leaf water content and the presence of crystalline surface salts, pubescence, and the amount and structure of surface waxes. Of particular ecological interest is the observation that the spectral properties of leaves tend to alter with environmental aridity. The environmental and seasonal changes in reflectivity with variation in leaf water content in *Atriplex* (Mooney et al., 1977) and with pubescence in *Encelia* (Figure 2.18) both help to minimise leaf temperatures in summer. The increased radiation absorption in winter may help to increase leaf photosynthesis both by raising leaf temperature when air temperatures are suboptimal and by increasing total absorbed PAR.

It is questionable whether the primary effect of many of the radiative high-temperature avoidance mechanisms described above is to avoid direct thermal damage or to minimise transpirational water loss. Any mechanism lowering T_ℓ lowers the water vapour pressure inside the leaf and tends to reduce water loss and conserve moisture as well as tending to increase the ratio of photosynthesis to water loss (particularly at supraoptimal temperatures). The efficiency of water use is discussed in more detail in Chapter 10. It is worth noting, however, that water conservation often tends to have higher priority than minimising leaf temperature (at least by means of evaporative cooling).

Cold environments

Many arctic and alpine species have 'cushion' or 'rosette' habits where the plant forms a dense canopy within a few centimetres of the ground surface. This gives rise to a high boundary layer resistance as air movement is inhibited within the canopy and the whole plant is within the layer of markedly reduced wind speed. Coupled with efficient radiation absorption, this enables the temperatures of leaves and flowers to be 10°C or more above air temperature (Geiger, 1950). It is even possible that the dense pubescence on the lower surfaces of the leaves of many alpine plants (e.g. *Alchemilla alpina*) may increase radiation absorption by reflecting transmitted light back into the leaf (Eller, 1977). The metabolic component of tissue energy balance can also significantly affect tissue temperature (Section 7.2.2): for example, spadix temperatures of certain Araceae can even be raised as much as 35°C above air temperature by thermogenic respiration in the early spring (Seymour *et al.*, 1983). Thermogenesis is, however, unlikely to be a significant factor in determining leaf temperature in most other plants even though there does appear to be enhanced operation of the potentially thermogenic alternative respiratory pathway in arctic plants.

In other species, particularly the large rosette plants of genera such as *Senecio* and *Lobelia*, that grow at high altitudes in the tropics, the meristematic tissues are protected from frost damage by the formation of 'night buds' where the adult outer leaves fold inwards by a nyctinastic movement (Beck *et al.*, 1982). The insulation provided by the adult outer leaves (that can tolerate freezing) is sufficient, especially when combined with a degree of supercooling and with the large thermal mass of the plants, to prevent the more sensitive tissues from freezing.

9.5.3 Thermal climate and plant response

Different features of the temperature regime are critical at different times of year. In winter, the minimum temperatures may determine what species survive, while the occurrence of blossom-damaging frosts in spring can be crucial for other species. Similarly, the sensitivity of different species to variation in growing-season temperature depends on how close the environment is to their natural optimum. For example, for a hypothetical yield response of the form of that in Figure 9.8 (for $B = 0$), a 5°C rise in temperature would double yield at 9°C, have no effect at 27°C and cause complete failure at 38°C.

Where the complete information required for crop or ecosystem modelling is not available, two simple methods, in addition to mean temperatures, are often used to characterise the temperature regime. The first is to use the concept of thermal time (see Section 9.3.1 above) and to calculate the total number of growing degree-days typically available during the season. Examples of the thermal time required for maturation of different crops range from 1500 °C d above 5°C for spring barley to about 4000 °C d above 10°C for rice. The other approximate approach is to determine the length of the growing season, assuming an appropriate minimum temperature for growth. A minimum of 5 or 6°C is often assumed for temperate crops and about 10°C for crops such as maize. Although these thresholds are not necessarily exact for any species, the length of period during which mean temperatures remain consistently above the appropriate threshold, called the 'length of the growing season', provides a useful simple measure of climate. The potential growing period is particularly important for grain and fruit crops that need a minimum period to complete a reproductive cycle. Growing season duration is most frequently a limiting factor in northern latitudes or at high altitude.

An example of temperature sensitivity

Both growing-season length and growing degree-days are particularly sensitive to small climatic shifts when mean temperatures are close to T_{base}. This is illustrated for Akureyri in Iceland in Figure 9.20. Here a small shift in temperature has a large effect on growing-season length and an even more dramatic effect on growing degree-days. For example, a 2.4°C reduction in temperature from the long-term mean decreases the growing season to 75% of normal and decreases the number of degree-days D to only 46% of

normal (Table 9.5). Changes of this magnitude would be disastrous if crop yields were affected proportionately. Although it is difficult to quantify the effect of temperature on crop yield using observational data alone, hay yields in Iceland for a period of 25 years are available and can be related to seasonal temperature using multiple regression techniques. As a first approach, Bryson (1974) noted that mean yields averaged 4.33 tonne ha^{-1} in the late 1950s, but only 3.22 tonne ha^{-1}, or 75% of that, in 1966 and 1967. In the same periods, corresponding

Table 9.5 Effects of changes in mean temperature at Akureyri, Iceland, on the growing season and growing degree-days above 5°C, where 'Clino' is the climatic normal temperature. (Data from Bryson, 1974.)

	Mean warm season temperature (°C)	Growing season length (days)		Growing degree-days	
		d	%	°C d	%
Clino	7.47	158	100	597	100
Clino – 1.0°C	6.47	144	91	443	73
Clino – 2.4°C	5.07	118	75	276	46

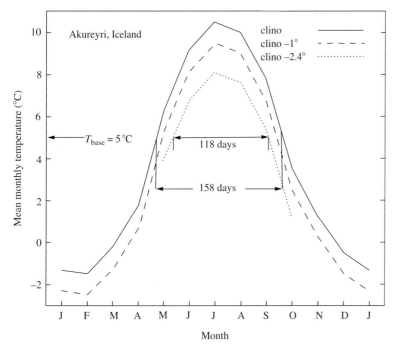

Figure 9.20 The effect of reduction of the annual mean temperature for Akureyri, Iceland, by 1°C or 2.4°C below the climatic 'normal' (clino) on growing season length (measured above 5°C). (After Bryson, 1974.)

mean warm-season temperatures were 7.65°C and 6.83°C. This yield reduction occurred even though more fertiliser was applied in the 1960s, so he attributed the effect to the lower temperature, the actual reduction corresponding approximately to that expected on the basis of the reduction in degree-days. Unfortunately this clear result is an oversimplification as other factors are also important so that the magnitude of the temperature effect is not as great if other combinations of years are selected. In fact more objective regression studies showed that mean cool-season temperature (which averaged −0.08°C, well below the assumed growth threshold) was a better yield predictor (perhaps because it measured winter kill of the grass), while other factors such as nitrogen application also contributed (H. G. Jones, unpublished).

9.5.4 Temperature and plant distribution

Although there are often clear physiological differences in temperature adaptation between species it is not entirely straightforward to explain why specific plants grow in particular thermal climates. One approach to identification of the environmental factors that determine the distribution of any species is to correlate the plant's natural distribution with the climate over the area where it is found; for example Woodward (1988) has demonstrated that a combination of the absolute minimum temperature (T_{min}) and the number of growing degree days (GDD) can provide a very useful predictor of plant distribution for many European species, with some species limited by the length of the growing season and others by the absolute minimum temperature. For example, using GDD defined with a base temperature of 0°C, *Koenigia islandica* in Europe only grows at sites where GDD <1800, but *Bromus* (= *Anisantha*) *sterilis* only grows at sites where GDD >2000, while *Tilia cordata* requires both $T_{min} > −40°C$ and $GDD > 2000$. Although tropical species clearly are not physiologically capable of growing in arctic areas, the question nevertheless arises as to why the more arctic species do not grow in warmer climates where there may be no physiological reason for their absence. Evidence that this exclusion is a result of competitive ability is nicely provided by experiments such as those of Woodward and Pigott (1975) on the alpine species, *Sedum roseum*, and the lowland species, *Sedum telephium*, which showed that when competition was eliminated, both species could grow equally well at the lowland site (though the lowland species could not succeed at the upland site – presumably because of poor cold tolerance). This implies that development of an ability to tolerate low temperatures confers some competitive disadvantage under warmer conditions. More sophisticated climatic correlation studies would use regression against a wider selection of climatic variables, including aspects of the water balance.

9.6 Sample problems

9.1 The net radiation absorbed by a leaf is 400 W m^{-2}. What are (i) the leaf temperature if $T_a = 25°C$, $h = 0.4$, $r_{aH} = 40$ s m^{-1}, $r_{ew} = 200$ s m^{-1}; (ii) the thermal time constant for this leaf if $\rho^* c_p^* = 2.7$ MJ m^{-3} and $\ell^* = 1$ mm; (iii) the time constant when the stomata are closed; (iv) the leaf temperature if the whole leaf is wet; (v) the corresponding thermal time constant?

9.2 Respiration rate of a fruit is 0.1 mg CO$_2$ m^{-2} s^{-1} at 13°C and 0.19 mg CO$_2$ m^{-2} s^{-1} at 19°C. What are (i) the Q_{10}, (ii) the activation energy?

9.3 The thermal time between flowering and maturity for a certain plant is constant at 600 degree-days above 6°C, while the rate of net photosynthesis (= 50 g m^{-2} day^{-1} for $k = 1$) is given by Eq. (9.16) with $B = −5$ and $T_{max} = 30°C$. Assuming a constant temperature over the period, plot the relationship between the net photosynthesis and temperature. What is the optimum temperature?

10

Drought and other abiotic stresses

Contents

In the previous chapter we considered temperature as one important abiotic factor affecting plant performance; in this chapter we discuss some other important environmental stresses, though concentrating on water deficits. Over large areas of the Earth's surface, lack of water is the major factor limiting plant productivity. The average primary productivity of deserts is less than 0.1 tonne $ha^{-1} yr^{-1}$; this is at least two orders of magnitude less than the productivity achieved when water is non-limiting. Even in relatively moist climates, such as that of southern England, drought lowers the yields of crops such as barley by an average of 10 to 15% each year, while the yields of more sensitive crops such as salads and potatoes may be reduced even further if unirrigated.

There are many possible definitions of drought and aridity that range from meteorological droughts defined in terms of the length of the rainless period, through definitions that allow for the water storage capacity of the soil and the evaporative demand of the atmosphere, to those that include some aspect of plant performance. In the following, drought is used to refer to any combination of restricted water supply (e.g. as a result of low rainfall or poor soil water storage) and/or enhanced rate of water loss (resulting from high evaporative demand) that tends to reduce plant productivity. A common meteorological quantification of aridity is the ratio between annual precipitation and annual total potential evapotranspiration, with values <0.05 representing hyper-arid climates, 0.05 to 0.20 representing arid and 0.2 to 0.5 representing semi-arid situations (UNEP, 1997).

The first part of this chapter outlines the effects of water deficits on individual plant processes, while the following section considers the various mechanisms that enable plants to grow or survive in dry environments. The final part of the chapter considers briefly some other important abiotic stresses and common features of plant responses to stress. Further aspects related to the breeding of crop plants tolerant of drought are discussed in Chapter 12. Because the main emphasis of this text is on the whole-plant physiological processes underlying plant response and adaptation to the environment, details of the biochemical and molecular processes involved are only briefly outlined and interested readers are referred to more specialist reviews.

10.1 Plant water deficits and physiological processes

Among the very many texts and reviews that discuss the physiology of plant responses to water deficits, the following contain particularly useful material (Jones *et al.*, 1989; Kramer & Boyer, 1995; Levitt, 1980) while the effects of water deficits on processes such as stomatal behaviour and photosynthesis have already been discussed in some detail in Chapters 4, 6 and 7. References to more specialist articles, especially to reviews outlining recent advances in understanding of the more molecular aspects of stress signalling and the biochemical and genetic pathways involved in plant responses, are provided at appropriate places in the text.

10.1.1 Effects of water deficits

A simple description of the relative sensitivity to water deficits of different processes is difficult, not only because of the large species differences and the marked capacity for acclimation to stress, as has already been seen for stomata and photosynthesis, but also because of the complexity of the ways in which water deficits have their effect. In particular, we have already seen that the effect of soil drying on processes such as shoot growth and photosynthesis may be mediated both by hydraulic signalling where soil water status directly affects shoot water status and by chemical signalling (involving ABA and pH) (see Sections 4.3.3 and 6.4.5), indicating that the water status of the responding organ is not all that has to be considered.

The most obvious effect of even mild stress is to reduce growth, with cell enlargement being particularly sensitive to water deficits (Hsiao, 1973). As was outlined in Chapter 4, turgor pressure (rather than total water potential) in the growing cells provides the driving force for cell expansion, but the actual rate of extension is controlled by variation in constants in the Lockhart equation: yield threshold (Y) and the extensibility ϕ (Eq. (4.16)). It is probable that, at least in those cases where changes in shoot growth have been observed in the absence of detectable

differences in shoot water status (e.g. Gowing *et al.*, 1990), modulation involves some form of signalling from the root to the shoot. Some synthesis of cell wall materials may continue during mild stresses that inhibit expansion growth. This can be manifest as 'stored growth' so that most of the growth lost during a short stress may be recovered after rewatering (Acevedo *et al.*, 1971). Cell division, though affected by water stress, is normally less sensitive than cell expansion.

In addition to simple growth inhibition, water deficits can greatly modify plant development and morphology (Table 10.1). For example, the differential sensitivity of roots and shoots (with root growth being less sensitive to water deficits) leads to large increases in the root to shoot ratio in drought (Sharp & Davies, 1989). Although overall root expansion is inhibited by drought, the apical few millimetres of roots are particularly insensitive to drought, with elongation occurring at the control rate in this zone even when the water potential is reduced to −1.6 MPa (Sharp *et al.*, 2004). Other effects on vegetative development include the reduction of tillering in grasses and the early termination of extension growth in perennials with the formation of dormant buds. Water deficits also increase the abscission of leaves and fruits, particularly after relief of stress. Not only does water stress decrease the size of leaves as a result of reduced cell expansion and cell division, but, at least in wheat, it leads to a reduction in the proportion of epidermal cells that form stomata and an increase in the number of epidermal trichomes or hairs (Quarrie & Jones, 1977).

Water deficits also affect reproductive development, with a period of water stress being required to stimulate floral initiation in some species such as lychee (Menzel, 1983). In other cases severe water stress can cause the early emergence of ready-differentiated floral buds (Jones, 1987b). Water deficits tend to advance flowering in annuals and to delay flowering in perennials. In wheat, for example, mild deficits can advance flowering by up to a week, though with corresponding decreases in the number of spikelets and in pollen fertility and grain set (Angus & Moncur, 1977). Stress treatments are

Table 10.1 Evidence for the involvement of abscisic acid (ABA) in plant responses to water deficit stress based on the similarity of responses to water deficits and to exogenous ABA application (collated largely from Addicott, 1983; Davies & Jones, 1991; Jones, 1981b).

Response	Water stress	ABA	
Short term			
Stomatal conductance	Decrease	Decrease	+++[a]
Photosynthesis	Decrease	Decrease (primarily stomatal effect)	+++
Membrane permeability	Increase/decrease	Increase/decrease	+
Ion transport	Increase/decrease	Increase/decrease	+
Long term: biochemical and physiological			
Specific *m*RNA and protein synthesis[b]	Increase	Increase	++
Proline and betaine accumulation	Increase	Increase	++
Osmotic adaptation[c]	Yes	Yes	+
Photosynthetic enzyme activity	Decrease	Decrease	+
Desiccation tolerance[d]	Increase	Increase	+
Salinity and cold tolerance	Induces	Induces	++
Wax production[e]	Increase	Increase	+
Long term: growth			
General growth inhibition	Yes	Yes	+++[a]
Cell division	Decrease	Decrease	+++
Cell expansion	Decrease	Decrease	+++
Germination	Inhibits	Inhibits	++
Root growth	Increase/decrease	Increase/decrease	++
Root/shoot ratio	Increase	Increase	++
Long term: morphology			
Production of trichomes	Increase	Increase	++
Stomatal index	Decrease	Decrease	++
Tillering in grasses[f]	Decrease	Decrease/increase	+
Conversion from aquatic to aerial leaf type	Yes	Yes	++
Induction of dormancy, terminal buds or perennation organs	Yes	Yes	++

Table 10.1 (*cont.*)

Response	Water stress	ABA	
Long term: reproductive			
Flowering in annuals	Often advanced	Often advanced	+
Flower induction in perennials	Inhibited	Inhibited	+
Flower abscission	Increased	Increased	+
Pollen viability	Decreased	Decreased	+
Seed set	Decreased	Decreased	+

[a] The strength of correlation is indicated as ranging from weak (+) to strong (+++).
Extra references:
[b] Heikkila *et al.*, 1984; Mundy and Chua, 1988; Cohen and Bray, 1990
[c] Henson, 1985
[d] Gaff, 1980; Bartels *et al.*, 1990
[e] Baker, 1974
[f] Hall and McWha, 1981

often used by plant breeders to speed up generation times as, for example, in the 'single seed descent' approach used by cereal breeders.

Just about every aspect of cellular metabolism and fine structure has been reported to be affected by water deficits. Particularly characteristic changes include: increases in rates of degradative as compared with synthetic reactions; decreased protein synthesis; increases in the concentrations of free amino acids (particularly proline, which may increase to as much as 1% of leaf dry matter in some species), glycine-betaine, di- and poly-amines (with osmotic stress especially giving rise to increases in putrescine: Smith, 1984) and sugars; all with corresponding changes in relevant enzyme activities. Many of these changes may be considered to be adaptive, but it is often difficult to distinguish changes that are a manifestation of cell or tissue damage from those that represent acclimation or responses to counter damage. For example, as with other types of environmental stress, water deficits tend to shift the cell redox potential to a more oxidised state and to increase concentrations of free radicals (both presumably

damaging changes), but levels of reducing agents such as glutathione, and free radical scavenging systems such as superoxide dismutase, which can counter the damage, both tend to increase (see e.g. Alscher & Cumming, 1990).

All the effects described above contribute to the general decrease in dry matter production and seed yield that is characteristic of drought. Although the effects on carbon assimilation per unit leaf area (or on other physiological processes) may be important, the dominant factor contributing to reduced productivity in drought is generally the reduction in leaf area. In addition, seed yield can be particularly sensitive to stress at critical periods of development (e.g. microsporogenesis: see Salter & Goode, 1967).

10.1.2 Mechanisms and the role of plant growth regulators

How even small water deficits can have such major metabolic and developmental consequences is still not clear, even though in recent years there have been great advances in our understanding of the molecular signalling mechanisms involved in drought responses. As was pointed out by Hsiao (1973), it is difficult to see how the possible effects of mild stress on water activity, macromolecular structure or the concentration of molecules in the cytoplasm can be primary stress sensors. The best evidence is for a sensor(s) responding to turgor pressure or possibly to cell size (Chapter 4). At least in growing cells, small changes in turgor may reduce cell expansion with the consequent build up of unused cell wall materials or other metabolites then affecting metabolism. There is also evidence that turgor can directly affect ion transport through the involvement of membrane stretch sensors or other turgor-dependent systems (Cosgrove & Hedrich, 1991; Kacperska, 2004). For example, osmosensitive or stretch-activated ion channels have been reported for both anions (Qi et al., 2004; Roberts, 2006) and cations such as Ca^{2+} (Zhang et al., 2007) and K^+ (Liu & Luan, 1998). Relative water content (RWC) is often a good proxy for turgor pressure and as it can be relatively easily measured it may even be more useful than total water potential in many cases as a measure of tissue stress. Most of the observed effects of water deficits are almost certainly secondary and result from the operation of specific plant regulatory responses.

Plant growth regulators (sometimes called phytohormones) are critically involved in the complex cellular and long-distance signalling involved in integration of plant responses to stresses including drought (see recent reviews such as those in Khan et al., 2012). Central to drought responses is the involvement of abscisic acid (ABA): as we have seen in Chapter 4, this growth regulator plays a major role in integration of whole-plant responses both to water stress and to a wide range of other environmental stresses including salinity and high temperatures (Wilkinson & Davies, 2002). Evidence for the involvement of ABA in stomatal closure in response to drought has been outlined in Chapter 6. Other particularly important observations include the fact that ABA concentrations rise rapidly in stressed plants (this rise tends to be a function of turgor pressure rather than total water potential), and the very close

correspondence between the responses to water deficits and to exogenously supplied ABA of a wide range of short- and long-term plant responses (Table 10.1). These observations, when combined with information obtained from the study of ABA deficient and ABA insensitive mutants (Chapter 6), provide compelling evidence that ABA does indeed have a general role as an endogenous plant growth regulator involved in plant adaptation to water deficits and other stresses, particularly in relation to long-distance signalling.

Major advances have been made in recent years in our understanding of the cellular and biochemical mechanisms involved in ABA metabolism and signal transduction (Zhu, 2002) and the identification of the ABA receptor (see Klingler et al., 2010). However, notwithstanding the evidence for ABA involvement in drought stress signalling at the cellular level there is also strong evidence for a number of important ABA-independent signalling pathways in drought and other stress responses (Shinozaki & Yamaguchi-Shinozaki, 2007; Zhu, 2002). For example some responses to osmotic stress appear to operate independently of ABA, for example through the dehydration responsive elements (DRE) first identified in 1994 (Yamaguchi-Shinozaki & Shinozaki, 1994). Unfortunately a significant proportion of the more molecular studies in the past that aimed to identify drought-responsive genes have used rather extreme desiccation treatments, often in combination with very limited (or no) information on the actual water status achieved (Jones, 2007 noted that approximately 50% of papers published between 2003 and 2005 in major plant journals on gene expression under water stress included no measure of water status). Future advances in understanding will critically depend on precise quantification of water status and the use of realistic stresses.

At the whole-plant level also, there is evidence for the involvement of many of the other groups of plant hormones. For example, although ethylene production is particularly stimulated by flooding, it is also stimulated by water deficits; this stimulation has been implicated in a number of observed responses including abscission of leaves and fruits, leaf epinasty, and also stomatal closure and decreased assimilation. Cytokinins have also been implicated in some drought-induced responses, such as leaf senescence and stomatal closure; for example Blackman and Davies (1985) suggest that these may be a consequence of the reduced supply of cytokinins in the transpiration stream that occurs in drought. There have also been suggestions (Cheong & Choi, 2003) of some involvement in drought responses of other groups such as the jasmonates, though these have been primarily implicated in responses to biotic stresses. Although water stress may affect levels of gibberellins and auxins, there is little evidence for these growth regulators having a major role in regulating stress responses.

10.2 Drought tolerance

The term drought resistance has long been used to refer to the ability of plants to survive drought, but I prefer *drought tolerance* to describe all mechanisms that tend to maintain plant survival or productivity in drought conditions. In an agricultural or horticultural context, a more drought-tolerant cultivar is one that has a higher yield of marketable product in drought conditions than does a less tolerant one. Many farmers and breeders also look for some degree of yield stability from year to year as a component of drought tolerance (see e.g. Fischer & Turner, 1978). In natural ecosystems, however, a drought-tolerant species is one that has the ability to survive and reproduce in a relatively dry environment. In this case, drought tolerance does not necessarily rely on a high productivity. It follows, therefore, that the mechanisms favouring drought tolerance in typical agricultural monocultures may be distinct from those that have evolved in natural ecosystems.

Plants that can exist in dry environments are called *xerophytes* (cf. *hygrophytes* in wet habitats and *mesophytes* in intermediate habitats). Although xerophytes occur naturally only in dry situations, this is not necessarily because they are truly xerophilous (drought liking) but rather because their competitive ability is greater than other species only in dry places.

Under careful husbandry, when competition from other species is eliminated, growth and production by these species is improved by increasing the availability of water.

There are many different ways in which plants may be adapted to dry conditions so that there is no simple set of morphological or physiological criteria that can be used to distinguish xerophytes. Many of the methods that have been proposed for classifying xerophytes (Levitt, 1980; Maximov, 1929) have been in terms of the different ecological niches occupied, using descriptions such as drought escaping, drought evading and drought enduring. Partly because of difficulties in attributing plants to one or other group, I think that the most useful approach is to concentrate on the mechanisms that contribute to drought tolerance, recognising that one plant may have several. These mechanisms may conveniently be classified into three main types: (i) *stress avoidance* – that is, those mechanisms that minimise the occurrence of damaging tissue water deficits; (ii) *stress tolerance* – that is, those physiological adaptations that enable plants to continue functioning in spite of plant water deficits; and (iii) *efficiency mechanisms* – that is, those that optimise the utilisation of resources, especially water. A useful subdivision of these groups is presented in Table 10.2, while some details are discussed below.

In addition to the constitutive expression of drought adaptations, all plants show an ability to *acclimate* to drought, of which stomatal regulation of water loss, osmoregulation and the general regulation of leaf area development and root growth are perhaps the most obvious. Indeed almost any character that favours adaptation to drought conditions can acclimate to a greater or lesser degree.

10.2.1 Avoidance of plant water deficits

Drought escape
A plant that rapidly completes its life cycle, or at least its reproductive cycle, can escape periods of drought and grow during periods of favourable soil moisture. This mechanism is typical of the desert ephemerals

Table 10.2 Drought tolerance mechanisms, showing some potential costs of the different tolerance strategies. Note that the different strategies can be either constitutive or may develop in response to drought.

	Examples	Costs
1. Avoidance of plant water deficits		
(a) Drought escape	Short growth cycle, dormant period, leaf drop	Short season (low potential yield)
(b) Water conservation	Small leaves, limited leaf area, stomatal closure, high cuticular resistance, reflective leaves, limited radiation absorption	May 'waste' some available water
(c) Efficient water uptake	Extensive, or deep or dense root system	Substantial carbohydrate needed for root growth
2. Tolerance of plant water deficits		
(a) Turgor maintenance	Osmotic adaptation, cells with low elastic modulus	Metabolic costs
(b) Desiccation tolerance	Production of intracellular compatible solutes, enzymes tolerant of low water status	Metabolic costs limit yield in good years
3. Efficiency mechanisms		
(a) Efficient use of available water	Stomatal closure, especially in afternoon	Low maximum rate
(b) Maximal harvest index	High proportion of dry mass in seed	Less adaptable

that can complete their life cycles from germination to seed maturation in as little as four to six weeks. (*Arabidopsis thaliana*, the plant beloved of molecular biologists because of its short life cycle and small genome, is one of these.) A similar, though less extreme adaptation is found in many crop plants, where the most drought-tolerant cultivars, at least in environments with a marked dry season, are frequently those that flower and mature the earliest, thus avoiding the worst of the dry season. Many annual plants even show a dynamic response of this type, flowering earlier than usual if they are subjected to water stress, as would occur as the soil dries out in a dry year. In general, this group does not rely on any other physiological mechanisms for surviving in dry climates.

Water conservation

Plant adaptations that limit the rate of water loss can prevent the development of detrimental plant water deficits in two ways. They can either conserve soil water for an extended period, thus maintaining soil (and plant) water potential suitably high over a sufficient period for seed ripening, or else a reduced transpirational flux can reduce the depression of ψ_ℓ that results from the frictional resistances in the transpiration pathway (see Chapter 4). Plants having mechanisms for restricting water loss have been termed water-savers (Levitt, 1980), and include many of the plants that are commonly thought of as xerophytes.

Particular adaptations that minimise transpiration include the following: a thick cuticle with a correspondingly low cuticular conductance (often less than 2 mmol m^{-2} s^{-1} (or 0.05 mm s^{-1}) in these plants); small leaves with a small total transpiring surface per plant (leaves may be reduced or even absent with some semi-desert plants); a high leaf reflectivity and other adaptations that minimise radiation absorption (see Chapters 2 and 9); a low stomatal conductance as achieved by stomatal closure or by very small, sunken or sparsely distributed stomata. A secondary effect of small leaves is that their high boundary-layer conductance results in leaf temperature closely following air temperature, thus avoiding the high leaf temperatures and the consequent large driving force for evaporation that can develop when stomata start to close in large-leaved plants. It is often suggested that a tomentum, or thick layer of leaf hairs or trichomes, can potentially help to conserve water, particularly if the hairs are dead, but in practice the effect on the total leaf diffusive resistance is usually rather small (for example, even a tomentum 1 mm thick has a diffusive resistance (see Eq. (3.21)) equal to $\ell/D = 1/24.2 = 0.041$ s mm^{-1} (or 1.025 m^2 s mol^{-1}), which is small in comparison with the stomatal resistance, which can be about two orders of magnitude larger when the stomata are closed). Leaf hairs probably have greater significance for radiation balance and for the balance between water loss and assimilation. The role of these and other morphological characters that might be expected to favour water conservation will be discussed in more detail in the section on efficiency mechanisms.

CAM plants, with their reversed stomatal cycle and thick cuticles, are particularly effective at limiting water loss under stress conditions. For example, a specimen of *Echinocactus* has been reported to lose less than 30% of its weight in six years when maintained without water (Maximov, 1929). In addition to these characters that have obvious direct implications for water conservation, many 'xeromorphic' plants have an extreme development of structural tissues such as schlerenchyma and collenchyma and the presence of spines and other protective structures. The latter have probably evolved not as drought tolerance mechanisms, but as characters reducing grazing damage and therefore enhancing competitive ability in environments where little other vegetation may be present.

Although some of the characters that reduce water loss are fairly constant irrespective of environmental conditions (such as thick cuticle or stomatal morphology), most respond to drought in some degree. For example, wax development, cuticle thickness, hair development and leaf reflectivity have all been reported to increase with increasing drought thus better fitting the plant for the environment. Similarly, regulation of leaf area in response to changing water availability is perhaps the most important and most general aspect of drought avoidance and acclimation to drought. Control of leaf area includes both reductions in leaf area expansion as the water supply declines and, as the severity of stress increases, the stimulation of leaf senescence and leaf fall so as to tailor the leaf-surface area to the available water. Rather than simply being a direct response to decreasing water availability, most leaf death in response to drought (or other environmental stresses) involves a tightly regulated form of *programmed cell death* (PCD) similar to apoptosis in animals that lets the plant efficiently recycle nutrients. Coordination of the various events in senescence involves a complex range of sensing and signalling processes including signalling involving hormones (cytokinins and ABA), reactive oxygen species and changes in gene expression (Munné-Bosch & Alegre, 2004).

Effective water uptake

Many plants that are successful in dry habitats have no specific adaptations for controlling water loss but rely on the development of a very deep and extensive root system that can obtain water from a large volume of soil or from a deep water table. Many desert shrubs (e.g. mesquite, *Prosopis juliflora*) have deep root systems, while functional roots of *Quercus fusiformis* have been found 22 m deep in limestone caves in Texas (Jackson *et al.*, 1999). Polunin (1960) even quotes Rubner as saying that roots of tamarisk were detected as deep as 50 m during excavation of the

Suez Canal. Root development is very plastic, with more of a plant's resources (e.g. carbohydrate) being put into root growth relative to the shoot in dry conditions. In some cases the absolute root growth may even be enhanced. Plants that have an effective water supply system often behave as water spenders (Levitt, 1980) and do not limit their rate of transpiration. The high costs of root growth mean that those plants that develop large root systems may not be as competitive or productive as smaller rooted plants when water is freely available.

Many plants can also take up significant water by interception of fog or dew or by interception of rainfall. Although it is often assumed that intercepted rainfall rapidly evaporates, especially in trees with their effective coupling to the atmosphere, there is good evidence that it can be absorbed by leaves and contribute substantially to water balance in dryland ecosystems (Breshears *et al.*, 2008).

10.2.2 Tolerance of plant water deficits

There are many mechanisms that contribute to the ability of plants to maintain physiological activity as tissue water content or ψ falls. In recent years a large number of genes with potential roles in the tolerance of tissue water deficits have been described (Ingram & Bartels, 1996) with their roles largely being attributable to one or both of the groups of physiological mechanisms outlined below.

Turgor maintenance

Turgor maintenance by means of increases in cell solute concentration (i.e. lowering ψ_π – see Eq. (4.7)) in response to water stress is widespread, occurring in leaves, roots and reproductive organs of many species (Morgan, 1984; Zhang *et al.*, 1999). This process, often called *osmotic adjustment* or osmoregulation, is probably the most important mechanism for maintaining physiological activity as ψ falls as a result of drought or salinity stress. The term osmotic adjustment is usually reserved to describe an active process where there is a net increase in osmotically active solutes in the cell; this should be distinguished

from any passive concentration effect that occurs as cells lose water and shrink.

Although both inorganic ions (especially K^+ and Cl^-) and organic solutes can be involved in osmotic adjustment of cells, changes in organic solutes including sugars, sugar alcohols, amino acids and various organic acids tend to be much more important. The most important of the organic solutes are those highly soluble uncharged molecules, known as *compatible solutes*, that interfere least with cellular metabolism. Examples of such compatible solutes include simple sugars (such as trehalose, fructose and glucose), sugar alcohols and cyclitols (including glycerol, mannitol, sorbitol, quercitol, etc.), amino acids (especially proline) and quaternary ammonium compounds such as glycine-betaine. The hydroxyl groups on these solutes facilitate hydrogen bonding with water and thus tend to act as osmoprotectants stabilising critical enzymes and other cellular structures. In saline environments especially, Na^+ and Cl^- can be exploited as vacuolar solutes, but it is necessary that they are maintained at low levels in the cytoplasm if cell metabolism is not to be damaged. In general, diurnal changes in ψ_π are rather small, with osmotic adjustment usually taking days or even months. Although the capacity for osmotic adjustment has been positively related to yield in some crops, there are cases where species that have a large capacity for osmoregulation are actually more sensitive to drought than are non-osmoregulating species (Quisenbery *et al.*, 1984); this suggests that there may be a 'cost' associated with osmoregulation.

Full turgor maintenance occurs if the decrease in ψ_π equals any decrease in total ψ, so turgor remains constant (see Figure 10.1). Even partial turgor maintenance, where $(d\,\psi_\pi/d\,\psi) < 1.0$, can be advantageous as it will defer the time at which turgor reaches zero. Many species show at least partial turgor maintenance, particularly where drought is imposed slowly, while full turgor maintenance is often observed over only a limited range of ψ. Some examples of full turgor maintenance over a limited range of ψ and of partial turgor maintenance are shown in Figure 10.2.

Any passive concentration effect as cell water decreases is greatest for a given fall in ψ for those plants such as succulents with relatively elastic walls as shown in Figure 10.3. The possible contribution of this passive adjustment to turgor maintenance is illustrated in Figure 10.1 for cells with differing *bulk modulus of elasticity* (ε_B). An important consequence of any passive turgor maintenance is that measurements of osmotic adjustment need to compare

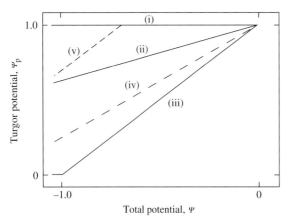

Figure 10.1 Schematic representation of the relationship between turgor potential (ψ_p, normalised to 1.0 when $\psi = 0$) and total water potential (ψ) for (i) full turgor maintenance where the change in osmotic potential completely compensates for the fall in ψ (i.e. $d\psi_\pi/d\psi = 1$); (ii) partial turgor maintenance ($0 < (d\psi_\pi/d\psi) < 1$); (iii) no turgor maintenance with constant solute concentration (extremely rigid walls); (iv) passive turgor maintenance with solutes concentrating as a result of cell shrinkage with elastic cell walls; (v) an example where turgor maintenance is only achieved over a limited range of ψ.

plants at the same water status (either at the turgor loss point or at full hydration and full turgor), otherwise some component of the passive adjustment will be incorrectly included in the results. The magnitude of osmotic adjustment has been reported to vary both within and between species, with the amounts for typical crop plants ranging from 0.1 to 1.7 MPa, though xerophytes may show up to 4 MPa osmotic adjustment and halophytes as much as 10 MPa (for sources see Morgan, 1984; Zhang *et al.*, 1999).

The extreme development of structural tissues in many xeromorphic plants that was noted above results in inextensible cells with a large modulus of elasticity (ε_B). This is an additional factor often associated with the ability to survive low water potentials as it allows the cells to withstand high turgor pressures (Figure 10.3), thus permitting high osmotic concentrations with the consequent ability to maintain turgor down to very low values of ψ. The associated constancy of cell volume may also be important in maintenance of physiological activity over a wide range of ψ without a need for osmoregulation or compatible solutes. It has also been suggested that a decrease in cell size, as commonly occurs in droughted plants, can also help in maintaining turgor through corresponding increases in tissue ε_B (Cutler *et al.*, 1977).

The opposite extreme is found in water storage tissues of succulent species. In this case ε_B is very small. As a result large changes in cell volume occur for relatively small changes in ψ (Figure 10.3): that is, there is a large tissue capacitance. This large

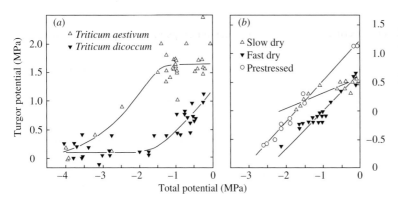

Figure 10.2 Examples of relationships between turgor potential and leaf water potential for (a) two wheat genotypes and (b) sorghum plants: ▼ previously well-watered and dried rapidly; △ previously well-watered and dried slowly; ○ previously stressed to −1.6 MPa and dried rapidly. (Data from Turner & Jones, 1980.)

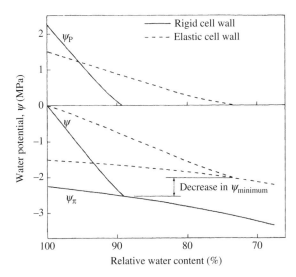

Figure 10.3 Höfler diagrams for cells with elastic (ε_B is small) or rigid (ε_B is large) cell walls, illustrating the potentially lower minimum ψ for positive turgor with rigid cell walls and the smaller volume changes for a given change of ψ.

capacitance acts to increase the time constant for water exchange (cf. Eq. (4.33)), and may help buffer against rapid environmental changes. It has been estimated that decreasing ε_B from a high value of 40 MPa to a low one of 4 MPa would increase the half-time for water exchange from 0.6 to 6 min (Tyree & Karamanos, 1981). It is, however, difficult to envisage how changes over this time scale might be ecologically significant. Equally, the amount of water stored is generally small in relation to potential evaporation rates but, where the stomata are closed so that rates of water loss are slow, storage may be adequate to maintain turgor for quite long periods.

The generality of osmotic adjustment has led to it being a major target for the genetic engineering of drought tolerance (Valliyodan & Nguyen, 2006; Wang et al., 2003; Yang et al., 2010). With increased understanding of the metabolic pathways involved in the synthesis of key compatible solutes, it is becoming feasible to engineer the metabolic pathways to increase concentrations of compatible solutes and even to link these to stress-induced promoters (Shinozaki & Yamaguchi-Shinozaki, 2007) so that any disadvantages of constitutive expression can be

overcome. Many examples of transgenic plants with enhanced synthesis of osmotic solutes such as proline have been reported (see Jiang & Wang, 2011; Yang et al., 2010) though it has proved hard to show that any associated improvements in drought tolerance (where found) are related to osmotic adjustment (Molinari et al., 2007), and in many cases it is even hard to demonstrate any actual quantifiable improvement in drought tolerance (Yang et al., 2010).

Desiccation tolerance and metabolism

Survival is as important as the ability to continue functioning at low water potential. This survival ability, or desiccation tolerance, has often been measured in terms of the environmental water potential or tissue water content that causes 50% cell death (Levitt, 1980). Usually tissue is 'equilibrated' with air at a given humidity and ψ (calculated according to Eq. (5.14)), though in many published examples where short exposure periods were used, vapour equilibrium may not have been attained. There are large differences between plant species and between tissues within plants in their ability to tolerate tissue water deficits.

'Resurrection plants'. There is an extreme capacity to survive severe desiccation in many poikilohydric lower plants such as algae, lichens and bryophytes (Oliver et al., 2000; Proctor, 2000). Similarly seeds and pollen grains of higher plants can often tolerate extreme desiccation. There are, however, relatively few higher plant species in which the vegetative tissues can tolerate severe desiccation and recover rapidly on rewatering (Farrant, 2000). These *resurrection plants* are particularly characteristic of granitic rock outcrops in the tropics (Porembski & Barthlott, 2000) though they do occur in other regions: a well known example is *Myrothamnus flabellifolia*, whose air-dry leaves start to respire within half an hour of re-watering and show the first signs of photosynthesis within 6.5 h. Lower plants, such as the moss *Tortula ruralis*, may recover more rapidly. The mechanisms involved in conferring the extreme desiccation tolerance found in lichens, algae and bryophytes appear to differ from those in higher

plants, though in all cases mechanical stabilisation of membranes and proteins during desiccation is critical. The remarkable desiccation tolerance of resurrection plants appears to involve a range of mechanisms including: the ability to maintain subcellular structure and physiological integrity as water is lost; the tolerance and detoxification of oxidative stress brought about by unregulated metabolism as desiccation progresses; and an ability to repair damage rapidly on rehydration (as appears to be the case in the resurrection plant *Borya nitida*).

Stabilisation of macromolecular structure as water is lost is primarily dependent on the presence of an adequate complement of compatible solutes and other molecular chaperones, while effective damage-repair mechanisms are also critical. It is common in resurrection species for water-filled vacuoles to be replaced during desiccation by a number of small vacuoles filled with non-aqueous materials and for 'vitrification' of the cytoplasm to occur (Farrant, 2000). An important protective mechanism during early stages of desiccation is the minimisation of the production of reactive oxygen species (ROS) by the downregulation of photosynthesis, whether this is achieved by breaking down chlorophyll or by masking the chlorophyll so electron transport is inhibited. Extreme desiccation tolerance is commonly associated with an ability to tolerate other environmental stresses such as high temperature and UV-B radiation. In vascular resurrection plants recovery is usually only possible where the initial dehydration occurs slowly.

Non-resurrection plants. Although there may be lessons to be learned from resurrection plants for the development of desiccation-tolerant crop plants, the mechanisms involved in resurrection and crop plants appear to be qualitatively different, so there are likely to be major trade-offs if resurrection-type mechanisms are to be incorporated into crop plants. Nevertheless even within crops there are clear differences between and within species in their ability to tolerate tissue water deficits. There has been rapid progress recently in the elucidation of

drought-responsive genes and signalling pathways in plants (Yang *et al.*, 2010) that one might expect to relate to drought tolerance mechanisms, though it is difficult to distinguish those responses that are positively adaptive from those that are merely manifestations of damage. As we would expect, the more limited tolerance of tissue water deficits in conventional plants involves a range of contributory mechanisms.

1. Compatible solutes and molecular chaperones as discussed above have been found to be particularly effective at protecting cytoplasmic proteins and cell membranes from desiccation, even when they increase to high concentrations during osmotic adjustment or during tissue desiccation. To be effective they need to be compartmented in the cytoplasm. The main molecules involved as compatible solutes have been listed above. An enhanced production of *late embryogenesis abundant* (LEA) proteins is also a characteristic response to tissue drying and although their production does not contribute significantly to osmotic adjustment, it is likely that they contribute to tolerance at least partly by stabilising proteins and subcellular structures. They are extremely hydrophilic and as a result bind water, sequester ions and maintain protein and membrane structure, acting as molecular chaperones (Bray, 1997). Similar mechanisms involving compatible solutes and cold stress proteins as osmoprotectants, are also involved in freezing tolerance, where dehydration is also a cause of injury (see Chapter 9).

2. Antioxidant systems are also critical in the cellular tolerance of water deficits as production of reactive oxygen species and *oxidative stress* is commonly a factor in desiccation injury with increasing evidence that chemical injury in the form of oxidative damage (especially that caused by oxygen radicals) is an important factor in desiccation injury (Arias *et al.*, 2011; Procházková & Wilhelmová, 2011). Mitochondria and especially chloroplasts are major sites of ROS production.

The superoxide radical (O_2^-) and other oxygen radicals can be produced by a number of reactions in cells including autoxidation of a number of reduced compounds, and the 'Mehler' reaction in chloroplasts where O_2 rather than CO_2 becomes the ultimate acceptor for electron transport (as can occur when assimilation is blocked by water stress; this may also be involved in photo-inhibitory damage: see Chapter 7). Once formed, O_2^- undergoes further reduction to form the very damaging hydroxyl radical ($^\bullet OH$), which can cause lipid peroxidation and the formation of hydrogen peroxide. Plants contain a number of antioxidant mechanisms that protect against the production of oxygen radicals including: (i) water-soluble reductants such as thiol-containing compounds (e.g. glutathione) and ascorbate; (ii) fat-soluble vitamins such as α-tocopherol and β-carotene; and (iii) enzymic antioxidants such as catalase and superoxide dismutase. Reactive oxygen species metabolism also appears to play a critical role in regulation of many stress responses.

3. Transcriptional and post-transcriptional regulation. Complex networks of genes and associated molecular signalling mechanisms are involved in sensing drought stress and in controlling the developmental and physiological responses underlying the 'avoidance' strategies outlined above as well as the induction of stress tolerance by osmoregulation and the development of desiccation-tolerant biochemistry. Drought, as well as other environmental stresses, induces the expression of many transcriptional regulators that affect downstream metabolism; the relevant signalling pathways include both those involving ABA signalling and ABA-independent signalling using the dehydration-responsive elements (*DRE*) genes and other second-messenger and metabolic cascades. Details of these signalling pathways may be found in appropriate texts and reviews (Shinozaki & Yamaguchi-Shinozaki, 2007; Taiz & Zeiger, 2010; Valliyodan & Nguyen, 2006; Wang *et al.*, 2003; Yang *et al.*, 2010).

10.2.3 Efficiency mechanisms

Any mechanism enhancing survival in drought conditions tends to decrease the potential dry matter productivity. For example, total photosynthesis would be decreased by stomatal closure, by leaf rolling, by decreases in leaf area or even by reductions in growing season length. The 'ideal' plant for any environment involves a compromise between water conservation and productivity mechanisms with the *optimum* balance depending on the aridity of the environment. In addition to such optimal expression of stress-avoidance characters, drought tolerance is favoured by any mechanisms that improve either the efficiency with which a limited supply of water is used for photosynthesis or the efficiency of subsequent conversion of photosynthate into reproductive structures (or yield in the case of crops).

Water use efficiency

Following common usage, the ratio of net assimilation to water loss will be given the general term water use efficiency (WUE), though it is, of course, not a true efficiency as it does not have a maximum value of unity. The precise definitions of assimilation or water use differ between authors: for example, water loss may be in mass units or molar units and assimilation may be expressed in terms of net CO_2 exchange ($\mathbf{P}_n$ in either molar or mass units), dry matter growth or yield (Y_d) or economic yield (Y). Total net daily CO_2 uptake may be converted into dry matter using a value of 0.61 to 0.68 g dry matter per g CO_2 (see Chapter 7), though for photosynthetic data for single leaves it is also necessary to allow for respiratory losses at night and from other tissues. Economic yield may be obtained from Y_d by multiplication by the harvest index (see below and Chapter 12). Unfortunately, many published estimates of Y_d have neglected root dry matter, which may vary from less than 20% of plant dry mass when ample water is available, to more than 50% in dry conditions.

Similarly water loss may be expressed in terms of either water transpired ($\mathbf{E}_\ell$) or total evaporation ($\mathbf{E}$).

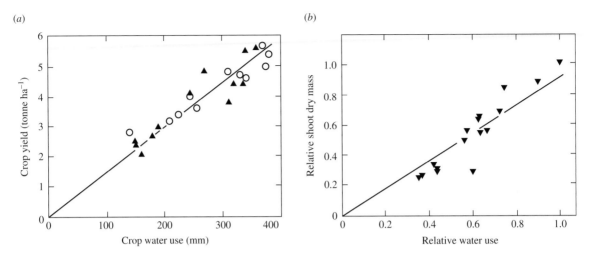

Figure 10.4 (*a*) Grain yield as a function of water used under a range of irrigation treatments for barley in 1976 (○) and wheat in 1979 (▲) in South-Eastern England (data from: Day *et al.*, 1987; Innes & Blackwell, 1981). (*b*) Relative shoot dry mass as a function of relative water use for cowpea in California in 1976 (after Turk & Hall, 1980). The water use efficiency (WUE) for any point is given by the slope of a line joining that point to the origin, so that the lines join points of equal WUE.

The yield per unit transpiration (transpiration efficiency; TE = Y/E_ℓ) is a better measure of plant performance than yield per unit evaporation (Y/E), though the latter is ecologically important because a significant proportion of total evaporation occurs directly from the soil.

Because of the sensitivity of evaporation to the environment (particularly the atmospheric vapour pressure deficit, *D*), WUE varies markedly from place to place and from year to year. Many more recent results are similar to those of N. M. Tulaikov, with 'Beloturka' wheat in Russia (see Maximov, 1929): WUE (in terms of dry matter) in one year (1917) ranged from 2.13×10^{-3} at Kostychev to 4.22×10^{-3} in the relatively cool and humid climate at Leningrad, while at one site values ranged from 1.74×10^{-3} to 3.31×10^{-3} over the years 1911 to 1917.

In contrast, there is extensive evidence that for any one species, the seasonal WUE in one environment can be surprisingly constant over a wide range of treatments (sowing date, planting density, water supply, nutrition, etc.). Some examples of the effects of different irrigation regimes (and total water use) on grain yields of wheat and barley in southern England

are shown in Figure 10.4(*a*). The close similarity between the years is somewhat fortuitous, but all treatments lie close to a straight line through the origin, implying nearly constant WUE. Similar constancy of WUE in terms of dry matter production by cowpea (*Vigna unguiculata*) is shown in Figure 10.4(*b*). In many cases where WUE is not constant over a range of treatments, deviations can be attributed to differing amounts of soil evaporation.

In an early analytical attempt to explain site-to-site variability of WUE, de Wit (1958) recognised that water loss from plants (E_ℓ) was a dominant feature and showed that for dry climates:

$$Y/E_\ell = k/E_o \qquad (10.1)$$

where E_o was the mean daily free water evaporation and *k* is a parameter that was relatively constant for any species. This relationship broke down only if growth was seriously 'nutrition limited', if the water available was excessive, or in humid regions (where Y/E_o was approximately constant). There remained, however, large differences between species, as is clearly illustrated by the well-known results in Table 10.3. Although absolute values of WUE are

Table 10.3 Values of water use efficiency (Y_d/E_ℓ) for some potted plants grown at Akron, Colorado (rearranged from Maximov, 1929, after data from Shantz & Piemeisel, 1927).

	Y_d/E_ℓ ($\times 10^3$)
C_4 plants	
Cereals	2.63–3.88
millet cvs	2.72–3.88
sorghum cvs	2.63–3.65
maize cvs	2.67–3.34
Other C_4 gramineae	2.96–3.38
Other C_4	2.41–3.85
Range for C_4 plants	2.41–3.88
C_3 plants	
Cereals	1.47–2.20
wheat cvs	1.93–2.20
oat cvs	1.66–1.89
rice cvs	1.47
Other C_3 gramineae	0.97–1.58
Other C_3 crops	1.09–2.65
alfalfa cvs	1.09–1.60
pulses	1.33–1.76
sugar beet	2.65
Native C_3 plants	0.88–1.73
Range for C_3 plants	0.88–2.65

variable (depending on weather), differences between species are very consistent and are clearly related to photosynthetic pathway. CAM plants have the highest dry matter efficiencies (e.g. 20×10^{-3} to 35×10^{-3} for *Agave* and pineapple: Joshi *et al.*, 1965; Neales *et al.*, 1968) followed by C_4 plants, which are in

turn approximately twice as efficient as C_3 plants. Table 10.3 illustrates that differences between species with one photosynthetic pathway are no larger than differences between cultivars of one species. Interestingly, there is no obvious relationship between WUE and aridity of habitat. The analysis of WUE will be expanded in Section 10.3 below.

Effective use of available water (EUW)

In many situations, efficiency alone is a poor competitive strategy because greater efficiency is often associated with a slow rate of water use. Such 'efficient' plants might then lose some of their potentially available water to faster growing competitors. This is, however, less of a problem in typical agricultural or horticultural monocultures. Nevertheless, a related point that is often forgotten in discussion of efficiency mechanisms is that it is no use just having a high efficiency (e.g. of water use) if water supply is non-limiting. Other things being equal, productivity in dry environments is likely to be greater for a plant whose behaviour tends to *maximise assimilation in relation to the amount of water available* than for a plant that simply has a high ratio of assimilation to water evaporated. Blum (2009) has suggested that plant breeders and agronomists should shift their emphasis from improving WUE to the improvement of their effective use of water (EUW). This would include strategies that maximise the amount of available water (e.g. through minimising soil evaporation and run-off).

Enhanced harvest index

For agricultural crops effective drought tolerance will also be enhanced by any mechanism that ensures that a higher proportion of photosynthate is devoted to yield production. For example cereals with a higher *harvest index* (= the fraction of dry matter in the harvestable grain) will produce higher yields for the same amount of photosynthesis. Similar considerations will apply in natural ecosystems. Improvements in harvest index were a key component of the advances in crop yield made during the 'green revolution' in the 1960s.

10.3 Further analysis of water use efficiency

10.3.1 Simple models based on leaf gas exchange

The analysis originally developed by Bierhuizen and Slatyer (1965) can be used to explain variation in WUE. Using equations described earlier for leaf gas exchange (see Chapter 7), and using molar units (though here expressing driving forces in terms of partial pressure, rather than mole fraction as used previously), one can write for instantaneous transpiration:

$$\mathbf{E}_\ell^m = \frac{(e_\ell - e_a)}{P(r_a + r_\ell)} \tag{10.2}$$

where e_ℓ is shorthand for the saturation vapour pressure at leaf temperature, and for instantaneous assimilation:

$$\mathbf{P}_n^m = \frac{(p_a' - p_i')}{P(r_a' + r_{\ell}')} = \frac{(p_a' - p_x')}{P(r_a' + r_{\ell}' + r_i')} \tag{10.3}$$

where p_x' is an 'internal CO_2 concentration' usually taken as equal to Γ and the intracellular resistance (r_i') is the sum of the mesophyll and biochemical components (see Section 7.6.2). Combining Eqs. (10.2) and (10.3) gives:

$$\frac{\mathbf{P}_n^m}{\mathbf{E}_\ell^m} = \frac{(p_a' - p_x')(r_a + r_\ell)}{(e_\ell - e_a)(r_a' + r_\ell' + r_i')} \tag{10.4}$$

Since p_x' can be assumed constant, WUE at any instant for a given environment is proportional to $(r_a + r_\ell)/(r_a' + r_\ell' + r_i')$. This ratio is larger for C_4 than for C_3 plants because of the larger r_i' (and often smaller r_ℓ) in C_3 plants and explains the observed differences between plants with different photosynthetic pathways (Table 10.3).

This resistance approach is not altogether satisfactory because of non-linearity in the CO_2 response curve, but a useful simplification can be made using the observation that the ratio of gas-phase to liquid-phase resistances is often nearly constant over a range of conditions (nutrition, age,

irradiance, etc.) so that p_i'/p_a' is approximately constant at near 0.3 for C_4 plants and 0.7 for C_3 plants (see Chapter 7). For example as irradiance decreases, both r_i' and r_ℓ' increase. Combining Eqs. (10.2) and (10.3) gives:

$$\frac{\mathbf{P}_n^m}{\mathbf{E}_\ell^m} = \frac{(p_a' - p_i')(r_a + r_\ell)}{(e_\ell - e_a)(r_a' + r_\ell')} \tag{10.5}$$

from which the resistances can he eliminated (because the CO_2 resistances $(r_a' + r_\ell') \simeq 1.6 (r_a + r_\ell)$ – see Chapter 3), to give:

$$\frac{\mathbf{P}_n^m}{\mathbf{E}_\ell^m} = \frac{p_a'[1 - (p_i'/p_a')]}{1.6 (e_\ell - e_a)} \tag{10.6}$$

where p_i'/p_a' is constant, this can be written:

$$\mathbf{P}_n^m/\mathbf{E}_\ell^m = k_m/(e_\ell - e_a) \simeq k_m/D \tag{10.7}$$

where k_m is a constant that depends on species. Because leaf temperature is not usually known (for calculation of e_ℓ) the leaf–air vapour pressure difference can be approximated by D. In agronomic studies, where one is concerned with dry matter yield, it is more usual to express Eq. (10.7) in terms of mass fluxes of CO_2 so that:

$$\mathbf{P}_n/\mathbf{E}_\ell = (M_C/M_W)(\mathbf{P}_n^m/\mathbf{E}_\ell^m) = 2.44 (\mathbf{P}_n^m/\mathbf{E}_\ell^m) \tag{10.8}$$

Equation (10.7) emphasises the role of leaf–air vapour pressure difference in determining WUE in different climates. The importance of this can be illustrated by studies of the effect of irrigation on productivity using small plots: in a monsoon climate where rainfall affects D, the effect of rain on production may be larger than the effect of an equivalent amount of irrigation; similarly the effects on yield when water supply is controlled using rainshelters will not be the same as equivalent amounts of rain falling in the dry season.

When extrapolating this approach (based on leaf gas exchange) to the crop or community level it is necessary to take account of a number of additional factors such as evaporation from the soil and diurnal and seasonal variation in stomatal aperture and vapour pressure deficit, as discussed below.

10.3.2 Extrapolation to long term

Integration of Eq. (10.7) over daily, or longer, periods requires several additional assumptions. Most important is estimation of the proportion of $\mathbf{P}_n$ lost in respiration. The simple conversion from CO_2 to dry mass given above is appropriate if net CO_2 exchange is available for the whole plant for the full 24 h. Where only daytime $\mathbf{P}_n$ is available, one needs to allow for growth respiration (with a conversion efficiency $\simeq 0.53$ g dry mass per g CO_2) and maintenance respiration (about 15 to 30% of $\mathbf{P}_n$) giving about 0.37 to 0.45 g dry mass per g CO_2 overall (see Chapter 7).

A second problem is estimation of $(e_\ell - e_a)$. We have already seen that it is necessary to approximate this term as D (use of a model that allows for leaf–air temperature differences is described in the next section). Second, it is difficult to estimate an appropriate daily mean for D, though Tanner (1981) has approximated the integrated daytime saturation deficit (D^*) as 1.45 times the mean of the average saturation deficit calculated at minimum and maximum temperatures.

Assuming that appropriate mean values for the leaf–air vapour pressure difference are available, Eq. (10.6) can be integrated to give a total WUE over the life of the plant (WUE*) by including a term (ϕ_C) to represent the losses of CO_2 not associated with assimilation through the stomata (respiration losses at night and from the roots), and a term (ϕ_W) for the losses of water vapour from the soil or through the cuticle:

$$\text{WUE}^* = \frac{p_a'[1 - (p_i'/p_a')](1 - \phi_C)}{1.6 D^*(1 + \phi_W)} \tag{10.9}$$

10.3.3 Carbon isotope discrimination

The basis of the discrimination (Δ) between the two stable isotopes of carbon (^{13}C and ^{12}C) was described in Chapter 7. It was shown there (Eq. (7.39)) that Δ is a function of p_i'/p_a' in C_3 plants. Rearranging Eq. (7.39) (and using vapour pressures) to give p_i'/p_a' in terms of Δ and substituting into 10.9 gives:

$$\text{WUE}^* = \frac{p_a'(0.030 - \Delta)(1 - \phi_C)}{1.6(0.0256)D^*(1 + \phi_W)} \tag{10.10}$$

This equation implies that WUE should decrease linearly with increasing Δ so that measurement of Δ might provide an indicator of differences in WUE for plants growing in a particular environment. A particular advantage of the technique should be that measurement of the carbon isotope composition of dry matter theoretically integrates $\mathbf{P}_n/\mathbf{E}_\ell$ over time.

Negative relationships between WUE and Δ of the plant dry matter as predicted by Eq. (10.10) have been found for a number of species both in pots and in the field (e.g. Figure 10.5), though field results tend to be more variable. These results suggest that measurements of Δ may provide a useful selection method for WUE, at least in C_3 species, especially since reasonably high heritabilities have been found for both water use efficiency and Δ (Hubick *et al.*, 1988). The use of carbon isotope discrimination has been successful in the breeding of wheat cultivars

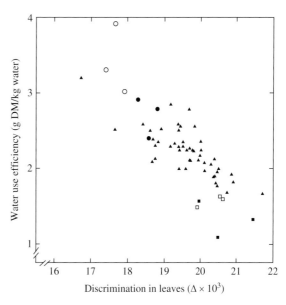

Figure 10.5 Water use efficiency (g dry matter/kg water) of whole peanut plants plotted against carbon isotope discrimination measured in dried leaf material. Solid symbols: well-watered plants; open symbols: water-stressed plants; Chico ($\square$, $\blacksquare$); Tifton-8 ($\circ$, $\bullet$): F2 progeny of Chico × Tifton-8 ($\blacktriangle$). (Data from Hubick *et al.*, 1988.)

such as Drysdale adapted for dryland conditions (Condon *et al.*, 2002), though it is interesting to note that such screening appears to work best when conducted under conditions of non-limiting water supply.

10.3.4 More sophisticated leaf models

Unfortunately the predictions from the simple models (e.g. Eqs. (10.4) or (10.9)) are not always accurate because they do not allow for differences between leaf and air temperature or for the precise shape of the photosynthesis response curves. Models that include a more complete description of leaf photosynthesis and of leaf energy balance are available (e.g. Cowan, 1977; Jones, 1976; Leuning, 1995; Thornley & France, 2007).

Although more detailed models are available, the main effects are well illustrated by the model shown

in Figures 10.6 and 10.7. As expected from Eq. (10.4), stomatal closure tends to increase the instantaneous WUE at the expense of absolute production (Figure 10.6). The magnitude of the cuticular resistance, r_c, (or in crop models the soil evaporation) is crucial. As r_c decreases, the general advantage of stomatal closure disappears so that an optimum stomatal resistance becomes apparent (Figure 10.6(*b*)). This effect occurs because the long liquid-phase component in this pathway to the chloroplast means that CO_2 uptake through the cuticle is negligible even where water loss may be significant. Although WUE usually tends to increase as stomata close, it is possible in some cases when the ratio of r_i'/r_a' is less than a critical value (e.g. in C_4 plants with large leaves) for WUE to *increase* with increasing stomatal aperture. This prediction has been confirmed experimentally for a crop (Baldocchi *et al.*, 1985).

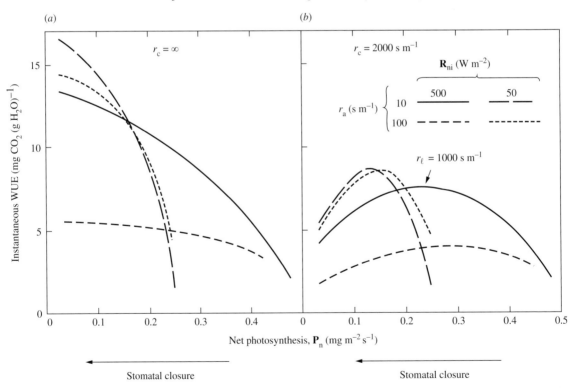

Figure 10.6 Variation of instantaneous water use efficiency (WUE) with photosynthesis rate (which increases as stomata open) for C_3 leaves (*a*) assuming an infinite cuticular resistance (r_c); (*b*) assuming an r_c of 2000 s m^{-1} (after Jones, 1976). The right-hand end of each curve corresponds to a leaf resistance of 50 s m^{-1}, and the left-hand end to 6400 s m^{-1}. The value of r_ℓ' giving maximum WUE for $\mathbf{R}_{ni} = 500$ W m^{-2} and $r_a' = 10$ s m^{-1} is shown by the arrow.

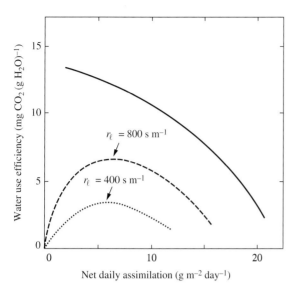

Figure 10.7 Variation of daily water use efficiency (WUE) with photosynthetic rate for C_3 leaves with $\mathbf{R}_{ni} = 500$ W m^{-2} and $r_a' = 10$ s m^{-1} for different amounts of respiration at night (**R**, mg m^{-2} s^{-1}). —— $\mathbf{R} = 0$; ---- $\mathbf{R} = 0.14\,\mathbf{P}_n + 0.05$; ····· $\mathbf{R} = 0.14\,\mathbf{P}_n + 0.15$. The arrows show values of r_ℓ' giving maximum WUE.

Figure 10.6 also illustrates the complexity of effects of alterations in leaf boundary layer resistance (e.g. resulting from leaf size, hairiness or exposure). Although increasing r_a, as one might get with larger leaves, normally leads to decreased WUE, particularly at high light or when stomata are open, the reverse can be true when stomata are closed. There are also interactions with the photosynthetic pathway in terms of the optimum combination of stomatal and boundary layer characters for any particular environment.

Although it can be instructive to calculate the stomatal resistance that gives the maximum instantaneous WUE (as in Figure 10.6), of more relevance to plant growth is the resistance giving maximum WUE over daily or longer periods. This resistance is smaller because of the need to make up for respiration at night (see above). Figure 10.7 shows how this optimum resistance changes as respiration increases. Further details of this model and its extension to canopies are discussed by Jones (1976).

It is worth reiterating at this point the earlier comment that maximum production for the amount

of water available is likely to be a more generally advantageous strategy than maximum WUE (i.e. it is no use maximising WUE if some water is left unused). In such cases, the optimum stomatal aperture may be between fully open and that giving maximum WUE. Calculation of this optimal behaviour is discussed in the next two sections (similar calculations for processes such as leaf area development are also possible). One application of these studies is to provide information on the relative importance of different drought tolerance mechanisms, by comparing these predictions with the behaviour of real plants. A very close similarity is unlikely as evolution is determined by a plant's overall fitness, which is the summation of many factors over the whole life cycle. An alternative application of these studies is in plant breeding (see Chapter 12). The 'shorthand' terminology used in these sections and elsewhere should be interpreted in a non-teleological sense: for example, an 'optimistic plant' (see below) simply describes a pattern of response without implying purpose.

10.3.5 Optimal stomatal behaviour

Although at any one time there may be an optimum stomatal aperture giving maximum instantaneous WUE, optimum use of water over a period involves optimal distribution of stomatal opening as the environment changes. Clearly it is more efficient to restrict periods of open stomata and rapid photosynthesis to those times when potential evaporation is low, particularly in the morning. Therefore, stomatal closure during midday and in the afternoon, as commonly observed in water-stressed plants, will tend to improve WUE.

Cowan and Farquhar (Cowan, 1977; Cowan & Farquhar, 1977) developed a model to quantify the optimal behaviour for stomata in a changing environment. Starting from the premise that the optimal behaviour is that where the *average* evaporation rate is minimal for a given average rate of assimilation, they present a theorem that defines the optimal behaviour. The criterion used is equivalent to that suggested above of maximising

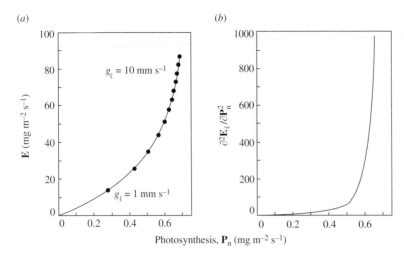

Figure 10.8 (*a*) A typical relationship between $\mathbf{E}_\ell$ and $\mathbf{P}_n$ for a C_3 plant in a constant environment calculated using the model of Jones (1976). The relative increase in $\mathbf{E}_\ell$ (i.e. $\Delta\mathbf{E}_\ell/\mathbf{E}_\ell$) for any change in g_ℓ is always greater than the corresponding relative increase in $\mathbf{P}_n$ (i.e. $\Delta\mathbf{P}_n/\mathbf{P}_n$). (*b*) The second derivative $\partial^2\mathbf{E}_\ell/\partial\mathbf{P}_n{}^2$.

the total assimilation for a given total water use and may also be framed in these terms.

As we have seen, for the majority of plants and environments $\mathbf{E}_\ell$ is relatively more sensitive to changes in stomatal conductance than is $\mathbf{P}_n$ (Figure 10.8). Mathematically this can be stated as: $\partial^2\mathbf{E}_\ell/\partial\mathbf{P}_n{}^2$ is everywhere > 0. For this particular case, the optimal stomatal behaviour is that which maintains:

$$(\partial\mathbf{E}_\ell/\partial g_\ell)/(\partial\mathbf{P}_n/\partial g_\ell) = \partial\mathbf{E}_\ell/\partial\mathbf{P}_n = \lambda \qquad (10.11)$$

where λ is a constant depending on the average rate of assimilation required (or water available). If the amount of water available is limited, the constant λ is small. Using the terminology of economics, the optimal conductance is that which maintains the marginal cost of a change in conductance equal to the marginal *benefit*, so the ratio λ is small if water is 'expensive'. In practice this means that, as the environment changes during a day, the stomata adjust to keep the ratio of the sensitivities of evaporation and assimilation to infinitesimal changes in g_ℓ at a constant value. Putting this another way, it means that the stomata operate in such a way as to keep the assimilation rate at that position on a curve relating $\mathbf{P}_n$ and $\mathbf{E}_\ell$ where the slope is a constant value, irrespective of changes in the shape of this curve as the environmental conditions change. It can be more convenient to use the reciprocal of λ (i.e. λ^{-1}) as this represents the marginal WUE (Hari *et al.*, 1986).

Figure 10.9 shows, for typical diurnal variation in climatic variables, daily trends in $\mathbf{E}_\ell$ and $\mathbf{P}_n$ for different constant values of λ. As the amount of available water, and therefore λ, decreases there is an increasing tendency towards midday stomatal closure. Midday stomatal closure is, in fact, a well-known phenomenon in water-stressed plants, and this analysis suggests that this behaviour may have evolved as a way of maximising WUE in drought conditions. Although the behaviour that maintains λ constant gives a higher value of WUE over a day than any constant stomatal conductance, surprisingly, perhaps, the difference is not large particularly at high rates of water use.

The conclusion that $\partial\mathbf{E}_\ell/\partial\mathbf{P}_n$ must remain constant over time for the optimal average WUE of a single leaf can be extended to a complete canopy, where it can be shown that optimal WUE is obtained if the stomata operate so that $\partial\mathbf{E}_\ell/\partial\mathbf{P}_n$ is constant over all the leaves in the canopy. Evidence that $\partial\mathbf{E}_\ell/\partial\mathbf{P}_n$ is in fact maintained approximately constant, as required by this hypothesis, has been provided for plants in controlled environments (Farquhar *et al.*, 1980a) and in the field (see Manzoni *et al.*, 2011). Consistent with optimal behaviour, it has been found that λ is reasonably constant during a day in many situations, and, if one takes account of non-stomatal water losses, its value tends to decrease as expected as water availability decreases.

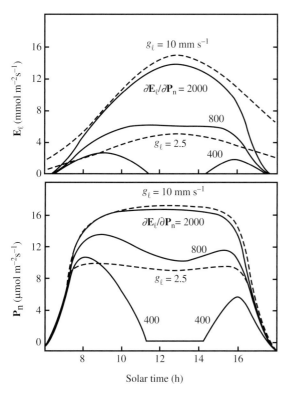

Figure 10.9 Optimal time courses for transpiration (E_ℓ) and assimilation (P_n) during a typical day for various magnitudes of $\partial E_\ell / \partial P_n$ (solid lines). These curves show that midday stomatal closure is an optimal behaviour when water is limited and $\partial E_\ell / \partial P_n$ is small. Also shown are E_ℓ and P_n for constant values of leaf conductance (g_ℓ) (dashed lines) (after Cowan & Farquhar, 1977).

Under extreme drought, however, there are indications that as photosynthetic biochemistry is inhibited the value of λ can increase somewhat (Manzoni *et al.*, 2011).

The strategy of maintaining $\partial E_\ell / \partial P_n$ constant can be shown to be not the most efficient, but the most inefficient stomatal behaviour if $\partial^2 E_\ell / \partial P_n^2 < 0$. This relationship between P_n and E_ℓ may occur in C_4 plants (particularly if r_a is large as occurs with large leaves or low wind speeds) but is rare for C_3 plants. In this case the most efficient behaviour is one where the stomata are either fully closed or fully open, with the period open being appropriate to use all the available water.

10.3.6 Optimal patterns of water use in an unpredictable environment

Even for a plant growing in a particular location with a known climate, it is difficult to define the most efficient pattern for water use during the season. This is because the various weather components (rainfall, temperature, etc.) usually vary from year to year both in amount and in the way in which they are distributed through the season, and hence are unpredictable. Although it is possible, at least in principle, to use techniques such as those outlined above to determine the best patterns of stomatal behaviour or leaf area development if the exact patterns of rainfall and other weather factors are known for the whole season, this information is never available to the plant (or to the farmer) at the beginning of the season. Therefore any optimal pattern for water use during the season must take account of the probability of future rainfall. Cowan (1982, 1986) has extended his earlier model for stomatal behaviour to take account of future rainfall probabilities and concluded that one would expect the stomatal conductance (and hence the assimilation rate) to decrease with increasing soil water deficit because this behaviour reduces the probability of plants completely drying the soil profile and then dying. Another related conclusion is that the optimal stomatal conductance declines during the season for an annual crop, even if the environment remains constant. The reasons for these results can be understood by means of the following, rather simplistic, analysis of the consequences of uncertainty in the pattern of future rainfall.

The pattern of water use that is adopted by any plant is inevitably a compromise that depends on the climate. This is because a plant that is over 'optimistic', for example, and produces a large transpiring leaf area on the assumption that more rain will fall, may not be able to produce any seed at all in the odd dry year. On the other hand, a relatively 'pessimistic' plant that ensures the production of some seed, even though no extra rain falls (e.g. by initiating reproductive growth early), may not be able to respond adequately in the wetter than average year. Clearly the more variable the climate, the more difficulty there will be in finding a type of behaviour that maintains the productivity close to the retrospectively determined 'ideal' behaviour in most years.

Desert ephemerals tend to be relatively pessimistic, behaving as water spenders, while perennials can be more optimistic since they have reserves to survive the odd bad year. In an agricultural situation, different farmers may have differing definitions of their ideal crops. A peasant farmer, for example, may be most concerned to avoid starvation in any year and might therefore prefer a crop variety that reliably produces some yield every year (e.g. Figure 10.10(a)), even though it has a low yield potential in the good years. A farmer with greater resources, on the other hand, may prefer a cultivar that has a greater long-term average yield, even though yield may be very low in dry years (Figure 10.10(b)).

The effect of climate and climatic variability on the productivity of plants with different types of water use behaviour (optimistic, pessimistic and responsive or non-responsive to changes in soil moisture as shown in Figure 10.11) has been investigated by the use of a Monte Carlo modelling technique. Figure 10.12 illustrates the effects of these stomatal behaviours on the probability distributions of total

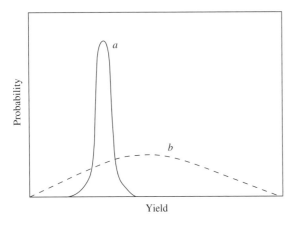

Figure 10.10 Hypothetical yield probability distributions over many years at one site for two contrasting crop varieties: (a) stable but low yield (——); (b) more variable variety with higher mean yield (– – – –).

assimilation over a period for a particular climate with moderately variable rainfall. It is apparent from Figure 10.12 that, although relatively pessimistic and optimistic behaviours may produce the same average yield over many years, the more pessimistic behaviour is likely to lead to very low yields much less frequently. An alternative model, based on trade-offs between (i) maximum growth rate and the water potential at which growth stops and (ii) maximum growth rate and rate of water use to investigate strategies that plants can adopt to maximise seasonally averaged growth rate in uncertain rainfall environments has been proposed by Sambatti and Caylor (2007). These authors concluded that enhanced drought tolerance in terms of a lowered water potential at which growth stops only becomes beneficial when the interval between rainfall events falls below the rate of soil water depletion.

Although desert ephemerals may be regarded as pessimistic in terms of their overall water use pattern, as they have a very short life cycle whatever the water supply (and hence a low potential yield), their stomatal behaviour may be relatively optimistic. This illustrates that the optimal stomatal behaviour depends on all aspects of the environment and of the plant life cycle.

Figure 10.12 also shows the effects of one type of stomatal behaviour where the plants respond to the amount of water available. Interestingly, such responsive behaviour is very little better, on average, than the best 'constant' behaviour shown in Figure 10.12(a), but it is more adaptable to a wide range of climates. Stomatal closure in response to increasing water deficit (Figure 10.11(d)) as found in most plants, provides a good general compromise. It is an optimistic behaviour that permits achievement of high productivity in a wet year, while preventing serious damage in short drought periods. Only in a long drought would a more conservative or pessimistic behaviour be better, and then only for annual species.

Optimisation of stomatal behaviour is only one of the plant responses involved in maximising assimilation with limited water availability. In a particularly interesting extension of Cowan's analysis

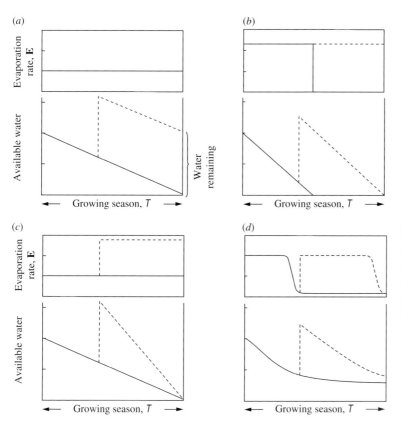

Figure 10.11 Four classes of stomatal behaviour and their effects on water use. The solid lines represent the time course of evaporation rate (upper half of each figure) or water remaining in the soil (lower half) if no rain falls. The dotted lines illustrate the consequences of rain part way through the season. (*a*) Stomata fixed to use initial soil water by the end of season (*pessimistic non-responsive*); (*b*) stomata fixed to use water faster than justified on the basis of the initial water in the soil (*optimistic non-responsive*); (*c*) stomata respond to keep rate of water use at that required to use all currently available water by end of season (*pessimistic responsive*); (*d*) stomata initially optimistic, but with ability to close preventing complete desiccation (after Jones, 1980).

to include the optimal allocation of carbon to the synthesis of leaves and roots, Givnish (1986) demonstrated that the following summary relationship is optimal:

$$r_i'/r_\ell' = f/(1-f) \tag{10.12}$$

where f is the proportion of carbon allocated to leaves as compared with roots. This remarkable relationship suggests that the ratio of r_i' (important for determining assimilation) to r_ℓ' (important for determining water loss) should be equal to the ratio between the allocation of carbon for synthesis of new leaves and the allocation to new roots.

10.3.7 Antitranspirants

As mentioned above (e.g. Eq. (10.4)), there is a tendency for WUE to improve as stomata close. This has provided the impetus for many attempts over the

years to improve crop WUE by the application of antitranspirants (see e.g. Solárová *et al.*, 1981). The main types of antitranspirant are as follows.

1. *Compounds that close stomata*, such as ABA, phenyl mercuric acetate and decenylsuccinic acid. Unfortunately these compounds are often ineffective, or too expensive, or else toxic. Even when they do close stomata, improvements in WUE are often not observed in the field.

2. *Film-forming compounds*, such as silicone emulsions or plastic films, are also rarely very effective or long lasting. Unfortunately these compounds are less permeable to CO_2 than to H_2O (see Table 10.4), so their application tends to reduce WUE. Equally important perhaps is that they are all even less permeable to O_2. The fact that no 'ideal' antitranspirant materials are known that are more permeable to CO_2 than to H_2O is hardly surprising in

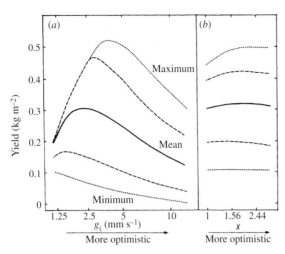

Figure 10.12 Examples of how the yield probability distribution for a particular climate changes depending on whether stomatal conductance is optimistic or pessimistic. The mean yield over a large number of simulations (500) is given by the solid line, the dotted lines give the highest and lowest yields observed and the dashed lines give the 5% tails. (a) The yield distribution with a range of fixed stomatal apertures; (b) the results if the stomata can respond to the amount of water available in such a way that the stomatal aperture is set at *x* times the value required to finish all the currently available water by the end of the season (after Jones, 1981c).

view of the greater molecular mass of CO_2 and the similar polar nature of the two molecules.

3. *Reflecting materials.* Increasing leaf reflectivity by application of materials such as kaolinite can help to lower leaf temperature and hence decrease E_ℓ. Although reflection in the PAR is increased more than in the infrared, at high irradiances (above those saturating P_n), this may not be too much of a disadvantage, so that useful overall improvements in WUE can be obtained.

Although film-forming antitranspirants have proved to be of value for minimising water loss and improving plant survival, especially of high-value perennials during transplanting, in only a very few cases have they proved beneficial for improving WUE. The lack of success in field crops is not surprising when one considers the degree of coupling of the crop to the atmosphere (Section 5.3.3). Both single plants

and plants in controlled-environment chambers are closely coupled to the environment so that **E** (and hence WUE) is determined by stomatal conductance, as the area of crop grown increases the effective boundary layer conductance decreases so that **E** becomes increasingly decoupled and hence insensitive to antitranspirants. This explains why the results of experimental tests on single plants (and in controlled environments) do not readily extrapolate to the field.

10.3.8 Why is WUE so conservative?

Although the instantaneous ratio P_n/E_ℓ can vary over a wide range, it remains to be explained why WUE over a whole season (after allowance is made for the humidity deficit) is so remarkably constant for one species (e.g. Figure 10.4). The major factor is almost certainly that the long-term mean is primarily determined when the stomata are fully open, because transpiration and photosynthesis when the stomata are closed, or nearly so, are small proportions of the total. Cowan and Farquhar (1977) have pointed out that, particularly when stomata are fairly wide open, the advantage for WUE of the optimal pattern of stomatal behaviour over that obtained by a constant aperture throughout the day can be relatively small (of the order of 10%). The other factor is the tendency of p_i' to be relatively stable in the long term so that Eq. (10.7) is nearly true (i.e. r_ℓ' adjusts as r_i' changes maintaining $(r_a + r_\ell)/(r_a' + r_\ell' + r_i')$ more nearly constant).

10.4 Irrigation and irrigation scheduling

The increasing worldwide shortage of water is leading to a renewed emphasis on improved methods for irrigation scheduling and for water application that make best use of available water. For example precision application by trickle irrigation or mini-sprinklers is slowly replacing coarser methods such as flood irrigation and rain-gun application. Indeed there is now interest in developing precision agriculture approaches such as *site specific*

Table 10.4 Permeabilities of various materials to water vapour, carbon dioxide and oxygen at 25°C unless stated otherwise (data from Brandrup & Immergut, 1975).

	Permeability ratio	Permeability ($\times 10^{14}$) (cm^2 s^{-1} Pa^{-1})		
	(H_2O/CO_2)	H_2O	CO_2	O_2
Polychloroprene (Neoprene G)	3.5	6825	194	30
Polypropylene	5.8	383	66.2[a]	17.3[a]
Polyethylene (density 0.914)	7.1	675	94.5	21.6
Polystyrene	11.4	9000	78.8	19.7
Gutta percha	14.4	3825	266	46.2
Natural rubber	14.9	17 180	1148	175
Polyvinylidene chloride (Saran)	16.3	3.75	0.23[a]	0.040[a]
Butyl rubber	21.3	825[b]	38.7	9.75
Polyvinyl chloride	1750	2060	1.18	0.34
Nylon 6	1770	1330	0.75[a]	0.29[a]
Cellulose (Cellophane)	404 000	14 300	0.035	0.016

[a] $= 30°C$; [b] $= 37.5°C$.

management (SSM) where the technology allows a variable rate of application to take account of varying requirements across a field. In order to make the most effective use of precision application approaches it is necessary to have corresponding precision scheduling approaches. Conventionally irrigation scheduling has aimed to maintain soil water close to or above *field capacity* (the soil water content after natural drainage for two days after saturation); this has usually been achieved either by direct measurement of soil moisture content (using any of the range of sensors from neutron probes, time-domain reflectometry (TDR), capacitance sensors or tensiometers (see e.g. Boyer, 1995; Hillel, 2004; Kramer & Boyer, 1995)) or by the use of soil water balance calculations based on the difference between irrigation plus precipitation (minus any run-off and deep drainage) and evaporation (Allen *et al.*, 1998). Although the water balance method is quite reliable, and the dominant

approach worldwide, it is subject to cumulative errors and is most effective when it is combined with soil moisture measurements to regularly reset the baseline.

The increased availability of sensor information on soil or plant water status is opening up the development of automated approaches to irrigation scheduling. Control strategies adopted may be based on either *open loop* controllers where the amount of irrigation applied is based on the calculated water loss (e.g. from calculated ET) over the previous time interval, or *closed-loop* controllers that depend on a feedback control responding to some aspect of the response (e.g. soil moisture status, stomatal aperture or stem diameter) (Romero *et al.*, 2012). The problem with the simpler open-loop systems is that small errors in calibration can lead to accumulating errors in the amount of water applied and hence in increasing deviations from the target soil moisture,

Table 10.5 A summary of the main methods available for plant 'stress' sensing and for irrigation scheduling, indicating their main advantages and disadvantages (abbreviated from Jones, 2007).

	Advantages	Disadvantages
1. Soil water measurement	Many are easy to use in practice; many are precise and can be automated	Soil heterogeneity requires many sensors; often not representative of root ψ (depends on evaporative demand); many sensors measure only a small soil volume
(a) Soil water potential (tensiometers, psychrometers, etc.)	Gives measure of water availability	Does not directly indicate 'how much' water to apply; tensiometers restricted to high ψ, other sensors often difficult to use
(b) Soil water content (gravimetric, neutron probe, capacitance/TDR)	Indicate 'how much' water to apply; widely used to recalibrate water balance	Limited replication
2. Soil water balance calculation (requires estimates of **E** and precipitation)	Easy to apply in principle and widely used; indicate 'how much' water to apply; gives average for crop	Not as accurate as direct measurement; depends on good estimates of crop coefficients; needs regular recalibration
3. Plant 'stress' sensing	Measures the plant response directly; integrates soil and environmental effects; potentially very sensitive	Does not indicate 'how much' water to apply; calibration needed to derive 'control' thresholds; largely at research/development stage
(a) Tissue water status		Leaf water status subject to homeostatic regulation so not very sensitive (isohydric plants); subject to short-term, environmentally determined fluctuations
Visible wilting	Easy to detect	Not precise; often appears after damage
ψ- pressure chamber	Widely used reference technique with $\psi_{\text{pre-dawn}}$ and ψ_{st} being particularly useful	Labour intensive; impossible to automate; instruments such as the leaf patch pressure sensor need calibration
ψ- psychrometer (and pressure probe)	Thermodynamically based measures	Can be automated; difficult to use in field
Tissue water content (RWC, leaf thickness, stem or fruit diameter)	Easy to automate; commercial morphometric systems available	Limited replication as expensive and often difficult to use
Xylem cavitation	Sensitive to increasing stress	Cavitation rate depends on stress history

Table 10.5 (*cont.*)

	Advantages	Disadvantages
(b) Physiological responses	Potentially more sensitive than measures of leaf water status	Often require sophisticated equipment; do not indicate 'how much' water is required so need calibration to determine control thresholds
Stomatal conductance	Very sensitive in isohydric species; porometry provides a reference method for research; thermal remote sensing can be used remotely and scaled to substantial areas of crop	Less sensitive in anisohydric species; porometry labour intensive; canopy temperature requires calibration for environment (e.g. reference surfaces)
Sap flow sensors	Measure of stomatal conductance; can be automated	Expensive and needs skilled operator; needs correction for environmental demand
Growth rate	Can be very sensitive	Instrumentation delicate and expensive; primarily laboratory system

while feedback systems continually correct any deviations. More sophisticated control can be achieved using proportional controllers.

10.4.1 Plant-based irrigation scheduling

Because growth and many plant physiological responses depend on plant water status (which not only depends on soil moisture but is also controlled by environmental demand and hydraulic flow resistances from the soil to plant tissues), it can be argued that the use of plant-based stress sensing has advantages as an alternative signal for driving irrigation scheduling decisions (Jones, 1990, 2004b, 2007). The main approaches available for detecting plant water deficits and plant responses that are suitable for irrigation scheduling (together with their advantages and disadvantages) are summarised in Table 10.5. Unfortunately many of these 'physiological' measures, such as measurement of stem or pre-dawn water potentials (ψ_{st} and $\psi_{pre-dawn}$), or of fruit or stem diameter changes are labour intensive and often

require specialist training and equipment. There are many publications reporting the use for irrigation scheduling of indicators based on measurements of sap flow, leaf thickness, fruit shrinkage or trunk diameter variation, or with the leaf patch-clamp pressure sensor (Zimmermann *et al.*, 2008), but the outputs are often difficult to calibrate and tend to depend not only on water status but also on age and environmental conditions (Fernández *et al.*, 2011a, 2011b; Huguet *et al.*, 1992). Maximum sensitivity can be obtained when these measurements are normalised against well-watered crops in the same environment, but this is not always feasible. Nevertheless useful data can be obtained for research purposes, even though these techniques are not generally used for irrigation scheduling in commercial situations.

Another potential plant-based measure of water stress is stomatal closure, as this is one of the most sensitive plant responses to water deficits, and this has been proposed as another indirect measure of stress for irrigation scheduling. Although measurement of g_s by porometer is labour intensive and difficult to automate and therefore impractical for scheduling, g_s can readily be detected by remote sensing techniques as variation in canopy temperature (Section 6.3.5). As discussed in the next section this allows canopy-scale estimates, which show particular promise for irrigation scheduling.

10.4.2 Crop water stress index and related measures

Remote sensing of canopy temperature as a measure of stomatal closure (as a proxy indicator of 'stress') has been widely proposed (see reviews by Jones, 2004a; Jones & Vaughan, 2010; Maes & Steppe, 2012). As a first step in accounting for the variation in canopy temperature caused by factors other than stomatal opening, Jackson et al. (1977) normalised data by defining a 'stress degree day' calculated as the difference between canopy and air temperatures at a fixed time of day; this could be integrated over time to give a measure of plant stress. The next critical advance was made by Idso, Jackson and colleagues (Idso, 1981; Jackson et al., 1981) who improved the environmental normalisation by correcting for the further variation caused by differences in air vapour pressure deficit (D). Their *Crop Water Stress Index* (CWSI) assumed (see Figure 10.13) that a plot of ($T_\ell - T_a$) against D for well-watered crops is linear with ($T_\ell - T_a$) decreasing solely as D increases; this line they termed a 'non-water-stressed baseline'. The value of ($T_\ell - T_a$) for stressed crops (with the stomata only partly closed) at any D, is intermediate between the non-water-stressed baseline value (T_{base}) and a potential maximum (T_{max}) obtained when stomata are completely closed (assumed constant across all D – see sample point x in Figure 10.13). An index in the range 0 to 1 is then calculated as:

$$\text{CWSI} = (T_x - T_{base})/(T_{max} - T_{base}) \qquad (10.13)$$

with larger values being indicative of increasing water deficits. A theoretical analysis of the approach (Jackson, 1982) shows that the CWSI is inversely related to crop transpiration, rather than to stomatal conductance. Crop Water Stress Index is actually a measure of $(\mathbf{E_p} - \mathbf{E})/\mathbf{E_p} = (1 - \mathbf{E}/\mathbf{E_p})$ where $\mathbf{E_p}$ is the potential transpiration from the well-watered crop.

Although this approach works well when measurements are restricted to near midday in hot dry climates (such as in Arizona, USA, where it was developed), there are difficulties in its more general application. The problems include: (i) in more temperate climates humidity deficits are smaller so that a given absolute error in temperature measurement leads to a larger relative error (Figure 10.13(b)); (ii) irradiance varies substantially with time of day and in many climates with cloud cover so that there can be major effects on actual canopy temperature; and (iii) variation in wind speed (and in canopy roughness) also affects leaf temperature (see Eqs. (9.5) and (9.6)). Although differences between crops and even between growth stages can be eliminated by the use of empirically determined non-water-stressed baselines, other factors such as plot size and environmental coupling can be important, and critically the clear sky conditions necessary for reliable application are rare in many climates, therefore alternative approaches such as the use of physical reference surfaces as described below have been developed.

Use of reference surfaces

As was indicated in Section 6.3.5 one can reformulate Eq. (10.13) by replacing the empirical value for the non-water stressed baseline (which implies a finite stomatal conductance) by the temperature of a freely transpiring water surface and replacing the value of the upper line with that of a non-transpiring surface with the same radiative properties. Although these temperature extremes can be obtained by calculation from Eqs. (9.5) or (9.6) where appropriate meteorological data exist, it is often more convenient simply to measure the temperatures of 'wet' and 'dry' physical reference surfaces with similar optical properties to the real leaves (Jones, 1999). The use of physical reference surfaces has the advantage that it automatically eliminates errors due to calibration of the sensor and it avoids the need for accurate estimation of the net radiation absorbed that is required for calculation.

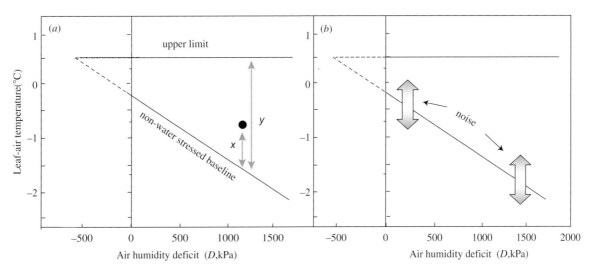

Figure 10.13 (a) A schematic illustration of the calculation of Idso's crop water stress index (CWSI), showing the dependence on the atmospheric humidity deficit (D) of the non-water stressed baseline and the upper limit. The CWSI for a canopy whose temperature and humidity deficit is given by the indicated point is calculated as x/y. (b) An illustration showing that for a given absolute error in temperature measurement, the relative error increases as the humidity deficit decreases, as tends to occur in temperate climates.

Using reference surfaces one can define an index that is a direct analogue of CWSI (I_{CWSI}) as:

$$I_{CWSI} = (T - T_w)/(T_d - T_w) \qquad (10.14)$$

where T_w is the temperature of a wet reference surface where the surface resistance to water vapour is zero, and T_d is the temperature of a non-evaporating surface where the surface resistance to water vapour is infinite. In many cases it is more useful to use an index that is directly proportional to stomatal conductance (I_{gs}). Rearranging Eq. (6.22), we see that the stomatal conductance ($= 1/r_\ell$) can be estimated as:

$$g_\ell = (1/(r_{aW} + (s/\gamma)r_{HR})) \left(\frac{T_d - T}{T - T_W} \right) \qquad (10.15)$$

where $(T_d - T)/(T - T_w)$ represents the conductance index, I_{gs}. (Note that this index becomes somewhat unstable when leaf or canopy temperature approaches that of the wet reference.) Although the stress indices (I_{CWSI} and I_{gs}) may be useful for relative studies it is often useful to be able to estimate stomatal conductance (or resistance) directly using Eq. (10.15). In this equation the first term on the right-hand side represents a scaling factor that depends only on the magnitude of the boundary layer resistance and on temperature.

Choice of reference surfaces

For leaf-scale studies the most convenient references that mimic the shortwave absorptance of leaves is to use actual leaves that have either been sprayed with water ($r_\ell = 0$) or whose transpiration has been stopped by covering with petroleum jelly ($r_\ell = \infty$). When scaling up to larger scales larger surfaces are required, but it is critical that they have similar aerodynamic and shortwave radiative properties to the canopy. Although there are advantages to having both wet and dry reference surfaces, it can be shown that estimates of g_ℓ are almost as good using only dry reference surfaces supplemented by humidity measurements (Leinonen et al., 2006). Further discussion of the use of thermal sensing for irrigation scheduling and detailed derivation of the relevant equations may be found elsewhere (Guilioni et al., 2008; Maes & Steppe, 2012).

In applying these 'stress' indices it is important to remember that, in spite of their name, they are not measures of stress, rather they indicate crop, and specifically stomatal, response. It is also important to remember that stomata can close for other reasons than drought (e.g. too much water would also lead to stomatal closure). In such cases irrigation scheduling using canopy temperature would tend to over-irrigate.

10.4.3 Deficit irrigation and partial rootzone drying

Historically irrigation has usually aimed to maximise growth, so surplus irrigation water was commonly applied and it was not necessary to have high-precision irrigation scheduling systems. With increasing shortages of water, however, there is now an increasing need for irrigation strategies that can conserve available water. *Deficit irrigation* (DI) strategies that do not fully replace all evaporative losses are now becoming widespread, at least partly because they lead to enhanced WUE as a result of the stomatal closure that occurs in such situations. A further benefit for many crops, especially fruit crops, is that the slight water deficits achieved by this means tend to enhance the production of grain or fruit at the expense of vegetative growth, and for fruits also to have improved quality in terms of sugar content. A difficulty with implementation of DI systems is that they require very precise monitoring and control of soil or plant water status in order to avoid any excessive deficits with their serious consequences for the crop.

We saw in Section 4.3.3 that when plants are grown in split-root systems with some roots maintained in wet soil but others allowed to dry, signals from the roots in drying soil to the shoots led to water-conservation responses such as stomatal closure and altered shoot development (Davies & Zhang, 1991) with no depression of the shoot water status. This result has been applied by Dry and colleagues (see e.g. Stoll *et al.*, 2000) to the development of practical irrigation systems where adequate water to maintain shoot water status is applied to only part of the root system with the rest of the roots being allowed to dry out slowly (sending drought signals to the shoots). When the drying side of the root gets too dry to signal effectively, the irrigation is switched from the 'wet' to the 'dry' side of the root system and the formerly wet side allowed to dry out (see Figure 10.14). This system

has become known as *partial rootzone drying* (PRD) and has been reported to have substantial benefits over comparable DI treatments in some, but not all, situations for grapevine and other crops. Although many putative 'tests' of PRD have unfortunately omitted adequate control treatments, a meta-analysis of published studies where comparable amounts of water were applied showed that PRD increased yield over DI in 6 out of 15 studies, with other agronomic benefits such as increased fruit size or quality also being noted in most cases and with no case where DI was better (Dodd, 2009).

The reasons for the variable results are not entirely clear, though it would be expected that PRD would only be beneficial in cases where yield is based on reproductive growth (i.e. not in vegetative crops such as lettuce). One advantage of PRD is that it uses less water than full irrigation that wets all of the root zone. In addition, in comparison with DI, precision scheduling should be much less critical as long as wet soil is maintained around some of the root volume. The development of PRD appears to be a good example where an understanding of the physiology underlying a plant environmental response has led to improved agronomic practice.

10.5 Other abiotic stresses

Many other environmental stresses can critically affect plant growth and reproduction. Space prevents us here going into detail of these, so we will only outline briefly some of the key abiotic stresses. Some, such as damage by ultraviolet radiation (see Sections 8.2.4 and 8.4.6), and temperature (Chapter 9) have already been discussed. The interested reader can find much relevant information in appropriate textbooks and reviews (e.g. Khan *et al.*, 2012; Pessarakli, 2011).

10.5.1 Flooding and waterlogging

An excess of water can be stressful for plants because of the resulting lack of oxygen reaching the roots or other submerged parts. Plants adapted to low oxygen

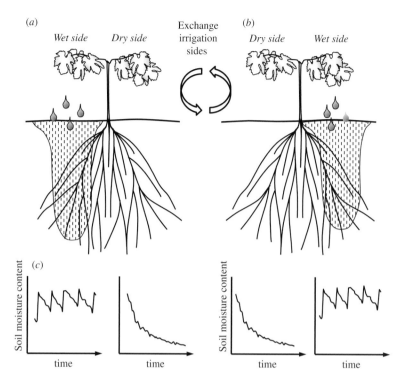

(a)

Wet side   Dry side

Exchange
irrigation
sides

(b)

Dry side   Wet side

Figure 10.14 Illustration of the principle of partial rootzone drying: (*a*) Irrigation is first applied to one side of the root system, maintaining the soil water content close to optimal (lower graphs) while the other side of the root system dries out; when a critical water deficit is reached on the dry side, irrigation is then switched to the other side allowing the first side to dry out (*b*). (Adapted from Stoll, 2000.)

(c)

Soil moisture content

time   time

Soil moisture content

time   time

environments including water plants such as water lilies have a range of morphological and functional adaptations (Bailey-Serres & Voesenek, 2008). Morphological adaptations to a low oxygen (*hypoxic*) environment include the characteristic development of longitudinally connected air spaces (*aerenchyma*) in the root parenchyma that transports oxygen down to submerged tissues; this aerenchyma development, which can reach 55% of the cross-sectional area of roots and submerged rhizomes, is often enhanced in response to flooding. In some groups of plants such as mangroves, root aeration is also facilitated by the development of lenticel-covered roots called pneumatophores that have a large volume of aerenchyma. Another common strategy is for plants maintain surface gas films or plastron layers over leaves (as are also found in many aquatic insects), maintained by hydrophobic surfaces or leaf hairs. These thin but extensive layers facilitate gas diffusion (both of O_2 in the dark and CO_2 in the light) to the submerged leaf or root by enlarging the effective gas-water interface beyond that just at the surface of

stomata or lenticels (Pedersen *et al.*, 2009). The various responses to submergence such as enhanced shoot growth as found in flooding-tolerant species such as rice have been recognised as the *low oxygen escape syndrome* (LOES).

There is some evidence that the limited air supply that can be transported down to roots by diffusion can be supplemented by forced ventilation processes driven by pressure gradients (Colmer, 2003; Dacey, 1980). It has been suggested that the positive tissue pressures required to drive flow can be generated either directly by exposure of stems to wind or as a consequence of pressurisation resulting from gradients in humidity or of temperature differences driving flow from younger to older leaves (see Box 10.1). Both wind-driven suction through a plant by a 'Venturi' effect and positive pressure forcing of air through submerged rhizomes have been documented for aquatic macrophytes such as *Phragmites australis*, while many species use humidity- or temperature-driven pressurisation (Colmer, 2003).

Box 10.1 | Aeration in aquatic plants

There are several potential mechanisms that can lead to the generation of pressure gradients and consequent forced aeration of roots and rhizomes in aquatic plants (*a*) A pressure gradient can be generated by a 'Venturi' effect (the reduction in fluid pressure when flow is enhanced at a constriction) caused by rapid air flow over tall broken culms, as has been reported for *Phragmites australis*. (*b*) Alternatively wind pressure could directly force air into shoots, for example through the stomata, generating a positive pressure. (*c*) Positive pressures can also be generated in leaves or shoots by processes of humidity-driven diffusion and thermal diffusion. These processes generate a pressure difference across any microporous partition with pores up to 0.3 μm (though some pressurisation can occur with pores as large as 1 to 3 μm) as such pores offer significantly more resistance to mass flow than to diffusion of gas molecules. For example for a water lily leaf, where the intercellular spaces are saturated with water vapour and the outside air is drier, the vapour pressure difference ($e_i - e_o$) is balanced by a concentration difference of other atmospheric gases, which then diffuse inwards and pressurise the intercellular spaces as long as internal humidity is maintained by a supply of water to balance any water vapour losses. This process leads to a potential static pressure difference ($\Delta P = e_i - e_o$). This humidity-driven pressurisation can be supplemented by a thermal effect arising from any heating of the inside partition (as for example in sunlight) both through the increased internal humidity resulting from the increases in e_s, and from an analogous process of thermal diffusion otherwise known as thermo-osmosis (see discussion and references in Colmer, 2003).

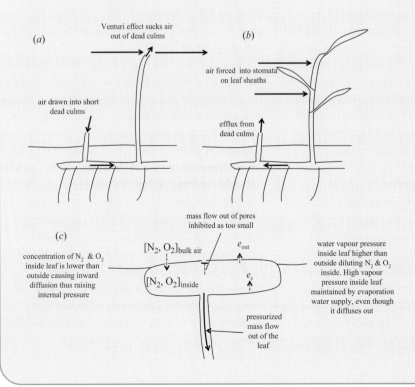

(*a*) Venturi effect sucks air out of dead culms

air drawn into short dead culms

(*b*) air forced into stomata on leaf sheaths

efflux from dead culms

(*c*)

concentration of N_2 & O_2 inside leaf is lower than outside causing inward diffusion thus raising internal pressure

mass flow out of pores inhibited as too small

$[N_2, O_2]_{bulk\ air}$　e_{out}

$[N_2, O_2]_{inside}$　e_s

pressurized mass flow out of the leaf

water vapour pressure inside leaf higher than outside diluting N_2 & O_2 inside. High vapour pressure inside leaf maintained by evaporation water supply, even though it diffuses out

Metabolic adaptations to flooding and anaerobiosis include the shift of metabolism towards the anaerobic (but inefficient) generation of ATP by glycolysis and fermentation of the resulting pyruvate to regenerate NAD^+ while producing ethanol. Though ethanol is toxic, it readily diffuses out of cells, while the alternative fermentation of pyruvate to the less toxic lactate is limited by its effect on cytoplasmic acidification. Immediate responses to flooding or anoxia in mesophytes include the shift of metabolism and substantial decreases in root/stem hydraulic conductance resulting, in some species such as tomato, in the induction of leaf epinasty (downward leaf curvature) and stomatal closure. These both act to partially offset effects of the reduced conductance. Although several hormones have been implicated, the most important component of the long-distance signalling within flooded plants is 1-aminocyclopropane-1-carboxylic acid (ACC), which is transported from the roots in the xylem and oxidised to ethylene (C_2H_4) in the leaves by ACC oxidase, which itself increases in response to flooding (Jackson, 2002). The enhanced export from the roots of ACC results both from an inhibition of ACC oxidase by low O_2 concentrations in the roots and by an increase in ACC synthesis. The ethylene released is directly involved in stimulation of cell expansion in the adaxial petiole cells leading to epinastic curvature and in overall inhibition of leaf growth. Ethylene also appears to be a key component of responses to mechanical wounding whether caused by wind, grazing or insect attack.

10.5.2 Salinity stress

Salinity is a major problem worldwide with more than 800 million hectares being salt affected, both from natural accumulation of salts and from salinisation induced by irrigation under conditions where excess salts are not leached away from the root zone. Salinity affects plants in three ways: first there is the reduction in water potential caused by the osmotic effect of the solutes and the resulting reduced availability of soil water, and then there are the direct toxic effects of the salts such as Na^+, and third the associated nutrient imbalances.

There are substantial differences between and within species in their salinity tolerance, with the NaCl concentrations in the water supply that inhibit growth by 50% ranging from <100 mM in rice to >400 mM in halophytes (salt tolerant plants) such as *Atriplex* species (saltbushes) (Munns & Tester, 2008). Further information on mechanisms of salinity response and tolerance in plants and on the signalling pathways that control the various observed responses may be found in appropriate texts and reviews (Munns, 2002; Munns & Tester, 2008; Pessarakli, 2011; Taiz & Zeiger, 2010; Zhu, 2002). There are three key mechanisms of salinity tolerance.

1. *Tolerance of osmotic stress.* Osmotic adjustment can allow plants to continue to take up water and maintain turgor as salinity increases. This allows plants to continue growing and photosynthesising normally even as the water potential declines, though often with an initial transient inhibition. With increasing salt concentrations, growth, stomatal aperture and photosynthesis are inhibited as with drought.
2. *Exclusion of Na^+.* Prevention of Na^+ uptake by exclusion at the roots ensures that toxic concentrations of salts do not accumulate in the sensitive tissues, especially the leaf blades. This needs to be associated with adaptations that allow plants to maintain turgor in spite of the lowering of the water potential.
3. *Tissue tolerance of high salt concentrations.* High concentrations of Na^+ in cytoplasm above about 30 mM are toxic, so tolerance requires compartmentation of Na^+ and Cl^- into the vacuoles or their disposal by transfer to older leaves as they die. Many halophytes, at least among the dicots, show a characteristic succulent morphology resulting from a salt-induced increase in cell size as a result of the increased vacuolar volume that follows as salts are partitioned there and water flows in. Another common adaptation is the active excretion of salts from the transpiration

stream by specialised glands in the leaves, thus slowing net uptake by the leaf tissues. Compartmentation of salts into vacuoles also requires the accumulation of compatible solutes (e.g. sucrose, proline and glycine-betaine) in the cytoplasm to balance the osmotic potential of the sequestered salts.

10.5.3 Mechanical stress

Mechanical perturbations such as those caused by insect and herbivore damage as well as by wind are among the environmental stresses that plants must tolerate. A number of the characteristic responses shown by plants were documented by Darwin in the nineteenth century (Darwin, 1890). Plants detect and respond rapidly to mechanical stimuli using a complex network of second messenger and hormone pathways leading to the expression of a range of defence-related genes and alterations in growth including decreased shoot expansion coupled with increased radial growth. *Thigmomorphogenesis* (Jaffe, 1973) is the term used to describe the response to touch as might occur in windy environments and during herbivory (see review by Chehab *et al.*, 2009).

Plants growing in windy environments show many characteristic adaptations including dwarfing and the development of reaction wood to help withstand the force of the wind (see Section 11.1 for further discussion of wind and its effects on plants). The commonest thigmomorphogenic responses in response to wind are the slowly developing reductions in shoot elongation and the corresponding increases in radial expansion, though a number of other phenomena such as enhanced abiotic and biotic stress resistance and enhanced senescence are often seen (Biddington, 1986; Braam, 2005). In addition to these irreversible morphogenetic responses to touch, there are a number of very rapid reversible (nastic) responses such as the leaf folding by *Mimosa pudica* in response to a light touch, and the very rapid closure of the Venus fly trap's leaf when its trigger hairs are stimulated. These latter responses involve very rapid changes in turgor pressure resulting from ion loss from motor cells in response to intercellular electrical signals.

Although the actual mechanism of touch sensing is uncertain, membrane stretch receptors may be involved. In any case, early steps in the response include the gating of Ca^{2+} channels, altered action potentials and production of ROS, while almost all the major groups of plant hormones have been implicated, with particular emphasis on ABA, ethylene and jasmonates. Following the initial discovery of touch-inducible genes in the calcium signalling pathway (Braam & Davis, 1990) many further genes have been identified, several of which are calmodulin-related or encode calmodulin (a calcium binding protein), thus implicating Ca^{2+} signalling.

10.5.4 Cross tolerance and common responses

It is now clear that there is much commonality between the signalling pathways and physiological responses involved in plant responses to different abiotic and biotic stresses. In many cases this is not surprising as, for example, drought, freezing and salinity stress all have the common feature of lowering the water potential, while many stresses lead to reduced growth or stomatal closure. It is also commonly found that exposure to one stress not only leads to development of tolerance to further exposure to that stress, but also to unrelated stresses. For example cold hardening also enhances subsequent drought tolerance.

The complex signal transduction cascades involved in stress responses converge at a number of points with many key components that are common to many response pathways (Harrison, 2012; Kacperska, 2004; Pastori & Foyer, 2002). This convergence leads to significant cross-talk between the different stress response pathways. Particularly important features that are common to many stress responses include the involvement of the ABA signalling pathway (e.g. Table 10.1), calcium signalling and ROS production. For example, even mechanical stimulation can enhance drought tolerance, probably through the stimulation of ABA production (Biddington & Dearman, 1985). Many of the critical molecular switches that regulate gene expression in response to stress are also common to many stress response pathways. For example transcription factors in the highly conserved NAC superfamily have been implicated in both the ABA-dependent and ABA-independent pathways involved in response to abiotic stresses such as drought, salinity and cold, and in both the jasmonate and salicylic acid pathways in response to biotic stresses such as pathogens, herbivory and wounding (Puranik *et al.*, 2012). Similarly, intercellular signalling in response to biotic and abiotic stresses often involves transmission of action potentials through the controlled gating of anion channels (Roelfsema *et al.*, 2012).

11

Other environmental factors: wind, altitude, climate change and atmospheric pollutants

Contents

This chapter considers a number of related aspects of the aerial environment that have not been adequately treated elsewhere in this book – these include wind and the effects of altitude, climate change and the 'greenhouse effect' and their implications for plant growth, and the effects of atmospheric pollutants. All these areas bring together principles that have been introduced earlier. Further details of these and other features of the microenvironment of plants are discussed by Geiger (1965), Grace *et al.* (1981), Campbell and Norman (1998), Garratt (1992), and Monteith and Unsworth (2008) while information on the scientific consensus on aspects of climate change may be found in the fourth consensus report of the Intergovernmental Panel on Climate Change (Core-Writing-Team *et al.*, 2008; Solomon *et al.*, 2007).

11.1 Wind

Not only is wind directly involved in heat and mass transfer by forced convection (see Chapter 3), but it is important to plants in many other ways including the dispersal of pollen and seeds and other propagules, and in shaping vegetation, either directly or,

particularly at coastal sites, by means of transported sand or salt (Grace, 1977).

11.1.1 Measurement and variability

Wind is very variable both in direction and velocity. In general wind speeds tend to be greater during the day than at night (Figure 11.1), largely as a result of the convection processes set up by solar heating of the Earth's surface during the day. Standard meteorological estimates of wind speed are obtained at 10 m above the surface where wind speeds only rarely fall below 1 m s^{-1}. For instance, mean annual wind speed is greater than 4.5 m s^{-1} over most of Britain, with the highest values occurring on mountains and near coasts. However, as seen in Chapter 3, the wind speed decreases rapidly as one approaches a surface (e.g. Figure 3.7), so that wind speeds near and within vegetation tend to be much slower than at 10 m. At agrometeorological sites wind speeds are commonly recorded at 2 m above the ground.

The various methods available for measuring wind are described by Grace (1977) and in texts on meteorological instrumentation (e.g. Brock & Richardson, 2001). For meteorological recording the rotating-cup anemometer is widely used, usually in conjunction with a directional vane. Rotating cups tend to be insensitive to low wind velocities as they have a substantial starting inertia, though sensitive types can be used within crop canopies to measure wind speeds as low as 0.15 m s^{-1}. For studies of smaller scale, more rapid velocity fluctuations that are required for eddy covariance studies of heat and mass transfer (see Figure 3.9), the standard instrument is the sonic anemometer. This measures wind speed, based on the time of flight of ultrasonic pulses between pairs of transducers oriented to obtain one-, two- or three-dimensional flow. Hot-wire anemometers provide a cheap alternative for small scale turbulence studies: these are based on the fact that the cooling of a heated fine wire or small thermistor is dependent on wind velocity. A number of other types of anemometer are used for special purposes including windmill-type anemometers and Pitot pressure tubes.

Although mean wind speeds are commonly reported, some processes such as damage are more dependent on maximum gust speeds than on average speeds.

11.1.2 Wind and evaporation

Increases in wind speed decrease the boundary layer resistance (see Chapter 3); this generally causes evaporation rate to increase. If, however, leaf temperature is significantly above air temperature (for example, in conditions of high irradiance with moderate stomatal closure: see Figure 9.1), increasing wind speed can decrease evaporation (Figure 11.2). This is because the increased heat loss lowers leaf temperature and therefore the water vapour pressure in the leaf. The resulting reduction in driving force for evaporation can be greater than the reduction in

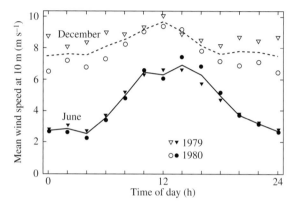

Figure 11.1 Average hourly mean values of wind speed at 10 m for June and December for two years at East Malling, Kent. The mean standard deviation of hourly values is 2.8 m s^{-1} for June and 4.6 m s^{-1} for December.

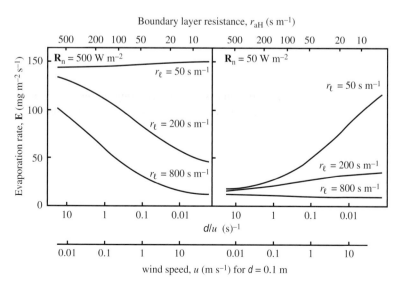

Figure 11.2 Calculated dependence of evaporation rate for a flat leaf (Eq. (5.26) with $D = 1$ kPa) on the boundary layer resistance (r_{aH}) or on the parameter d/u (see Eq. (3.31)) for various r_ℓ and for high or low net radiation ($\mathbf{R}_n$). The equivalent wind speeds for a leaf with characteristic dimension $d = 0.1$ m are also shown.

the transfer resistance. This effect follows from the evaporation equation (Eq. (5.26)) where the boundary layer resistance occurs in both numerator and denominator.

It can be shown by differentiation of Eq. (5.26) with respect to r_a that E is independent of r_a (and hence wind speed) when:

$$r_\ell = \rho_a c_p D[(s/\gamma) + 1)]/(s(R_n - G)) \qquad (11.1)$$

Under such conditions, $T_\ell = T_a$. If r_ℓ is less than the value given by this equation, E increases with increasing wind speed, but when r_ℓ exceeds this critical value, E decreases with increasing wind speed. For a free water surface, where the surface resistance term is zero, E always increases with wind speed (u), roughly in proportion to $\sqrt{u}$. The effect of boundary layer resistance on evaporation rate in a range of conditions is illustrated in Figure 11.2.

High wind speeds can also affect evaporation by indirect effects on either the stomatal or cuticular resistances. There are conflicting reports on the effect of wind on the stomata (Grace, 1977). In many cases it is likely that any closure in response to increased wind speed is simply a feedback or feedforward response (see Chapter 6) to increased evaporation rate. Species tolerant of more exposed habitats may have stomata that are less responsive to wind speed, though it is very difficult to generalise. For example, both *Rhododendron ferrugineum* and *Pinus cembra* occur above the treeline in the European Alps, but the former is found mainly in more sheltered microsites.

Using potted plants in a wind tunnel, it was apparent that *Rhododendron* stomata were the more sensitive to raised wind speed (Caldwell, 1970), though no such differential sensitivity could be detected in the field (A. Cernusca, unpublished data). It may be that the wind tunnel results were atypical for a number of reasons, including the difficulty of growing the plants in pots, problems of natural endogenous rhythms obscuring responses, and differences in leaf size with consequent differences in r_a (Ch. Körner, personal communication). Cuticular resistances, on the other hand, may be decreased by wind. This probably results from damage to the cuticle as a result of leaves flexing in the wind and rubbing against each other.

11.1.3 Dwarfing, deformation and mechanical damage

Plants developing in windy environments show a range of characteristics from stunted growth and deformation to actual breakage. At sites with a predominant prevailing wind, it is a common observation that plants often show a marked asymmetry, with growth being very much reduced to windward (Figure 11.3). There can be several reasons for this including 'wind training', where the branches are oriented downwind as a result of constant wind pressure. The reduced growth on the windward side can be attributed to both physiological and mechanical effects. These include, for example, enhanced desiccation and a more extreme temperature regime for exposed tissues

(a)

(b)

Figure 11.3 Typical effects of wind on shrubs and trees showing both wind-pruning and wind-training. (a) Mixed beech woodland at Gruinart on Islay, and (b) *Sorbus* at Ballyvaughan, County Clare, Ireland.

(particularly in alpine environments), and direct damage or damage by materials such as salt or sand transported by the wind. All of these may preferentially kill buds on the windward side. These effects also lead to a marked tendency for the whole-plant stature to be smaller in more exposed situations. Ecological advantages of dwarf plants in cold environments have already been discussed in Chapter 9. For example, the reduced wind speeds in the vicinity of cushion plants enable the tissues to be much warmer than the air in high irradiance conditions.

Even quite moderate wind velocities can affect growth. Often a major factor is the impaired plant water status that results from the increased evaporation rate frequently associated with increased wind speed (see above). Where stomata are closed, cuticular damage resulting from high wind speeds can also lead to increased water loss. In many cases, however, no decline of ψ_ℓ is observed at elevated wind speeds. In alpine environments the decreased leaf temperature, and hence decreased metabolic rate that results from the more efficient sensible heat transfer at high wind speeds, can be more important than any effects on water status. In arid environments, however, desiccation effects tend to be dominant.

As we saw in Section 10.5.3 the shaking effect of wind can be an important factor reducing growth. For example, Neel and Harris (1971) demonstrated that growth of *Liquidambar* could be reduced to 30% of normal when the plants were 'shaken' for only 30 s each day; while less extreme results have been reported for many other species. Thigmomorphogenic responses, including the production of shorter and sturdier plants and the production of flexure- or compression-wood in trees in response to bending, provide an important means of adaptation to windy environments. In spite of the discovery of touch-stimulated genes the detailed mechanisms of these thigmomorphogenic responses are still unclear, though they do not appear to be consistently related to impaired water status (as might be expected, for instance, if shaking induced cavitation in the xylem vessels).

11.1.4 Lodging

An important problem in high winds is that the forces exerted on the plants can lead to structural failure, either the uprooting of whole plants (particularly common with trees) or else the breaking or buckling of stems. This process of plants being laid flat by the wind is called lodging (or windthrow in trees). In addition to the obvious importance in forestry, lodging can significantly reduce yield in cereals, where the commonest form of lodging is buckling of the stem. The lodging of cereals can decrease the harvestable yield either by effects on photosynthesis (e.g. as a result of either poor light penetration in the compressed canopy or stress caused by the damaged conducting system) or else by making the grain difficult to harvest.

The occurrence of lodging depends on the forces exerted on the plant by wind, rain, etc., on the height from the ground at which they act, and on the strength of the stem (Grace, 1977; Niklas, 1992). In a cereal, for example, the force due to the wind acts primarily on the head of the plant and induces a *torque* (**T**) or turning moment that increases down the stem and causes bending. The torque at a point is the product of the force $\times$ the perpendicular distance of its line of action from that point (see Figure 11.4), so it has units of N m (or J). Thus the total torque at the base of a stem is:

$$\mathbf{T} = \Sigma(\mathbf{F}_i\, h_i) \tag{11.2}$$

where h_i is the height of the i-th part of the plant, and $\mathbf{F}_i$ is the horizontal force due to the wind acting on that part. Once a significant bending occurs, there is an additional turning moment due to gravity $= \Sigma(m_i\, x_i\, g)$ where x_i is the displacement from the vertical of a part of mass m_i and g is the acceleration due to gravity. Therefore the total torque is:

$$\mathbf{T} = \Sigma(\mathbf{F}_i\, h_i + m_i\, x_i\, g) \tag{11.3}$$

The torque induces stresses (S) that cause deformation (strain), in this case stem bending, but this is resisted by the bending-resistance moment of the stem. The maximum bending-resistance moment is called the

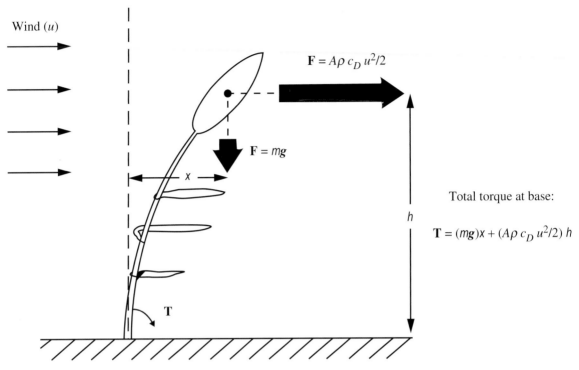

Wind (u)

$F = A\rho\, c_D\, u^2/2$

$F = mg$

x

h

Total torque at base:

$T = (mg)x + (A\rho\, c_D\, u^2/2)\, h$

T

Figure 11.4 Illustration of forces' action on a cereal ear in the wind showing that the total torque (**T**) acting is the sum of the force due to gravity (resulting from any displacement from the vertical) and that due to the force of the wind.

stem strength. The magnitude of the stress for a cylindrical stem is related to the torque by:

$$S = \mathbf{T}d/2I \qquad (11.4)$$

where d is the stem diameter, and I is the *moment of inertia*. I depends on the shape of the stem cross-section, being equal to $\pi d^4/64$ for a solid cylinder and $\pi(d^4 - d_i^4)/64$ for a hollow cylinder with internal diameter d_i. The maximum stress that can be withstood without irreversible deformation occurring is called the elastic limit.

The amount of bending is proportional to S (and hence **T**) and inversely proportional to the flexural rigidity or stem stiffness, given by the product $\varepsilon_Y\, I$, where ε_Y is a measure of linear elasticity of the stem material called Young's modulus (with units of pressure). A large value of $\varepsilon_Y\, I$ represents a stiff stem. A very stiff stem, as in trees, will transfer the torque operating on it to the root system and may thus promote root failure.

The actual force exerted by the wind (**F**) results largely from form drag, which depends on the shape and size of the object (cf. skin friction – see Section 3.2.3). 'Streamlined' objects minimise turbulence and form drag in a flowing fluid (Figure 11.5), while a 'bluff' body such as a cube is subject to greater form drag as a result of the turbulence and also the lowered pressure that occurs in the object's wake. The maximum potential rate of transfer of momentum for a completely bluff body is equal to $0.5\rho u^2$ (Eq. (3.17)). In practice, some of the fluid slips round the sides of the object so that total force applied to the object can be given by:

$$\mathbf{F} = c_f 0.5\rho u^2 A \qquad (11.5)$$

where c_f is a dimensionless form drag coefficient dependent on u, ρ is the air density and A is the area projected in the direction of the wind. For a cylinder, form drag contributes 57% of the total drag at a Reynolds number (see Section 3.3.1) of 10 (e.g. for

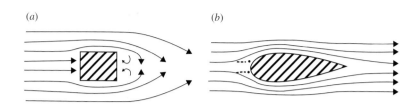

Figure 11.5 Flow around (a) a bluff body and (b) a streamlined body. The dotted lines show the trajectories of dense objects impacting on the streamlined body.

$d = 0.001$ m, $u = 0.15$ m s^{-1}) increasing to 97% at a Reynolds number of 10 000 (e.g. $d = 0.1$ m, $u = 1.5$ m s^{-1}). The drag on a flat leaf oriented parallel to the flow is small and largely a result of skin friction, but when normal to the flow, drag is much greater and form drag predominates. In most cases it is convenient to combine the skin friction and form drag and to define an overall drag coefficient, c_D, which for leaves is typically between 0.03 and 0.6. For spheres and for cylinders normal to the flow, c_D is typically between 0.4 and 1.2 for Reynolds numbers between 10^2 and 10^5 (Monteith & Unsworth, 2008).

For many plants increasing wind speed forces the foliage to align with the streamlines so that the value of the drag coefficient, c_D decreases in a way that is dependent on the rigidity of the foliage. Mayhead (1973) has presented data on drag coefficients for a number of British forest trees: at wind speeds that are liable to cause windthrow (above about 30 m s^{-1}) c_D for a variety of conifers ranged between 0.14 for the very supple Western hemlock (*Tsuga heterophylla*) to 0.36 for the much more rigid Grand fir (*Abies grandis*). At a wind speed of 10 m s^{-1} the corresponding values of c_D for the two species were above 0.3 and 0.8 respectively.

If the airstream contains denser particles such as raindrops, these can increase the force exerted on the object by impacting on it, partly because the denser particles have greater inertia and consequently less tendency to be diverted in the streamlines round the plant (Figure 11.5). It is possible to calculate the relative forces exerted by raindrops and wind on, for example, a cereal ear. The force due to wind is given by Eq. (11.5): assuming a wind velocity of 5 m s^{-1} at 2 m, the average velocity at the level of the ears might be $\simeq 1.5$ m s^{-1} (Eq. (3.36)) and assuming $c_D = 0.4$, this gives, for an ear of 0.01 m^2 cross-sectional area, a

continuous force $\mathbf{F} = 0.4 \times 1.2 \times (1.5)^2 \times 0.01/2 = 5.4 \times 10^{-3}$ N. For a large enough object, the separate impulses of individual raindrops are equivalent to a steady force given by the mass of water impacting each second × velocity. Assuming that the mean horizontal velocity of the drops is 5 m s^{-1} this gives, for a low rate of precipitation of 1 mm h^{-1} ($= 0.28$ g m^{-2} s^{-1}), $\mathbf{F} = $ (mass m^{-2} s^{-1}) $\times u \times A = 0.28 \times 10^{-3} \times 5 \times 0.01$ N $= 1.4 \times 10^{-5}$ N. Even for a high rate of precipitation of 25 mm h^{-1}, $\mathbf{F} = 3.5 \times 10^{-4}$ N, which, surprisingly, is more than an order of magnitude smaller than the force of the wind.

The tendency to lodge will also be enhanced if the frequency of turbulence in the wind corresponds to a natural resonant frequency of the plant, or if the stem is weakened by disease or moisture absorption. Such dynamic responses have been shown to be particularly important for storm damage in forests. Work on the dynamic behaviour of trees has been reviewed by Mayer (1987).

A reduction in the tendency to lodge has been an objective for some years in cereal breeding programmes. The turning moment may be reduced by decreasing the size of the ear, by improving its streamlining or by decreasing its height. The latter possibility has provided a major impetus for breeding the dwarf cereals that are increasingly grown throughout the world. The original problem was that the increased levels of nitrogen fertiliser that were applied to increase yield also increased stem height and therefore the tendency to lodge. Reduction of stem height using dwarfing genes conferred an improved tolerance of higher levels of nitrogen.

An alternative approach to reduction of lodging might be to increase the stem strength, for example by increasing I either by increasing the material in the stem or preferably by increasing the stem diameter (as hollow stems have a higher I for any given mass of material). Increases in the elastic limit could also be advantageous.

11.1.5 Shelter

Shelter belts have been used since prehistoric times to reduce wind damage to crops and livestock. It has been known empirically for many years that erection of windbreaks, whether natural belts of trees or artificial screens of slatted wood or plastic mesh, can have a very important mollifying effect in windy environments, resulting in greatly improved plant growth and yield, particularly where drought is also a problem. As well as having direct effects on crop performance, shelter belts can also improve soil water distribution by collecting snow during the winter, so that the melt water is more evenly distributed than would otherwise be the case.

Micrometeorological effects of shelter

Reviews of some micrometeorological aspects of shelter may be found in McNaughton (1989) and Jensen (1985). Although the detailed effects of a windbreak depend on its orientation and on the turbulence characteristics of the wind, typical airflow patterns behind a thin solid barrier such as a wall can be sketched as in Figure 11.6(a). In this case there is a large recirculating eddy downwind of the barrier. As the porosity of the windbreak increases, however, the recirculating eddies retreat downwind, becoming smaller and more intermittent until they finally vanish when the porosity increases above about 30%. As a result porous windbreaks are more effective than solid windbreaks. Characteristic patterns of variation in relative wind velocity with distance behind the barrier and with porosity are shown in Figure 11.6(b), demonstrating that the effect can extend downwind to about 25 × the height of the barrier.

A triangular region that extends between the barrier and a line to the ground at about $8h$ behind the barrier (where h is the height of the barrier) is termed the 'quiet zone'; this remains similar in shape as barrier porosity increases. In this region, not only is

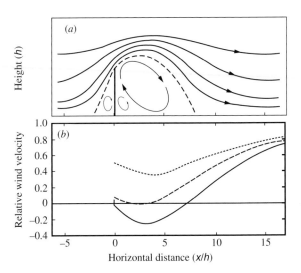

Figure 11.6 (a) Typical wind-flow patterns (expanded vertical scale) caused by a windbreak of low porosity and height (h) $\simeq 100\ z_o$ standing normal to the wind flow, showing the recirculating eddy immediately behind the barrier. The horizontal distance is represented as multiples of the barrier height and the broken line indicates the separation streamline. (b) Corresponding horizontal profiles of mean wind velocity at a height equal to $0.25h$ with barriers having porosities equal to zero (———), 0.3 (------) or 0.5 (.............) (McNaughton, 1989).

mean wind speed reduced but so is the size of the turbulent eddies. Behind and above this region lies a wake zone where turbulence is enhanced. Transport processes therefore tend to be depressed in the quiet zone (where the crop becomes relatively decoupled from the overhead airstream), and enhanced in the wake.

The reduced rates of turbulent transfer immediately behind shelter tend to lead to increased air and surface temperatures during the day, as incoming radiation may be less easily dissipated than in the open field. At night, shelter may have the opposite effect on temperature, partly by preventing the break up of inversions near the ground. It is difficult to generalise concerning the effects of shelter on air humidity and canopy evaporation because of the complexities of the energy balance (see Chapter 5), but it is usually observed that the increased canopy boundary layer resistance in the quiet zone behind the shelter reduces the rate of removal of water vapour, thus leading to a build up of humidity and a concomitant decrease in crop evaporation. As a result of this reduced evaporation rate, leaf water potentials are commonly higher than in exposed sites.

Radiation is little affected by north–south oriented shelter belts, since any shading in the morning or evening is partially compensated for by increased reflection off the shelter belt. An east–west oriented shelter belt, however, tends to reduce net radiation receipt to its north (in the northern hemisphere), the distance of influence depending on height as well as on solar elevation (itself a function of latitude and time of year).

Effects of shelter on plants

In addition to shading effects, shelter belts can be harmful in that they use significant water and nutrients. For this reason, species like alder, that have nitrogen-fixing capability, are often planted as shelter species. Another problem is that shelter belts may harbour both pests (including birds as well as insects) and diseases. In general, however, crop yields are usually markedly improved by shelter. Grace (1977) reviewed more than 95 experiments conducted at many sites with a wide range of species. In these experiments the average yield benefit of shelter was 23%, with fewer than a quarter of the cases giving less than 10% benefit.

The improved yields usually result from raised temperatures or from better water status that follows from decreased potential evaporation. This can lead to higher stomatal conductances and more rapid photosynthesis. Another beneficial effect of shelter is protection of soils from wind erosion. Further discussion of shelter effects may be found in texts by Grace (1977) and Rosenberg *et al.* (1983).

11.1.6 Dispersal of pollen, seeds and other propagules

Many species have evolved to take advantage of the wind to aid cross-fertilisation and dispersal (Daubenmire, 1974). *Anemophily* (wind pollination) is widespread (e.g. in the Gramineae and in conifers), particularly among species of cool and cold climates. Pollination efficiencies of different types of floral structure in relation to patterns of wind flow can be studied by means of scale models in a wind tunnel. In using scale models, it is necessary to adjust the wind speed to maintain the value of the Reynolds number ($= ud/v$ – Chapter 3) equal to that appropriate for the original floral structure in its natural environment (i.e. wind flow must decrease as dimensions increase).

There are several types of wind dispersal (*anemochory*). These include having minute disseminules that can be easily blown long distances (e.g. the minute seeds of the Orchidaceae, as well as the spores of fungi and bryophytes). Larger seeds may have hairy parachutes (as in the Asteraceae) or wings (as in many tree species). In addition, some species (e.g. Chenopodiaceae) have papery fruits that readily roll along the ground, while others, such as tumble weed, have vegetative propagules that roll in the wind.

11.2 Altitude

Altitudinal variation of climate causes many visible changes in the composition of vegetation and in the growth habit of individual plants. Plants growing at high elevations often exhibit many characteristic morphological and physiological features including dwarf, compact habit and small, narrow or densely pubescent leaves (Daubenmire, 1974; Körner, 1999; Larcher, 1995). In addition to altitude, local topography plays a major role in determining the microclimate at any site in the mountains. For example aspect and slope affect the shortwave irradiance (and hence soil temperatures – see e.g. Figure 2.11), while topography also affects the pattern of wind flow and the degree of shelter and, on a large scale, can even have marked effects on precipitation (high to the windward and low to the leeward of a mountain range). In this section, however, the effects of altitude will be considered only in relation to level sites. Further information on mountain climates and microclimates may be found in the classic text by Geiger (1965) and in Barry (2008) while useful discussion of their impacts on plant life may be found in the excellent book on alpine plant life by Körner (1999).

Table 11.1 Estimated average values of different climatic factors at 600 m and 2600 m above sea level in the Central European Alps (from data compiled by Körner & Mayr, 1981). Summer refers to June, July and August.

Height above sea level		600 m	2600 m
Atmospheric pressure	$(10^4$ Pa)	9.46	7.40
Average wind speed (summer)	(m s^{-1})	1	4
Average water vapour pressure (summer)	$(10^2$ Pa)	14.7	6.9
Maximum water vapour pressure deficit (July)	$(10^2$ Pa)	~20	~8
Annual number of days with fog		0–10	80
Annual sum of precipitation (±30%)	(mm)	900	1800
Average air temperature (July)	(°C)	18	5
Annual average air temperature	(°C)	8	−3
Relative global radiation on clear days (summer)	(%)	100	120
Relative global radiation on overcast days (summer)	(%)	100	260
Number of clear days (summer)		10	5
Number of sunshine hours (July)		200	160
Evaporation from low vegetation (July)[a]	(mm day^{-1})	4–6	3–4
Number of days with snow cover		80	280

[a] Ch. Körner, personal communication.

Some typical climatic data for different altitudes are summarised in Table 11.1. The decreasing pressure and increasing wind speed with increasing altitude are probably the most fundamental physical effects of altitude; most other effects follow from these.

11.2.1 Pressure

The atmospheric pressure over the altitudinal range of botanical interest may be derived. For a thin layer of atmosphere of thickness dz, the downward pressure (force per unit area) due to this layer alone (dP) is given by the mass per unit area multiplied by the acceleration due to gravity (g). Since mass per unit area equals the mass per unit volume (the density, ρ) multiplied by the thickness of the layer:

$$dP = \rho\,dz\,g \tag{11.6}$$

substituting from Eq. (3.6) this gives:

$$dP = (PM_A/\mathscr{R}T)dz\,g \tag{11.7}$$

Integrating this from $z = 0$ (sea level), where $P = P°$, to any altitude z, gives the pressure at that altitude (P_z) as:

$$P_z \simeq P°\exp(-M_A\,g\,z/\mathscr{R}T) \tag{11.8}$$

where T is the mean temperature over the altitude range. The observed mean pressures at 600 and 2600 m in the Central Alps (Table 11.1) are close to the values predicted by Eq. (11.8), though note that actual values also depend on factors such as air humidity.

11.2.2 Temperature

As a result of the reduction of atmospheric pressure with height, there is a marked tendency for temperature to decrease. The reason for this is that as a parcel of air rises, it tends to expand as the pressure decreases. This expansion requires the performance of work. If there is no heat exchange with the environment, i.e. the system is *adiabatic*, the energy for this expansion is extracted from the air itself, thus lowering its temperature. The rate at which air temperature changes with altitude is called the lapse rate. For dry air this theoretical temperature gradient, the *dry adiabatic lapse rate*, is approximately equal to $0.01°C\ m^{-1}$. The adiabatic lapse rate for wet air, saturated with water vapour, is much smaller, ranging from 0.003 to $0.007°C\ m^{-1}$, depending on air temperature. The smaller lapse rate in wet air arises because condensation liberates some heat that partially compensates for the air expansion.

In natural situations, temperatures decrease on average by about $0.006°C\ m^{-1}$ (e.g. Table 11.1), though quite large deviations do occur. Even negative lapse rates may be found on occasions in valleys in winter when temperature inversions occur. The general temperature reduction with altitude means that cold tolerance becomes an increasingly important factor determining plant distribution at high elevations (e.g. Körner, 1999). The general temperature reduction with altitude also affects the growing season; for example, in Britain the decrease can be 12 to 15 days per 100 m rise, or approximately 5% per 100 m rise.

It is important to recognise that plant surface temperatures can diverge substantially from air temperatures recorded at local meteorological stations (2 m above the ground). Dwarf plants and cushion plants particularly are insulated from the prevailing wind by the surface boundary layer and Körner (1999) has summarised a number studies showing that maximum leaf temperatures for a range of species in alpine systems in the arctic and the alps, respectively, averaged as much as 17.4°C and 24.3°C above air temperature. Such temperature excesses have important implications not only for evaporation, but for photosynthesis and plant growth.

11.2.3 Partial pressure and molecular diffusion coefficients

As the total air pressure decreases with altitude, the partial pressures of its component gases (including N_2, CO_2 and O_2) decrease in proportion. On average, water vapour partial pressure decreases rather more rapidly because of the lowered water-holding capacity of air and the resulting condensation as air temperature falls. These changes all affect the driving forces for diffusion to and from plant leaves. Similarly, the changes of pressure and temperature with altitude affect diffusion coefficients according to Eq. (3.18). Although the effects of temperature and pressure are in opposite directions, the pressure effect is dominant so that the diffusion coefficient, D, increases with altitude.

11.2.4 Relative humidity and precipitation

Although absolute water vapour pressure decreases with altitude, relative humidity tends to increase with increasing altitude as a result of adiabatic cooling. In such cases the vapour pressure deficit tends to decrease. This results in a general tendency for precipitation (and cloudiness) to increase with altitude. Detailed trends depend, however, on latitude and local topography (particularly in relation to prevailing wind direction) so there is no unique relation between altitude and precipitation (see Figure 11.7). It is worth noting that although the relative humidity may even decrease with altitude in semi-arid mountain regions, this cannot be taken to imply that evaporation increases because the absolute vapour pressure deficit still decreases in such cases as the temperature falls. In more humid areas where the air vapour pressure deficit is small towards mountain summits, this does not necessarily imply that there is no evaporation because the leaves can heat up substantially under high insolation leading to significant leaf–air vapour pressure differences.

Table 11.2 Daily totals of diffuse and total shortwave radiation on a horizontal surface in June at three altitudes in the European alps (data from Dirmhirn, 1964).

Altitude (m)	Clear sky			Overcast
	Diffuse (MJ m^{-2} day^{-1})	Total (MJ m^{-2} day^{-1})	Diffuse/total (%)	Diffuse (= total) (MJ m^{-2} day^{-1})
500	4.2	28.9	14.5	6.5
1500	3.3	32.6	10.0	10.3
3000	2.6	34.9	7.4	16.9

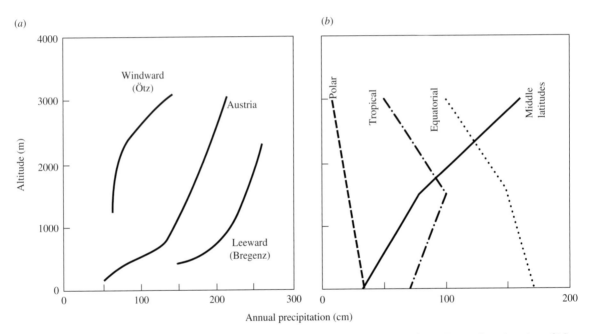

Figure 11.7 Altitudinal profile of mean annual precipitation (*a*) in the Austrian alps, and (*b*) in different climatic regions. (After Lauscher, 1976.)

11.2.5 Radiation

On clear days, the incoming solar radiation increases with altitude as the attenuating air mass decreases (Table 11.2; see also Chapter 2). Remembering that the air mass at any altitude is proportional to P/P^o, the theoretical altitude dependence of $\mathbf{I}_{S(dir)}$ on a clear day (ignoring changes in atmospheric water vapour with altitude) can be determined using Eqs. (11.8) and (2.11) (but see Gates, 1980 for details).

Prediction of altitudinal variation in total (global) shortwave radiation ($\mathbf{I}_S$) is complicated by the variation in cloudiness with altitude (see entries for fog, overcast days and sunshine hours in Table 11.1). In many regions cloud cover and fog tend to increase along with the altitudinal increases in relative humidity. The effects

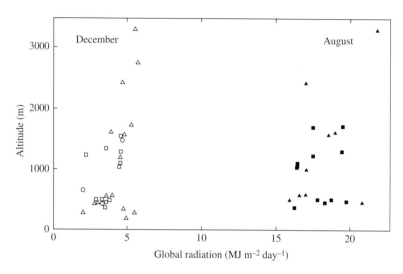

Figure 11.8 Altitudinal variation of mean daily global radiation for August and December 1980 on a horizontal surface for different sites: ○ central Switzerland; △▲ eastern Switzerland; □ ■ western Switzerland. (Data from Valko, 1984.)

of this are typified by the report (Yoshino, 1975) that sunshine duration in the Japanese mountains fell from around 2200 h yr^{-1} at heights up to 500 m to about 1300 h yr^{-1} at 1500 m, though on the highest peaks, sunshine duration increased (and fog decreased) with further increases in elevation. The increased cloudiness and smaller air mass at high elevations tend to balance out in the long term so that global radiation receipts may be relatively insensitive to altitude (Figure 11.8), though with a tendency to increase slightly with altitude (Körner, 1999). With clear skies, diffuse radiation decreases with increasing altitude, but conversely in overcast conditions diffuse radiation (and global radiation) increases with altitude (Table 11.2).

Ultraviolet

Radiation increases with altitude are particularly marked in the UV (Caldwell, 1981): for example Dirmhirn (1964) quotes a 4.8-fold increase of UV-B wavelengths (280 to 315 nm) in the solar beam on rising from 200 m to 3000 m in winter. In this context it is also worth noting that depletion of the stratospheric ozone layer as a result of atmospheric pollutants such as chlorofluorocarbons is also likely to be biologically important (Anon, 1989), so both effects will be considered together.

Even quite small changes in UV-B can have quite large effects on plant functioning (Anon, 1989;

Jenkins, 2009), with effects such as stomatal closure and the inhibition of growth and photosynthesis often, but not always, being observed. Assessment of the biological effects of UV-B is not simple; not only is there much variability between species and varieties in their sensitivity and their ability to acclimate (for example, by the synthesis of screening pigments such as flavonoids), but there is also great variability of natural UV-B irradiance with location and time of year, and the unnatural spectral distribution of UV-B from the supplementary lights often used (which results in the need to use an appropriate weighting function) causes further problems (see Section 8.4.6). Physiological effects of additional UV-B are often apparent for increases in UV-B irradiance of only 50 to 100 mW m^{-2} (equal to an additional daily dose of between about 1500 and 5000 J m^{-2} and simulating a 15 to 25% depletion of the ozone layer). The effects on photosynthesis have been intensively studied, and it is clear that exposure to UV can lead to rapid (50% within 3 h) downregulation of genes for a number of important photosynthetic proteins (Jordan, 2011).

11.2.6 Effects on plants

Gas exchange

The changes of photosynthesis and transpiration with altitude are difficult to predict, partly because of the complexity of the physical changes and partly because of biological responses to these changes. In this context it is necessary to distinguish physical effects of altitude (such as the effect of P on diffusion coefficients) from physiological and anatomical changes (such as alterations in stomatal dimensions and aperture).

The advantages of molar units for conductance are particularly apparent in studies of altitude effects on plant gas exchange. If one uses standard units (mm s^{-1}) for diffusive conductance (see Chapter 3) where $g_\ell = D/\ell$, and substitutes for D from Eq. (3.18), it follows that g_ℓ would tend to increase with altitude (as pressure decreases) even with constant stomatal dimensions. Taking as an example the July means in Table 11.1, it is apparent that g_ℓ would increase by 18% between 600 and 2600 m (i.e. $100 \times (9.46/7.40) \times (278/291)^{1.75}$) as a result of these purely physical effects. With molar units (mol m^{-2} s^{-1}), on the other hand, where $g_\ell = PD/\ell\mathscr{R}T$, the direct pressure dependence cancels the pressure dependence of D, so that the net effect of moving between 600 and 2600 m is a decrease in g_ℓ of only about 3% ($100 \times (278/291)^{0.75}$). The use of molar units does not, however, eliminate the need to consider the purely physical effects of altitude, because the fact that the boundary layer conductance, g_a, is proportional to $D^{2/3}$ means that the effect of pressure on g_a differs from that on g_ℓ. Making the appropriate substitutions it can be shown that, at constant wind speed, g_a would increase by c. 11% over the same attitudinal range, while the corresponding change in the molar conductance g_a would be a decrease of c. 9%. In addition to these effects, the altered driving forces with changed temperature and pressure and the altered g_a as a result of the wind speed gradient must also be considered.

Applying these results to evaporation, the fairly small physical effects on diffusion conductances are found to be much less important than the larger effects due to changes in stomatal dimensions and frequency, the increase in g_a resulting from increased wind speed, and especially the tendency for vapour pressure deficit to decrease with increasing altitude. The typical summer vapour pressures in Table 11.1,

for example, correspond to a decreased water vapour pressure deficit, D, of approximately 70% from 0.59 kPa at 600 m to 0.18 kPa at 2600 m, with the corresponding water vapour mole fraction deficit ($D_{rw} = x_{Ws} - x_{Wa}$) decreasing by about 61%. Frequently this effect is dominant and tends to decrease potential evaporation with altitude (Barry, 2008). A further factor to consider is radiation, as this is also an important determinant of $\mathbf{E}$ (Eq. (5.26)). Although $\mathbf{I}_S$ increases strongly with height in clear weather, the mean value is less sensitive to altitude. There is also evidence that, even allowing for the physical effects discussed above, stomatal conductance tends to increase with altitude (Körner & Mayr, 1981). This may result partly from the stomatal humidity response, where stomata tend to open more with the smaller humidity deficits at the higher altitude, though high-altitude species also tend to have a greater stomatal frequency, particularly on the upper leaf surface (Körner, 1999; Körner & Mayr, 1981). Smith and Geller (1979) have incorporated many of these factors into a model for predicting effects of altitudinal variation on $\mathbf{E}$.

An altitude correction for potential evaporation that has been found to be appropriate for Scotland is that monthly potential evaporation decreases by approximately 21 mm (100 m)$^{-1}$ in summer and 8 mm (100 m)$^{-1}$ in winter (Smith, 1967). The corresponding figures for England vary rather less at 17 and 12 mm (100 m)$^{-1}$ respectively. In spite of the tendency for stomatal density and maximum leaf diffusive conductance to increase with altitude, Körner (1999) reported that actual evapotranspiration on snow-free days was independent of altitude across a range of alpine grassland sites near Innsbruck (Austria) and averaged approximately 2.3 mm d^{-1} for sites between 580 m and 2530 m. In spite of this the seasonal values ranged from nearly 700 mm year^{-1} at lower altitudes to 210 to 250 mm year^{-1} at the upper grassland limits as a result of differing growing season lengths. On sunny days evapotranspiration rates typically reached 4.5 to 5 mm d^{-1} (with slightly lower values at the highest altitudes), though actual values were very dependent on the local vegetation types and exposure.

The effects of altitude on photosynthesis are just as complex as those on evaporation, with the details in any region depending on the actual lapse rates and radiation profiles so that no general predictions can be made. In general, alpine plants tend to have thicker leaves (lower specific leaf areas) than lowland plants; this can lead to lower photosynthetic rates per unit area and lower relative growth rates, when grown at similar temperatures (Atkin *et al.*, 1996). Another point to note is that the partial pressure of CO_2 consistently decreases with altitude, thus also tending to decrease photosynthesis, though temperature and radiation changes are again dominant. Although there can be quite large differences between plant species in their respiration rate, in their Q_{10} for respiration, and in their degree to which respiration acclimates to changing temperatures, there seems to be rather little evidence for consistent differences between arctic or alpine and lowland species in their respiration rates when measured at the same temperatures (Atkin & Tjoelker, 2003). It is also worth noting that alpine plants tend to have lower relative growth rates than more temperate species.

Plant form

The tendency for evaporation rate to decrease with altitude, together with evidence that the importance of leaf water potential as a rate limiting factor decreases with altitude (e.g. Körner & Mayr, 1981),

suggest that the xeromorphy commonly found in plants at high altitudes is not primarily related to the aerial environment, and may even be unrelated to water stress. For this reason the term *schleromorphy* (from the Greek word for hard) is perhaps a better term to describe the characteristic plant type. Although the dwarfing at high altitude has an environmental component, it is primarily a genetically determined adaptation to prevailing conditions, including high wind speeds, low temperatures, etc.

Treelines

The transition between trees and dwarf shrub vegetation in mountains is characteristically rather sudden with a narrow transition zone, within which many species show characteristically stunted and gnarled growth (*Krummholz*). The close relationship of the treeline to the climatically determined snow line at different latitudes argues for a climatic control of the treeline (Figure 11.9), though there is substantial variation within any region. Low temperature extremes do not appear to be a factor determining treelines, rather treelines depend on temperatures during the growing season when growth is determined. Although it has often been suggested that the level of the true climatically determined treeline (Figure 11.9) corresponds with that where the temperature of the warmest month is less than about 10°C, Körner (2012) has argued that across latitudes

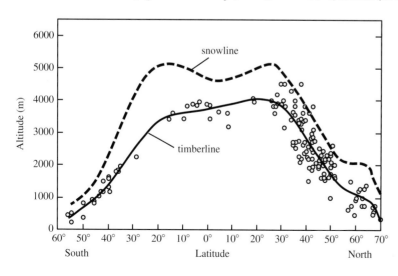

Figure 11.9 The latitudinal distribution of timberline and snowline, where the lines are fitted by eye to the data points; for clarity these points have been omitted for the snow line (redrawn from Körner, 1999).

the best approximation is to an average growing season temperature of close to 6.5°C (varying from 5.5°C in subtropical biomes to 7.4°C in warm temperate climates). The actual length of the growing season at the treeline is, in contrast, rather variable across the globe. The close correlation of treeline location with summer temperature suggests that an important factor may be the increased potential for raising tissue temperature above that of the air as the height of the vegetation, and hence the boundary layer conductance, decreases. The sharpness of treelines may be enhanced by frost killing of unhardened foliage or by frost drought (where trees cannot restrict transpiration losses to a level that can be replaced by water uptake from still-frozen soil, especially in late winter: Grace, 1989). Körner (2012) provides a detailed discussion of the factors determining treeline location.

11.3 Climate change and the 'greenhouse effect'

There is currently much concern about the reality and possible consequences of 'global warming' as a result of the build up of CO_2 and other so-called greenhouse gases in the Earth's atmosphere. There is now unequivocal evidence that global climates have been warming increasingly rapidly since pre-industrial times with the rate averaging 0.074°C/decade between 1906 and 2005, and 0.13°C/decade between 1956 and 2005 (Core-Writing-Team et al., 2008). Climate change is an emotive topic that has led to much misleading information being promulgated so it is important to assess critically any sources used and evidence for any statements made. A very useful summary of current scientific thinking on the topic is provided by Houghton (2009) while the publications from the Intergovernmental Panel for Climate Change, and especially their fourth assessment report, provide a useful source that outlines the current scientific consensus (www.ipcc.ch). It is clear from these and other sources that although there is still much uncertainty around the science of climate change prediction and the quantification of

anthropogenic effects, human activities and especially the production of *greenhouse gases* are contributing substantially to climate warming. Because there is so much information available elsewhere, here we only outline briefly some of the main considerations relating directly to plants and their interaction with the environment.

The main components of the energy balance of the Earth–atmosphere system are illustrated in Figure 11.10, which shows the situation at equilibrium when the solar radiation absorbed by the Earth–atmosphere system equals the outgoing thermal radiation. The term *greenhouse effect* is used to describe the partial trapping of the thermal radiation by the lower 10 to 15 km of the atmosphere (the troposphere). This is a major factor in making life on Earth possible as it increases the surface temperature by about 34°C above what would occur in the absence of an atmosphere. The tropospheric temperature increases towards the Earth's surface so that thermal radiation emission from atmospheric gases decreases with height (as a function of temperature according to Eq.(2.4)). This results in the upward thermal radiation from the top of the atmosphere being less than that emitted from the Earth's surface. At the actual average surface temperature ($T_s \simeq 288$ K) the thermal radiation emitted from the Earth is c. 390 W m^{-2}, but the actual longwave radiative flux from the top of the atmosphere is only c. 235 W m^{-2} (Figure 11.10), implying a trapping of 155 W m^{-2}. (The surface temperature of the Earth that would occur in the absence of an atmosphere can be determined by substituting the value of the radiative loss (equal to the solar radiation absorbed) of 235 W m^{-2} into Eq. (2.4): this gives a surface temperature of only c. 254 K.) Even an increase in trapping of 1 W m^{-2} would significantly increase the equilibrium surface temperature.

Any changes in the amount of energy trapped lead to climate change, with greater trapping leading to equilibrium increases in surface temperature. Alterations in the amount of energy trapped in the Earth–atmosphere system can result from a range of climate drivers that include (i) changes in the solar

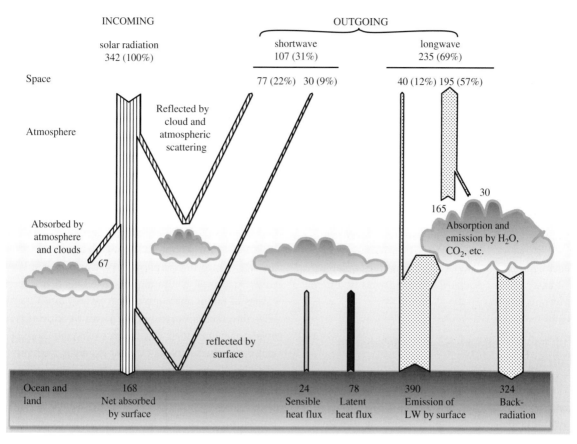

Figure 11.10 Schematic diagram of the global average of the various energy fluxes within the climate system. Incident shortwave is illustrated on the left-hand side of the diagram, while longwave (IR) fluxes are illustrated on the right-hand side. In addition energy transfer from the Earth's surface to the atmosphere by sensible heat (C) and latent heat (λE) are indicated at the centre. All fluxes are expressed in W m^{-2}. The system is at equilibrium with total incoming shortwave radiation (342 W m^{-2}) balanced by the outgoing, while 492 W m^{-2} are absorbed and emitted by the Earth's surface. (Data from Le Treut *et al.*, 2007.)

radiation input; (ii) changes in the amount of solar radiation reflected back to space (for example as a result of changes in cloudiness or surface albedo); (iii) changes in aerosol concentrations (for example sulphates and carbon particulates) that can affect both shortwave and longwave radiation fluxes; and (iv) changes in the absorption of thermal radiation by greenhouse gases in the greenhouse effect. It is convenient to define changes in energy trapping in terms of changes in the energy balance at the tropopause, so that the impact of different gases or other influences on climate change (such as volcanic

emissions or changes in solar input) are conveniently expressed in terms of their *radiative forcing* (RF), which is defined (with some *caveats*) as the change in net irradiance at the tropopause that results from the change. The value of RF is linearly related to the resulting global mean equilibrium temperature change (ΔT_s) once the new equilibrium atmospheric temperature profile is achieved, by a climate sensitivity parameter (λ).

It is important to distinguish the direct effects of radiative forcing from the net effects after the various biogeochemical feedbacks and adjustments that occur (see below). The direct effect of the different climate drivers such as changes in greenhouse gas concentrations, aerosols and solar irradiance (expressed as radiative forcing) provides a useful means for comparing the effects of different drivers. When discussing greenhouse gases it is necessary to note that not only do different gases have different efficiencies in absorbing thermal radiation, but that emission of a given amount today will have different long-term climatic effects as a result of their very different lifetimes in the atmosphere. This overall impact has been encapsulated in what is known as the *global warming potential* (GWP). The GWP compares the integrated radiative forcing over a specified period (e.g. 100 years) from a unit mass pulse emission and is a way of comparing the potential climate change associated with emissions of different greenhouse gases.

11.3.1 Greenhouse gases and radiative forcing

The absorption of outgoing longwave radiation is largely by the greenhouse gases, of which water vapour is the main natural absorber; this absorber is the main contributor to the present warming of the Earth's surface above what is expected without an atmosphere. Notwithstanding the importance of water vapour, however, the main anthropogenic effects on global surface temperatures are determined by changes in the atmospheric concentrations of CO_2 and other greenhouse gases because there is little direct effect of human activity on atmospheric water vapour (though it plays an important role in some of the climatic feedbacks). Although increasing atmospheric CO_2 is the major contributor to warming, the combined effect of a number of trace gases, principally methane, nitrous oxide and the chlorofluorocarbons (CFCs) and hydrofluorocarbons (HFCs), though present at concentrations that are two to six orders of magnitude lower than CO_2, can rival the effect of CO_2 because, per molecule, they absorb infrared radiation much more strongly, particularly in the critical 8 to 12 μm waveband. Details of global

and regional emissions of various pollutants are available from the Emission Database for Global Atmospheric Research (EDGAR, see http://edgar.jrc. ec.europa.eu).

Values of the estimated atmospheric concentrations of different long-lived greenhouse gases in 1750 and 2005, the molar efficiencies with which they absorb thermal radiation (radiative efficiency, expressed as $W m^{-2} ppb^{-1}$), their respective total radiative forcing between 1750 and 2005 ($W m^{-2}$), their lifetimes in the atmosphere (y) and their global warming potentials are tabulated in Table 11.3. Note that radiation absorption by trace gases (such as CFCs) does not necessarily follow the logarithmic relation expected from Beer's law (Chapter 2), being approximately linearly related to concentration because of their low concentration. It is also important to note that the effects of minor gases are not necessarily directly related to their absorption coefficients because of overlap between their absorption bands and those of major absorbers such as CO_2: for example, methane and nitrous oxide can lose about half of their potential trapping by such overlap.

Note that the total greenhouse gas forcing of around 2.63 $W m^{-2}$ is significantly offset by other anthropogenic changes including emission of aerosols such as sulphate, organic carbon and particulates from burning both fossil fuels and biomass that act to mitigate climate change. The direct effects of aerosols are primarily to affect shortwave radiation exchanges with the direction of the effect depending on whether they primarily scatter or absorb the incoming shortwave. Their indirect effects are primarily on the amount and radiative properties (and hence their albedo) of clouds as well as on their lifetimes. Table 11.4 also shows values for the radiative forcing that can be attributable to other anthropogenic and natural processes such as changes in solar radiation. Volcanic eruptions are intermittent events that have had major impacts on paleoclimates, while in recent years the eruption of Mount Pinatubo in 1991 was readily seen as an abrupt decrease in the rate of increase of atmospheric CO_2. This has been attributed to enhanced photosynthetic CO_2 uptake resulting

Table 11.3 Examples of the radiative efficiency, radiative forcing estimated (forcing between 1750 and 2005), lifetime and global warming potential for a 100-year timescale (GWP) for some sample greenhouse gases. For further details including the magnitudes of potential errors in the various terms see Solomon *et al.* (2007).

	Concentration (mol/mol[c])		Radiative efficiency (W m^{-2} ppb^{-1})	Radiative forcing (W m^{-2})	Lifetime[a] (y)	GWP[b]
	1750	2005				
CO_2	275 ppm	379 ± 0.65 ppm	1.4×10^{-5}	1.66	2 (−100[d])	1
CH_4	700 ppb	1774 ± 1.8 ppb	3.7×10^{-4}	0.48	12	25
N_2O	275 ppb	319 ± 0.12 ppb	3.03×10^{-3}	0.16	114	295
CFC-12 (CCl_2F_2)	0	538 ± 0.18 ppt	0.32	0.17	100	10 900
CCl_4		93 ± 0.17 ppt	0.13	0.012	26	1400
CF_4		74 ± 1.6 ppt	0.10	0.0034	50 000	7390
SF_6		5.6 ± 0.038 ppt	0.52	0.0029	3200	22 800

[a] Lifetime is defined as the reciprocal of fraction that is removed each year.
[b] Relative to CO_2.
[c] Concentrations expressed in volume terms as parts per million (vpm), parts per billion (ppb) or parts per trillion (ppt).
[d] CO_2 does not have a specific lifetime because its removal involves a number of exchanges between atmosphere, oceans and vegetation with different time constants ranging from rapid exchanges to slow redistribution.

from an increased diffuse radiation in clear days, which arose from enhanced scattering by the emitted aerosols compensating for a decrease in the direct beam radiation (Gu *et al.*, 2003).

11.3.2 Carbon cycle and carbon dioxide

Over the last 600 000 years or so the atmospheric concentration of CO_2 has oscillated between about 200 vpm and 280 vpm, with the value during the interglacial warm periods (including the 2000 years or so before AD 1800) being fairly stable at around 280 ± 15 vpm. It is only in the past 200 years that the concentration has increased outside this range to the present value of about 395 vpm, though with the marked seasonal variation attributed to

variations in photosynthesis as has been seen in the well-known Mauna Loa records (www.esrl.noaa.gov/gmd/ccgg/trends). The present rate of increase (c. 2 vpm per year) is unprecedented. Similarly, ice-core data suggest that methane concentrations oscillated between the pre-industrial value of 400 ppb and 700 ppb for the previous 600 000 years, while N_2O concentrations oscillated between about 220 and 275 ppb over the same period (Solomon *et al.*, 2007).

Carbon dioxide
Estimates of the amounts of carbon in different reservoirs and of the fluxes between them, are given in Table 11.5. The major non-natural sources of CO_2 are the burning of fossil fuels and the manufacture of

Table 11.4 Summary of radiative forcing (RF) for different climate change drivers (data from Solomon et al., 2007).

	RF (W m^{-2})	5–95% uncertainty range (W m^{-2})
Anthropogenic forcings		
Long-lived greenhouse gases	2.63	±0.26
CO_2	1.66	1.49 to 1.83
CH_4	0.48	0.43 to 0.53
N_2O	0.16	0.14 to 0.18
Halocarbons	0.34	0.31 to 0.37
Stratospheric ozone	−0.05	±0.1
Tropospheric ozone	0.35	−0.1 to 0.3
Surface albedo	−0.1	−0.4 to 0.2
Total aerosol		
Direct effect	−0.5	−0.9 to −0.1
Cloud albedo	−0.7	−1.1 to 0.4
Natural forcing		
Solar irradiance	0.12	0.06 to 0.18

cement. In the 1990s these processes released about 6.4 Gt yr^{-1} to the atmosphere, though the current rate may be as much as 7.2 Gt yr^{-1}. To this must be added the CO_2 release resulting from land use change, especially that resulting from tropical deforestation, giving a total anthropogenic emission of about 8.0 Gt yr^{-1} in the 1990s. Of this emission, about 3.2 Gt yr^{-1} is stored in the atmosphere leading to the observed rate of increase of atmospheric CO_2 concentration, which in the early 2000s reached 4.1 Gt yr^{-1}. A significant portion of the remainder (2.2 Gt yr^{-1}) is stored in the oceans, leaving a residual of about 2.6 Gt yr^{-1}, which is taken up by terrestrial ecosystems in the growth of replacement vegetation on cleared land, as a result of various land management practices and as a result of fertilising effects of the increased atmospheric CO_2 and N deposition. Note that the individual photosynthetic and respiratory fluxes are substantially larger than the anthropogenic changes, and nearly balance in the oceans and on land. Although the general trends are clear, it is difficult to predict accurately future changes in atmospheric CO_2, because the rate of uptake by the oceans, for example, is not well understood.

The role of forests in the global carbon balance is somewhat controversial. Although burning and clearance of tropical forests releases large quantities of CO_2 it does not necessarily follow that planting trees will reduce CO_2 build up in the atmosphere. In the long term the amount of carbon stored in trees and their products will tend to a steady value that depends on the species, its management and the uses of the wood products (Thompson & Matthews, 1989). By making assumptions about the fate of carbon (burning, early decomposition, or preservation for centuries before decay) it is possible to calculate the total carbon accumulated (Figure 11.11). The main point to note is that one might expect that eventually a steady state will be reached, the level of which (and the time taken to reach it) will depend on the use of the timber. Where the trees are long-lived and/or the product goes into a long-life use (such as building material with a life of hundreds of years) many hundreds of tonnes ha^{-1} will accumulate (and it will take hundreds of years to attain the steady state). If, on the other hand, the product is allowed to decay rapidly, smaller amounts will accumulate (and a lower steady state will be attained rapidly). Planting more woodland, therefore, is only likely to be beneficial to the global carbon

Table 11.5 Estimates of the magnitudes of the various carbon reservoirs and sinks for the 1990s from the fourth assessment of the IPCC (Solomon *et al.*, 2007). This shows both the current values and the amount that can be attributable to anthropogenic processes. Note that one Gtonne equals 10^{12} kg, or one petagramme (Pg).

	Existing values	Anthropogenic changes
Carbon reservoirs	(Gtonne)	(Gtonne)
CO_2 in atmosphere	662	165
Carbon in marine biota	3	
Carbon in soil and terrestrial biomass	2261	-39
Carbon in fossil fuel resources	3700	-244
Carbon in surface ocean (top 75 m)	918	18
Carbon in medium and deep ocean and sediment	37 350	100
Net fluxes	(Gtonne y^{-1})	(Gtonne y^{-1})
Atmospheric increase		3.2 ± 0.1
CO_2 fossil fuel emissions		6.4 ± 0.4
Atmosphere to ocean flux of CO_2		2.2 ± 0.4
Land to atmosphere flux of CO_2		1.0 ± 0.6
CO_2 release due to land use change		1.6 (0.5 to 2.7)
Net CO_2 flux into residual land sink		2.6 (-4.3 to -0.9)
Photosynthesis and respiration	(Gtonne y^{-1})	(Gtonne y^{-1})
Terrestrial gross primary productivity	120	2.6
Terrestrial respiration	119.6	
Ocean gross primary productivity	70	2.2
Ocean respiratory loss	70.6	

balance in the period before attainment of the steady state, or where the product is used to substitute for fossil fuel. It is currently uncertain whether old-growth mature tropical forests constitute a continuing carbon sink (Baker *et al.*, 2004). Grassland could even be more effective at lowering net CO_2 emissions than forests (because of a potentially higher net primary productivity) if used as a biomass source to replace fossil hydrocarbons or if some of the produce is prevented from decay. Unfortunately the achievement of high productivity by grassland or any other system requires a high input of nitrogen, but denitrification of fertilisers to N_2O is one of the reasons for increasing atmospheric N_2O, which may itself be damaging to the environment.

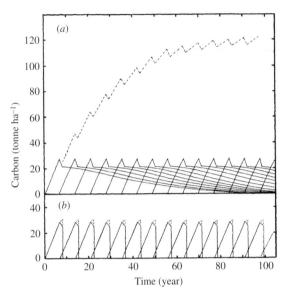

Figure 11.11 Modelled time course of carbon accumulation in a poplar coppice system, showing the importance of final use of the wood (based on calculations of Thompson & Matthews, 1989). The carbon accumulated in each rotation is given by the series of solid curves, while the total accumulated carbon at any time is shown by the dashed line. When the product is converted into a relatively long-lived product such as medium density fibreboard (*a*), the total carbon in trees and in the product would be expected to rise slowly to a steady-state value of 120 tonne ha^{-1} after about 100 years. In contrast, where the product is rapidly used as a biomass source for energy (*b*), the maximum amount of carbon locked up in the system would be less than 30 tonne ha^{-1}, and this would be reached in less than ten years.

Methane

As we have seen, global concentrations of methane were also relatively stable until about 150 years ago, and have increased substantially, though are now rising much less rapidly than the 1% per year reached during the 1970s and early 1980s. The reasons for the changes in methane concentration are not clear, though its concentration depends on the balance between biogenic sources (rice paddies, ruminant animals, biomass burning, and emission from swamps and marshes including the tundra regions) and its removal in the atmosphere by reaction with the photochemically produced hydroxyl radical. There is a potential strong positive feedback effect of

any global warming, which could be amplified if it leads to significant releases of CH_4 from warming tundra.

Nitrous oxide and other greenhouse gases

Atmospheric N_2O is increasing at about 0.26% per year, possibly as a result of either increased microbial nitrification and denitrification of agricultural fertilisers or combustion, though substantial quantities are produced by tropical forest soils and by outgassing of microbially produced N_2O in coastal oceanic waters. Together with the CFCs (especially CCl_3F (CFC-11) and CCl_2F_2 (CFC-12)) these trace gases probably make a contribution to global warming of a similar order to that of CH_4 (Table 11.4), particularly when one takes account of their long residence times in the atmosphere.

The other main climatic effect of CFCs on destruction of the tropospheric ozone layer, with the consequential increased transmission of damaging UV radiation to the Earth's surface, led to international agreement (the Montreal Protocol) to restrict their emission (Anon, 1987), but their long lifetimes (c. 120 yr for CFC-12) means that the effects of existing and known future emissions of these gases will take a long time to reverse. The replacements (hydrohalocarbons) have shorter atmospheric lifetimes, so that the cumulative global warming potential from their release is less even though they are effective absorbers of infrared radiation. The perfluorinated compounds (such as carbon tetrafluoride, CF_4) and sulphur hexafluoride are highly radiatively efficient and have extremely long lifetimes reaching tens of thousands of years so are essentially permanent, though it is worth noting that about half of the CF_4 emissions are from natural sources.

11.3.3 Effects on climate and feedbacks

The simple energy balance shown in Figure 11.10 indicates that increasing the absorption of thermal radiation in the atmosphere should lead to global surface warming. Climate, however, is the long-term result of an extremely complex and dynamic system of atmospheric motion that involves a large number of feedbacks that modify any simple prediction based on direct radiative forcing. Increasing temperature, for example, would be expected to increase the amount of water vapour in the atmosphere by increasing evaporation and by increasing the water-holding capacity of the air; this in turn affects atmospheric absorption of thermal radiation and the formation of clouds with their major effects on the radiation balance.

Unfortunately there is still substantial uncertainty in our quantification of the ways in which changing global temperatures and changes in aerosols might affect cloud cover and type and their resulting impacts on climate. The further feedback effects are complex as the net effect of increased cloud cover depends on whether the enhanced reflection of solar radiation (leading to cooling) is greater or less than the increased trapping of upward thermal radiation from the surface (leading to warming); this balance depends critically on cloud altitude and type.

The polar ice caps make a major contribution to the albedo of the Earth, so changes in their extent would significantly affect the amount, and distribution, of solar radiation absorbed as does the amount of deposition onto the snow of dusts and soot from biomass burning. Similarly changes in temperature can affect evaporation and the frequency of drought with consequent effects on vegetation growth, which may itself then lead to altered surface albedo and changed energy balance. Other aspects of the *vegetation feedbacks* have been discussed by Woodward et al. (1998). These include the fact that the stomatal closure stimulated by increasing CO_2 concentrations tends to lead to extra warming, but the final effect depends more on any resulting longer term changes in vegetation growth and structure and, for example, the balance between increasing leaf area index (with lower albedo leading to warming and increased evaporation leading to cooling) at high latitudes and decreasing vegetation cover (higher albedo and reduced evaporation) in subtropical regions. The fertilisation effect of enhanced CO_2 on

enhanced vegetation growth and the resulting enhanced carbon sink has been calculated (Woodward et al., 1998) to have a potential negative feedback effect on temperature of 0.7°C by 2100, partially compensating for other positive forcings.

Although many general circulation models (GCMs) simulating the behaviour of the atmosphere on a global scale have been developed that attempt to take account of atmospheric dynamics, there are still large uncertainties, with the temperature rises predicted for a doubling of CO_2 ranging between about 2 and 5 K. More importantly for plant production, there are likely to be large regional differences, with the largest temperature increases occurring near the poles, while effects on rainfall and the water balance are likely to be at least, if not more, important than changes in temperature. The difficulties involved in predicting effects on climate change have been discussed in detail by Houghton (2009).

In spite of the uncertainties, however, there are strong suggestions that associated with global warming will be changes in climatic variability and the probability of extremes, while there are also likely to be strong regional differences in climatic responses. For example warming trends are stronger nearer the poles.

11.3.4 Consequences of global warming for agriculture and natural ecosystems

There are a number of effects to consider. First, there are the direct effects of increased CO_2 concentration on plant performance. The most obvious are the effects on photosynthesis through CO_2 fertilisation (countered by any enhanced stomatal closure), and on evaporation and water use efficiency as a result of stomatal closure. These are reasonably well understood, though most research has concentrated on the effects of short-term changes in CO_2 concentration and ignored the longer term adaptations that can occur. Second, there are the effects of altered climate (temperature, humidity, rainfall, etc. and their seasonal distribution).

Effect of elevated carbon dioxide concentration on photosynthesis

There have been thousands of studies of the effects of increased CO_2 concentrations on photosynthesis; many involved measurements in greenhouses or controlled environment chambers, but they provide rather unnatural environments and their limited space allows only limited replication of small plants, as do experiments in open-top chambers (OTCs; see Figure 11.12). Even the largest OTCs (Medhurst et al., 2006), which can be 3.5 m diameter and 9.0 m high, can only contain one or two mature trees and edge effects can be problematic with no natural canopy closure or natural rooting. Furthermore plants in such chambers are completely uncoupled from the environment with the resulting difficulties of interpreting results (see Section 5.3.3). Even with effective air conditioning, the chamber environment cannot fully mimic the natural environment, so it is always necessary to compare plants in OTCs with and without enhanced CO_2 supply. It is also worth remembering that as in all cases where plants are grown in pots, their use tends to inhibit plant photosynthesis and growth (Poorter et al., 2012) and may thus give misleading results as

has been reported for many earlier studies of long-term CO_2 exposure.

A potentially better approach is the use of free-air carbon dioxide enrichment (FACE) experiments (Figure 11.12) that can allow CO_2 enrichment on 8 to 30 m diameter areas of naturally-growing plants (see e.g. Long et al., 2004) over long periods. Free-air carbon dioxide enrichment rings are equipped with sophisticated sensing and control systems that release CO_2 from pipes on the upwind side at a rate to maintain a specified elevated CO_2 concentration within the ring, though their operation can be problematic under still conditions as can occur at night when there is no air flow across the rings. In all cases, however, it is necessary to distinguish between the short-term effects of altered CO_2 concentration on photosynthesis and the effects of long-term exposure (for which there is much less information).

Differences in the short-term photosynthetic responses to CO_2 of plants having different photosynthetic pathways were outlined in Chapter 7. The net CO_2 fixation rate is often close to CO_2 saturation in C_4 plants at normal atmospheric concentrations of CO_2, so rather little enhancement of assimilation is achieved by increasing the CO_2

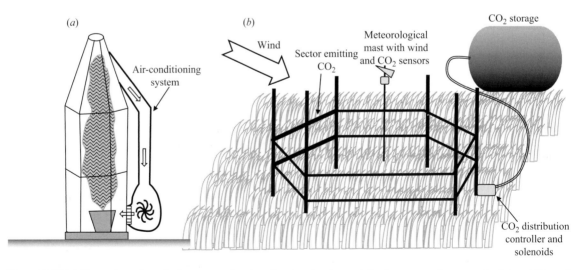

Figure 11.12 (a) Illustration of a typical open-top chamber showing the air-conditioning system that controls air humidity and temperature by recycling air through a condensing heat exchanger, and a CO_2 injection system to maintain the chosen concentration, and (b) a typical FACE system showing the CO_2 storage and supply system with distribution to, and release from, an active sector controlled on the basis of measured wind speed, direction and CO_2 concentration.

concentration. On the other hand, photosynthesis in C_3 plants tends to continue to increase with increasing CO_2 above 380 vpm (e.g. Figure 11.13). This response is partly the direct stimulation as a result of increased availability of CO_2 at the carboxylation site, and partly because CO_2 competitively inhibits photorespiratory loss by competing with O_2 and shifting Rubisco function towards carboxylation.

With long-term exposure to elevated CO_2 as in FACE experiments, however, it is frequently observed

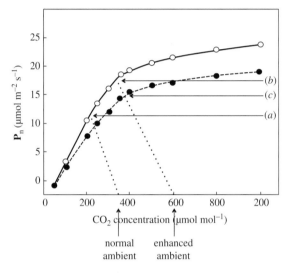

Figure 11.13 Typical responses of photosynthesis to CO_2 concentration (expressed as a mole fraction) in C_3 plants. The solid line shows the photosynthetic response to intercellular space CO_2 concentration (x_i') for plants grown with an ambient concentration (x_a') equal to 360 vpm. The corresponding curve for plants grown for several months under 600 vpm is shown as a dashed line. The diagonal dotted lines represent the 'supply functions' determined by stomatal conductance and the horizontal arrows indicate the operating points for plants (a) grown and measured at ambient CO_2, (b) grown at ambient and measured at enhanced CO_2 (no acclimation) and (c) grown at ambient and measured at enhanced CO_2 (with acclimation). These show the slight enhancement in net photosynthesis after long-term exposure to elevated CO_2, in spite of reductions in stomatal conductance and both the Rubisco-limited and the RuBP-regeneration limited rates. Data points shown for illustration are for *Lolium perenne* grown for several months at 360 or 600 μmol mol^{-1} (data from Ainsworth *et al.*, 2003).

that the short-term increases in net photosynthesis, as obtained from a CO_2 response curve, tend to be partly lost as the plants acclimate to elevated CO_2 as shown for the example in Figure 11.13. Although such acclimation is not always observed, the acclimation often involves both downregulation of Rubisco activity and sometimes a decreased photosynthetic capacity at light saturation; these may partly offset any gain from decreased photorespiration (Sage *et al.*, 1989). Nevertheless, meta-analyses of many FACE experiments (Long *et al.*, 2004; Nowak *et al.*, 2004; see Table 11.6) have shown that a general enhancement is maintained both of acclimated light-saturated photosynthesis, and of daily integrals of photosynthesis and of dry-matter growth and seed yield when CO_2 concentrations were increased to between 550 and 700 vpm over long periods. Perhaps surprisingly, even C_4 crop species have averaged a 20% stimulation of assimilation after long-term exposure to elevated CO_2. This perhaps results from the improved water status that follows from reduced stomatal aperture and improved water use efficiency. Among C_3 plants, trees show the largest mean response (47% enhancement), with crop plants and other herbaceous dicots (forbs) showing much smaller stimulation averaging about 15%. These stimulations in net assimilation occur in spite of consistent stomatal closure and downregulation of Rubisco (also reflected in the reduction in $V_{c,max}$), with a somewhat smaller reduction in maximum electron transport rate reflecting a lowered capacity for regeneration of RuBP. Interestingly, stomatal closure that occurs tends to maintain x_i/x_a relatively constant (Figure 11.13 and Table 11.6).

The net growth responses to elevated CO_2 depend both on alterations to photosynthesis and on a number of morphological responses to elevated CO_2 that include increases in leaf thickness (decreased leaf area per mass), effects on dry-matter partitioning, leaf-cell packing and stomatal distribution, together with the tendency for leaf number to increase. Indeed the benefits of CO_2 enrichment for crop growth have been well established over many years for a wide range of greenhouse crops (Mortensen, 1987) though,

Table 11.6 Meta-analysis of the effects of long-term elevation of atmospheric CO_2 concentration in FACE experiments (expressed as percentage increases) on various photosynthetic and growth parameters in different types of plant. Data represent mean values and 5 to 95% confidence range (Long *et al.*, 2004) for all except the ecosystem data (Nowak *et al.*, 2004), which are means and 25 to 75% percentiles.

	C_3 plants (% increases)	C_4 plants (% increases)
Light saturated photosynthesis (P_{sat})		
Average	34 (31.3 to 37.7)	10 (1 to 19)
Trees	47 (42 to 52)	
Forbs	15 (3.5 to 26.5)	
Legumes	21 (13 to 29)	
Grasses	36 (29.5 to 42.5)	-2 (-13 to 9)
Stomatal conductance (g_s)	-20 (-22 to -18)	-24 (-37 to -14.5)
c_i/c_a	-2 (-4 to 0)	
Rubisco	-19.5 (-32 to -7)	
$V_{c,max}$ (maximum carboxylation rate)	-13.5 (-15.7 to -11.3)	
J_{max} (maximum electron transport rate)	-5.5 (-8 to -3)	
Quantum yield	12 (3.3 to 20.7)	
Leaf area per unit dry mass (SLA)	-7 (-9.5 to -4,5)	3 (-7 to 6)
Leaf number	12.5 (6.5 to 18.5)	
Dry matter production	20 (17 to 23)	3 (-4.5 to 7)
Ecosystems	Above-ground production	Below-ground production
Bog	2.6 (-12.5 to 20)	10.5 (6 to 14)
Forest	23.5 (7 to 31)	35 (13 to 50.5)
Grassland	12.5 (5 to 24)	16 (11.5 to 23)

because CO_2 is often only supplemented in daylight hours, full acclimation may not occur and the results are not necessarily applicable to a future higher CO_2 environment outdoors. The great variability in the enhancement of growth rate by elevated CO_2 in different species implies that changing atmospheric CO_2 concentrations are likely to have large effects on the relative competitive ability of species and hence on plant distribution.

Effects on water use efficiency

An important effect of rising CO_2 concentration is expected to be an increase in WUE, because raising CO_2 tends to close stomata without decreasing P_n.

Morphological and anatomical changes (e.g. stomatal density) as a consequence of altered CO_2 would also influence WUE. Changes in WUE are likely to be particularly important in the main agricultural regions of the world where alterations in water availability are probably more crucial for plant production than are changes in mean temperatures in any CO_2-induced climate change.

Modelling of agricultural and ecological consequences

The general effects of temperature, water availability and other aspects such as climate variability have been discussed in other chapters, and the basic information is available to predict the effects of a given climate change on natural vegetation and on crop production. Their use for predicting the consequences of climate change is limited by the quality of the climatic predictions, which for any particular area cannot yet provide reliable predictions of the exact combinations of weather conditions that are likely to occur. Although simple models of plant growth can be used where there is a clear single most limiting environmental constraint, there is still a long way to go in developing models that fully take into account the seasonal variation in weather and its variability. This is essential if real progress is to be made, as environmental extremes such as frosts and droughts are probably the main constraints on crop and vegetation distribution patterns.

When considering the implications of climate change it is necessary to remember that climate will also affect many other factors and their interactions, such as the population and virulence of pests and pathogens. There have been many attempts to estimate the effects of climate change on agricultural and horticultural production, but they inevitably still involve a lot of uncertainty even though it is clear that it is likely that the best areas to grow specific crops may shift (Parry *et al.*, 2007). For example it is likely that the optimal zone for crops as diverse as temperate cereals and those perennial crops that require winter chilling (Sunley *et al.*, 2006) may shift towards the poles.

11.4 Atmospheric pollutants

In the past few decades, there has been increasing recognition of the importance of atmospheric pollutants for plant growth and distribution. Acidic pollutants are well known to lead to soil and freshwater acidification, while the widespread reporting in the 1980s of forest decline in Europe (especially silver fir, Norway spruce and beech) and in Eastern North America (red spruce) led to a major effort to improve our understanding of the mechanisms involved. These studies have emphasised the complexity of the plant responses involved and the interactions of pollutants with other environmental factors such as temperature and drought.

Many atmospheric constituents can have detrimental effects on plants. These include the various nitrogen oxides (NO_x), sulphur dioxide (SO_2) and other acidic gases predominantly arising from combustion of fossil fuels, photochemical oxidants such as ozone (O_3) and peroxyacetyl-nitrate (PAN), important constituents of photochemical smog, and, particularly in confined environments such as glasshouses, a range of highly toxic compounds such as di-*n*-butyl phthalate (used as a plasticiser).

Plants themselves can be important sources of many diverse *biogenic volatile organic compounds* (BVOCs) that play a role in protection against thermal and oxidative stresses and in signalling in defence against herbivores. These can influence atmospheric chemistry and the production or destruction of pollutants such as ozone (Loreto & Schnitzler, 2010). Interactions of BVOCs with atmospheric pollutants are complex with, for example, terpenes removing ozone in the absence of other pollutants though, in the presence of NO_x, they can enhance ozone production. Vegetation emits around 800×10^{12} g C yr^{-1} of these compounds globally (Fowler *et al.*, 2009), with about half of this total emission comprising the terpene *isoprene* (2-methyl-1,3-butadiene); among other BVOCs are various monoterpenes (commonly emitted by conifers), sesquiterpenes and other volatiles such as alcohols, aldehydes and ketones. Isoprene emissions mostly originate from recent photosynthetic metabolites and can reach 20% of the carbon fixed in photosynthesis, with rates of between 0.2 and 1 µg C m^{-2} s^{-1} having been reported for isoprene emission from leaves of the grass *Arundo donax* (Hewitt *et al.*, 1990b). Although some species (e.g. tobacco and sunflower) do not normally emit terpenes, many others, especially trees such as poplar, can emit large quantities, either directly from the mesophyll or from storage organs such as the resin ducts in conifers or the specialised leaf glands in the Lamiaceae. The rates of emission depend on differences in rates of synthesis and on physico-chemical processes determining their storage in and release from storage pools, with emission often being stimulated by biotic and abiotic stress, especially high temperature (Loreto & Schnitzler, 2010). The overall emission of BVOCs is of the same order as global methane emissions. It remains unclear what the evolutionary forces are determining why only some plants emit isoprene, with the trait having evolved and disappeared many times during evolution (Monson *et al.*, 2013). The presence of this trait probably results from trade-offs between potential benefits (possibly an enhanced tolerance of high temperature stress) and costs (including the associated carbon and energy loss) interacting with the long-term stability of environmental niches favouring emission.

11.4.1 Uptake and deposition processes

Uptake of pollutants occurs via a number of routes: dry deposition (which refers to the absorption of gases and capture of particles by plants), wet deposition in precipitation (rain or snow) and 'occult' deposition in intercepted cloud, fog or mist.

Gases

Fluxes of important gaseous pollutants such as SO_2, HNO_3, HCl, O_3 and NH_3 can be determined using standard micrometeorological techniques (Chapter 3) or by the exposure of plants or natural communities to pollutants in enclosures where gas fluxes can

be measured. The usual mass transfer theory is applicable, and it is possible to identify transfer resistances corresponding to transfer in the canopy boundary layer (r_A) and a 'surface' component (r_L) that comprises the parallel paths of uptake through stomata, directly through the leaf cuticle and to other surfaces (such as the soil). That is: $r_L = 1/(r_s^{-1} + r_c^{-1} + r_{soil}^{-1})$, where r_s, r_c and r_{soil} respectively are resistances of the pathways through the stomata, cuticle and soil. With the multiple sinks and pathways for a pollutant in a canopy (see Figure 5.4) it is not generally possible completely to separate the atmospheric and surface processes into two series resistances. If the concentration at the sink can be assumed to be zero, the total resistance to uptake (Σr_X) of a pollutant gas X is given by:

$$\Sigma r_X = r_{AX} + r_{LX} = 1/V_d = c_X/\mathbf{J}_X \qquad (11.9)$$

where c_X is the concentration of X in the air, $\mathbf{J}_X$ is the flux, and V_d is known as the *deposition velocity* (with units of m s^{-1}). This term is widely used in pollutant studies and is simply the reciprocal of the total uptake resistance (the conductance); this can also be regarded as the flux normalised for atmospheric concentration.

The surface resistance for uptake of reactive gases such as HNO_3 and HCl is normally considered to be negligible (i.e. these pollutants deposit on leaf and soil surfaces at rates determined by r_A, which itself depends on atmospheric turbulence). Deposition of these gases is therefore primarily dependent on their atmospheric concentrations and on atmospheric transfer processes. The resistance for NH_3 is negligible for wet surfaces but can become appreciable when the vegetation is dry where uptake is both through stomata and the cuticle.

For gases such as SO_2 and O_3, uptake depends on dissolution or reaction on wet leaf surfaces or within the substomatal cavity, and dry deposition is significantly determined by the canopy stomatal resistance, which can be up to 90% of the total resistance to uptake. Although non-stomatal deposition dominates for SO_2, for ozone typically around 30 to 70% of uptake is through stomata.

Deposition of these gases, therefore, tends to be greatest either during daylight when the stomata are open, or when the surface is wet. The damaging effects of ozone depend on its uptake through stomata where it generates damaging reactive oxygen species in the mesophyll; where it is only deposited on external surfaces of vegetation it causes relatively little damage. The actual rate of SO_2 uptake depends on the chemistry and especially the pH of the surface water film, which itself is determined partly by the balance between NH_3 and SO_2 uptake (see Flechard et al., 1999 for a more detailed discussion of the controls of deposition). Indeed soil acidification by SO_2 is substantially enhanced in the presence of NH_3. In addition, because the rate of deposition of these gases is not very sensitive to r_A, deposition of these gases is normally less dependent on canopy type (i.e. forest versus short crop) than is the case for HNO_3 and HCl.

Several gases including NH_3, NO and NO_2 can be absorbed or emitted by vegetation, though fluxes of NO tend to be predominantly away from the surface as a result of denitrification in anaerobic soil. The complex nature of ammonia sources and sinks in natural environments, with for example major sources arising from animal waste, potentially leads to substantial local advection, making flux measurement problematic. With these bi-directional fluxes it is useful to define a *canopy compensation point* (similar to that used for CO_2 exchange) as the concentration at which no net exchange takes place.

Wet deposition and in particles

The deposition of pollutants as particles and in droplets is a major mechanism for pollutant deposition. The mechanisms of deposition, which are equally applicable to transport of pollutants and of dusts, spores and bacteria, depend on particle size, changing from predominantly Brownian diffusion for particles less than about 1 μm, to impaction and increasingly sedimentation, with increasing size (Chamberlain & Little, 1981). Much of the particulate sulphate and nitrogen in the atmosphere, as well as

lead from car exhausts, occurs as aerosols of between 0.1 and 1 μm diameter. Mist and fog droplets, on the other hand, average about 20 μm, and dust particles up to 100 μm.

Sedimentation is the downward movement of particles under the influence of gravity. The rate of this settling, the sedimentation velocity (v_s) increases with the square of the particle diameter and is given by *Stokes law* for particles with diameters in the range $0.1 < d_p < 50$ μm and ρ about 1000 kg m^{-3} as:

$$v_s = (d_p{}^2 \boldsymbol{g} \rho_p)/(18\rho_a \boldsymbol{v}) \tag{11.10}$$

where ρ_p and ρ_a are the densities of the particle and of air, $\boldsymbol{g}$ is the acceleration due to gravity and $\boldsymbol{v}$ is the kinematic viscosity of air. Even for particles as large as 30 μm and with a density of 1000 kg m^{-3}, Eq. (11.10) predicts sedimentation velocities of only about 27 mm s^{-1} ($=((30 \times 10^{-6})^2 \times 9.8067 \times 1000)/(18 \times 1.2 \times 15.1 \times 10^{-6})$). This is much less than typical velocities of turbulent eddies in the atmosphere, except within very dense canopies, so it is clear that sedimentation is only important for large dust particles and water droplets.

When air flows round obstacles, entrained particles tend to continue in nearly straight lines because of their inertia leading to the potential for impaction. The impaction efficiency on an object in an airstream can be defined as the number of particles striking the obstacle divided by the number that would have passed through the space occupied by it if the object had not been there. The impaction efficiency is related to the *Stokes number*, which is equal to the stopping distance of the particle divided by the effective diameter of the obstacle (where the stopping distance is the horizontal distance travelled by a particle in still air when given an initial velocity u and is approximated by $v_s u/g$). Impaction efficiency therefore increases with wind speed, with particle size and with decreasing size of obstacle.

The particulate sulphate and nitrate in the atmosphere is only very inefficiently intercepted by vegetation because of the small size of the particles; transfer of these particles is therefore predominantly by Brownian motion, with some impaction and interception by micro-roughness elements such as leaf hairs. This size of particle tends to stick firmly once attached. Wind-tunnel studies suggest deposition velocities of less than 1 mm s^{-1} to forest with wind speeds as high as 5 m s^{-1} for 0.5 μm particles. This is much less than the corresponding rates of uptake of gaseous sources of S and N such as SO_2, NH_3 and HNO_3 (Fowler et al., 1989).

Sulphate or nitrate aerosols may act as nucleation centres for the formation of fog or cloud droplets, especially during the formation of orographic cloud when air is cooled as it rises over mountain ranges. Cloud droplets therefore tend to contain quite high concentrations of sulphate and nitrate. Because typical droplet sizes in orographic cloud are of the order of 20 μm in diameter (Fowler et al., 1989) they are collected relatively efficiently by impaction and sedimentation processes on vegetation, with

deposition velocities of perhaps 20 to 50 mm s^{-1} for moorland and up to 200 mm s^{-1} for forest in the atmospheric conditions that apply in upland Britain. These deposition rates are one to two orders of magnitude greater than for the original aerosols.

This *occult deposition* is a particularly important means of pollutant deposition at higher elevations where vegetation may be bathed in cloud for long periods. Because the concentrations of major ions in cloud water often exceed those in rainwater by two- to three-fold (up to six-fold), deposition of pollutant chemicals in cloud can in certain circumstances be more important than deposition in rainfall. Cloud deposition is greater onto aerodynamically rough surfaces such as forest, than onto crops or moorland, and is much greater than the deposition of small aerosol particles. Cloud deposition also has hydrological significance as an important

Table 11.7 Representative rates of sulphur and nitrogen deposition (kg ha^{-1} y^{-1}) as different forms in different environments (data from Fowler et al., 1989; Goulding et al., 1998; Irving, 1988).

	Dry		Wet	Occult	Total
	Gas	Particle			
Sulphur					
Kielder (UK)					
coniferous forest	3.1	–	13	6.5	22.6
cotton grass	3.1	–	13	1.3	17.4
Oak Ridge (USA)					
deciduous forest	7.8 ± 1.3	2.1 ± 0.3	8.5 ± 2.0	–	>18.4
Nitrogen (NO_3^- and NH_4^+)					
Kielder (UK)					
forest	13.5	–	8	1.9	23.4
cotton grass	4.0	–	8	0.4	12.4
Rothamsted (C. England)					
winter cereal (1995)	31.5	2.9	9.0		43.3

source of water for plants in montane environments. The effects of factors influencing the deposition of different pollutants are illustrated by the data in Table 11.7. The data for the Kielder Forest area (a large area of afforestation in northern England at about 300 m above sea level) show particularly well the differential effect of forest (15 m trees) as compared with moorland on deposition. The presence of trees increases sulphur input by about 30% (as a result of increased cloud-water interception), but increases nitrogen input by about 90% (as a result of both the increased cloud-water interception and the increased dry deposition rate that occurs for gases such as HNO_3 and NH_3). The actual rates of deposition at any site depend on local pollutant concentrations. For example, the wet deposition rates of nitrate and ammonium deposition at Rothamsted in central England have decreased substantially from a peak of about 18 kg N ha^{-1} y^{-1} in 1980 since when a number of factors such as a switch from coal to natural gas have led to a halving of the rate (Goulding et al., 1998).

11.4.2 Effects of pollutants on vegetation

The effects of pollutants depend not only on the sensitivity of the plants, but on the concentration and on the duration of exposure. Short episodes at high concentration may be more damaging than continuous exposure to lower concentrations that give rise to the same long-term time-averaged concentration. Lichens and bryophytes are often particularly sensitive to atmospheric pollution and have been used to provide a sensitive bioassay of pollution in many situations (e.g. Henderson, 1987), though it is often difficult to relate these results to the specific chemical problem that exists in any situation.

The effects of pollutants on plants can be direct as a result of uptake by leaf tissues, but in many cases the influences may be through processes such as soil acidification or even more subtle effects on pathogens or competitors. Research has emphasised effects on carbon assimilation and partitioning, water relations and nutrition. Detailed discussion and examples may

be found in many books and reviews (e.g. Bell & Treshow, 1992; Mathy, 1988; Omasa et al., 2005; Wellburn, 1994). In general there are large differences between species, and even cultivars, in their sensitivities to different atmospheric pollutants, with typical thresholds for damage by SO_2, for example, being of the order of >50 ppb (volume parts per 10^9) for the appearance of chronic effects (e.g. chlorosis and premature senescence) with long-term exposure, >180 ppb for episodic exposure and >500 ppb for the appearance of acute damage such as leaf necrosis. Ozone, on the other hand, can cause yield losses at 50 to 100 ppb, concentrations that are much closer to the normal background.

Methods of study

Pollution effects on vegetation are most commonly studied in fumigation chambers with more or less sophisticated environmental control, which may be either of the closed or of the 'open top' type. In some cases open-field fumigation systems, similar to the FACE systems described above, are used: these have the advantage of maintaining natural microclimatic conditions (McLeod et al., 1985).

Interactions

Pollutants interact both with each other and with biotic and abiotic stresses in their effects on plants in a complex manner. As an example of the type of effect that is commonly observed, Ting and Dugger (1968) reported that exposure of bean leaves to a combination of 0.3 vpm O_3 and 0.4 vpm SO_2 for four hours had no apparent detrimental effect, while exposure to the O_3 alone caused heavy damage. This has been explained in terms of stomatal closure in response to the SO_2 preventing entry of the O_3, though, in other experiments, low levels of SO_2 have been reported to cause stomatal opening (Unsworth et al., 1972), and hence potentially enhanced damage.

The effects of ozone on plant cells provide a good example of the complexity of the reactions involved in response to pollutants. Ozone can damage cell membranes by reacting directly either with proteins (oxidising cysteine and methionine residues) or with the unsaturated fatty acids. Ozone is also quite soluble in water and may decompose in aqueous cell compartments to produce highly reactive and damaging radicals such as hydroxyl ($OH^{\bullet}$), peroxy ($HO_2^{\bullet}$) and the superoxide ion ($O_2^{-\bullet}$). In addition to these direct effects, however, the impact may be enhanced by the formation of highly reactive hydroperoxides on reaction with gaseous alkenes produced within the leaf (Hewitt *et al.*, 1990a). These biogenic alkenes include ethylene (induced in response to a number of stresses) as well as isoprene or monoterpenes in appropriate species. The effect may even be amplified by positive feedback as tissue damage releases even more alkene.

An example of another important type of interaction occurs where exposure to pollutants enhances sensitivity to other stresses. Typical of this effect is an experiment where growth of *Phleum pratense* seedlings after exposure to a mixture of SO_2 and NO_2 at concentrations up to 90 vppb for 40 days was compared under well-watered and drought conditions (Wright *et al.*, 1986). Fumigation had little effect on the subsequent relative growth rate over a 23-day period when the plants were maintained well watered. When, however, water was withheld over the same period, the relative growth rate of the non-fumigated control plants decreased by about 50%, but the inhibition was much greater for the previously fumigated plants. This enhanced sensitivity to water deficit was related to a greater water use by the fumigated plants.

Pollutants such as O_3 and acidic mists can also markedly enhance sensitivity to frost (e.g. Wolfenden & Mansfield, 1990), and this appears to be a factor in some types of forest decline. There is, for example, strong evidence that the enhanced sensitivity of spruce to frost damage that is caused by pollutants results from a delay in the onset of hardening in autumn, rather than from an effect on the final degree of hardiness. Exposure to pollutants during the dormant period can also enhance winter damage, possibly by lowering the efficiency of control of water loss by the stomata or the cuticle. Another important type of interaction that should not be forgotten, but is relatively poorly understood, is that caused by the effects of pollutants on pests and disease organisms.

12 Physiology and crop yield improvement

Contents

This chapter introduces some of the ways in which information of the type discussed in earlier chapters can be applied to the improvement of crop yields. Farm yields have been improving over hundreds of years, though the rate of increase has been particularly rapid in the last 70 years or so (Figure 12.1(*a*)). These yield increases have resulted both from the introduction of new varieties and from advances in crop management (agronomy), including both the widespread use of fertilisers, herbicides, pesticides and fungicides, and improvements in machinery and irrigation. In addition to their increased yield potential, the new varieties that have been developed by plant breeders often incorporate improved pest or disease resistance and the ability to benefit from increased levels of fertiliser application; in cereals this is partly because newer dwarf genotypes are resistant to lodging. The 'green revolution' in the 1960s and 1970s was based on both the incorporation of dwarfing genes that conferred resistance to lodging and improved the harvest index (HI; the proportion of dry matter that is in harvestable yield) and on the introduction of photoperiod insensitivity genes that allowed the improved crops to be grown over a wide range of environments.

It is useful when discussing yield increases to distinguish between *potential yields* achieved with optimal agronomy under experiment station conditions, and *farm yields* as obtained by typical farmers and reported in national yield statistics (Figure 12.1(*b*)). An indication of the contribution made by new varieties to improvements in potential yield can be obtained from direct comparisons of yields of old and new varieties in one trial, while estimates of the relative contribution of breeding and improved management to yield increases can be obtained from statistical analyses of historic yield data from variety trials where any one variety was tested over several years. This latter approach allows one to partition any observed yield increases into a genetic component and a time component that includes changes in agronomy and changes in other factors such as climate and effects of changing disease or pest influence (see Mackay *et al.*, 2011). The yield gap between farm yields and potential yields is a useful measure of these non-genetic factors. A study based on the national average wheat yields in the UK (Figure 12.2(*a*)) indicated that agronomy and breeding made approximately similar contributions to the doubling of national cereal yields over the 30 years to 1978, while more recent analyses (Figure 12.2(*b*)) have shown that almost all the advance in cereal yields in the UK since then has apparently been due to variety (Mackay *et al.*, 2011). In contrast, genetic and agronomic advances

(a)

(b)

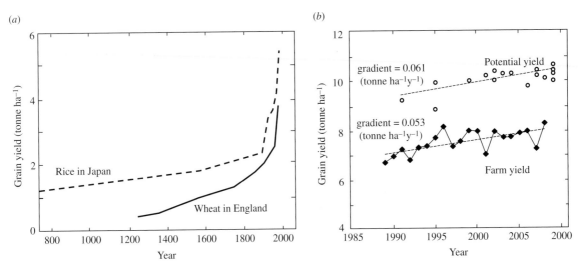

Figure 12.1 (a) Historical trends in grain yields of rice in Japan and of wheat in England (from data collected by Evans, 1975). (b) Recent trends in potential yield and farm yield for wheat in the UK (data from Fischer & Edmeades, 2010).

(a)

(b)

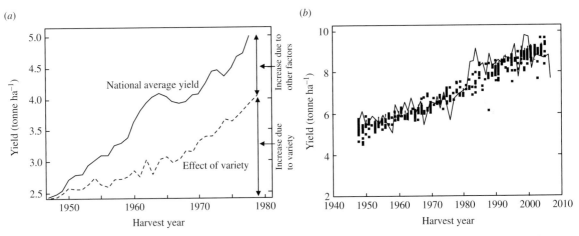

Figure 12.2 Analyses of contribution of variety improvement to increasing yields of winter wheat for England and Wales from national list and recommended list trials. (a) Variation of the five-year moving average of national yields 1947–1978 where the area below the dashed line represents the contribution of variety to the overall increase in yield (data from Silvey, 1981). (b) Trends over the period 1948–2007 where the continuous line represents the trend in yield over that period and the squares represent the effect of changing varieties over that period with each symbol indicating the year of introduction of a variety (data from Mackay et al., 2011).

have continued to contribute almost equally to yield increases of forage maize and sugar beet in the UK. Unfortunately it is not possible to partition precisely the contributions to yield improvement because of what are called genotype × environment (G×E) interactions, where different varieties respond

differently (e.g. the newer semi-dwarf cereal varieties have higher harvest indices and can respond better to high nitrogen input than can tall varieties that become liable to lodging). For example, Bingham et al. (2012) showed, for a sample of UK barley varieties released between 1931 and 2005 and grown in common trials, that though potential yields had increased by 72% over the period when grown with 110 kg N ha^{-1} supplemental nitrogen, the increase was only 40% when grown without supplemental nitrogen. The newer varieties apparently use nitrogen more efficiently.

In some crops, breeding has had a much smaller impact than in cereals. For example, some of the major varieties of both dessert and culinary apple currently grown in the UK were discovered in the nineteenth century. Another consideration is that the long-term trend of yield improvements for any one crop can vary greatly between different parts of the world. Maize yields, for example, nearly doubled in the USA between the 1960s and 1980s, increasing approximately linearly from about 4.1 tonne ha^{-1} in 1961 to about 7.3 tonne ha^{-1} in 1989 (largely as a result of the introduction of F1 hybrid varieties). In contrast, yields in middle Africa remained nearly constant over the same period at about 0.8 tonne ha^{-1} (derived from data of the Food and Agriculture Organization of the United Nations – see http://faostat.fao.org/site/567/default.aspx#ancor). The low yields and limited increases in developing countries relate more to problems related to growing conditions and agronomy than to variety. Fischer and Edmeades (2010) have reviewed the more recent progress in yields and the contribution of breeding for a range of crops grown in different environments. The FAO data show that by 2010 grain maize yields had increased to about 9.7 tonne ha^{-1} in the USA but to only about 1 tonne ha^{-1} in middle Africa.

An important question in relation to attempts both to identify the environmental factors such as radiation, temperature or water, that limit the yield of any crop and to overcome these constraints by breeding or management, is how closely the current yields approach the theoretical maximum. Although potential yields are still increasing by around 1% per annum (Fischer & Edmeades, 2010) there is some evidence that cereal potential yields have been levelling off in recent years, with some crops such as sorghum in the USA showing indications of a plateau being approached. The fact that there is still ample

scope for improvement, at least on a farm scale, is implicit from the observation that average yields are often only about half the best yields (6 tonne ha^{-1} compared with approximately 12 tonne ha^{-1} for wheat in the UK). The latter value is in fact approaching the potential yield calculated on the basis of the observed behaviour of a wheat crop in terms of leaf area development and the assumption that virtually all the dry matter produced during the 40-day grain filling period goes into the grain (Austin, 1978).

An alternative estimate of the potential yield can be obtained using the type of efficiency calculation described in Chapter 7, where it was shown that 3% efficiency of conversion of incident solar radiation to dry matter should be attainable. If one assumes an input of approximately 3 GJ of solar radiation during the life of a winter wheat crop in the UK, this corresponds to a production of about 5100 g dry matter m^{-2}, or about 23 tonne ha^{-1} of grain (assuming a harvest index of 45%). Even if one only considers the 2 GJ available during the summer (for a spring crop), 15 tonne ha^{-1} should be possible. It is likely, however, that yields of this order could only be achieved by quite radical alterations in the phenology of the wheat crop. Even higher yields are theoretically possible in higher radiation environments. In practice, average yields achieved by farmers (farmers' yields) tend to be substantially less than the potential yields obtainable with optimal agronomy as achieved in some experiment station trials; this 'yield gap' is commonly between about 20% and 80% (Lobell et al., 2009), with the highest yield gaps occurring in low rainfall environments. Interestingly these authors have also shown that agricultural crops commonly have lower net primary productivity (NPP) than adapted natural ecosystems.

Discussion of yield processes in crops and the role of physiology in increasing crop yields may be found in the classic books by Evans (1975) and Milthorpe and Moorby (1979). Other useful discussions may be found in a number of more recent texts (e.g. Fitter & Hay, 2001; Hall, 2001; Sadras & Calderini, 2009). The recent handbooks on wheat breeding published by the

International Maize and Wheat Improvement Centre in Mexico are particularly useful sources of relevant guidance (Pask *et al.*, 2011; Reynolds *et al.*, 2011).

There are many areas where physiology and an understanding of plant interactions with the aerial environment have made important contributions to crop production. For example, the understanding of photoperiodic control of flowering has enabled the development of widely adapted photoperiod-insensitive varieties (e.g. in cereals) that can be grown successfully over a wide geographic range, and has allowed the artificial control of flowering in horticulture. Similarly, knowledge of the natural control of plant growth and development has led to the successful use of plant growth regulators in fruit production (see Luckwill, 1981) and in propagation in vitro, while understanding of root–shoot signalling in plants has led to improvements in irrigation (Chapter 10). Some other potential applications of physiology to yield improvement are discussed in this chapter.

12.1 Variety improvement

The traditional approach of plant breeders to improving crop performance is to choose, as parents, plants known to complement each other for the required characters such as high yield or disease resistance, to cross them and then select those progeny that have the required combined desirable attributes of both parents. Since the first edition of this book appeared, much has changed in the tools available to plant physiologists and plant breeders, with the whole process of plant breeding having been transformed as a result of the rapid advances in molecular genetics and information technology that now allow substantial shortcuts to be made. As we have seen in previous chapters, molecular biology and what is known as functional genomics have greatly enhanced our ability to unravel the genetic mechanisms controlling plant development and stress responses and to identify and incorporate in new lines advantageous characters. Although genetic transformation and insertion of novel genes

into crops is now feasible and used in many parts of the world for specific purposes where single genes may be useful, such as the development of herbicide resistant crops or the insertion of single major resistance genes, most traits that are of interest to breeders have a much more complex basis. In such cases the newer molecular tools complement the more traditional approaches of crossing and selection that have been used by breeders over many years.

The many breeding strategies available and the practical aspects of plant breeding are discussed in detail in plant breeding texts such as those by Allard (1999) and Acquaah (2012), while the newer molecular techniques and their potential applications in plant breeding are also treated in a number of texts (e.g. Xu, 2010). Modern plant breeding now combines conventional approaches of crossing and selection that have been used by breeders over many years with the tools provided by genomics and other high-throughput technologies and bioinformatics.

12.1.1 Molecular genetics

The rapid advances in molecular genetics since the 1980s have substantially enhanced our understanding of the genetic composition of crop plants with complete genome sequences now being available for several crop and model species. These new technologies now allow rapid high-throughput analysis (so-called 'omics' studies) of gene sequences and of transcript, protein and metabolome profiles at particular stages of development or in response to specific environmental stresses (genomics, transcriptomics, proteomics and metabolomics). These tools have provided a range of molecular markers, including single nucleotide polymorphisms (SNPs), that can be used for genetic mapping, and marker-assisted selection (MAS) for the evaluation of germplasm and the analysis of gene function. The costs of large-scale genotyping and even gene sequencing are continuing to decrease rapidly making them more feasible for use in breeding programmes so that the use of MAS is increasing. Although gene isolation and insertion into crop plants through 'genetic manipulation' is now straightforward for many crops (Xu, 2010), the challenges remain, however, for complex characters such as stress tolerance, which are not only multigenic, but also whose optimal expression varies in a complex way with the exact sequence of environmental conditions. In such situations it becomes necessary to adopt a holistic systems approach to the identification of optimal genotypes.

Mapping and quantitative trait loci (QTLs)

Large numbers of molecular markers spread across the genome are rapidly becoming available for all major crops. Mapping studies provide powerful tools for identifying genes determining traits of interest, and the possibility to identify from studies of specific bi-parental crosses the genomic segments that are statistically associated with specific phenotypic characters (*quantitative trait loci*, or QTLs). Genomic and bioinformatic tools thus offer new opportunities for dissecting quantitative traits into their components. New advances in association genetics are improving our ability to identify QTLs in more complex populations, while fine-mapping can be used to identify the genes and even the specific nucleotides giving rise to the observed QTLs. The QTLs identified (or at least any closely linked markers) can then be used in marker-assisted selection for improved genotypes, or else, where the underlying gene sequences can be identified, these can potentially be incorporated directly into new transgenic varieties. Unfortunately mapping is not always reliable as it is difficult to map genes in chromosomal areas where recombination is rare.

Molecular markers that are either within genes, or at least closely linked to the appropriate QTLs, can be used to supplement or even replace phenotypic selection particularly now that selection for these markers is becoming increasingly cost effective. Indeed marker-assisted selection can be particularly powerful where phenotypes are difficult to identify, perhaps because they are only expressed at a late stage of development or are subject to genotype × environment interactions. Other problems with

particular crops include the fact that there may be a long period between making a cross and being able to test the progeny. In fruit trees, for example, it may be several years before any fruit is produced that can be tested. Even in annual crops, the number of plants tested would be greatest if all characters could be tested using seedlings.

Transformation and gene function

Sequencing and the identification of specific genes conveying particular beneficial characters and their subsequent cloning opens up opportunities for their direct insertion into new genotypes using the tools of genetic engineering. Even more importantly, however, modification of gene expression, whether by means of transformation, mutagenesis or other opportunities to modify the expression of specific existing genes, for example through gene silencing by techniques such as RNA interference (RNAi), provides a powerful tool for determining the function of particular genes through what is known as *reverse genetics*. This reverse genetics approach provides perhaps the most powerful tool we have had for the elucidation of gene pathways and the identification of the functions of specific genes, and is probably the most useful application of modern genomics technologies. These techniques are particularly useful in the dissection of complex networks of interacting genes that control the multigenic characters that are characteristic of stress response and adaptation.

Genetic engineering approaches have already led to the successful insertion of a number of genes into crop plants including those conferring herbicide resistance and insect resistance. In addition to these traits, others with a relatively simple genetic basis involving single dominant genes (e.g. flowering time, stem height and even salt tolerance) may be quite simple to incorporate in new transgenic varieties. Although uptake of transgenic crops in Europe has been relatively slow because of regulatory restrictions, approximately 160 million hectares of transgenic crops were being grown elsewhere in the world in 2011 (around half in developing countries) with the largest areas being for soybean, maize and cotton (www.isaaa.org/resources/publications/briefs/43/ executivesummary/default.asp). The most widely used transgenes are those that confer tolerance of the herbicide glyphosate, and those crops containing *Bacillus thuringiensis* (Bt) toxin genes to confer insect resistance. Other transgenic crops have included those with genes to slow the softening process in tomatoes or to improve nutritional content by enhancement of vitamin A content (golden rice).

Much greater challenges are involved in the development of transgenic crops requiring multiple transgenes as would be required for most stress tolerance (see Section 12.3.1, below), though an increasing proportion of transgenic crops have stacked traits with more than one transgene. In a few cases single genes may confer stress tolerance: for example the salinity resistance in wheat that has been associated with chromosome 4 of the D genome and which has been related to efficient exclusion of Na^+ has been localised to the Kna1 locus (Gorham *et al.*, 1990; Dubcovsky *et al.*, 1996) and there are hopes that other single genes could be isolated and incorporated into new cultivars to improve crop yield or environmental tolerance. In most cases, however, improvement of crop adaptation is likely to require the identification and insertion of complexes of genes (Tuberosa & Salvi, 2006). The complexity of adaptive and compensatory mechanisms in whole plants means that any individual modification will almost certainly require additional changes for its optimal expression, while the insertion of new biosynthetic pathways may involve the insertion of many genes.

12.1.2 Phenotyping platforms

The rapid advances in genomics and DNA sequencing, and in the analysis of the transcriptome, proteome and metabolome, have not yet been fully matched by advances in technology for identifying complex phenotypic characteristics such as growth rate, yield and stress tolerance. It has been notoriously difficult, for example, for breeders relying on visual selection methods to screen for characters such as yield.

Similarly, most physiological measures such as measurements of growth, photosynthesis, stomatal behaviour, water status and rates of water use tend to be labour intensive and suitable for comparing only tens to at most a hundred or so genotypes. Breeders, however, must be able to evaluate or screen very large numbers of progeny. As a result there has been substantial investment in the past ten years in the development of high-throughput *phenotyping platforms*, as breeders and researchers critically need the tools to screen large populations for specific advantageous characters.

The technology is developing rapidly with sophisticated automated phenotyping platforms being established in many laboratories. Pioneering facilities include those in Jülich (www.fz-juelich.de/), Montpellier (www1.montpellier.inra.fr/ibip/lepse/ english/ressources/index.htm), Canberra and Adelaide (www.plantphenomics.org.au), with an increasing number becoming available. These facilities generally combine controlled environment facilities for growing uniform plant material with systems for weighing and imaging individual potted plants. These systems can either involve robotic systems for moving cameras to individual plants, or else the use of automated plant handling conveyor-belt systems that transfer potted plants as required to weighing and imaging stations. The imaging systems used may include multiple-view cameras or laser scanners to determine plant structure and to follow growth; thermal cameras to measure temperature as an indicator of stomatal behaviour; fluorescence imagers to study photosynthesis; and visible/near infrared reflectance (multispectral or hyperspectral) sensors to provide information on biochemical composition of the tissues, or potentially on canopy development using vegetation indices (Furbank & Tester, 2011; Montes *et al.*, 2007). Weighing sensors provide direct information on water use. Other specialised sensors such as nuclear magnetic resonance (NMR) imagers and X-ray computed tomography (CT) may potentially also be incorporated in some cases, especially for the study of root system architecture, though cheaper systems using imaging of roots in transparent media may be more generally useful (Iyer-Pascuzzi *et al.*, 2010).

The use of multicolour fluorescence imagers (Langsdorf *et al.*, 2000) and sensors such as the Multiplex® (Force-A, Centre Universaire Paris Sud, Orsay, France; www.force-a.fr) allow one to monitor separately changes in fluorescence in the blue (440 nm), green (525 nm), red (685 nm) and far red (740 nm) where the shorter wavelengths represent fluorescence from phenolic compounds such as ferulic and chlorogenic acids in the cell walls. Changes in the ratios of between the fluorescence at different wavelengths can be used as powerful indicators of plant response to abiotic and biotic stresses (Langsdorf *et al.*, 2000; Jones & Vaughan, 2010) and for phenotyping (Bürling *et al.*, 2013).

It is, however, worth adding a word of caution here as simply collecting more data on plants will not of itself necessarily help in crop improvement; it is also necessary to have clear hypotheses as to what character (e.g. aspect of structure as determined from three-dimensional imaging or functional response such as stomatal behaviour) may be of value. Although one might think that selection of the appropriate phenotype is relatively straightforward when, for example, assessing stomatal behaviour by thermal imaging, it is often not clear whether one should be selecting for genotypes with high conductance (and hence high photosynthesis) or low conductance and hence enhanced water conservation, or indeed whether a high sensitivity to water deficit is more appropriate. Indeed the optimum depends on the environment within which the plant is to be grown; however in conventional glass house phenotyping platforms the results are more likely to depend on factors such as pot size and the interaction between plant size and pot size and not give useful information about the relative behaviour of genotypes with unconstrained roots in the field. Interpretation and use of morphological and growth data from three-dimensional imaging will be even more speculative.

It follows, therefore, that even more useful for plant breeders would be systems that allow the phenotyping of thousands of distinct genotypes in the natural field environment. Approaches such as the use of airborne or other unmanned aerial vehicles (UAV) sensors are particularly suitable for the study of canopy temperature as a measure of stomatal conductance, while imaging from mobile platforms such as 'cherry pickers' or balloons can give good coverage of breeding nurseries for thermal imagery (Jones *et al.*, 2009). The greatest flexibility and the highest resolution is provided by highly mobile imaging stations (White *et al.*, 2012) that move through the crop ('phenomobiles'; see Figure 12.3) or using sensors mounted on centre-pivot irrigators (O'Shaughnessy & Evett, 2010). Developments are rapid in this area with new systems continuing to become available.

High-throughput phenotyping, together with the developing techniques of informatics and systems modelling, transcriptomics and metabolomics, combined with machine-learning and multivariate statistical analyses of time-series response data are starting to provide tools for identification of complex regulatory networks involved in many stress responses (Beal *et al.*, 2005). Nevertheless there is still some way to go before this approach produces results that can be applied generally. An important step in this approach will then be the validation and testing of any candidate regulatory hub genes identified. It should then be possible to test any conclusions by using a reverse genetic approach involving the generation or screening of loss-of-function mutants in the appropriate genes.

12.1.3 Rapid screening tests

Rapid screening tests where a strong selection pressure is applied artificially have been successfully used in many cases, especially in breeding for pest and disease resistance. Similarly, seedling survival tests can be used for screening large numbers of plants for tolerance of cold, heat, drought or salt stress. Physiological knowledge has also proved valuable in selecting plants with appropriate daylength or vernalisation requirements for particular environments.

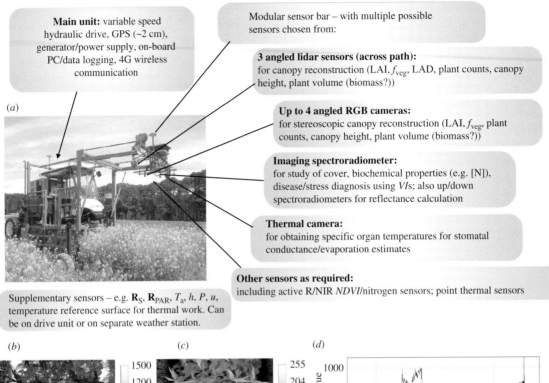

Main unit: variable speed hydraulic drive, GPS (~2 cm), generator/power supply, on-board PC/data logging, 4G wireless communication

Modular sensor bar – with multiple possible sensors chosen from:

3 angled lidar sensors (across path): for canopy reconstruction (LAI, f_{veg}, LAD, plant counts, canopy height, plant volume (biomass?))

(a)

Up to 4 angled RGB cameras: for stereoscopic canopy reconstruction (LAI, f_{veg}, plant counts, canopy height, plant volume (biomass?))

Imaging spectroradiometer: for study of cover, biochemical properties (e.g. [N]), disease/stress diagnosis using *VIs*; also up/down spectroradiometers for reflectance calculation

Thermal camera: for obtaining specific organ temperatures for stomatal conductance/evaporation estimates

Other sensors as required: including active R/NIR *NDVI*/nitrogen sensors; point thermal sensors

Supplementary sensors – e.g. R_S, R_{PAR}, T_a, h, P, u, temperature reference surface for thermal work. Can be on drive unit or on separate weather station.

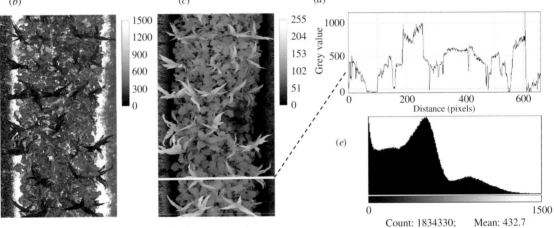

(b) *(c)* *(d)*

(e)

Count: 1834330;   Mean: 432.7

Figure 12.3 (*a*) Illustration of a mobile phenotyping platform (Phenomobile II from the Australian High Resolution Plant Phenomics Centre, Canberra; see www.plantphenomics.org.au/HRPPC). The sensor bar may be adapted to use any of a wide range of sensors depending on experimental objectives; these are likely to include sensors that give information on: canopy structure (such as stereo photography or Lidar); those that give information on canopy biochemical composition (e.g. broadband R/NIR sensors or spectroradiometers that can be used to derive various VIs as estimators of properties such as nitrogen or water content), and thermal cameras or sensors that give information on evaporation and stomatal conductance. Typical Lidar images (at 660 nm) obtained from movement of the line scanner over mixed crops of maize and oilseed rape are shown as either an intensity map in (*b*) or as a height map in (*c*). The height transect across the line marked in (*c*) is shown in (*d*) and (*e*) shows a histogram of height for the area shown in (*c*). (Unpublished data with the kind permission of R. Furbank, J. Jimenez-Berni, D. Deery and X. Sirault.)

Attributes for which screening can be carried out fall into four classes, only some of which are suitable for the phenotyping platforms currently being developed.

1. *Morphological and anatomical.* These include characters such as plant height, leaf size or stomatal frequency. In practice, these are the easiest and most widely used by breeders and are readily selected for using current phenotyping platforms.
2. *Compositional.* Screening for grain composition, such as protein or lysine content, is widespread. Only a limited number of characters such as pigment composition are suitable for rapid screening using hyperspectral reflectance, with precision being degraded on moving from single leaves to whole canopies. Included in this class is screening for hormone content, such as concentration of abscisic acid as a test for drought tolerance (see below).
3. *Process rates.* For example, these might include processes such as photosynthesis, respiration or vernalisation.
4. *Process control and functional responses.* It is generally more difficult to screen for differences in the control of water status or of processes such as growth, transpiration or photosynthesis as this would normally require assessment of *responses* of a character such as stomatal aperture to altered conditions and at least double the number of observations.

For many characters it is necessary to test the performance of the selected genotype at a crop scale, because as we have seen earlier (e.g. Chapter 5 for sensitivity of evaporation and water use efficiency to stomatal behaviour), the performance of the genotype depends on its interaction with neighbours and with the environment itself. Models can, however, provide a useful basis for predicting behaviour and limiting wasted effort on the development of inappropriate ideotypes. To be of value, a useful screening test must satisfy a number of criteria: (i) the character must be easier to assess than yield itself; (ii) there must be correlation (preferably causal) between the character and yield in the field (though sometimes this may only occur when combined with another character); (iii) the test should be simple, rapid, cheap and preferably capable of being used on seedlings at any time of year; and (iv) there must be heritable variation in the character. It is worth noting that a test involving a single measurement is more likely to be useful than the more complex tests usually required for estimates of responses.

Many other physiological tests have been proposed as a means of screening for particular characters though they have rarely been successful in yield improvement, often because of the compensation effects already discussed. A typical example of a physiological mass screening test is the compensation point test that was tried as a means of identifying mutants with the C_4 pathway (Section 7.9.4). The failure of this particular physiological screen and of other similar simplistic tests, as well as the generally very limited success obtained with attempts to insert so-called key or rate-limiting genes into cultivars by transformation, generally result from failure to recognise the complexity of plant growth and yield production process with the many compensating processes that occur. In the 1980s there was much interest in selecting cultured protoplasts for tolerance of stresses such as osmotic stress or salinity, because of the opportunities that this system provides for applying high selection pressures. The aim was that selected cell lines could be used directly for the clonal regeneration of improved genotypes, but again these studies have not yet lived up to expectation, partly because of the very different tolerance mechanisms commonly found in intact plants as compared with isolated cells (Rains, 1989).

12.1.4 Definition of the ideotype

In principle, a detailed knowledge of the mechanisms of yield production should enable us to short circuit the traditional empirical breeding approach of simply selecting those plants with the highest yield. In particular one could pin-point those processes that most limit yield so that effort can be concentrated on them and so perhaps even indicate ways of

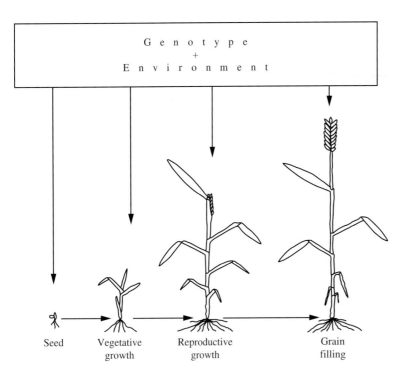

Figure 12.4 Final grain yield for cereal integrates the effects of genotype and environment over the whole growth cycle.

overcoming the limitations. Crop physiology uses mathematical models together with information on processes determining yield to determine the optimal responses at any stage and hence define the ideal plant or *ideotype* (Donald, 1968). A next stage would be to help devise rapid screening techniques that could be used by breeders. The objectives of physiology in a breeding programme can be summarised: (i) to define the ideal plant 'ideotype' for a particular situation, and (ii) to devise rapid screening procedures or techniques for applying high selection pressure.

The determination of the final yield of any crop is a complex process that depends on the cumulative effects of environment over the whole life interacting with genotypically determined developmental sequences (Figure 12.4). For grain crops the life cycle can be divided into the vegetative and reproductive phases, where the amount of growth during the vegetative phase sets a limit to the yield achievable.

The early attempts to gain an insight into the way that environment and plant genotype acted to determine cereal yields were based on yield

component analysis (Engledow & Wadham, 1923). Engledow and Wadham suggested that 'Theoretically the procedure would be to find out the plant characters which control yield per acre and by a synthetic series of hybridizations to accumulate into one plant form the optimum combination of characters'. In this approach the total yield per unit ground area can be treated as the product of several individual components; i.e. for a cereal:

$$\text{Yield} = \text{plants}\,\text{m}^{-2} \times \text{ears}\,\text{plant}^{-1} \times \text{grains}\,\text{ear}^{-1}$$
$$\times \text{mass}\,\text{grain}^{-1}$$

The idea was that effects on these different components could then be studied separately, but of course they all interact, with increasing planting density, for example, leading to the compensatory response of fewer and smaller ears per plant. There tends to be a hierarchy with grain mass being most stable (showing high heritability and small effect of environment), while ear size is less stable and the number of ears per plant is even less stable with the lowest heritability (Sadras & Slafer, 2012). Nevertheless where such compensatory responses can be shown to be incomplete, advances in yield should be obtainable.

Later, the techniques of growth analysis were adopted (e.g. Watson, 1958), with a change of emphasis from a leaf area basis (as in the net assimilation rate: Chapter 7), to a unit ground area basis (giving a crop growth rate, CGR). Using growth analysis, the dynamics of partitioning of carbohydrate could be analysed in more detail. Realisation of the importance of photosynthesis in productivity led to emphasis on leaf area index and its time integral, the leaf area duration (LAD). More recently there has been an emphasis on improving photosynthesis, though as we shall see below (Section 12.3.3) rather little of the yield advance seen since the beginning of the twentieth century has been attributable to enhanced leaf photosynthesis.

Additionally it became clear that the economic yield depended not only on the total dry matter production, but also on the harvest index. The importance of the concept of harvest index was recognised as long as a hundred years ago by the barley breeder, E. S. Beaven (see Donald & Hamblin, 1976), though he called it a 'migration coefficient'. The importance of harvest index in varietal improvements is well illustrated by a comparison of the wheat varieties Little Joss and Holdfast, that were widely grown in the UK in the late 1940s, with varieties such as Maris Huntsman, Maris Kinsman and Hobbit, that were available in the 1970s (see Table 12.1). Improvements in harvest index also played a major part in yield improvements of rice, wheat and maize during the green revolution. A large proportion of the increased yield in the modern semi-dwarf varieties is attributable to the increased harvest index, which itself is related to the smaller requirement for dry matter in the shorter stems. Similarly, yield responses to environmental factors such as water supply or nitrogen fertilisation partly depend on altered harvest index, as illustrated in Figure 12.5

Table 12.1 Yield characteristics of some winter wheat cultivars (data from Austin, 1978).

Cultivar	Date released	Height to base of ear (cm)	Relative yield (%)	Harvest index (%)
Little Joss	1908	130	100	30
Holdfast	1935	112	94	31
Maris Huntsman	1970	95	148	40
Maris Kinsman	1975	82	145	38
Hobbit	1975	67	166	45

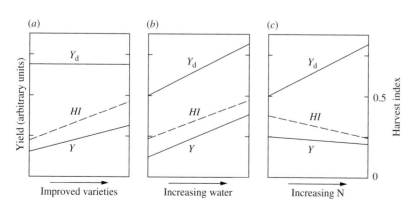

Figure 12.5 Contribution of changes in total dry matter production (Y_d) and harvest index (*HI*) to economic yield (*Y*): (*a*) with improved varieties (largely increased *HI*); (*b*) increasing water (both Y_d and *HI* increase); (*c*) increasing nitrogen fertilisation (Y_d increases but *HI* decreases). (After Donald & Hamblin, 1976.)

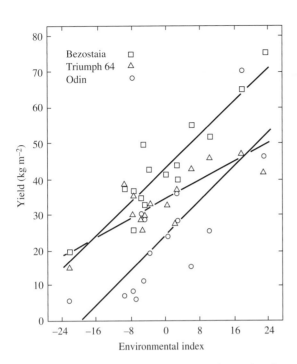

Figure 12.6 Typical genotype × environment interactions in wheat yields shown for a multisite international yield trial. Yield is plotted against an environmental index (the site mean yield as averaged over all varieties) and the regression lines for each of three genotypes shown. Although there is no G×E interaction for Bezostaia and Odin across these environments, the response of Triumph 64 is very different, with yields being much more stable across environments. (After Stroike & Johnson, 1972.)

Perhaps the greatest difficulty with attempts to define an ideal plant is that most plants have a great capacity for yield compensation. Donald's (1968) original ideotype for wheat included short strong stems with few erect leaves, large ears and a single culm, though it is now clear that there may be other ideotypes that are equally good. For example, it might be thought from study of the yield components that simply increasing ear number per plant would be a good way of increasing yield. Unfortunately such a simple approach does not necessarily work because the size of each ear is likely to decrease to compensate. In fact quite divergent strategies can result in similar yields. For example, two-row barley varieties tend to have many relatively small ears but they produce yields similar to those of the six-row

varieties, which have relatively few large ears; see for example the UK Recommended List of winter barley for 2012/13 (www.hgca.com). Compensation can occur at all stages of the life cycle. For example, if some seeds fail to germinate, neighbouring plants may compensate by producing more shoots and a greater leaf area.

Another problem is that the relative performance of different cultivars depends on the environment in which they are grown. A typical example of such genotype × environment interaction is shown in Figure 12.6. Usually cultivars with higher potential yield tend to be most affected by stress (in this case drought) with high-yield potential generally incompatible with high stress tolerance (see e.g. Sadras & Calderini, 2009). Treatment of genotype × environment interactions may be found in classical papers by Yates and Cochran (1938), Finlay and Wilkinson (1963), and Eberhardt and Russell (1966). These approaches assume that the relationship between genotype and environment is linear but advances in multivariate statistics have shown that the relationships are frequently more complex and techniques such as the additive main effects and multiplicative interactions (Gauch, 1992) have been developed to provide a fuller description. Further information may be found in Xu (2010).

12.2 Modelling and determination of crop ideotype

The use of models in environmental plant physiology was introduced in Chapter 1 and several examples where modelling techniques have been used to investigate the consequences of changes in plant morphology or of physiological response in different environments have already been discussed. These included the models for water use efficiency (Chapter 10), which can be used to determine optimum combinations of leaf size and stomatal behaviour for particular environments, and radiation transfer models that suggest an optimum leaf angle for maximal radiation interception (Chapter 2).

Such models can be used in attempts to explain plant distribution, while a similar approach may be adopted to determine crop ideotypes for maximum yield in certain environments. In this section a very simple example of optimisation is used to illustrate the principles of how modelling can be used in the determination of a crop ideotype. Similar optimisations can be applied to many other combinations of characters.

12.2.1 Timing of switch from vegetative to reproductive development

In an interesting example of the use of optimisation theory, Cohen (1971) showed that for any given environment, the maximum grain yield will be achieved by a plant that switches suddenly from a vegetative phase, where all available photosynthate is used for vegetative growth, to a reproductive phase where all resources are used for grain growth (see also Paltridge & Denhom, 1974). Many plants, for example cereals such as wheat, in fact approximate this type of response, where flowering on all shoots is approximately synchronous, occurring after vegetative growth ceases. On the other hand, there are many species, for example most legumes, that have an 'indeterminate' flowering pattern. That is, they continue to grow and produce flowers while the earliest formed seeds are ripening. This indeterminate behaviour is probably an adaptation to a relatively unpredictable environment with some seed being produced even if the growing season is very short, without sacrificing the potential for further seed production if water supply is maintained.

A question that might face a plant breeder breeding new plants for a particular environment, or an agronomist choosing a suitable genotype, is 'When is the optimum time for this vegetative to floral switch?'. To answer this one has first to set up a model for the crop, and then find the optimum by an appropriate method; this may be analytical or graphical or it may require an iterative solution (Thornley & France, 2007).

To start with a very simple model, one can assume that any photosynthate can go either into grain growth or into expansion of leaf area and that there is a fixed growing season of T days available (for ease of calculation assume $T = 10$ 'days', though a more realistic number can easily be substituted). If it is further assumed that the mass of photosynthate (m, in CO_2 equivalents) that gives one unit of leaf area (A) equals b, it follows that in the vegetative phase when all photosynthate is diverted into leaf area:

$$dA = dm/b \tag{12.1}$$

The value of b depends on leaf thickness and the amount of photosynthate that is required for supporting structures such as stems and roots. Similarly one might assume that the rate of production of photosynthate ($\mathbf{P} = dm/dt$) is proportional to the leaf area (though see below), i.e.:

$$\mathbf{P} = aA \tag{12.2}$$

where a is a constant. It is also necessary to assume an initial leaf area (A_o). On this basis various models can be derived, as follows.

Discontinuous model

Perhaps the simplest approach is to assume that increases in leaf area only occur at night, so that the growth in leaf area over the season is given by the discontinuous curve in Figure 12.7. If the vegetative to floral switch occurs early in the season there will be a long period available for filling grain, but the rate of grain filling will be slow because of the small leaf area available (dotted line in Figure 12.7(a)). If, however, the switch occurs later, the area (and consequently photosynthetic rate) is larger but the time available shorter.

The total photosynthate available for grain filling is given by the product of the photosynthetic rate for the appropriate area × time left. If the area after t days is denoted by A_t, the potential yield of grain dry matter (Y_d again in CO_2 equivalents) is given by:

$$Y_d = a A_t (T - t) \tag{12.3}$$

The optimum time for the switch is that which gives a maximum Y_d and may be determined by solving

(a)   (b)

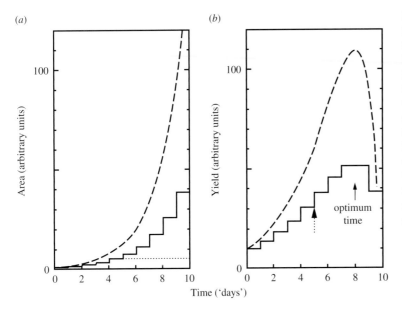

Figure 12.7 Yield models for a cereal (see text). (*a*) Increase in leaf area with time for the discontinuous model (solid line) or the continuous model (dashed line). The dotted line illustrates the time course for leaf area for the case where the switch to flowering occurs after five 'days' for the discontinuous model (and hence there is no further increase in leaf area). (*b*) Dependence of final yield for the two models on the time of the switch from vegetative to floral development. The dotted arrow indicates the yield corresponding to a flowering switch after five 'days'.

Eq. (12.3) for all t between 0 and T, as done in Figure 12.7(*b*). It is clear from this that for the particular values chosen ($A_o = 1$, $a = 1$, $b = 2$), the optimum switch occurs after eight days.

Continuous model

If photosynthate can be used immediately to generate new leaf area according to Eq. (12.1), the continuous exponential growth curve shown dashed in Figure 12.7(*a*) is obtained. In this case, the rate of change of area can be written:

$$dA/dt = (1/b)dm/dt = \mathbf{P}/b \tag{12.4}$$

Substituting from Eq. (12.2):

$$dA/dt - (a/b)A = 0 \tag{12.5}$$

This is a first-order differential equation of the form already seen above (e.g. Eq. (4.31)), the solution being:

$$A = c\,e^{(a/b)t} \tag{12.6}$$

where c is an arbitrary constant. The value of c may be obtained by substituting particular known values for the variables (= boundary conditions), i.e. at $t = 0$, $A = A_o$, which gives:

$$c = A_o \tag{12.7}$$

Substituting back into Eq. (12.6), a function for the time dependence of A is given by:

$$A_t = A_o e^{(a/b)t} \tag{12.8}$$

As before, one can solve to find the optimum value for t by calculating Y_d at various values of t and obtaining the optimum by graphical interpolation (Figure 12.7(*b*)).

A more elegant solution is to make use of calculus. Equation (12.8) can be substituted into Eq. (12.3) to give:

$$Y_d = a\,A_o e^{(a/b)t}(T - t) \tag{12.9}$$

It can be seen from Figure 12.7(*b*) that at the maximum Y_d, the slope of the curve relating Y_d and t is horizontal (i.e. $dY_d/dt = 0$). Therefore the optimum value for t can be determined by differentiating:

$$dY_d/dt = a\,A_o e^{(a/b)t}[(a/b)T - (a/b)t - 1] \tag{12.10}$$

then setting (dY_d/dt) equal to zero, and solving for t. This gives $t = T - (b/a)$, which can easily be shown to be an optimum rather than a minimum. As b increases, the optimum time gets earlier.

More complex models

Although the above models are useful for illustrative purposes, they include many major simplifications, such as that involved in Eq. (12.2). In practice, a more appropriate equation takes account of the diminishing returns as leaf area increases beyond a leaf area index of about 3, so that Eq. (12.2) can be replaced by, for example, a rectangular hyperbola (see Chapter 7):

$$\mathbf{P} = \mathbf{P}_{max}A/(k + A) \qquad (12.11)$$

Unfortunately, the differential equations resulting from the use of this or other more realistic assumptions rapidly become too complicated to be solved analytically, so that solution of more realistic models generally requires the use of numerical procedures on a computer. Other refinements could include treatment of seasonal and daily changes in environmental conditions, adequate modelling of stomatal behaviour and so on.

12.2.2 Dynamic crop simulation models

A number of dynamic crop simulation models have been developed to allow formulation and testing of hypotheses about how environment affects yield and also the investigation of how changing specific phenotypic characters might be expected to affect yield in particular environments. The earliest models (de Wit, 1965) were based on our developing understanding of the role of light interception and photosynthesis in the control of productivity. More recently, models have been improved and enhanced so that they usually incorporate a collection of subsystems that describe individual crop processes such as development, light interception, photosynthesis, respiration, water use and dry matter partitioning; they also often allow investigation of important management processes such as sowing date, water management and soil nutrition. The more mechanistic models tend to be aimed at understanding the genetic and environmental control of yield, while other, often more empirical, models are often aimed at optimisation of crop management.

It is possible to use any of these models to investigate the consequences of changing any particular component (e.g. time of flowering) in isolation to determine the optimum value. This can clearly be used to derive crop ideotypes for specific environments.

Sensitivity analysis of these simulation models can be used to pin-point deficiencies in our understanding of processes controlling yield and to determine those yield-determining processes that are most amenable to alteration by breeding or management. Large-scale ecosystem models can also be used in a similar fashion to investigate the consequences of changes in biotic or physical environment.

Over the years a wide range of large-scale dynamic simulation models have been developed for particular crops such as sorghum (Arkin et al., 1975), wheat (AFRCWHEAT2; Porter, 1993); CERES Wheat (Ritchie et al., 1985; see http://nowlin.css.msu.edu/wheat_book) and cotton (GOSSYM; Whisler et al., 1986). More recently many separate crop models have been incorporated into decision support platforms such as DSSAT (http://dssat.net) and APSIM (McCowan et al., 1996; www.apsim.info/Wiki/APSIM-and-the-APSIM-Initiative.ashx) that in addition to incorporating any of a range of different crop models are supported by climatic, soil and crop management databases. Both these systems are free to download.

In spite of the greater apparent realism in many of these large simulation models, there is evidence that the greater levels of reductionism and complexity involved in the most complex models may often be counterproductive. This is both because increasing model complexity makes it more difficult to understand the interactions between components, and because the more complex models have a much larger requirement for empirical parameters and may even give less accurate predictions of outputs such as yield or water use than do simpler models (Sinclair & Seligman, 1996). For example, comparisons of 5 (Jamieson et al., 1998) or 13 (Goudriaan, 1996) different wheat models showed substantial variation in their predictions under common environmental assumptions (with the simpler models sometimes outperforming the more complex models). Therefore there is much to be said, especially for educational and research purposes, for the approach with which this section began, that is, of using the simplest possible model for the purpose, since this can frequently give a better and more comprehensible picture of the system behaviour than the more complex models with all their multiple interactions.

The simple 'trial and error' optimisation approach described above can be usefully applied to a wide range of problems. I have found that the use of exercises of this type based on extremely simple assumptions can provide instructive insights into optimisation and how different factors interact in determining plant yield. Suitable examples for class projects include investigation of optimal sowing density (balancing the cost of extra seed against the earlier canopy closure at high seed rates), determining optimal stomatal behaviour (see Chapter 10) or determining the optimal partitioning of photosynthate between roots and shoots when water is limited. The APSIM modelling framework, mentioned above, provides a particularly convenient and powerful tool for such evaluations, which can be applied to these or more complex simulations as it is now freely available and reasonably easy to use. It can readily be used to investigate a wide range of crops and the effects of different management practices and it can be applied to investigate the potential validity of different potential plant ideotypes.

The increasingly powerful genomics tools that are becoming available open up an extension of the ideotype approach to what has been termed 'breeding by design', where rather than concentrating on single target traits one can use the molecular information to select and transfer what we hope are optimal combinations of specific alleles into new cultivars where different characters are pyramided (Fischer, 2011). This approach is now being more widely adopted with applications to crops as diverse as maize (Chapius et al., 2012) and peach (Quilot-Turion et al., 2012). Unfortunately, as we have seen from the difficulties with obtaining good crop simulation models, we are still a long way from being able to make robust predictions of characters that can be pyramided successfully. Nevertheless a strong case can be made that progress will be best achieved by linking the new information from functional genomics to whole-crop physiology using systems modelling (Yin & Struik, 2008), bearing in mind that the effects of genetic alterations tend to diminish on moving

from the molecular through to the crop level. Although much progress has been made, it is important to avoid repeating past mistakes where a naive overoptimism on the incorporation of specific characters or genes has rarely been translated into new varieties.

12.2.3 Validation of the ideotype

An essential stage in development of an ideotype for breeders is its validation. In its simplest form this involves a positive correlation within a set of genotypes between expression of the character and yield under appropriate conditions. A better approach is to use isogenic lines, which differ only in the character under investigation. Unfortunately preparation of isogenic lines usually requires a complete breeding programme itself as it involves a long series of back-crosses, unless the character appears as a mutant within a commercial variety. For this reason rather few characters have been tested using isogenic lines. Examples include stomatal frequency (see below) and the uniculm (single stem) mutant in barley.

12.3 Examples of applications

In the past, many successful applications of environmental physiology have been largely 'explanatory', though there have been important applications in the area of crop management including the use of evaporation models in irrigation scheduling and light penetration models in the evaluation of pruning systems for fruit tree management. In this section some examples of physiological methods that have potential for inclusion in plant breeding programmes are examined.

12.3.1 Breeding for drought tolerance

Drought is a major factor limiting yields of crops in many areas, but irrigation is often either not possible or uneconomic, so that much effort has been (and is being) devoted to breeding drought-tolerant cultivars. Much useful information can be found on websites such as: www.plantstress.com and www.generationcp.org

while useful information on relevant genes in *Arabidopsis* including those related to stress tolerance may be found on http://rarge.gsc.riken.jp. There are many possible characters for drought tolerance (Chapter 10) but, as was seen there, the best combination depends on the crop, the climate and even on the farming system. A list of characters that have been suggested for inclusion in a particular drought-tolerance breeding programme for sorghum (a crop predominantly grown in semi-arid areas) is presented in Table 12.2. I have indicated with +++ those characters that with current breeding technology are

Table 12.2 Characters for drought tolerance of possible use in a sorghum improvement programme (modified and extended from Seetharama *et al.*, 1982).

	Techniques	Genetic variability	Prospects for breeding
Morphological/phenological			
1. Maturity	Visual	+++	+++
2. Developmental plasticity	Visual	+?	+?
3. Glossy leaves	Visual	+++	+++
4. Leaf number, size, shape	Visual	++	++
Physiological – constitutive			
5. Desiccation tolerance	Survival tests	+++	+++
6. Heat tolerance	Survival, ion leakage	+++	+++
7. High growth rates	Visual, growth analysis	++?	++
8. Low respiration/high photosynthesis	Gas exchange/chlorophyll fluorescence	++?	+?
9. Recovery after stress	Visual	+++	+++
10. Anatomical (e.g. stomatal density)	Microscope	++?	?
11. Root/shoot ratio	Growth analysis	++	+
12. Liquid phase resistance	Pressure chamber	?	?
13. Deep roots	Root boxes, herbicide tests	++	+?
Physiological – facultative			
14. Stomatal closure	Leaf temperature, porometer	++	+++
15. Leaf rolling	Visual	+++	++
16. Epidermal wax production	Visual, chemical analysis	++	++?
17. Leaf area increase	Visual	++	+?

Table 12.2 (*cont.*)

	Techniques	Genetic variability	Prospects for breeding
18. Leaf senescence	Visual	+++	+++
19. Remobilisation of stem reserves	^{14}C, growth analysis	++?	+++?
20. Relative increase in root growth	Root box, growth analysis	++	++
Physiological – metabolic			
21. Osmotic adjustment and synthesis of compatible solutes	Psychrometry, chemical analysis	++?	+?
22. Activation of antioxidant systems	Chemical analysis	?	?
23. Production of abscisic acid	Chemical analysis	++	?

The number of pluses indicates the variability that exists for any character and the possibility of breeding for it.

most likely to provide significant improvements in sorghum drought tolerance. A number of other screening tests, such as the use of chlorophyll *a* fluorescence (Chapter 7; Havaux & Lannoye, 1985) or carbon isotope discrimination (Chapter 10), have been proposed and have in the case of the use of ^{13}C discrimination already shown promise in the release of commercial varieties (Condon *et al.*, 2002). Although there is evidence that there may be significant genetic variation in many metabolic characters (e.g. respiration) or responsive characters (e.g. developmental plasticity or stomatal closure), their use is at present speculative and really needs improved screening techniques, though thermal imaging appears to provide a good system for high-throughput phenotyping for stomatal response (Jones *et al.*, 2009). The use of heat and desiccation tolerance tests applicable to large numbers of seedlings also holds substantial promise, but general validation in terms of correlation with yield under drought conditions is still awaited.

Although there is still much optimism that breeding for drought responsiveness can be of value in the future, it is interesting that almost all successes thus far in improving yield under water-limiting conditions have resulted from selection in stress-free conditions (Cattivelli *et al.*, 2008). In spite of this, our increasing understanding of QTLs related to yield under drought conditions (Tuberosa & Salvi, 2006) provides an important tool for breeders, both through marker-assisted selection and even potentially through the identification of the underlying genes and their use in genetic engineering (though the fact that they mostly relate to quantitative multigenic traits does limit their potential application).

Use of stomatal characters

Because of the central role that stomata have in the control of water loss, much effort has concentrated on the use of stomatal characters in breeding for drought tolerance (see e.g. Table 12.2, numbers 10 and 14). Physiological thinking, along the lines described in Chapter 10, has led to the widely held view that drought tolerance would be favoured by reduced rates of water use, both because this would increase water use efficiency and because it would conserve soil water for longer. This should be attainable by reducing stomatal conductance. Several approaches have therefore been tried by physiologists and breeders to select for stomatal characters (Jones, 1987a).

Stomatal frequency. Because conductance might be expected to decrease as stomatal frequency decreases

Table 12.3 Effect of selection for stomatal frequency per unit area on water use by two pairs of barley lines (data from Jones, 1977b, based on isogenic lines developed by D. C. Rasmussen).

	Whole plant		Flag leaf			
	Transpiration (g day^{-1})	Leaf area (cm^2)	Stomatal frequency (mm^{-2})	Pore length (μm)	Leaf area (cm^2)	Leaf conductance (mm^{-1} s^{-1})
High frequency lines:						
Minn. 92–43	73	329	83.4	16.7	24.9	4.6
CI 5064	75	412	97.6	17.7	18.3	4.6
Low frequency lines:						
Minn. 161–16	110	513	67.8	19.1	29.0	5.1
CI 4176	128	522	65.8	20.4	21.7	4.7

(Chapter 6), there have been many attempts to select for reduced numbers of stomata per unit area of leaf. There is a fair degree of genetic variation for stomatal frequency and some of these studies have been successful in their initial objective. Although there are reports that the reduced frequency can reduce water use or increase water use efficiency, it can also be disadvantageous. In an attempt to validate the idea that a reduced stomatal frequency per unit area should improve drought tolerance, isogenic lines for stomatal frequency were developed in barley. Surprisingly it was discovered that, contrary to expectations, the low-frequency lines transpired up to 6% faster per unit area than the high-frequency lines (Table 12.3). Further analysis showed that although stomatal frequency had been successfully reduced by selection, this had been offset by increases in pore size (so that leaf conductance was unchanged) and in total leaf area. In fact the increase in leaf area had dominated in its effect on transpiration (Table 12.3). Future attempts to breed for low stomatal frequency must therefore take account of the negative correlation between leaf (and cell) size and stomatal frequency.

Stomatal conductance. Direct measurements of stomatal conductance (or proxies such as canopy temperature) should, in principle, be better than component characters such as stomatal frequency. Unfortunately, although measurements of conductance can be made with porometers, biological and environmental variability (particularly in the field) means that large numbers of measurements are required to distinguish different genotypes; therefore a more promising approach seems to be to use thermal imaging, which can rapidly average data for many leaves and plants (Jones *et al.*, 2009), though there are indications that the increased canopy temperature that can occur as canopy height decreases may confound the use of canopy temperature as a simple measure of conductance (e.g. Giunta *et al.*, 2008).

Stomatal response. A potential disadvantage of selection for low stomatal frequency or low conductance is that it might limit potential photosynthesis where water is not limiting. Therefore it should, in theory, be better to select a plant that is effective at closing its stomata in response to drought (Table 12.2). Unfortunately it is more difficult to measure a response than a steady state because one needs to make at least twice as many measurements, while a complete interpretation may also require simultaneous measurements of leaf water potential

(ψ_ℓ). A further complication is that both mean conductance and mean response tend to change during ontogeny (Jones, 1979).

Abscisic acid production

Because of the difficulty of selecting directly for stomatal response and because it is known that the plant growth regulator abscisic acid (ABA) is both produced in stress and closes stomata, there has been much interest in selection for a high production of ABA in response to drought as this could be a good short-cut to drought-tolerant varieties. Unfortunately such an approach is clearly too naive because high levels of ABA may either indicate good stomatal control and therefore avoidance of stress (as in drought-tolerant varieties of sorghum and maize; e.g. Larqué-Saavedra & Wain, 1974), or they may occur *because* the plant is particularly stressed (as may be the case in wheat; Quarrie & Jones, 1979). Breeding experiments for a number of species have now confirmed the heritability of ABA production (see Quarrie, 1991 for a detailed discussion of breeding for different ABA accumulation capacity).

Genetic engineering and insertion of genes

Approaches to the identification of genes for insertion into plants that might confer drought tolerance include (i) selection of genes on the basis of our understanding of drought adaptation mechanisms (e.g. genes related to ABA metabolism or osmoregulation) and subsequent identification of the appropriate candidate genes; and (ii) their identification from physiological studies using microarrays to identify those genes whose expression is enhanced in response to drought (Cattivelli *et al.*, 2008; Valliyodan & Nguyen, 2006). Unfortunately it is difficult to identify the critical genes that favour drought tolerance from the latter approach alone, as the expression of many hundreds of genes is altered by stress, many of which are just indicators of damage rather than tolerance. Nevertheless many hundreds of papers have been published in the past 15 to 20 years reporting attempts to engineer plants with enhanced drought tolerance based on insertion or

overexpression of transcription factors or structural genes known to be induced by water deficit treatments. The genes tested have included a number of transcription factors as well as genes encoding enzymes that regulate the synthesis of compatible solutes (including proline and glycine-betaine as well as various sugars and sugar alcohols), late embryogenesis abundant (LEA) proteins and heat shock proteins (Shinozaki & Yamaguchi-Shinozaki, 2007). Most of the early emphasis was on protective compounds such as osmoprotectants, antioxidants and ROS scavengers (see Section 10.2.2).

More recently the approach has been widened to take account of the appropriate regulatory networks involved in response to stress. Effort has substantially shifted from attempts to insert or upregulate specific structural genes to the overexpression of transcription factors that can have much wider ranging impacts on the regulated expression of cascades of drought-responsive genes. Most of these transcription factors regulate their target gene expression through binding to the cognate *cis*-elements in the promoters of the stress-related genes. Particularly well characterised systems include the drought-responsive elements (DRE) recognised by DREB or CBF transcription factors (e.g. Yamaguchi-Shinozaki & Shinozaki, 1994), the abscisic acid (ABA)-responsive element (ABRE) recognised by bZIP domains (e.g. Uno *et al.*, 2000; Zhu, 2002) and the NAC transcription factors (e.g. Hu *et al.*, 2006). Although constitutive overexpression of DREBs can lead to undesirable traits such as stunted growth, when put under the control of drought-inducible promoters their insertion can at least improve survival under severe stress conditions (Morran *et al.*, 2011).

Many papers have reported that overexpression of such transcription factors can improve drought tolerance. Indeed the level of current interest in this field is indicated by the fact that in a search of the Web of Knowledge (http://wok.mimas.ac.uk) with the keywords 'drought, tolerance, transgenic' I identified at least 50 papers published within one six-month period in 2012 that reported effects of transgenes on 'drought tolerance'. Eighteen of these papers reported the effects of overexpression of transcription factors, while other major groups included those genes involved in cell or hormonal signalling (nine) and osmoregulation (seven) with effects on a large number of other processes from stomatal behaviour to root architecture also being reported. In most cases, however, the reported improvements in drought (or salt) tolerance with transgenic plants have only been demonstrated under rather artificial conditions that bear little relation to normal yield production in the field. Where tested under realistic conditions, however, the transgenic plants have not generally lived up to their initial promise, probably largely as a result of the complexity of the compensations that are involved in whole-plant responses in the field and because of the widely diverging natures of droughts in different environments. Nevertheless some cases, such as for the rice lines overexpressing NAC transcription factors (Hu *et al.*, 2006), the transgenics have been shown to give increased grain yield under both well-watered and drought conditions in the field.

12.3.2 Photosynthesis and crop yield

Historical yield trends have been dominated by improvements in harvest index, though other characters such as promotion of flowering time, increased grain number per unit area, altered leaf size and thickness, as well as reductions in height, have also contributed in some cases (Giunta *et al.*, 2008). Because photosynthesis is essential for biomass production it has been argued that breeding for increased photosynthesis should also be a route to increasing crop yield as there is wide variation between plants in leaf photosynthetic rates. Indeed there is evidence for several crops that increases in biomass production have contributed to historic yield increases (Fischer & Edmeades, 2010). Although there has been controversy as to whether yields are limited by 'source' processes (i.e. photosynthesis) or by 'sink' processes (e.g. capacity for grain growth), the general consensus is that both are important and co-limiting so that improving photosynthesis should contribute to improving yield.

With this background, many studies have demonstrated varietal or species differences in photosynthesis at various levels including activity of Rubisco, ^{14}C fixation by protoplasts, leaf slices or discs, or CO_2 fixation at the leaf level. There has, however, rarely been an indication from any of these studies of a clear positive association between leaf photosynthetic rate and crop yield. In fact the reverse is often the case. For example, Dunstone *et al.* (1973) in a good early study, the results of which have been confirmed several times since, showed that in wheat the highest photosynthetic rates tend to occur in small-leaved primitive diploid species, and the lowest rates in the high-yielding modern hexaploid varieties. It seems that total leaf area, and leaf area display, are much more crucial determinants of productivity. Indeed a study of Australian wheats released since 1958 has shown increases in canopy radiation use efficiency associated with changes in light-penetration profiles (Sadras *et al.*, 2012) even though there was no corresponding increase in leaf photosynthesis. These conclusions are supported by many studies showing a close relationship between dry matter productivity and leaf area or intercepted radiation (see Chapter 7).

Notwithstanding the failure so far consistently to relate crop yield to leaf photosynthesis, many workers are still optimistic, and it is probably true that such a relationship would hold *if other factors could be kept equal*. Indeed evidence from FACE CO_2 enrichment experiments does confirm that increasing photosynthesis can lead to enhanced crop yields (Section 11.3.3). As a result there is increasing optimism that manipulation of photosynthesis and light use efficiency, as already discussed in Section 7.9.4, has real potential as a route to enhancing crop yields in the future. For example there appears to be significant scope to improve the amounts, or more importantly the properties, of Rubisco, as there is substantial variability between species in both the rate constant for carboxylation (with reported values between 8.1 and 34 µM in a range of higher plants) and the specificity factor (with reported values varying between 70 and 120 in the same range of plants) (Parry *et al.*, 2011). Similarly there is scope to enhance the capacity for RuBP regeneration and the stability of Rubisco activase (Murchie *et al.*, 2009;

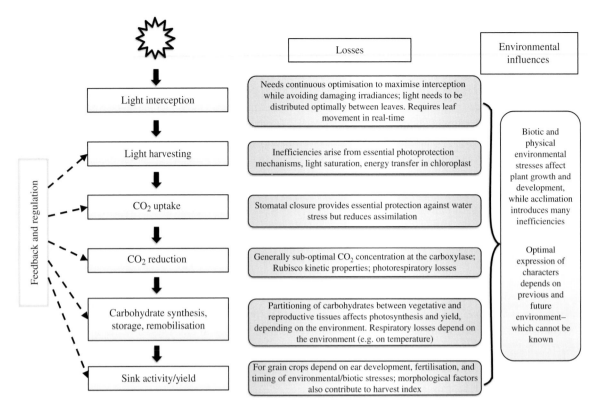

Figure 12.8 An illustration of some of the reasons why photosynthesis in C$_3$ plants rarely approaches its potential rate. Losses can occur at any of the stages indicated, both as a result of inbuilt inefficiencies (e.g. photorespiratory losses or the fact that it is not possible to fully optimise light interception across a range of irradiation conditions) or as a result of protective and regulatory mechanisms that avoid damage by environmental stresses (e.g. stomatal closure in response to drought). Feedback and regulatory processes may lead to compensations that may minimise the advantage of any single alteration.

Parry *et al.*, 2011). Other possibilities that could be targeted for increasing photosynthesis include other approaches to the inhibition of photorespiration and improvement of the efficiency of light conversion by reductions in the energy loss through non-photochemical quenching (NPQ). Note, however, that any such inhibition of NPQ would need to be linked with alterations in canopy structure to ensure that leaf orientation is such that no leaves are exposed to the damaging excessive radiation that NPQ protects against.

A step change in photosynthesis, and more importantly in water use efficiency, could potentially be achieved if it proved possible to insert the C$_4$ photosynthetic pathway into C$_3$ crops; although this is a very speculative endeavour significant effort is being devoted to this approach (http://c4rice.irri.org).

There has been recent speculation that overall there might be scope to increase photosynthesis by up to 50% as compared with current crops (Zhu *et al.*, 2010), though I suspect that such a figure is very overoptimistic, not least because this will require coordinated manipulation of many interlinking regulatory pathways each of which interact in as-yet unknown ways. In practice photosynthesis tends to operate well below its potential rate because of a complex array of inbuilt inefficiencies and of a range of protective and regulatory mechanisms that act to balance growth and survival (see Figure 12.8), so achievements of an improved crop yield must target all of these protective and regulatory processes and not just the photosynthetic system itself.

APPENDICES

APPENDIX 1

Units and conversion factors

The International System of Units (SI) is followed, though in some cases multiples of base units (such as mm) are used when they give more appropriate sized numbers. The following table defines the derived units in terms of SI base units and gives some useful conversions to other units in common use.

Quantity	SI base units	SI derived units	Equivalent forms of SI units	Equivalents in other units
Mass	1 kg			= 2.2046 pounds
Length	1 m			= 3.2808 feet
Time	1 s	(or min, h, etc.)		
Temperature	1 K			= 1°C
Electric current	1 A			
Amount of substance	1 mol			
Energy	$1 \text{ kg m}^2 \text{ s}^{-2}$	joule (J)	N m	= 10^7 erg = 0.2388 calorie
Force	1 kg m s^{-2}	newton (N)	J m^{-1}	= 10^5 dyne
Pressure	$1 \text{ kg m}^{-1} \text{ s}^{-2}$	pascal (Pa)	$\text{N m}^{-2}; \text{J m}^{-3}$	= 10 dyne cm^{-2} = 10^{-5} bar = 0.9869×10^{-5} atmosphere = 7.5×10^{-3} mm Hg
Power	$1 \text{ kg m}^2 \text{ s}^{-3}$	watt (W)	J s^{-1}	= 10^7 erg s^{-1}
Electric charge	1 s A	coulomb (C)		
Electric potential difference	$1 \text{ m}^2 \text{ kg s}^{-3} \text{ A}^{-1}$	volt (V)	J C^{-1}	
Kinematic viscosity	$1 \text{ m}^2 \text{ s}^{-1}$			= 10^4 stokes
Dynamic viscosity	1 Pa s			= 10 poise

APPENDIX 2

Mutual diffusion coefficients for binary mixtures containing air or water at 20°C

Values of D in air may be corrected for temperature (giving values with less then 1% error over the range 0 to 45°C) by multiplying by $(T/293.2)^{1.75}$ (data for solutes from Weast (1969), others mostly from Monteith and Unsworth (2008).

	Symbol	In air $(mm^2\ s^{-1})$	In water $(mm^2\ s^{-1})$
Water	D_W	24.2	0.0024^a
Carbon dioxide	D_C	14.7	0.0018
Oxygen	D_O	20.2	0.0020
Heat	D_H (= thermal diffusivity)	21.5	0.144
Momentum	D_M (= kinematic viscosity, v)	15.1	1.01
Sucrose (0.38% solution)		–	0.52×10^{-3}
Glycine (dilute)		–	1.06×10^{-3}
$CaCl_2$ (10 mol m^{-3})		–	1.12×10^{-3}
NaCl (10 mol m^{-3})		–	1.55×10^{-3}
KCl (10 mol m^{-3})		–	1.92×10^{-3}

[a] Coefficient of self-diffusion.

APPENDIX 3

Some temperature-dependent properties of air and water

Density of dry air (ρ_a), density of air saturated with water vapour (ρ_{as}), psychrometer constant ($\gamma = P\,c_p/0.622\,\lambda$), latent heat of vaporisation of water (λ), radiative resistance ($r_R = \rho c_p/4\varepsilon\rho T^3$), the factor converting conductance in units of mm s^{-1} to mmol m^{-2} s^{-1} ($g/g = g^m/g = P/\mathscr{R}T$) and kinematic viscosity of water (v).

T (°C)	ρ_a (kg m^{-3})	ρ_{as} (kg m^{-3})	γ (Pa K^{-1})	λ (MJ kg^{-1})	r_R (s m^{-1})	g/g (mmol m^{-2} s^{-1}/mm s^{-1})	v (mm^2 s^{-1})
−5	1.316	1.314	64.6	2.513	304	44.8	
0	1.292	1.289	64.9	2.501	282	44.0	1.79
5	1.269	1.265	65.2	2.489	263	43.2	
10	1.246	1.240	65.6	2.477	244	42.5	1.31
15	1.225	1.217	65.9	2.465	228	41.7	
20	1.204	1.194	66.1	2.454	213	41.0	1.01
25	1.183	1.169	66.5	2.442	199	40.3	
30	1.164	1.145	66.8	2.430	186	39.7	0.80
35	1.146	1.121	67.2	2.418	174	39.0	
40	1.128	1.096	67.5	2.406	164	38.4	0.66
45	1.110	1.068	67.8	2.394	154	37.8	

APPENDIX 4

Temperature dependence of air humidity and associated quantities

T (°C)	e_s (Pa)	c_{sW} (g m^{-3})	s (Pa °C)	ε	T (°C)	e_s (Pa)	c_{sW} (g m^{-3})	s (Pa °C)	ε
−5	421 (402)[a]	3.41	32	0.50	20	2337	17.30	145	2.20
−4	455 (437)[a]	3.66	34	0.53	21	2486	18.34	153	2.31
−3	490 (476)[a]	3.93	37	0.57	22	2643	19.43	162	2.44
−2	528 (517)[a]	4.22	39	0.60	23	2809	20.58	170	2.56
−1	568 (562)[a]	4.52	42	0.65	24	2983	21.78	179	2.69
0	611	4.85	45	0.69	25	3167	23.05	189	2.84
1	657	5.19	48	0.74	26	3361	24.38	199	2.99
2	705	5.56	51	0.78	27	3565	25.78	210	3.15
3	758	5.95	54	0.83	28	3780	27.24	221	3.31
4	813	6.36	57	0.88	29	4005	28.78	232	3.48
5	872	6.79	61	0.94	30	4243	30.38	244	3.66
6	935	7.26	65	1.00	31	4493	32.07	257	3.84
7	1002	7.75	69	1.06	32	4755	33.83	269	4.02
8	1072	8.27	73	1.12	33	5031	35.68	283	4.22
9	1147	8.82	78	1.19	34	5320	37.61	297	4.43
10	1227	9.40	83	1.26	35	5624	39.63	312	4.65
11	1312	10.01	88	1.34	36	5942	41.75	327	4.86
12	1402	10.66	93	1.42	37	6276	43.96	343	5.09
13	1497	11.35	98	1.49	38	6626	46.26	357	5.33
14	1598	12.07	104	1.58	39	6993	48.67	376	5.58
15	1704	12.83	110	1.67	40	7378	51.19	394	5.84
16	1817	13.63	117	1.77	41	7780	53.82	413	6.11
17	1937	14.48	123	1.86	42	8202	56.56	432	6.39
18	2063	15.37	130	1.97	43	8642	59.41	452	6.68
19	2196	16.31	137	2.07	44	9103	62.39	473	6.98

[a] Saturation vapour pressure over ice.

Calculation of saturation vapour pressure over water

The table above shows saturation water vapour pressure (e_s), saturation water vapour concentration (c_{sw}), slope of saturation vapour pressure curve (s) and the ratio of the increase of latent heat content to increase of sensible heat content of saturated air ($\varepsilon = s/\gamma$).

The value of $e_{s(T)}$ (the saturation vapour pressure of moist air over water) is approximated over the normal range of environmental temperatures by the following version of the Magnus equation (equation from the *CR-5 Users Manual 2009–12* from Buck Research – www.hygrometers.com modified from Buck, 1981):

$$e_{s(T)} = f(a \exp(bT/(c + T))$$

where T is in °C, $e_{s(T)}$ is in Pa, and the empirical coefficients are: $a = 611.21$, $b = 18.678 - (T/234.5)$, $c = 257.14$. The term $f = 1.00072 + (10^{-7} P (0.032 + 5.9 \times 10^{-6} T^2) \simeq 1.001)$ is an enhancement factor that corrects for the slight departure of the behaviour of water in air from that of a pure gas.

APPENDIX 5

Thermal properties and densities of various materials and tissues at 20°C

Approximate values selected principally from Herrington (1969), Weast (1969), Leyton (1975) and Edwards *et al.* (1979).

	Specific heat capacity, c_p (J kg^{-1} K^{-1})	Thermal conductivity, k (W m^{-2} K^{-1})	Density, ρ (kg m^{-3})
Air	1010	0.0257	1.204
Aluminium	896	237.0	2710
Cellulose	2500	–	1270–1610
Glucose	1260	–	1560
Plant leaves	3500–4000	0.24–0.57	530–910
Seasoned oak wood	2400	0.21–0.35	820
Fresh red pine wood	1960–3130	0.15–0.38	360–490
Polyethylene (high density)	2090	0.33	960
Polyvinyl chloride	1050	0.092	1714
Clay soil: dry	890	0.25	1600
Clay soil: wet (40% water)	1550	1.58	2000
Peat: dry	300	0.06	400[a]
Peat: wet (40% water)	1100	0.50	1160[a]
Water	4182	0.59	998.2[b]

[a] From www.simetric.co.uk/si_materials.htm.
[b] Rising to a maximum of 1000 kg m^{-3} at 4°C.

APPENDIX 6

Physical constants and other quantities

Constant	Value
Acceleration due to gravity (at sea level, latitude 45°) (g)	9.8067 m s^{-2}
Avogadro's number	6.022×10^{23} particles mol^{-1}
Gas constant ($\mathscr{R}$)	8.3144 J K^{-1} mol^{-1}
Planck's constant (h)	6.6261×10^{-34} J s
Solar constant ($\mathbf{I}_{pA}$)	1366 W m^{-2}
Speed of light *in vacuo* (c)	2.99792458×10^{8} m s^{-1}
Stefan–Boltzmann constant (σ)	5.6703×10^{-8} W m^{-2} K^{-4}
Molar volume of ideal gas at 0°C, 100 kPa	2.27106×10^{-2} m^{3} mol^{-1}
Molar volume of ideal gas at 0°C, 101.3 kPa	2.241×10^{-2} m^{3} mol^{-1}
Molecular mass of air (M_A)	28.964×10^{-3} kg mol^{-1}
Specific heat of air (c_p)	1012 J kg^{-1} K^{-1}
Water – dielectric constant (at 20°C) ($\mathscr{D}$)	80.2
Water – dynamic viscosity (at 20°C) ($\eta = \rho \mathbf{v}$)	1.008×10^{-3} N s m^{-2} (= Pa s)
Water – latent heat of fusion	334 kJ kg^{-1} or 6.01 kJ mol^{-1}
Water – partial molal volume (at 20°C) ($\overline{V}_W$)	18.05×10^{-6} m^{3} mol^{-1}
Water – surface tension against air (at 10°C) (σ)	74.2×10^{-3} N m^{-1}
Water – surface tension against air (at 20°C) (σ)	72.8×10^{-3} N m^{-1}
Water – surface tension against air (at 30°C) (σ)	71.2×10^{-3} N m^{-1}

APPENDIX 7

Solar geometry and radiation approximations

Useful relationships for calculating irradiance for modelling purposes may be derived from spherical geometry. These include the following (expressing all angles in degrees):

Solar elevation

The solar elevation at any site is given by:

$$\sin \beta = \cos \theta = \sin \lambda \sin \delta + \cos \lambda \cos \delta \cos h \quad (A7.1)$$

where β is the solar elevation above the horizontal, θ is the zenith angle of the sun (the complement of β), λ is the latitude of the observer, δ is the angle between the sun's rays and the equatorial plane of the Earth (solar declination) and is a function only of the time of year (see Table A7.1), h is the hour angle of the sun (the angular distance from the meridian of the observer) and is given by $15 (t - t_o)$ where t is the time in hours and t_o is the time at solar noon. Unfortunately the time of solar noon varies during the year by an amount that is given by 'the equation of time' (Table A7.1). In the Western hemisphere, the standard time at local apparent noon = 12.00 – (equation of time) – $4 \times$ (longitude in degree). As an example of the method of calculation, standard time at local apparent noon at New York (74°W) on 1 February would be 07 h 12.3 min GMT (i.e. 12 h + 17.3 – 296 min), which is equal to a local time (Eastern Standard Time) of 12 h 12.3 min because Eastern Standard Time is 5 h before the Greenwich standard. A number of useful solar calculators are now available on the web (e.g. at www.susdesign.com/sunangle).

Daylength

The daylength (N), that is the number of hours that the sun is above the horizon, may be obtained by solving equation A7.1 for $\beta = 0$.

This gives the hour angle of the sun, h, at sunrise or sunset as:

$$\cos h = -\tan \lambda \tan \delta \quad (A7.2)$$

so that the daylength in hours equals $2\,h/15$.

Angle between any surface and the sun

This is given by:

$$\cos \xi = [(\sin \lambda \cos h)(-\cos \alpha \sin \chi) - \sin h(\sin \alpha \sin \chi) \\ + (\cos \lambda \cos h) \cos \chi] \cos \delta + [\cos \lambda (\cos \alpha \sin \chi) \\ + \sin \lambda \cos \chi] \sin \delta \quad (A7.3)$$

where ξ is the angle between the sun's rays and the normal to the surface, χ is the zenith angle ($=$ slope) of the surface, α is the azimuth or aspect of the surface (measured east from north). This equation can be used to calculate irradiance at sloping sites or on leaves of any orientation (see e.g. Figure 2.9).

An example of application

Estimate the direct irradiance on a horizontal surface at solar noon on 1 April at a site at sea level and 45°N latitude.

Substituting in Eq. (A7.1) gives:

$\sin \beta = \sin 45 \sin 4.1 + \cos 45 \cos 4.1 \cos 0 = 0.756$

Using Eq. (2.11) with $m = 1/\sin \beta$ (Eq. (2.10)), and assuming an atmospheric transmittance of 0.7, gives:

$\mathbf{I}_{S(dir)} = 1370 \times 0.7^{1.32} \times 0.756 = 646\,\text{W m}^{-2}$

Distance to the sun

The value of the irradiance incident at the top of the atmosphere varies by up to about 3% as a result of seasonal variation in the distance between the Earth and the sun.

Table A7.1 Solar declination (δ, degree) and the equation of time (e, min) on the first day of each month.

Month	δ	e	Month	δ	e
January	−23.1	−3	July	+23.2	−4
February	−17.3	−14	August	+18.3	−6
March	−8.0	−13	September	+8.6	0
April	+4.1	−4	October	−2.8	+10
May	+14.8	+3	November	−14.1	+16
June	+21.9	+2	December	−21.6	+11

For simulation purposes δ may be obtained from:

$\delta = -23.4 \cos[360(t_d + 10)/365]$

where t_d is the number of the day in the year.

APPENDIX 8

Measurement of leaf boundary layer conductance

As an alternative to the use of equations such as Eqs. (3.31) to (3.33) for estimating g_a it is often better to measure leaf boundary conductance more directly. This is particularly true for very irregular-shaped leaves or for leaves in gas exchange chambers. Three main approaches are available, based on measurements of either the evaporation rate or the energy balance of model leaves (see Brenner & Jarvis, 1995).

From evaporation rate

The boundary layer conductance to water loss, g_{aW} may be determined directly from the rate of evaporation (**E**) from a 'wet' model (having no surface resistance analogous to cuticular or stomatal components) of the same dimensions and surface characteristics and exposed in the same situation, by using Eq. (5.20):

$$g_{aW} = \mathbf{E}/[(0.622\rho_a/P)(e_{s(Ts)} - e_a)] \tag{A8.1}$$

where $e_{s(Ts)}$ is the saturation vapour pressure at 'leaf' temperature and e_a is the water vapour pressure in the bulk air.

Adequate models can be made from wet blotting paper, though the exact surface characteristics of real leaves may be difficult to mimic. **E** is usually estimated gravimetrically, but can also be estimated by gas-exchange techniques in a leaf chamber (see Chapter 6) if the purpose is to determine g_a inside a particular gas-exchange cuvette.

From Newton's law of cooling

Another approach is to measure the heat transfer properties of model leaves in a radiation environment where net radiation is zero and where there is no evaporative cooling (e.g. aluminium model leaves in the dark). Real leaves can also be used if evaporation is prevented by covering the surface with a material such as petroleum jelly to prevent transpiration. In either case the only significant mode of energy exchange with the environment is by 'sensible' heat transfer (that is convection and conduction), and the rate of heat loss is proportional to the leaf–air temperature difference (ΔT) as predicted by Eq. (3.29). This is an example of Newton's law of cooling and provides a convenient method for estimation of g_{aH}.

The technique is to follow the time course of the change of 'leaf' temperature (T_ℓ) after the model has been heated above air temperature. Leaf temperature approaches air temperature as shown in Figure A8.1. For small temperature differences one can ignore the small contribution of differences in longwave radiation balance so one can write the instantaneous rate of heat loss per unit area by the 'leaf' as the rate of change of T_ℓ multiplied by its thermal capacity per unit area, i.e.:

$$\mathbf{C} = -\rho^* c_p^* \ell^* (dT_\ell/dt) = -\rho^* c_p^* \ell^* (d\Delta T/dt) \tag{A8.2}$$

where ρ^*, c_p^* and ℓ^* are, respectively, the density, specific heat and thickness of the 'leaf', and ΔT is the air–leaf temperature difference. Where there is no other significant form of energy exchange (i.e. net radiation is zero), this can be equated to the rate of sensible heat loss given by Eq. (3.29) to give the differential equation:

$$(\rho^* c_p^* \ell^*)(d\Delta T/dt) + g_{aH}\rho_a c_p \Delta T = 0 \tag{A8.3}$$

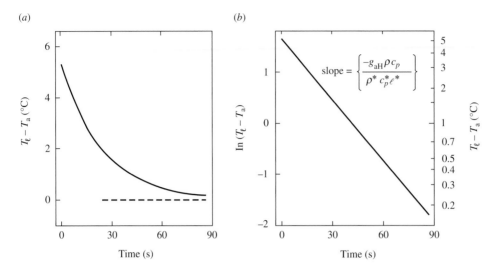

(a)

(b)

$$\text{slope} = \left\{\frac{-g_{aH}\rho c_p}{\rho^* c_p^* \ell^*}\right\}$$

Figure A8.1 Estimation of g_{aH} for an aluminium leaf model ($\ell^* = 0.001$ m, $c_p^* = 899$ J kg^{-1}, $\rho^* = 2.702 \times 10^3$) from a cooling curve. The time course of 'leaf'–air temperature difference ($T_\ell - T_a$) in (a) is transformed by plotting $\ln(T_\ell - T_a)$ against time in (b). The slope in this example is -0.04 s^{-1} = $-g_{aH}\rho_a c_p/\rho^* c_p^* \ell^*$. Substituting values for ρ^*, c_p^*, ℓ^*, ρ_a and c_p gives $g_{aH} = 0.04 \times 0.001 \times 899 \times 2702/1010 \times 1.204 = 80$ mm s^{-1}.

where ρ_a and c_p are the density and specific heat of air. Solution of Eq. (A8.3) and rearrangement gives:

$$g_{aH} = (\rho^* c_p^* \ell^* / (t_2 - t_1)\rho_a c_p) \ln (\Delta T_1 / \Delta T_2) \quad \text{(A8.4)}$$

where t_1 and t_2 are two times when the temperature differences are given by ΔT_1 and ΔT_2. The value of g_{aH} can then be determined from the slope of a plot of t against $\ln(\Delta T)$ as shown in Figure A8.1.

Using heated leaf replicas

A third approach that is particularly appropriate for estimation of g_{aH} for leaves in natural environments is to use heated leaf replicas (Brenner & Jarvis, 1995). Leaf replicas can be made using electrical heating tape or resistance wire sandwiched between sheets of brass or aluminium to spread the heat output evenly. One can estimate g_{aH} by comparing the temperatures (e.g. measured using thermocouples) of a heated replica and a similarly exposed unheated replica

(using subscripts 'h' and 'u' to denote heated and unheated leaves).

Using Eq. (3.29), the energy balance of each leaf in the steady state is given by:

$$\mathbf{R}_{n-h} + P_e = \rho_a c_p (T_h - T_a) g_{aH} \quad \text{(A8.5)}$$

$$\mathbf{R}_{n-u} = \rho_a c_p (T_u - T_a) g_{aH} \quad \text{(A8.6)}$$

where P_e is electrical heat output ($= I^2 R$; where I is the current and R is the electrical resistance). Since any incoming radiation for the two replicas will be the same, the difference in net radiative heat loss ($\mathbf{R}_{n-h} - \mathbf{R}_{n-u}$) depends only on longwave emission and can be approximated using the linearisation in Eq. (5.7) as:

$$\mathbf{R}_{n-h} - \mathbf{R}_{n-u} = 4\varepsilon\sigma T_u^3 (T_u - T_h) \quad \text{(A8.7)}$$

Therefore one can subtract Eq. (A8.6) from (A8.5) to give, after rearrangement, the following estimate of g_{aH}:

$$g_{aH} = (1/\rho_a c_p)([P_e/(T_h - T_u)] - 4\varepsilon\sigma T_u^3) \quad \text{(A8.8)}$$

APPENDIX 9

Derivation of Equation (9.9)

When the leaf temperature is at the equilibrium temperature for a particular environment ($T_\ell = T_e$):

$$\mathbf{S} = 0 = \mathbf{R}_n - \mathbf{C} - \lambda \mathbf{E} \tag{A9.1}$$

From Eq. (5.10) it follows that:

$$\mathbf{R}_n = \mathbf{R}_{ni} - \rho_a c_p (T_e - T_a)/r_R \tag{A9.2}$$

and from Eq. (9.3) that:

$$\mathbf{C} = \rho_a c_p (T_e - T_a)/r_{aH} \tag{A9.3}$$

and from Eq. (9.4) that:

$$\lambda \mathbf{E} = (0.622 \, \rho_a \lambda/P)(e_{s(Te)} - e_a)/(r_{aW} + r_{\ell W}) \tag{A9.4}$$

This can be expanded, making use of Eq. (5.21) to give:

$$\lambda \mathbf{E} = (0.622 \rho_a \lambda/P)(D + s(T_e - T_a))/(r_{aW} + r_{\ell W})$$
$$= (\rho_a c_p/\gamma)(D + s(T_e - T_a))/(r_{aW} + r_{\ell W}) \tag{A9.5}$$

Substitution of Eqs. (A9.2), (A9.3) and (A9.5) into Eq. (A9.1) gives:

$$0 = \mathbf{R}_{ni} - \rho_a c_p (T_e - T_a)[(1/r_{HR}) + (s/\gamma(r_{aW} + r_{\ell W}))]$$
$$- \rho_a c_p D/[\gamma(r_{aW} + r_{\ell W})] \tag{A9.6}$$

If, however, $T_\ell \neq T_e$:

$$\mathbf{S} = \mathbf{R}_{ni} - \rho_a c_p (T_\ell - T_a)[(1/r_{HR}) + (s/\gamma(r_{aW} + r_{\ell W}))]$$
$$- \rho_a c_p D/[\gamma(r_{aW} + r_{\ell W})] \tag{A9.7}$$

so substracting Eq. (A9.6) from (A9.7) gives:

$$\mathbf{S} = \rho_a c_p (T_e - T_\ell)[(1/r_{HR}) + (s/\gamma(r_{aW} + r_{\ell W}))] \tag{A9.8}$$

which can be substituted into Eq. (9.8) to give Eq. (9.9).

APPENDIX 10

Answers to sample problems

2.1 (i) (a) The total shortwave radiation absorbed is the sum over all wavebands 0.3–3.0 μm of $\alpha \mathbf{I}$, i.e. $(0.85 \times 450) + (0.20 \times 380) + (0.65 \times 70) = 504$ W m^{-2}.

(b) The shortwave absorption coefficient is the ratio of the total shortwave radiation absorbed (504 W m^{-2}) to the total incident shortwave $(450 + 380 + 70 = 900$ W m$^{-2}) = 0.56$.

(c) As there is no sensible heat or latent heat exchange, the energy absorbed $(\mathbf{I}_{S(absorbed)} + \mathbf{I}_{L(absorbed)}) =$ the thermal radiation given off (which from Eq. (2.4) $= \varepsilon \sigma T^4$). $\mathbf{I}_{L(absorbed)} = \varepsilon \sigma\, T_{environment}{}^4$. Assuming $\varepsilon = 1$, substituting 293 K for $T_{environment}$, and rearranging gives $\mathbf{I}_{S(absorbed)} = \sigma(T^4 - 293^4)$. Therefore with further rearrangement, $T^4 = \mathbf{I}_{S(absorbed)}/(\sigma + 293^4) = 504$ (W m^{-2})/5.6703 × 10^{-8} (W m^{-2} K^{-4}) $+ 293^4$ (K^4), so that taking the fourth root gives leaf temperature as 357 K or 84°C.

(ii) Sensible heat exchange has not been allowed for.

2.2 (i) The net radiation absorbed is the difference between that absorbed $(\alpha \mathbf{I}_{Sd} + \alpha \mathbf{I}_{Su} + \mathbf{I}_{Ld} + \mathbf{I}_{Lu})$ and the thermal radiation emitted from the two sides of the leaf, i.e. $((0.5 \times 500) + (0.3 \times 0.5 \times 500) + (\sigma\, 268^4) + (\sigma\, 297^4) - 2\, \sigma\, 293^4) = 223$ W m^{-2}.

(ii) We have assumed that the emissivities equal 1.

2.3 (i) From Eq. (2.1), $\mathbf{E} = hc\lambda = (6.6262 \times 10^{-34}\,\text{J s}) \times (2.998 \times 10^8\,\text{m s}^{-1})/(500 \times 10^{-9}\,\text{m}) =$

3.97×10^{-19} J photon^{-1} for green light. A similar calculation for infrared gives 0.993×10^{-19} J photon^{-1}.

(ii) The wavenumber $= \lambda^{-1}$ (in cm^{-1}), so for green light the wavenumber is $1/(500 \times 10^{-7}\,\text{cm}) = 20\,000$ cm^{-1}, and for infrared light it is given by $1/(2000 \times 10^{-7}\,\text{cm}) = 5000$ cm^{-1}.

(iii) Since energy per photon is proportional to $1/\lambda$, the number of photons per unit energy is proportional to λ, so that there are 4× as many photons at 2000 nm as for equal energy at 500 nm.

2.4 (i) The fraction of ground area sunlit (which for opaque leaves is $\mathbf{I}/\mathbf{I}_o$) equals e^{-L} with a horizontal leaved canopy (Eq. (2.16)) so that (a) for $L = 1$, this equals $e^{-1} = 0.368$, and (b) for $L = 5$, this equals $e^{-5} = 0.0067$.

(ii) For a horizontal leaved canopy, the sunlit leaf area index $(L_{sunlit}) = 1 - e^{-L}$ (Eq. (2.17)), so the corresponding values for L_{sunlit} are 0.632 and 0.993.

(iii) For randomly oriented leaves, $L_{sunlit} = (1 - e^{-\mathcal{k}L})/\mathcal{k}$, where $\mathcal{k}_{,} = 0.5\,\text{cosec}\beta = 0.5/\sin\beta$. For $\beta = 40°$, $\mathcal{k} = 0.5/0.6428 = 0.7779$. Therefore for $L = 1$, $L_{sunlit} = (1 - \exp(-1 \times 0.7779))/0.7779 = 0.695$. Similarly for $L = 5$, $L_{sunlit} = 1.259$.

2.5 $\mathbf{I}/\mathbf{I}_o = 0.25$, which, from Eq. (2.18) $= e^{-\mathcal{k}L}$, so $L = -[\ln(0.25)]/\mathcal{k}$, and using Table 2.5 to calculate $\mathcal{k}$: (i) for $\mathcal{k} = 1/(2\sin(60))$, $L = 2.401$; (ii) for $\mathcal{k} = 1$, $L = 1.386$.

3.1 (i) From Eq. (3.20), $\mathbf{J}_W = (24.2 \times 10^{-6}\,\text{mm}^2\,\text{s}^{-1}) \times ((17.3 - 11)\,\text{g m}^{-3})/(0.1\,\text{m}) = 1.525 \times 10^{-3}$ g m^{-2} s^{-1} = 1.525 mg m^{-2} s^{-1}.

(ii) From Eq. (3.21), $g_W = D_W/\ell$ (24.2 mm^2 s^{-1})/ (100 mm) = 0.242 mm s^{-1}.

(iii) $J_W^m = J_W^m/M_W = (1.525/18) = 0.085$ mmol m^{-2} s^{-1}.

(iv) From Appendix 3, $g_W = 41 \times g_W = 9.92$ mmol m^{-2} s^{-1}.

3.2 (i) (a) If one assumes that the characteristic dimension is $0.9 \times$ diameter, and that for a leaf g_{aH} is $1.5\times$ the value given by Eq. (3.31), $g_{aH} = 1.5 \times 6.62$ ((1 m^{-1})/ $(0.9 \times 0.02$ m)$^{0.5}$ = 74 mm s^{-1}.

(b) From Table 3.2, $g_{aW} = 1.08 \times 74 = 79.9$ mm s^{-1}.

(c) $g_{aM} = 0.8 \times 74 = 59.2$ mm s^{-1}.

(d) δ for momentum is given by $2 \times D_M/g_M = (2 \times 15.1$ mm^2 s^{-1})/(59.2 mm s^{-1}) = 0.51 mm.

(ii) For heat the surface is now the top of the tomentum, so g_{aH} is still 74 mm s^{-1}; for water vapour, on the other hand, there is an extra resistance approximately equal to $\ell/D_W = (1$ mm)/(24.2 mm^2 s^{-1}), so by the rules for two resistances in series $g_{aW} = 1/[(1/24.2) + (1/79.9)] = 18.6$ mm s^{-1}; for momentum the conductance is still unchanged at 59.2 mm s^{-1}.

(iii) The Reynolds number $(ud/\nu) = (1 \times 10^3$ mm s$^{-1}) \times (0.9 \times 20$ mm)/(15.1 mm^2 s^{-1}) = 1192, which is in the range where a laminar boundary layer breaks down.

3.3 (i) If one assumes that $d = 0.64$, $h = 0.512$ m and $z_o = 0.13 h = 0.104$ m, and substitutes into Eq. (3.37), $u_* = (4 \times 0.41)/\ln[(2 - 0.512)/ 0.104] = 0.616$ m s^{-1}.

(ii) Also substituting into Eq. (3.37), $u_{0.8} = (0.616/0.41) \ln[(0.8 - 0.512)/0.104] = 1.53$ m s^{-1}.

(iii) Substituting in Eq. (3.42), $\tau = (1.204$ kg m$^{-3}) \times (0.616$ m s$^{-1})^2 = 0.457$ kg m^{-1} s^{-2}.

(iv) From Eq. (3.45), $g_{AM} = (0.616$ m s$^{-1})^2/ (4$ m s$^{-1}) = 0.95$ m s^{-1}.

4.1 (i) (a) From Eq. (4.1) (assuming that $\alpha = 0$, $T = 20°$C), $h = [2 \times (7.28 \times 10^{-2}$ N m$^{-1}) \times$ cos 0°]/[0.5 × 10^{-3} (m) × 998.2 (kg m^{-3}) × 9.8 (m s^{-2})] = 0.0298 m = 2.98 cm.

(b) 2.98 cm (it still rises to the same height, but further along the capillary).

(c) 2.98 cosα (cm) = 1.92 cm.

(d) As for (a), but substituting 0.5 μm for 0.5 mm, gives 29.8 m.

(ii) Pressure required to prevent capillary rise = $(2\sigma \cos\alpha)/r = 0.291$ MPa.

4.2 (i) $\psi_p = \psi - \psi_\pi = -1 + 1.5$ MPa = 0.5 MPa.

(ii) The solute concentration decreases from c_s to $c_s/1.25$, therefore since ψ_π is proportional to c_s (Eq. (4.8)), the new value of $\psi_\pi = \psi_{\pi o}/1.25 = -1.5/1.25$ MPa = -1.2 MPa.

(iii) The new $\psi_p = -0.5 + 1.2$ MPa = 0.7 MPa.

(iv) If ψ is linearly related to volume, when ψ reaches 0, the volume will be $1.5 \times V_o$. The initial relative water content of the cell will equal initial volume/turgid volume = $1/1.5 = 0.666$ (assuming the whole volume is water).

(v) The easiest way to determine ε_B is graphically giving the value at full turgor as 2 MPa.

4.3 (i) From Eq. (4.25), $J_v = (0.1 \times 10^{-3}$ (m))$^2/ [8 \times (1.008 \times 10^{-3}$ (N s m$^{-2})) \times 1$ (m)] $\times (5 \times 10^3$ Pa) = 6.2004×10^{-3} m s^{-1}. The volume flow per pipe therefore = $J_v \times \pi r^2 = 1.95 \times 10^{-10}$ m^3 s^{-1}.

(ii) From Eq. (4.23), $L = 6.2004 \times 10^{-3}$ (m s$^{-1}) \times 1$ (m)/(5 × 10^3 (Pa)) = 1.24×10^{-6} m^2 s^{-1} Pa^{-1}; from Eq. (4.24), $L_p = 1.24 \times 10^{-6}$ m s^{-1} Pa^{-1}; $R = 1/L_p = 8.06 \times 10^{-5}$ Pa s m^{-1}.

(iii) Rearranging Eq. (4.25) and substituting appropriate values gives $r^2 = (6.2004 \times 10^{-3}$ (m s$^{-1})) \times (8 \times 1.008 \times 10^{-3}$ (N s m$^{-2}))/ (1 \times 10^3$ (Pa)) = 5.00×10^{-8} m^2, therefore $r = 2.24 \times 10^{-4}$ m, and $d = 0.448$ mm.

4.4 (i) From Eq. (4.28):
(a) = (1.1 (MPa))/(0.1 × 10^{-6} (m^3 m^{-2} s^{-1})) = 1.1×10^7 MPa s m^{-1}.

(b) $= (1.1\ (\text{MPa}))/[(0.1/10) \times 10^{-6}\ (\text{m}^3\ \text{plant}^{-1}$
$\text{s}^{-1})] = 1.1 \times 10^8\ \text{MPa s m}^{-3}$.

(c) $= (1.1\ (\text{MPa}))/[30 \times (0.1/10) \times 10^{-6}$
$(\text{m}^3\ \text{m}^{-2}\ \text{s}^{-1})] = 3.67 \times 10^6\ \text{MPa s m}^{-1}$.

(ii) Assuming that all resistances stay the same, and that ψ_{soil} stays at –0.1 MPa, the potential drop across the plant will double (as flow per plant has doubled) to give $\psi_\ell = -2.3$ MPa.

(iii) If half the shoots are removed, the shoot resistance doubles so that the total plant resistance $= R + R/2 = 3R/2$, therefore the potential drop across the plant $= 1.5 \times 1.1 = 1.65$ MPa, and $\psi_\ell = -1.75$ MPa.

5.1 (i) $e_s = 4243$ Pa (Appendix 4)

(ii) $e = 0.4 \times 4243\ \text{Pa} = 1697$ Pa.

(iii) $c_W = 0.4 \times 30.38 = 12.15\ \text{g m}^{-3}$ (using Appendix 4).

(iv) $D = 4243 - 1697\ \text{Pa} = 2546$ Pa.

(v) From Figure 5.2 (see Eq. (5.18)), $T_{\text{wb}} = 20°C$.

(vi) $T_d = 15°C$ (i.e. the temperature from Appendix 4 at which e_s equals 1697 Pa).

(vii) From Eq. (3.7), $m_W = 12.15\ (\text{g m}^{-3})/$
$(1.164 - [(1.164 - 1.145) \times 0.4]\ \text{kg m}^{-3} = 1.05 \times 10^{-2}$.

(viii) For $P = 101.3$ kPa, $x_W = 1697/(101.3 \times 10^3)$
$= 1.68 \times 10^{-2}$.

(ix) From Eq. (5.14), $\psi = [8.3144\ (\text{J K}^{-1}\ \text{mol}^{-1}) \times$
$303\ (\text{K})/(18.05 \times 10^{-6}\ (\text{m}^3\ \text{mol}^{-1}))]\ln(0.4) = -127.9$ MPa.

5.2 (i) Assuming that $\varepsilon = 1$ and substituting in Eq. (5.6), $R_{\text{ni}} = 430 + 4 \times (5.6703 \times 10^{-8}) \times (295^4 - 292^4)\ \text{W m}^{-2} = 447\ \text{W m}^{-2}$.

(ii) From Eq. (5.9), $g_R = (4 \times 5.6703 \times 10^{-8} \times 292^3)/(1012 \times 1.204) = 4.64 \times 10^{-3}\ \text{m s}^{-1} = 4.64\ \text{mm s}^{-1}$.

5.3 (i) Substituting into Eq. (5.26):

(a) When the surface is wet $g_A = g_W$, so that $\mathbf{E}$ (for forest) $= [145\ \text{Pa K}^{-1} \times 400\ \text{W m}^{-2} + 1.204\ \text{kg m}^{-3} \times 1010\ \text{J kg}^{-1}$
$\text{K}^{-1} \times 0.2\ \text{m s}^{-1} \times 1000\ \text{Pa}]/[2.454 \times$

$10^6\ \text{J kg}^{-1} \times (145\ \text{Pa K}^{-1} + (66.1\ \text{Pa K}^{-1} \times 0.2/0.2))] = 0.581\ \text{g m}^{-2}\ \text{s}^{-1}$.
Similarly for short grass, but substituting $g_A = 0.01\ \text{m s}^{-1}$ gives $\mathbf{E} = 0.135\ \text{g m}^{-2}\ \text{s}^{-1}$.

(b) $g_W = ((g_A)^{-1} + (g_L)^{-1})^{-1}$ so that for forest ($g_A = 0.2\ \text{m s}^{-1}$) and where $g_L = 0.03\ \text{m s}^{-1}$, $g_W = ((0.2)^{-1} + (0.03)^{-1})^{-1} = 0.02609\ \text{m s}^{-1}$. A similar calculation for grassland gives $g_W = 0.0075\ \text{m s}^{-1}$.
Substitution of these values into Eq. (5.26) gives $\mathbf{E}$ of 0.188 g m^{-2} s^{-1} and 0.123 g m^{-2} s^{-1} respectively for forest and grass.

(ii) The Bowen ratio, $\beta = \mathbf{C}/\lambda\mathbf{E}$. Making use of the energy balance (Eq. (5.1)) and remembering that $\mathbf{M}$ and $\mathbf{S}$ are zero at the steady state, enables one to write $\beta = ((\mathbf{R_n} - \mathbf{G}) - \lambda\mathbf{E})/\lambda\mathbf{E}$. Using the answers from (i) above enables one to write for wet forest, for example, $\beta = (400 - 2454 \times 0.581)/(2454 \times 0.581) = -0.72$.
Corresponding answers for the other cases are 0.2, –0.133 and 0.33.

5.4 (i) From Eq. (5.32), $\Omega = (2.20 + 1)/(2.20 + 1 + 15/5) = 0.516$.

(ii) From Eq. (5.33) the relative reduction in transpiration $(d\mathbf{E}/\mathbf{E}) = (1 - \Omega)(dg_\ell/g_\ell) = 0.484 \times 0.5 = 0.242$.

(iii) Eq. (5.33) is only strictly true for small changes in g_ℓ, so expressing the changes with respect to the initial $\mathbf{E}$ and g_ℓ is only an approximation.

6.1 (i) $r_{\text{aW}} = (c_{\text{W(leaf)}} - c_{\text{W(air)}})/\mathbf{E}_{\text{(blottingpaper)}} = (17.30\ \text{g m}^{-3} - 0.2 \times 17.30\ \text{g m}^{-3})/0.230\ \text{g m}^{-2}\ \text{s}^{-1} = 60.2\ \text{s m}^{-1}$.

(ii) The cuticular resistance, r_c, is given by the difference between the total resistance to water loss with the stomata closed and the boundary layer resistance (i.e. assuming that the final rate is achieved when the stomata close completely) $= [(17.30 \times 0.8\ \text{g m}^{-3})/(0.002\ \text{g m}^{-2}\ \text{s}^{-1})] - 60.2\ \text{s m}^{-1} = 6860\ \text{s m}^{-1}$.

(iii) For the initial rate of water loss, $r_{\ell W} = [(17.30 \times 0.8 \text{ g m}^{-3})/(0.08 \text{ g m}^{-2} \text{ s}^{-1})] - 60.2 \text{ s m}^{-1} = 112.8 \text{ s m}^{-1}$. Since the stomatal and cuticular resistances are in parallel, the stomatal resistance is given by $((r_{\ell W})^{-1} - (r_c)^{-1})^{-1} = 114.7 \text{ s m}^{-1}$.

6.2 (i) From Eq. (6.3), r_{sW} (for one surface) $= [10 \times 10^{-6} \text{ m} + (\pi \times 2.5 \times 10^{-6} \text{ m}/4)]/(200 \times 10^6 \text{ m}^{-2} \times \pi \times (2.5 \times 10^{-6} \text{ m})^2 \times 24.2 \times 10^{-6} \text{ m}^2 \text{ s}^{-1}) = 125.9 \text{ s m}^{-1}$; therefore for both surfaces, $r_{sW} = 125.9/2 = 62.9 \text{ s m}^{-1}$.

(ii) $g_{sW} = 1/(62.9 \text{ s m}^{-1}) = 0.0159 \text{ m s}^{-1} = 15.9 \text{ mm s}^{-1}$.

6.3 (i) As $T_\ell = T_a$ one can use Eq. (6.10). Assuming the $T_a = 25°C$, the molar volume of air $= 0.02241 \text{ m}^3 \text{ mol}^{-1} \times 298 \text{ K}/273 \text{ K} = 0.024462 \text{ m}^3 \text{ mol}^{-1}$, so that $u_e = (2 \times 10^{-6} \text{ m}^3 \text{ s}^{-1})/(0.024462 \text{ m}^3 \text{ mol}^{-1}) = 8.176 \times 10^{-5} \text{ mol s}^{-1}$, so that $g_W^{-1} = [[(1/0.35) - 1] \times 1.5 \times 10^{-4} \text{ m}^2/(8.176 \times 10^{-5} \text{ mol s}^{-1})]$, therefore $g_W = 0.293 \text{ mol m}^{-2} \text{ s}^{-1}$.

(ii) When the system is non-isothermal one needs to use Eq. (6.9). $x_{Ws} = e_{s(T\ell)}/P = (3167 \text{ Pa})/(1.013 \times 10^5 \text{ Pa}) = 0.03126$, $x_{Wo} = (0.35 \times 3565 \text{ Pa})/(1.013 \times 10^5) = 0.01232$, and $x_{We} = 0$, so that $g_W = (8.176 \times 10^{-5} \text{ mol s}^{-1} \times 0.01232)/(1.5 \times 10^{-4} \text{ m}^2 \times (0.03126 - 0.01232) \times (1 - 0.01232)) = 0.359 \text{ mol m}^{-2} \text{ s}^{-1}$.

6.4 (i) The stomatal response to D (in kPa) can be written as $g_\ell = 10(1 - (1/3)D)$, so that for $D = 1 \text{ kPa}$, $g_\ell = 6.6666 \text{ mm s}^{-1}$. In mass units, $E = (c_{Ws} - c_{Wa}) \times g_W$, so, using Eq. (5.11) to convert D to the corresponding water vapour concentration difference, one obtains (assuming a T of 293 K), $E = (2.17/293) \text{ g m}^{-3} \text{ Pa}^{-1} \times 1000 \text{ Pa} \times 0.006666 \text{ m s}^{-1} = 0.0494 \text{ g m}^{-2} \text{ s}^{-1}$.

(ii) Assuming a linear response of g_ℓ to ψ, g_ℓ can be written as $g_{\ell o} \times (1 + 0.5\psi)$, but as $\psi = -10E$, combining gives $g_{\ell o}(1-5E)$, which together with the humidity response gives g_ℓ (mm s^{-1}) $= 10(1 - (1/3)) \times (1-5E)$, therefore

$E = (2.17/293) \text{ g m}^{-2} \text{ Pa}^{-1} \times 1000 \text{ (Pa)} \times 0.006666 \text{ (m s}^{-1}) \times (1-5 E)$. Rearranging and solving for E gives $0.040 \text{ g m}^{-2} \text{ s}^{-1}$.

7.1 From Table 7.2:

(i) $F_v/F_m = 2.7/3.7 = 0.730$.

(ii) $F_q' = 3.2 - 1.2 = 2.0$.

(iii) $q_P = 2.0/2.3 = 0.870$.

(iv) $NPQ = (3.7/3.2) - 1 = 0.156$.

(v) Quantum yield $= (3.12 - 1.2)/3.2 = 0.625$.

7.2 (i) $u_e = (5 \times 10^{-6} \text{ m}^3 \text{ s}^{-1})/(0.0227107 \text{ m}^3 \text{ mol}^{-1} \times 296 \text{ K}/273 \text{ K}) = 2.031 \times 10^{-4} \text{ mol s}^{-1}$.

(ii) From Eq. (6.6b), $u_o = 2.031 \times 10^{-4} \text{ mol s}^{-1} \times (1 - 0.5/100)/(1 - 1.5/100) = 2.052 \times 10^{-4} \text{ mol s}^{-1}$.

(iii) The CO_2 concentration is 0.6 g m^{-3}, dividing by M_C ($= 44$) gives 0.01364 mol m^{-3}. The number of moles of gas per m$^3 = 1/(0.0227107 \text{ m}^3 \text{ mol}^{-1} \times 296/273) = 40.611 \text{ mol m}^{-3}$, therefore $x_e' = 0.01364/40.611 = 335.8 \times 10^{-6} \text{ mol mol}^{-1}$, or 335.8 ppm. (Alternatively one could use Eqs. (3.5) and (3.6).)

(iv) From Eq. (7.10), $P_m = [(2.031 \times 10^{-4} \text{ mol s}^{-1} \times 335.8 \times 10^{-6}) - (2.052 \times 10^{-4} \text{ mol s}^{-1} \times 335.8 \times 10^{-6} \times 450/600)](10 \times 10^{-4} \text{ m}^2) = 16.52 \text{ μmol m}^{-2} \text{ s}^{-1}$.

(v) From Eq. (6.9), $r_W = (10^{-3} \text{ m}^2)[(2.809/100) - (1.5/100)] \times [1 - (1.5/100)]/[(2.031 \times 10^{-4} \text{ mol s}^{-1}) \times (1.5/100 - 0.5/100)] = 6.348 \text{ m}^2 \text{ s mol}^{-1}$, therefore $g_W = 0.158 \text{ mol m}^{-2} \text{ s}^{-1}$, therefore from Table 3.2 and assuming largely still air $g' = (0.68/1.12) \times 0.158 \text{ mol m}^{-2} \text{ s}^{-1} = 0.096 \text{ mol m}^{-2}\text{s}^{-1}$.

7.3 Estimating slopes from Figure 7.19 gives ℓ'_g for the different methods:

(i) 7.5 m^2 s μmol^{-1}/(7.5 m^2 s μmol^{-1} + 7.0 m^2 s μmol^{-1}) $= 0.52$.

(ii) (21.5 μmol m^{-2} s^{-1} - 18.9 μmol m^2 s^{-1})/(21.5 μmol m^{-2} s^{-1}) $= 0.12$.

(iii) 7.5 m^2 s μmol^{-1}/(7.5 m^2 s μmol^{-1} + 22.1 m^2 s μmol^{-1}) $= 0.25$.

7.4 Since the relative masses of CO_2 and sucrose per mol of C are 44/30, and assuming an energy equivalent of 16 kJ g^{-1} in sucrose, this is equivalent to $16 \times 30/44$ kJ $(g\ CO_2)^{-1} = 10.9$ kg g^{-1}. Therefore:

(i) For May the efficiency of gross photosynthesis is $((288/7) \times 10.9$ kJ $(g\ CO_2)^{-1}$ m^{-2} day$^{-1})/(14.5 \times 10^3$ kJ m^{-2} day$^{-1}) = 3.1\%$. The corresponding figure for July is $((304/7) \times 10.9/(17 \times 10^3)) = 2.8\%$. Assuming that the energy in PAR is 50% of that in global radiation, the corresponding efficiencies in terms of incident PAR are 6.2% and 5.6% respectively.

(ii) For net productivity one needs to use the average energy per g in dry matter (17.5 kJ g^{-1}), so that the efficiency in May $= ((167/7) \times 17.5 \times (30/44))/$ $(14.5 \times 10^3) = 2\%$, and in July $((101/7) \times 17.5 \times (30/44))/(17 \times 10^3) = 1.0\%$, with the corresponding values in terms of PAR being 3.9% and 2%.

8.1 (i) From Eq. (8.4): (a) $\zeta = 0.904 \times 1.1 = 0.994$; (b) $\zeta = 0.904 \times 1.1 \times 0.08/0.35 = 0.227$.

(ii) Estimating from Figure 8.3, gives ϕ as 0.52 and 0.28.

9.1 (i) Substituting into Eq. (9.5):
$$T_\ell = 25°C + [(40 \times ((40/1.08) + 200)\ s\ m^{-1}$$
$\times 66.5$ Pa $K^{-1} \times 400$ W $m^{-2})/(1010$ J kg^{-1} K^{-1} $\times 1.204$ kg $m^{-3} \times (66.5$ Pa $K^{-1} \times (237.03$ s $m^{-1}) + (189$ Pa $K^{-1}) \times 40$ s $m^{-1}))] - [(40$ s m^{-1} $\times 0.6 \times 3167$ Pa$)/(66.5$ Pa $K^{-1} \times (237.03$ s $m^{-1}) + (189$ Pa $K^{-1}) \times 40$ s $m^{-1})] = 30.6°C$.

(ii) From Eq. (9.11), $\tau = (2.7 \times 10^6$ J $m^{-3} \times$ 1×10^{-3} m$)/[(1010$ J kg^{-1} $K^{-1} \times 1.204$ kg $m^{-3}) \times [((1/40) + (4 \times 5.6703 \times 10^{-8} \times 298^3/1010 \times 1.204)\ s\ m^{-1}) + (189$ Pa $K^{-1}/$ $(66.5$ Pa $K^{-1} \times 237.03$ s $m^{-1}))]] = 53$ s.

(iii) Similarly, when $r_{\ell W} = \infty$, $\tau = 74$ s.

(iv) Substituting $r_{\ell W} = 0$ into Eq. (9.5) gives $T = 20.7°C$.

(v) 20.8 s.

9.2 (i) From Eq. (9.20), $Q_{10} \approx (0.19/0.1)\exp(10/6) = 2.91$.

(ii) Rearranging Eq. (9.19): $E_a = \mathscr{R}T(T + 10) \times \ln(Q_{10}) = 75.4$ kJ mol^{-1}.

9.3 25°C.

REFERENCES

Aasamaa K, Sõber A (2011) Stomatal sensitivities to changes in leaf water potential, air humidity, CO_2 concentration and light intensity, and the effect of abscisic acid on the sensitivities in six temperate deciduous tree species. *Environmental and Experimental Botany* **71**, 72–8.

Abdou WA, Helmlinger MC, Conel JE *et al.* (2000) Ground measurements of surface BRF and HDRF using PARABOLA III. *Journal of Geophysical Research: Atmospheres* **106**, 11967–76.

Acevedo E, Hsiao TC, Henderson DW (1971) Immediate and subsequent responses of maize leaves to changes in water status. *Plant Physiology* **48**, 631–6.

Acharya BR, Assmann SM (2009) Hormone interactions in stomatal function. *Plant Molecular Biology* **69**, 451–62.

Acock B, Charles-Edwards DA, Fitter DJ *et al.* (1978) The contribution of leaves from different levels within a tomato crop to canopy net photosynthesis: an experimental examination of two canopy models. *Journal of Experimental Botany* **29**, 815–27.

Acock B, Grange RI (1981) Equilibrium models of leaf water relations. In *Mathematics and Plant Physiology* (eds Rose DA & Charles-Edwards DA), pp. 29–47. Academic Press, London.

Acquaah G (2012) *Principles of Plant Genetics and Breeding*, 2nd edn. Blackwell, Oxford.

Addicott FT, ed. (1983) *Abscisic Acid*. Praeger, New York.

Agam N, Berliner PR (2006) Dew formation and water vapor adsorption in semi-arid environments: a review. *Journal of Arid Environments* **65**, 572–90.

Ainsworth EA, Davey PA, Hymus GJ *et al.* (2003) Is stimulation of leaf photosynthesis by elevated carbon dioxide concentration maintained in the long term? A test with *Lolium perenne* grown for 10 years at two nitrogen fertilization levels under Free Air CO_2 Enrichment (FACE). *Plant, Cell & Environment* **26**, 705–14.

Alder NN, Pockman WT, Sperry JS, Nuismer S (1997) Use of centrifugal force in the study of xylem embolism. *Journal of Experimental Botany* **48**, 665–74.

Allard RW (1999) *Principles of Plant Breeding*, 2nd edn. John Wiley & Sons, Inc., New York.

Allaway WG, Austin B, Slatyer RO (1974) Carbon dioxide and water vapour exchange parameters of photosynthesis in a Crassulacean plant *Bryophyllum diagremontiana*. *Australian Journal of Plant Physiology* **1**, 397–405.

Allen RG, Pereira LS, Howell TA, Jensen ME (2011) Evapotranspiration information reporting: I. Factors governing measurement accuracy. *Agricultural Water Management* **98**, 899–920.

Allen RG, Pereira LS, Raes D, Smith M (1998) *Crop Evapotranspiration: Guidelines for Computing Crop Water Requirements. FAO Irrigation and Drainage Paper 56*. FAO Land and Water Division, Rome, Italy.

Allen RG, Pruitt WO, Raes D, Smith MAH, Pereira LS (2005) Estimating evaporation from bare soil and the crop coefficient for the initial period using common soils information. *Journal of Irrigation and Drainage Engineering* **131**, 14–23.

Allen RG, Tasumi M, Trezza R (2007) Satellite-based energy balance for mapping evapotranspiration with internalized calibration (METRIC) – model. *Journal of Irrigation and Drainage Engineering* **133**, 380–94.

Alscher RG, Cumming JR (1990) *Stress Responses in Plants: Adaptation and Acclimation Mechanisms*. Wiley-Liss, New York.

Amthor JS (1989) *Respiration and Crop Productivity*. Springer-Verlag, New York.

Amthor JS (2000) The McCree-de Wit-Penning de Vries-Thornley respiration paradigms: 30 years later. *Annals of Botany* **86**, 1–20.

Angeles G, Bond B, Boyer JS *et al.* (2004) The cohesion-tension theory. *New Phytologist* **163**, 451–2.

Angus JF, Moncur MW (1977) Water stress and phenology in wheat. *Australian Journal of Agricultural Research* **28**, 177–81.

Anon (1964) *Mean Daily Solar Radiation, Monthly and Annual.* US Department of Commerce, Washington.

Anon (1980) *Solar Radiation Data for the United Kingdom 1951–1975.* Meteorological Office, Bracknell.

Anon (1987) *Montreal Protocol on Substances that Deplete the Ozone Layer.* United Nations Environment Programme, Nairobi, Kenya.

Anon (1989) *Environmental Effects Panel Report (ISBN 92 807 1245 4).* United Nations Environment Programme, Nairobi, Kenya.

Ansari AQ, Loomis WE (1959) Leaf temperatures. *American Journal of Botany* **46**, 713–17.

Arias DG, Piattoni CV, Guerrero SA, Iglesias AA (2011) Biochemical mechanisms for the maintenance of oxidative stress under control in plants. In *Handbook of Plant and Crop Stress* (ed. Pessarakli M), pp. 157–90. CRC Press, Boca Raton, FL.

Arkin GF, Vanderlip RL, Ritchie JT (1975) A dynamic grain sorghum growth model. *Transactions of the American Society of Agricultural Engineers* **19**, 622–30.

Armond PA, Mooney HA (1978) Correlation of photosynthetic unit size and density with photosynthetic capacity. *Carnegie Institution Year Book* **77**, 234–7.

Árnadóttir J, Chalfie M (2010) Eucaryotic mechanosensitive channels. *Annual Review of Biophysics* **39**, 111–37.

Aronson EL, McNulty SG (2009) Appropriate experimental ecosystem warming methods by ecosystem, objective, and practicality. *Agricultural and Forest Meteorology* **149**, 1791–9.

Atkin OK, Botman B, Lambers H (1996) The causes of inherently slow growth in alpine plants: an analysis based on the underlying carbon economies of alpine and lowland *Poa* species. *Functional Ecology* **10**, 698–707.

Atkin OK, Macherel D (2009) The crucial role of plant mitochondria in orchestrating drought tolerance. *Annals of Botany* **103**, 581–97.

Atkin OK, Scheurwater I, Pons TL (2007) Respiration as a percentage of daily photosynthesis in whole plants is homeostatic at moderate, but not high, growth temperatures. *New Phytologist* **174**, 367–80.

Atkin OK, Tjoelker MG (2003) Thermal acclimation and the dynamic response of plant respiration to temperature. *Trends in Plant Science* **8**, 343–51.

Atkins PW, de Paula J (2009) *Physical Chemistry*, 9th edn. Oxford University Press, Oxford.

Aubinet M, Vesala T, Papale D, eds (2012) *Eddy Covariance: a Practical Guide to Measurement and Data Analysis.* Springer, Berlin.

Austin RB (1978) Actual and potential yields of wheat and barley in the United Kingdom. *Agricultural Development and Advisory Service Quarterly Review* **29**, 76–87.

Azzari G, Goulden ML, Rusu RB (2013) Rapid characterisation of vegetation structure with a microsoft Kinect sensor. *Sensors* **13**, 2384–98.

Bae G, Choi G (2008) Decoding of light signals by plant phytochromes and their interacting proteins. *Annual Review of Plant Biology* **59**, 281–311.

Bailey-Serres J, Voesenek LACJ (2008) Flooding stress: acclimations and genetic diversity. *Annual Review of Plant Biology* **59**, 313–19.

Bainbridge R, Evans GC, Rackham O (1968) *Light as an Ecological Factor.* Blackwell, Oxford.

Baker EA (1974) The influence of environment on leaf wax development in *Brassica oleracea* var. *gemmifera. New Phytologist* **73**, 955–66.

Baker NR (2008) Chlorophyll fluorescence: a probe of photosynthesis in vivo. *Annual Review of Plant Biology* **59**, 89–113.

Baker TR, Phillips OL, Malhi Y *et al.* (2004) Increasing biomass in Amazonian forest plots. *Philosophical Transactions of The Royal Society of London, Series B* **359**, 353–65.

Baldocchi DD, Verma SB, Rosenberg NJ, Blad BL, Specht JE (1985) Microclimate plant architectural interactions: influence of leaf width on the mass and energy exchange of a soybean canopy. *Agricultural and Forest Meteorology* **35**, 1–20.

Baldridge AM, Hook SJ, Grove CI, Rivera G (2009) The ASTER spectral library version 2.0. *Remote Sensing of Environment* **113**, 711–15.

Ball JT, Woodrow IE, Berry JA (1987) A model predicting stomatal conductance and its contribution to the control of photosynthesis under different environmental conditions. In *Proceedings of VII International Photosynthesis Congress* (ed. Biggins J), pp. 221–34. Martinus Nijhoff, Dordrecht.

Ballaré CL (2009) Illuminated behaviour: phytochrome as a key regulator of light foraging and plant anti-herbivore defence. *Plant, Cell & Environment* **32**, 713–25.

Balling A, Zimmerman U (1990) Comparative measurements of the xylem pressure of Nicotiana plants by means of the pressure bomb and pressure probe. *Planta* **182**, 525–8.

Bangerth F (1979) Calcium related physiological disorders of plants. *Annual Review of Phytopathology* **17**, 97–122.

Baret F, Guyot G (1991) Potentials and limits of vegetation indices for LAI and APAR assessment. *Remote Sensing of Environment* **35**, 161–73.

Baroli I, Price GD, Badger MR, von Caemmerer S (2008) The contribution of photosynthesis to the red light response of stomatal conductance. *Plant Physiology* **146**, 737–47.

Barrs HD (1968) Determination of water deficits in plant tissues. In *Water Deficits and Plant Growth* (ed. Kozlowski TT), pp. 235–368. Academic Press, New York and London.

Barry RG (2008) *Mountain Weather and Climate*, 3rd edn. Cambridge University Press, Cambridge.

Bartels D, Schneider K, Terstappen G, Piatkowski D, Salamini F (1990) Molecular cloning of abscisic acid-modulated genes which are induced during desiccation of the resurrection plant *Ceratostigma plantagineum*. *Planta* **181**, 27–34.

Barton CVM, Ellsworth DS, Medlyn BE *et al.* (2010) Whole-tree chambers for elevated atmospheric CO_2 experimentation and tree scale flux measurements in south-eastern Australia: The Hawkesbury Forest Experiment. *Agricultural and Forest Meteorology* **150**, 941–51.

Bastiaanssen WGM, Menenti M, Feddes RA, Holtslag AAM (1998) A remote sensing surface energy balance algorithm for land (SEBAL). 1. Formulation. *Journal of Hydrology* **213**, 198–212.

Bates LM, Hall AE (1981) Stomatal closure with soil moisture depletion not associated with changes in bulk water status. *Oecologia* **50**, 62–5.

Beakbane AB, Mujamder PK (1975) A relationship between stomatal density and growth potential in apple rootstocks. *Journal of Horticultural Science* **50**, 285–9.

Beal MJ, Falciani F, Ghahramani Z, Rangel C, Wild DL (2005) A Bayesian approach to reconstructing genetic regulatory pathways with hidden factors. *Bioinformatics* **21**, 349–56.

Beck E, Senser M, Scheibe R, Steiger H-M, Pongratz P (1982) Frost avoidance and freezing tolerance in Afroalpine 'giant rosette' plants. *Plant, Cell & Environment* **5**, 212–22.

Becker P, Meinzer FC, Wullschleger SD (2000) Hydraulic limitation of tree height: a critique. *Functional Ecology* **14**, 4–11.

Beljaars ACM, Bosveld FC (1997) Cabauw data for the validation of land surface parameterization schemes. *Journal of Climate* **10**, 1172–93.

Bell CJ, Rose DA (1981) Light measurement and the terminology of flow. *Plant, Cell and Environment* **4**, 89–96.

Bell JNB, Treshow M, eds (1992) *Air Pollution and Plant Life*, 2nd edn. Wiley-Blackwell, Chichester, UK.

Bentley RE, ed. (1998) *Handbook of Temperature Measurement Vol. 1. Temperature and Humidity Measurement*. Springer-Verlag, Singapore.

Bergmann DC, Sack FD (2007) Stomatal development. *Annual Review of Plant Biology* **58**, 163–81.

Berk A, Bernstein LS, Anderson GP *et al.* (1998) MODTRAN cloud and multiple scattering upgrades with application to AVIRIS. *Remote Sensing of Environment* **65**, 367–75.

Bidabé B (1967) Action de la température sur l'évolution des bourgeons de pommier et comparaison de méthodes de contrôle de l'époque de floraison. *Annales de Physiologie Végétale* **9**, 65–86.

Biddington NL (1986) The effects of mechanically-induced stress in plants: a review. *Plant Growth Regulation* **4**, 103–23.

Biddington NL, Dearman JA (1985) The effects of mechanically-induced stress on water loss and drought resistance in lettuce, cauliflower and celery seedlings. *Annals of Botany* **56**, 795–802.

Bierhuizen JF, Slatyer RO (1965) Effect of atmospheric concentration of water vapour and CO_2 in determining transpiration–photosynthesis relationships of cotton leaves. *Agricultural Meteorology* **2**, 259–70.

Bilger W, Schreiber U, Lange OL (1984) Determination of leaf heat resistance: comparative investigation of chlorophyll fluorescence changes and tissue necrosis methods. *Oecologia* **63**, 156–62.

Bingham IJ, Karley AJ, White PJ, Thomas WTB, Russell JR (2012) Analysis of improvements in nitrogen use efficiency associated with 75 years of spring barley breeding. *European Journal of Agronomy* **42**, 49–58.

Biscoe PV, Gallagher JN, Littleton EJ, Monteith JL, Scott RK (1975a) Barley and its environment. IV. Sources of assimilate for the grain. *Journal of Applied Ecology* 12, 295–318.

Biscoe PV, Scott RK, Monteith JL (1975b) Barley and its environment. III. Carbon budget of the stand. *Journal of Applied Ecology* 12, 269–93.

Björkman O, Badger MR, Armond PA (1980) Response and adaptations of photosynthesis to high temperatures. In *Adaptation of Plants to Water and High Temperature Stress* (eds Turner NC & Kramer PJ), pp. 233–49. Wiley, New York.

Björkman O, Boardman NK, Anderson JM *et al.* (1972a) The effect of light intensity during growth of *Atriplex patula* on the capacity of photosynthetic reactions, chloroplast components and structure. *Carnegie Institution Year Book* 71, 115–35.

Björkman O, Demmig B (1987) Photon yield of O_2 evolution and chlorophyll fluoresecence characteristics at 77 K among vascular plant of diverse origins. *Planta* 170, 489–504.

Björkman O, Gauhl E, Nobs MA (1970) Comparative studies of *Atriplex* species with and without β-carboxylation photosynthesis and their first generation hybrid. *Carnegie Institution Year Book* 68, 620–33.

Björkman O, Ludow MM, Morrow PA (1972b) Photosynthetic performance of two rainforest species in their native habitat and analysis of the gas exchange. *Carnegie Institution Year Book* 71, 94–102.

Björkman O, Mooney HA, Ehleringer JR (1975) Photosynthetic responses of plants from habitats with contrasting thermal environments. *Carnegie Institution Year Book* 74, 743–8.

Björkman O, Nobs MA, Berry JA *et al.* (1973) Physiological adaptation to diverse environments: approaches and facilities to study plant responses to contrasting thermal and water regimes. *Carnegie Institution Year Book* 72, 393–403.

Blackburn GA (1998) Spectral indices for estimating photosynthetic pigment concentrations: a test using senescent tree leaves. *International Journal of Remote Sensing* 19, 657–75.

Blackman FF (1905) Optima and limiting factors. *Annals of Botany* 19, 281–95.

Blackman PJ, Davies WJ (1985) Root to shoot communication in maize plants of the effects of soil drying. *Journal of Experimental Botany* 36, 39–48.

Bliss D, Smith H (1985) Penetration of light into soil and its role in the control of seed germination. *Plant, Cell & Environment* 8, 475–83.

Blum A (2009) Effective use of water (EUW) and not water-use efficiency (WUE) is the target of crop yield improvement under drought stress. *Field Crops Research* 112, 119–23.

Blümel K, Chmielewski F-M (2012) Shortcomings of classical phenological forcing models and a way to overcome them. *Agricultural and Forest Meteorology* 164, 10–19.

Boardman NK, Björkman O, Anderson JM, Goodchild DJ, Thoree SW (1975) Photosynthetic adaptation of higher plants to light intensity: relationship between chloroplast structure, composition of the photosystems and photosynthetic rates. In *Proceedings of the Third International Congress on Photosynthesis* (ed. Avron M), pp. 1809–27. Elsevier, Amsterdam.

Bohrer G, Mourad H, Laursen TA *et al.* (2005) Finite element tree crown hydrodynamics model (FETCH) using porous media flow within branching elements: a new representation of tree hydrodynamics. *Water Resources Research* 41, W11404. doi:11410.11029/12005WR004181.

Bond DM, Dennis ES, Finnegan EJ (2011) The low temperature response pathways for cold acclimation and vernalization are independent. *Plant, Cell & Environment* 34, 1737–48.

Boote KJ, Jones JW, Pickering NB (1996) Potential uses and limitations of models. *Agronomy Journal* 88, 704–15.

Bourgeois G, Jenni S, Laurence H, Tremblay N (2000) Improving the prediction of processing pea maturity based on the growing-degree day approach. *HortScience* 35, 611–14.

Box GEP, Hunter WG, Hunter JS (2005) *Statistics for Experimenters: Design, Innovation and Discovery*, 2nd edn. Wiley, New York.

Boyer JS (1995) *Measuring the Water Status of Plants and Soils*. Academic Press Inc., London.

Boyer JS, James RA, Munns R, Condon TAG, Passioura JB (2008) Osmotic adjustment leads to anomalously low estimates of relative water content in wheat and barley. *Functional Plant Biology* 35, 1172–82.

Braam J (2005) In touch: plant responses to mechanical stimuli. *New Phytologist* 13, 373–89.

Braam J, Davis RW (1990) Rain-, wind-, and touch-induced expression of calmodulin and calmodulin-related genes in *Arabidopsis*. *Cell* 60, 357–64.

Brandrup J, Immergut EH, eds (1975) *Polymer Handbook*, 2nd edn. Wiley, New York.

Bray EA (1997) Plant responses to water deficit. *Trends in Plant Science* 2, 48–54.

Bréda NJJ (2003) Ground-based measurements of leaf area index: a review of methods, instruments and current controversies. *Journal of Experimental Botany* 54, 2403–17.

Brenner AJ, Jarvis PG (1995) A heated leaf replica technique for determination of leaf boundary layer conductance in the field. *Agricultural and Forest Meteorology* 72, 261–75.

Breshears DD, McDowell NG, Goddard KL *et al.* (2008) Foliar absorption of intercepted rainfall improves woody plant water status most during drought. *Ecology* 89, 41–7.

Brock FV, Richardson SJ (2001) *Meteorological Measurement Systems*. Oxford University Press, Oxford.

Brodersen CR, McElrone AJ, Choat B, Matthews MA, Shackel KA (2010) The dynamics of embolism repair in xylem: in vivo visualizations using high-resolution computed tomography. *Plant Physiology* 154, 1088–95.

Brooking IR (1996) Temperature response of vernalization in wheat: a developmental analysis. *Annals of Botany* 78, 507–12.

Brooks A, Farquhar GD (1985) Effect of temperature on the CO_2/O_2 specificity of ribulose-1,5-bisphosphate carboxylase/oxygenase and the rate of respiration in the light. *Planta* 165, 397–406.

Brown KW, Jordan WR, Thomas JC (1976) Water stress induced alterations of the stomatal response to decreases in leaf water potential. *Physiologia Plantarum* 37, 1–5.

Brown KW, Rosenberg NJ (1970) Influence of leaf age, illumination, and upper and lower suface differences on stomatal resistance of sugar beet (*Beta vulgaris*) leaves. *Agronomy Journal* 62, 20–4.

Brown RW, van Haveren BP (1972) *Psychrometry in Water Relations Research*. Utah Agricultural Experiment Station, Logan, Utah.

Bryson RA (1974) A perspective on climate change. *Science* 184, 753–60.

Buck AL (1981) New equations for computing vapor pressure and enhancement factor. *Journal of Applied Meteorology* 20, 1527–32.

Buckley TN (2005) The control of stomata by water balance. *New Phytologist* 168, 275–92.

Bunce JA (2006) How do leaf hydraulics limit stomatal conductance at high water vapour pressure deficits? *Plant, Cell & Environment* 29, 1644–50.

Burke MJ, Stushnoff C (1979) Frost hardiness: a discussion of possible molecular causes of injury with particular reference to deep supercooling of water. In *Stress Physiology of Crop Plants* (eds Mussell H & Staples RC), pp. 198–225. Wiley, New York.

Bürling K, Cerovic ZG, Cornic G *et al.* (2013) Fluorescence-based sensing of drought-induced stress in the vegetative phase of four contrasting wheat genotypes. *Environmental and Experimental Botany* 81, 51–9.

Businger JA (1975) Aerodynamics of vegetated surfaces. In *Heat and Mass Transfer in the Biosphere. I. Transfer Processes in the Plant Environment* (eds de Vries DA & Afgan NH), pp. 139–65. Scripta, Washington.

Bykov OD, Koshkin VA, Čatský J (1981) Carbon dioxide compensation of C_3 and C_4 plants: dependence on temperature. *Photosynthetica* 15, 114–21.

Caird MA, Richards JH, Donovan LA (2007) Nighttime stomatal conductance and transpiration in C_3 and C_4 plants. *Plant Physiology* 143, 4–10.

Calder IR (1976) The measurement of water losses from a forested area using a 'natural' lysimeter. *Journal of Hydrology* 30, 311–25.

Caldwell M, Teramura AH, Tevini M *et al.* (1995) Effects of increased solar ultraviolet radiation on terrestrial plants. *Ambio* 24, 166–73.

Caldwell MM (1970) Plant gas exchange at high wind speeds. *Plant Physiology* 46, 536–7.

Caldwell MM (1981) Plant responses to solar ultra violet radiation. In *Encyclopedia of Plant Physiology, Vol 12A: Physiological Plant Ecology I – Responses to the Physical Environment* (eds Lange OL, Nobel ES, Osmond CB, & Ziegler H), pp. 169–97. Springer-Verlag, Berlin.

Caldwell MM, Dawson TE, Richards JH (1998) Hydraulic lift: consequences of water efflux from the roots of plants. *Oecologia* 113, 151–61.

Camacho-B SE, Hall AE, Kaufmann MR (1974) Efficiency and regulation of water transport in some woody and herbaceous species. *Plant Physiology* **54**, 169–72.

Campbell GS, Norman JM (1998) *An Introduction to Environmental Biophysics*, 2nd edn. Springer, New York.

Campbell JB (2007) *Introduction to Remote Sensing*, 4th edn. Taylor and Francis, London.

Canny M (1997) Vessel contents during transpiration – embolisms and refilling. *American Journal of Botany* **84**, 1223–30.

Cardenas AC (1983) *A Pheno-climatological Assessment of Millets and Other Cereal Grains in Tropical Cropping Patterns.* MSc Thesis, University of Nebraska.

Carlson TN, Ripley DA (1997) On the relation between NDVI, fractional vegetation cover and leaf area index. *Remote Sensing of Environment* **62**, 241–52.

Carter GA (1991) Primary and secondary effects of water content on the spectral reflectance of leaves. *American Journal of Botany* **78**, 916–24.

Casa R, Jones HG (2005) LAI retrieval from multiangular image classification and inversion of a ray tracing model. *Remote Sensing of Environment* **98**, 414–28.

Casson SA, Franklin KA, Gray JE *et al.* (2009) Phytochrome B and PIF4 regulate stomatal development in response to light quantity. *Current Biology* **19**, 229–34.

Castellví F, Snyder RL (2009) Sensible heat flux estimates using surface renewal analysis: a study case over a peach orchard. *Agricultural and Forest Meteorology* **149**, 1397–402.

Cattivelli L, Rizza F, Badeck F-W *et al.* (2008) Drought tolerance improvement in crop plants: an integrated view from breeding to genomics. *Field Crops Research* **105**, 1–14.

Čermák J, Gašpárek J, De Lorenzi F, Jones HG (2007) Stand biometry and leaf area distribution in an old olive grove at Andria, southern Italy. *Annals of Forest Sciences* **64**, 491–501.

Čermák J, Kučera J (1981) The compensation of natural temperature gradient at the measuring point during the sap flow rate determination in trees. *Biologia Plantarum* **23**, 469–71.

Čermák J, Kučera J, Nadezhdina N (2004) Sap flow measurements with some thermodynamic methods, flow integration within trees and scaling up from sample trees to entire forest stands. *Trees: Structure and Function* **18**, 529–46.

Chamberlain AC, Little P (1981) Transport and capture of particles by vegetation. In *Plants and their Atmospheric Environment* (eds Grace J, Ford ED & Jarvis PG), pp. 147–73. Blackwell, Oxford.

Chapius R, Delluc C, Debeuf R, Tardieu F, Welcker C (2012) Resiliences to water deficit in a phenotyping platform and in the field: how related are they in maize? *European Journal of Agronomy* **42**, 59–67.

Chaves MM, Flexas J, Pinheiro C (2009) Photosynthesis under drought and salt stress: regulation mechanisms from whole plant to cell. *Annals of Botany* **103**, 551–60.

Chehab EW, Eich E, Braam J (2009) Thigmomorphogenesis: a complex plant response to mechano-stimulation. *Journal of Experimental Botany* **60**, 43–56.

Chen JM (1996) Optically-based methods for measuring seasonal variation of leaf area index in boreal conifer stands. *Agricultural and Forest Meteorology* **80**, 135–63.

Chen JM, Leblanc SG (1997) A four-scale bidirectional reflectance model based on canopy architecture. *IEEE Transactions on Geoscience and Remote Sensing* **35**, 1316–37.

Chen M, Chory J (2011) Phytochrome signaling mechanisms and the control of plant development. *Trends in Cell Biology* **11**, 664–71.

Cheong J-J, Choi YD (2003) Methyl jasmonate as a vital substance in plants. *Trends in Genetics* **19**, 409–13.

Cheung YNS, Tyree MT, Dainty J (1975) Water relations parameters on single leaves obtained in a pressure bomb, and some ecological interpretations. *Canadian Journal of Botany* **53**, 1342–6.

Choudhury BJ (1994) Synergism of multispectral satellite observations for estimating regional land surface evaporation. *Remote Sensing of Environment* **49**, 264–74.

Christie JM (2007) Phototropin blue-light receptors. *Annual Review of Plant Biology* **58**, 21–45.

Ciha AJ, Brun WA (1975) Stomatal size and frequency in soybean. *Crop Science* **15**, 309–13.

Cochard H, Cruiziat P, Tyree MT (1992) Use of positive pressures to establish vulnerability curves – further support for the air-seeding hypothesis and implications for pressure-volume analysis. *Plant Physiology* **100**, 205–9.

Cohen A, Bray EA (1990) Characterization of three mRNAs that accumulate in wilted tomato leaves in response to elevated levels of endogenous abscisic acid. *Planta* **182**, 27–33.

Cohen D (1971) Maximising final yield when growth is limited by time or by limiting resources. *Journal of Theoretical Biology* **33**, 299–307.

Cohen S, Fuchs M (1987) The distribution of leaf area, radiation, photosynthesis and transpiration in a Shamouti orange hedgerow orchard. I. Leaf area and radiation. *Agricultural and Forest Meteorology* **40**, 123–44.

Collatz J, Ferrar PJ, Slatyer RO (1976) Effects of water stress and differential hardening treatments on photosynthetic characteristics of a xeromorphic shrub, *Eucalyptus socialis* F. Muel. *Oecologia* **23**, 95–105.

Colmer TD (2003) Long-distance transport of gases in plants: a perspective on internal aeration and radial oxygen loss from roots. *Plant, Cell & Environment* **26**, 17–36.

Condon AG, Richards RA, Rebetzke GJ, Farquhar GD (2002) Improving intrinsic water-use efficiency. *Crop Science* **42**, 122–31.

Cooper JP (1970) Potential production and energy conversion in temperate and tropical grasses. *Herbage Abstracts* **40**, 1–15.

Cooper JP, ed. (1975) *Photosynthesis and Productivity in Different Environments*. Cambridge University Press, Cambridge.

Core-Writing-Team, Pachauri RK, Reisinger A, eds (2008) *IPCC 2007, Climate Change 2007: Synthesis Report. Contribution of Working Groups I, II and III to the Fourth Assessment Report of the Intergovernmental Panel on Climate Change*. IPCC, Geneva, Switzerland.

Cosgrove DJ (1986) Biophysical control of plant cell growth. *Annual Review of Plant Physiology* **37**, 377–405.

Cosgrove DJ (1999) Enzymes and other agents that enhance cell wall extensibility. *Annual Review of Plant Physiology and Plant Molecular Biology* **50**, 391–417.

Cosgrove DJ (2005) Growth of the plant cell wall. *Nature Reviews Molecular and Cell Biology* **6**, 850–61.

Cosgrove DJ, Hedrich R (1991) Stretch-activated chloride, potassium, and calcium channels coexisting in plasma membranes of guard cells of *Vicia faba. Planta* **186**, 143–53.

Coulson KL (1975) *Solar and Terrestrial Radiation: Methods and Measurements*. Academic Press, New York.

Covell S, Ellis RH, Roberts EH, Summerfield RJ (1986) The influence of temperature on seed germination rate in grain legumes. I. A comparison of chickpea, lentil, soybean and cowpea at constant temperatures. *Journal of Experimental Botany* **37**, 705–15.

Cowan IR (1977) Stomatal behaviour and environment. *Advances in Botanical Research* **4**, 117–228.

Cowan IR (1982) Water-use and optimization of carbon assimilation. In *Encyclopedia of Plant Physiology, New Series, Vol. 12B* (eds Lange OL, Nobel PS, Osmond CB, & Ziegler H), pp. 589–613. Springer-Verlag, Berlin, Heidelberg, New York.

Cowan IR (1986) Economics of carbon fixation in higher plants. In *On the Economy of Plant Form and Function* (ed. Givnish TJ), pp. 133–70. Cambridge University Press, Cambridge.

Cowan IR, Farquhar GD (1977) Stomatal function in relation to leaf metabolism and environment. *Symposium of the Society for Experimental Biology* **31**, 471–505.

Crank J (1979) *The Mathematics of Diffusion*, 2nd edn. Oxford University Press, Oxford.

Crombie DS, Milburn JA, Hipkins MF (1985) Maximum sustainable sap tensions in *Rhododendron* and other species. *Planta* **163**, 27–33.

Cussler EL (2007) *Diffusion: Mass Transfer in Fluid Systems*, 3rd edn. Cambridge University Press, Cambridge.

Cutler JM, Rains DM, Loomis RS (1977) The importance of cell size in the water relations of plants. *Physiologia Plantarum* **40**, 255–60.

Dacey JWH (1980) Internal winds in water lilies: an adaptation for life in anaerobic sediments. *Science* **210**, 1017–19.

Dainty J (1963) Water relations of plant cells. *Advances in Botanical Research* **1**, 279–326.

Darvishzadeh R, Skidmore A, Atzberger C, van Wieren S (2008) Estimation of vegetation LAI from hyperspectral reflectance data: effects of soil type and plant architecture. *International Journal of*

Applied Earth Observation and Geoinformation **10**, 358–73.

Darwin C (1890) *The Power of Movement in Plants*. William Clowes and Sons Ltd., London.

Darwin F, Pertz DFM (1911) On a new method of estimating the aperture of stomata. *Proceedings of the Royal Society of London, Series B* **84**, 136–54.

Daubenmire R (1974) *Plants and Environment: a Textbook of Plant Autecology*, 3rd edn. Wiley, New York.

Davies WJ, Jones HG, eds (1991) *Abscisic Acid: Physiology and Biochemistry*. Bios Scientific Publishers Ltd, Oxford.

Davies WJ, Wilkinson S, Loveys B (2002) Stomatal control by chemical signalling and the exploitation of this mechanism to increase water use efficiency in agriculture. *New Phytologist* **153**, 449–60.

Davies WJ, Zhang J (1991) Root signals and the regulation of growth and development of plants in drying soil. *Annual Review of Plant Physiology and Plant Molecular Biology* **42**, 55–76.

Day W, Legg BJ, French BK *et al.* (1987) A drought experiment using mobile shelters: the effect of drought on barley yield, water use and nutrient uptake. *Journal of Agricultural Science, Cambridge* **91**, 599–623.

de Pury DGG, Farquhar GD (1997) Simple scaling of photosynthesis from leaves to canopy without the errors of big-leaf models. *Plant, Cell & Environment* **20**, 537–57.

de Wit CT (1958) Transpiration and crop yields. *Verslagen van Landbouwkundige Onderzoekingen* **64**, 1–88.

de Wit CT (1965) *Photosynthesis of leaf canopies*. *Agricultural Research Report no. 663*. PUDOC, Wageningen.

Delieu TJ, Walker DA (1983) Simultaneous measurement of oxygen evolution and chlorophyll fluorescence from leaf pieces. *Plant Physiology* **73**, 534–41.

Demarsy E, Frankhauser C (2009) Higher plants use LOV to perceive blue light. *Current Opinion in Plant Biology* **12**, 69–74.

Demmig-Adams B, Adams III WW (1992) Photoprotection and other responses of plants to high light stress. *Annual Review of Plant Physiology and Plant Molecular Biology* **43**, 599–626.

Denmead OT (1969) Comparative micrometeorology of a wheat field and a forest of *Pinus radiata*. *Agricultural Meteorology* **6**, 357–71.

Denmead OT, Bradley EF (1987) On scalar transport in plant canopies. *Irrigation Science* **8**, 131–49.

Denmead OT, McIlroy IC (1970) Measurements of non-potential evaporation from wheat. *Agricultural Meteorology* **7**, 285–302.

Denmead OT, Shaw RH (1962) Availability of soil water to plants as affected by soil moisture content and meteorological conditions. *Agronomy Journal* **45**, 385–90.

Dennis ES, Peacock WJ (2007) Epigenetic regulation of flowering. *Current Opinion in Plant Biology* **10**, 520–7.

Dewar RC (2002) The Ball–Berry–Leuning and Tardieu–Davies stomatal models: synthesis and extension within a spatially aggregated picture of guard cell function. *Plant, Cell & Environment* **25**, 1383–98.

Dirmhirn I (1964) *Das Strahlungsfeld in Lebensraum*. Akademische Verlagsgesellschaft, Frankfurt-am-Main.

Dixon MA, Tyree M (1984) A new stem hygrometer, corrected for temperature gradients and calibrated against the pressure bomb. *Plant, Cell & Environment* **7**, 693–7.

Dodd IC (2009) Rhizosphere manipulations to maximize 'crop per drop' during deficit irrigation. *Journal of Experimental Botany* **60**, 2454–9.

Donald CM (1968) The breeding of crop ideotypes. *Euphytica* **17**, 385–403.

Donald CM, Hamblin J (1976) The biological yield and harvest index of cereals as agronomic and plant breeding criteria. *Advances in Agronomy* **28**, 361–405.

Doorenbos J, Pruitt WO (1984) *Guidelines for Predicting Crop Water Requirements. FAO Irrigation and Drainage Paper 24*. Food and Agriculture Organization of the United Nations, Rome.

Downs RJ, Hellmers H (1975) *Environment and the Experimental Control of Plant Growth*. Academic Press, London.

Dreyer E, Le Roux X, Montpied P, Daudet FA, Masson F (2001) Temperature response of leaf photosynthetic capacity in seedlings from seven temperate tree species. *Tree Physiology* **21**, 223–32.

Dubcovsky J, Maria GS, Epstein E, Luo MC, Dvořák J (1996) Mapping of the K^+/Na^+ discrimination locus Kna1 in wheat. *Theoretical and Applied Genetics* **92**, 448–54.

Dubois J-JB, Fiscus EL, Brooer FL, Flowers MD, Reid CD (2007) Optimizing the statistical estimation of the parameters of the Farquhar–von Caemmerer–Berry model of photosynthesis. *New Phytologist* **176**, 402–14.

Dunstone RL, Gifford RM, Evans LT (1973) Photosynthetic characteristics of modern and primitive wheat species in relation to ontogeny and adaptation to light. *Australian Journal of Biological Sciences* **26**, 295–307.

Dwyer LM, Stewart DW, Carrigan L *et al.* (1999) A general thermal index for maize. *Agronomy Journal* **91**, 940–6.

Eberhardt SA, Russell WA (1966) Stability parameters for comparing varieties. *Crop Science* **6**, 36–40.

Edwards DK, Denny VE, Mills AF (1979) *Transfer Processes*, 2nd edn. McGraw-Hill, New York.

Ehleringer JR (1980) Leaf morphology and reflectance in relation to water and temperature stress. In *Adaptation of Plants to Water and High Temperature Stress* (eds Turner NC & Kramer PJ), pp. 295–308. Wiley, New York.

Ehleringer JR, Björkman O (1977) Quantum yields for CO_2 uptake in C_3 and C_4 plants. *Plant Physiology* **59**, 86–90.

Ehleringer JR, Cerling TE, Helliker BR (1997) C-4 photosynthesis, atmospheric CO_2 and climate. *Oecologia* **112**, 285–99.

Ehleringer JR, Forseth I (1980) Solar tracking by plants. *Science* **210**, 1094–8.

Ehleringer JR, Hall AE, Farquhar GD, eds (1993) *Stable Isotopes and Plant Carbon–Water Relationships.* Academic Press Inc., San Diego, CA.

Eichinger WE, Parlange MB, Stricker H (1996) On the concept of equilibrium evaporation and the value of the Priestley–Taylor coefficient. *Water Resources Research* **32**, 161–4.

El Fadli KI, Cerveny RS, Burt CC *et al.* (2012) World Meteorological Organization assessment of the purported world record 58°C temperature extreme at El Azizia, Libya (13 September 1922). *Bulletin of the American Meteorological Society* http://dx.doi.org/10.1175/BAMS-D-12-00093.1.

El-Sharkawy M, Hesketh J (1965) Photosynthesis among species in relation to characteristics of leaf and CO_2 diffusion resistances. *Crop Science* **19**, 517–21.

Eller BM (1977) Leaf pubescence: the significance of lower surface hairs for the spectral properties of the upper surface. *Journal of Experimental Botany* **28**, 1054–9.

Ellmore GS, Ewers FW (1986) Fluid flow in the outermost xylem increment of a ring-porous tree. *American Journal of Botany* **73**, 1771–4.

Engledow FL, Wadham SM (1923) Investigations on the yield of cereals. Part I. *Journal of Agricultural Science, Cambridge* **21**, 391–409.

Evans GC (1972) *The Quantitative Analysis of Plant Growth.* Blackwell, Oxford.

Evans JR (1989) Photosynthesis and nitrogen relationships in leaves of C_3 plants. *Oecologia* **78**, 9–19.

Evans JR, Kaldenhoff R, Genty B, Terashima I (2009) Resistances along the CO_2 diffusion pathway inside leaves. *Journal of Experimental Botany* **60**, 2235–48.

Evans LT (1975) *Crop Physiology: Some Case Histories.* Cambridge University Press, Cambridge.

Evans LT, Wardlaw IF, Fischer RA (1975) Wheat. In *Crop Physiology* (ed. Evans LT), pp. 101–49. Cambridge University Press, Cambridge.

Falkowski PG, Raven JA (2007) *Aquatic Photosynthesis,* 2nd edn. Princeton University Press, Princeton, NJ.

Fanjul L, Jones HG (1982) Rapid stomatal responses to humidity. *Planta* **154**, 135–8.

Farquhar GD (1978) Feedforward responses of stomata to humidity. *Australian Journal of Plant Physiology* **5**, 787–800.

Farquhar GD, Cernusak LA (2012) Ternary effects on the gas exchange of isotopologues of carbon dioxide. *Plant, Cell & Environment* **35**, 1221–31.

Farquhar GD, Ehleringer JR, Hubick KT (1989) Carbon isotope discrimination in photosynthesis. *Annual Review of Plant Physiology* **40**, 503–37.

Farquhar GD, O'Leary MH, Berry JA (1982) On the relationship between carbon isotope discrimination and intercellular carbon dioxide concentration in leaves. *Australian Journal of Plant Physiology* **9**, 121–37.

Farquhar GD, Schultze E-D, Küppers M (1980a) Responses to humidity by stomata of *Nicotiana glauca* L. and *Corylus avellana* L. are consistent with the optimisation of carbon dioxide uptake with respect to water loss. *Australian Journal of Plant Physiology* **7**, 315–27.

Farquhar GD, von Caemmerer S, Berry JA (1980b) A biochemical model of photosynthetic CO_2 assimilation in leaves of C_3 species. *Planta* **149**, 78–90.

Farquhar TD, Sharkey TD (1982) Stomatal conductance and photosynthesis. *Annual Review of Plant Physiology* **33**, 317–45.

Farrant JM (2000) A comparison of mechanisms of desiccation tolerance among three angiosperm resurrection plant species. *Plant Ecology* **151**, 29–39.

Fell D (1997) *Understanding the Control of Metabolism.* Portland Press Ltd., London.

Fereres E, Goldhamer DA (2003) Suitability of stem diameter variations and water potential as indicators for irrigation scheduling of almond trees. *Journal of Horticultural Science & Biotechnology* **78**, 139–44.

Fernández JE, Rodriguez-Dominguez CM, Perez-Martin A *et al.* (2011a) Online-monitoring of tree water stress in a hedgerow olive orchard using the leaf patch clamp pressure probe. *Agricultural Water Management* **100**, 25–35.

Fernández JE, Torres-Ruiz JM, Diaz-Espejo A *et al.* (2011b) Use of maximum trunk diameter measurements to detect water stress in mature 'Arbequina' olive trees under deficit irrigation. *Agricultural Water Management* **98**, 1813–21.

Finlay KW, Wilkinson GN (1963) The analysis of adaptation in a plant breeding programme. *Australian Journal of Agricultural Research* **14**, 742–54.

Fischer RA (2011) Wheat physiology: a review of recent developments. *Crop & Pasture Science* **62**, 95–114.

Fischer RA, Edmeades GO (2010) Breeding and cereal yield progress. *Crop Science* **50**, S85–S98.

Fischer RA, Turner NC (1978) Plant productivity in the arid and semiarid zones. *Annual Review of Plant Physiology* **29**, 277–317.

Fiscus EL (1975) The interaction between osmotic- and pressure-induced water flow in plant roots. *Plant Physiology* **55**, 917–22.

Fisher MJ, Charles-Edwards DA, Ludlow MM (1981) An analysis of the effects of repeated short-term soil water deficits on stomatal conductance to carbon dioxide and leaf photosynthesis by the legume *Macroptilium atropurpureum cv. Siratro. Australian Journal of Plant Physiology* **8**, 347–57.

Fitter AH, Hay RKM (2001) *Environmental Physiology of Plants*, 3rd edn. Academic Press, London.

Fleagle RG, Businger JA (1980) *An Introduction to Atmospheric Physics*, 2nd edn. Academic Press, New York.

Flechard CR, Fowler D, Sutton MA, Cape JN (1999) A dynamic chemical model of bi-directional ammonia exchange between semi-natural vegetation and the atmosphere. *Quarterly Journal of the Royal Meteorological Society* **125**, 2611–41.

Flood PJ, Harbinson J, Aarts MGM (2011) Natural genetic variation in plant photosynthesis. *Trends in Plant Science* **16**, 327–35.

Florez-Sarasa ID, Bouma TJ, Medrano H, Azcon-Bieto J, Ribas-Carbo M (2007) Contribution of the cytochrome and alternative pathways to growth respiration and maintenance respiration in *Arabidopsis thaliana. Physiologia Plantarum* **129**, 143–51.

Fowler D, Cape JN, Unsworth MH (1989) Deposition of atmospheric pollutants on forests. *Philosophical Transactions of the Royal Society of London, Series B* **324**, 247–65.

Fowler D, Pilegaard K, Sutton MA *et al.* (2009) Atmospheric composition change: ecosystems–atmosphere interactions. *Atmospheric Environment* **43**, 5193–267.

Foyer CH, Bloom AJ, Queval G, Noctor G (2009) Photorespiratory metabolism: genes, mutants, energetics and redox signalling. *Annual Review of Plant Biology* **60**, 455–84.

Franklin KA (2008) Shade avoidance. *New Phytologist* **179**, 930–44.

Franklin KA, Quail PH (2010) Phytochrome function in *Arabidopsis. Journal of Experimental Botany* **61**, 11–24.

Franks F (1972) *Water: A Comprehensive Treatise, Vol. 1: The Physics and Physical Chemistry of Water.* Plenum Press, New York.

Franks PJ, Farquhar GD (1999) A relationship between humidity response, growth form and photosynthetic operating point in C_3 plants. *Plant, Cell & Environment* **22**, 1337–49.

Fu QS, Cheng LL, Guo YD, Turgeon R (2011) Phloem loading strategies and water relations in trees and herbaceous plants. *Plant Physiology* **157**, 1518–27.

Furbank RT (2011) Evolution of the C_4 photosynthetic mechanism: are there really three C_4 acid decarboxyation types? *Journal of Experimental Botany* **62**, 3103–8.

Furbank RT, Tester M (2011) Phenomics: technologies to relieve the phenotyping bottleneck. *Trends in Plant Science* **16**, 635–44.

Gaastra P (1959) Photosynthesis of crop plants as influenced by light, carbon dioxide, temperature, and stomatal diffusion resistance. *Mededelingen van de Landbouwhoogeschool te Wageningen* **59**, 1–68.

Gaff DF (1980) Protoplasmic tolerance of extreme water stress. In *Adaptation of Plants to Water and High Temperature Stress* (eds Turner NC & Kramer PJ), pp. 207–30. Wiley, New York.

Gamon JA, Peñuelas J, Field CB (1992) A narrow-waveband spectral index that tracks diurnal changes in photosynthetic efficiency. *Remote Sensing of Environment* **41**, 35–44.

Gardner MJ, Hubbard KE, Hotta CT, Dodd AN, Webb AAR (2006) How plants tell the time. *Biochemical Journal* **397**, 15–24.

Garner WW, Allard HA (1920) Effect of the relative length of day and night and other factors of the environment on growth and reproduction in plants. *Journal of Agricultural Research* **18**, 553–606.

Garratt JR (1992) *The Atmospheric Boundary Layer.* Cambridge University Press, Cambridge.

Garrigues S, Shabanov NV, Swanson K *et al.* (2008) Intercomparison and sensitivity analysis of Leaf Area Index retrievals from LAI-2000, AccuPAR, and digital hemispherical photography over croplands. *Agricultural and Forest Meteorology* **148**, 1193–209.

Gates DM (1980) *Biophysical Ecology.* Springer Verlag, New York.

Gauch HG (1992) *Statistical Analysis of Regional Yield Trials: AMMI Analysis of Factorial Designs.* Elsevier, Amsterdam.

Gay AP, Hurd RG (1975) The influence of light on the stomatal density in the tomato. *New Phytologist* **75**, 37–46.

Geiger DM (1950) *The Climate Near the Ground.* Harvard University Press, Boston.

Geiger DM (1965) *The Climate Near the Ground*, 4th edn. Harvard University Press, Cambridge, MA.

Genty B, Briantais J-M, Baker NR (1989) The relationship between the quantum yield of photosynthetic electron transport and quenching of chlorophyll fluorescence. *Biochimica et Biophysica Acta* **990**, 87–92.

Ghannoum O (2009) C_4 photosynthesis and water stress. *Annals of Botany* **103**, 635–44.

Gifford RM, Musgrave RB (1973) Stomatal role in the variability of net CO_2 exchange rates by two maize inbreds. *Australian Journal of Biological Sciences* **26**, 35–44.

Giunta F, Motzo R, Pruneddu G (2008) Has long-term selection for yield in durum wheat also induced changes in leaf and canopy traits? *Field Crops Research* **106**, 68–76.

Givnish TJ (1986) Optimal stomatal conductance, allocation of energy between leaves and roots, and the marginal cost of transpiration. In *On the Economy of Plant Form and Function* (ed. Givnish TJ), pp. 171–213. Cambridge University Press, Cambridge.

Goel NS, Strebel DE (1984) Simple beta distribution representation of leaf orientation in vegetation canopies. *Agronomy Journal* **76**, 800–2.

Gollan T, Passioura JB, Munns R (1986) Soil water status affects the stomatal conductance of fully turgid wheat and sunflower leaves. *Australian Journal of Plant Physiology* **13**, 459–64.

Gollan T, Turner NC, Schultze E-D (1985) The responses of stomata and leaf gas exchange to vapour pressure deficits and soil water content. III. In the schlerophyllous woody species *Nerium oleander.* *Oecologia* **65**, 356–62.

Gommers CMM, Visser EJW, St Onge KR, Voesenek LACJ, Pierik R (2013) Shade tolerance: when growing tall is not an option. *Trends in Plant Science* **18**, 65–71.

Goodwin SM, Jenks MA (2005) Plant cuticle function as a barrier to water loss. In *Plant Abiotic Stress* (eds Jenks MA & Hasegawa PM), pp. 14–36. Blackwell Publishing, Oxford.

Gorham J, Wyn-Jones RG, Bristol A (1990) Partial characterisation of the trait for enhanced K^+/Na^+ discrimination in the D genome of wheat. *Planta* **180**, 590–7.

Goudriaan J (1986) A simple and fast numerical method for the computation of daily totals of photosynthesis. *Agricultural and Forest Meteorology* 38, 249–54.

Goudriaan J (1996) Predicting crop yields under global change. In *Global Change and Terrestrial Ecosystems* (eds Walker BH & Steffen W), pp. 260–74. Cambridge University Press, Cambridge.

Goulding KWT, Bailey NJ, Bradbury NJ *et al.* (1998) Nitrogen deposition and its contribution to nitrogen cycling and associated soil processes. *New Phytologist* 139, 49–58.

Gowing DJG, Davies WJ, Jones HG (1990) A positive root-sourced signal as an indicator of soil drying in apple, *Malus* × *domestica* Borkh. *Journal of Experimental Botany* 41, 1535–40.

Grace J (1977) *Plant Response to Wind*. Academic Press, London.

Grace J (1981) Some effects of wind on plants. In *Plants and their Atmospheric Environment* (eds Grace J, Ford ED, & Jarvis PG), pp. 31–56. Blackwell, Oxford.

Grace J (1989) Tree lines. *Philosophical Transactions of the Royal Society of London Series B* 324, 233–45.

Grace J, Ford ED, Jarvis PG, eds (1981) *Plants and their Atmospheric Environment*. Blackwell, Oxford.

Granier A (1987) Evaluation of transpiration in a Douglas-fir stand by means of sap flow measurements. *Tree Physiology* 3, 309–19.

Grant DR (1970) Some measurements of evaporation in a field of barley. *Journal of Agricultural Science* 75, 433–43.

Grant RH (1999) Potential effect of soybean heliotropism on ultraviolet-B irradiance and dose. *Agronomy Journal* 91, 1017–23.

Grassi G, Magnani F (2005) Stomatal, mesophyll conductance and biochemical limitations to photosynthesis as affected by drought and leaf ontogeny in ash and oak trees. *Plant, Cell & Environment* 28, 834–49.

Green SR, Clothier BE, Jardine B (2003) Theory and practical application of heat pulse to measure sap flow. *Agronomy Journal* 95, 1371–9.

Grime JP (1979) *Plant Strategies and Vegetation Processes*. Wiley, Chichester, UK.

Grime JP (1989) Whole-plant responses to stress in natural and agricultural systems. In *Plants Under Stress* (eds Jones HG, Flowers TJ & Jones MB), pp. 157–80. Cambridge Unversity Press, Cambridge.

Gu L, Baldocchi DD, Wofsy SC *et al.* (2003) Response of a deciduous forest to the Mount Pinatubo eruption: enhanced photosynthesis. *Science* 299, 2035–8.

Gu L, Pallardy SG, Tu KB, Law BE, Wullschleger SD (2010) Reliable estimation of biochemical parameters from C_3 leaf photosynthesis–intercellular carbon dioxide response curves. *Plant, Cell & Environment* 33, 1852–74.

Guilioni L, Jones HG, Leinonen I, Lhomme JP (2008) On the relationships between stomatal resistance and leaf temperatures in thermography. *Agricultural and Forest Meteorology* 148, 1908–12.

Gusta LV, Burke MJ, Kapoor AC (1975) Determination of unfrozen water in winter cereals at subfreezing temperatures. *Plant Physiology* 56, 707–9.

Gusta LV, Fowler DB (1979) Cold resistance and injury in winter cereals. In *Stress Physiology of Crop Plants* (eds Mussell H & Staples RC), pp. 160–78. Wiley, New York.

Guy RD, Berry JA, Fogel ML, Hoering TC (1989) Partitioning of respiratory electrons in the dark in leaves of transgenic tobacco with modified levels of alternative oxidase. *Planta* 177, 171–80.

Guyot G, Phulpin T, eds (1997) *Physical Measurements and Signatures in Remote Sensing*. Balkema, Rotterdam.

Ha S, Vankova R, Yamaguchi-Shinozaki K, Shinozaki K, Phan Tran L-S (2012) Cytokinins: metabolism and function in plant adaptation to environmental stresses. *Trends in Plant Science* 17, 172–9.

Hack HRB (1974) The selection of an infiltration technique for estimating the degree of stomatal opening. *Annals of Botany* 38, 93–114.

Hales S (1727) *Vegetable Staticks: or, an Account of some Statistical Experiments on the Sap in Vegetables*. W. Innys and R. Manby; T. Woodward, London.

Hall AE (2001) *Crop Responses to Environment*. CRC Press, Boca Raton.

Hall AE, Kaufmann MR (1975) Stomatal response to environment with *Sesamum indicum* L. *Plant Physiology* 55, 455–9.

Hall AE, Schulze E-D, Lange OL (1976) Current perspectives of steady-state stomatal responses to environment. In *Water and Plant Life* (eds Lange OL, Kappen L & Schulze E-D), pp. 169–87. Springer-Verlag, Berlin.

Hall DO, Rao KK (1999) *Photosynthesis*, 6th edn. Cambridge University Press, Cambridge.

Hall HK, McWha JA (1981) Effects of abscisic acid on growth of wheat (*Triticum aestivum* L.). *Annals of Botany* 47, 427–33.

Hansen P (1970) ^{14}C-studies on apple trees. VI. The influence of the fruit on the photosynthesis of leaves and the relative photosynthesis of fruit and leaves. *Physiologia Plantarum* 23, 805–10.

Hapke B (1993) *Theory of Reflectance and Emittance Spectroscopy*. Cambridge University Press, Cambridge.

Hari P, Mäkelä A, Korpilahti E, Homberg M (1986) Optimal control for gas exchange. *Tree Physiology* 2, 169–75.

Harley PC, Baldocchi DD (1995) Scaling carbon-dioxide and water-vapor exchange from leaf to canopy in a deciduous forest. 1. Leaf model parametrization. *Plant, Cell & Environment* 18, 1146–56.

Harmer SL (2009) The circadian system in plants. *Annual Review of Plant Biology* 60, 357–77.

Harrison MA (2012) Cross-talk between phytohormone signalling pathways under both optimal and stressful environmental conditions. In *Phytohormones and Abiotic Stress Tolerance in Plants* (eds Khan NA, Nazar R, Iqbal N & Anjum NA), pp. 49–76. Springer, Heidelberg.

Havaux M, Lannoye R (1985) Drought resistance of hard wheat cultivars measured by a rapid chlorophyll fluorescence test. *Journal of Agricultural Science, Cambridge* 104, 501–4.

Heikkila JJ, Papp JET, Schultz GA, Bewley JD (1984) Induction of heat shock protein messenger RNA in maize hypocotyls by water stress. *Plant Physiology* 76, 270–4.

Henderson A (1987) Literature on air pollution and lichens XXV. *Lichenologist* 19, 205–10.

Henson IE (1985) Solute accumulation and growth in plants of pearl millet (*Pennisetum americanum* [L.] Leeke) exposed to abscisic acid or water stress. *Journal of Experimental Botany* 36, 1889–99.

Henzell RG, McCree KJ, van Bavel CHM, Schertz KF (1976) Sorghum genotype variation in stomatal sensitivity to water deficit *Crop Science* 16, 660–2.

Henzler T, Waterhouse RN, Smyth AJ *et al.* (1999) Diurnal variations in hydraulic conductivity and root pressure can be correlated with the expression of putative aquaporins in the roots of *Lotus japonicus*. *Planta* 210, 50–60.

Herbst M, Kappen L, Thamm F, Vanselow R (1996) Simultaneous measurements of transpiration, soil evaporation and total evaporation in a maize field in northern Germany. *Journal of Experimental Botany* 47, 1957–62.

Herrington LP (1969) *On Temperature and Heat Flow in Tree Stems. Bulletin 73*. Yale University, School of Forestry, New Haven, CT.

Hewitt CN, Kok GL, Fall R (1990a) Hydroperoxides in plants exposed to ozone mediate air pollution damage to alkene emitters. *Nature* 344, 56–8.

Hewitt CN, Monson RK, Fall R (1990b) Isoprene emissions from the grass *Arundo donax* L. are not linked to photorespiration. *Plant Science* 66, 139–44.

Hillel D (2004) *Introduction to Environmental Soil Physics*. Academic Press, San Diego.

Hocking PJ (1980) The composition of phloem exudate and xylem sap from tree tobacco (*Nicotiana glauca* Grah.). *Annals of Botany* 45, 633–43.

Holmes MG, Smith H (1975) The function of phytochrome in plants growing in the natural environment. *Nature* 254, 512–14.

Hotta CT, Gardner MJ, Hubbard KE *et al.* (2007) Modulation of environmental responses of plants by circadian clocks. *Plant, Cell & Environment* 30, 333–49.

Hough MN, Jones RJA (1997) The United Kingdom Meteorological Office rainfall and evaporation calculation system: MORECS version 2.0 – an overview. *Hydrology and Earth System Sciences* 1, 227–39.

Houghton JT (2009) *Global Warming: the Complete Briefing*, 4th edn. Cambridge University Press, Cambridge.

Hsiao TC (1973) Plant responses to water stress. *Annual Review of Plant Physiology* 24, 519–70.

Hu H, Dai M, Yao J *et al.* (2006) Overexpressing a NAM, ATAF, and CUC (NAC) transcription factor enhances drought resistance and salt tolerance in rice. *Proceedings of the National Academy of Sciences of the USA* 103, 12987–92.

Hubick KT, Shorter R, Farquhar GD (1988) Heritability and genotype × environment interactions of carbon isotope discrimination and transpiration efficiency in peanut (*Arachis hypogaea* L.). *Australian Journal of Plant Physiology* 15, 799–813.

Hughes JE, Morgan DC, Lambton PA, Black CR, Smith H (1984) Photoperiodic signals during twilight. *Plant, Cell & Environment* **7**, 269–77.

Huguet JG, Li SH, Lorendeau JY, Pelloux G (1992) Specific micromorphometric reactions of fruit trees to water stress and irrigation scheduling automation. *Journal of Horticultural Science* **67**, 631–40.

Hunt R, Causton DR, Shipley B, Askew AP (2002) A modern tool for classical growth analysis. *Annals of Botany* **90**, 485–8.

Hüsken D, Steudle E, Zimmermann U (1978) Pressure probe technique for measuring water relations of cells in higher plants. *Plant Physiology* **61**, 158–63.

Hutton JT, Norrish K (1974) Silicon content of wheat husks in relation to water transpired. *Australian Journal of Agricultural Research* **25**, 203–12.

Hyer EJ, Goetz SJ (2004) Comparison and sensitivity analysis of instruments and radiometric methods for LAI estimation: assessments from a boreal forest site. *Agricultural and Forest Meteorology* **122**, 157–74.

Hylton CM, Rawsthorne S, Smith AM, Jones AD (1988) Glycine decarboxylase is confined to the bundle-sheath cells of leaves of C_3-C_4 intermediate species. *Planta* **175**, 452–9.

Iba K (2002) Acclimative response to temperature stress in higher plants: approaches of gene engineering for temperature tolerance. *Annual Review of Plant Biology* **53**, 225–45.

Idso SB (1981) A set of equations for full spectrum and 8-μm to 14-μm and 10.5- μm to 12.5-μm thermal-radiation from cloudless skies. *Water Resources Research* **17**, 295–304.

Idso SB, Jackson RD, Ehrler WL, Mitchell ST (1969) A method for determination of infrared emittance of leaves. *Ecology* **50**, 899–902.

Ingram J, Bartels D (1996) Molecular basis of dehydration tolerance in plants. *Annual Review of Plant Physiology and Plant Molecular Biology* **47**, 377–403.

Innes P, Blackwell RD (1981) The effect of drought on water use and yield of two spring wheat genotypes. *Journal of Agricultural Science, Cambridge* **96**, 603–10.

Irmak S, Mutiibwa D, Irmak A *et al.* (2008) On the scaling up leaf stomatal resistance to canopy resistance using photosynthetic photon flux density. *Agricultural and Forest Meteorology* **148**, 1034–44.

Irving PM (1988) Overview of the US national acid precipitation assessment programme. In *Air Pollution and Ecosystems* (ed. Mathy P). D. Reidel, Dordrecht.

Islam MR, Sikder S (2011) Phenology and degree days of rice cultivars under organic culture. *Bangladesh Journal of Botany* **40**, 149–53.

Iwanoff L (1928) Zur Methodik der Transpirations-bestimmung am Standort. *Berichte der Deutschen Botanischen Gesellshaft* **46**, 306–10.

Iyer-Pascuzzi AS, Symonova O, Mileyko Y *et al.* (2010) Imaging and analysis platform for automatic phenotyping and trait ranking of plant root systems. *Plant Physiology* **152**, 1148–57.

Jackson JE, Palmer JW (1979) A simple model of light transmission and interception by discontinuous canopies. *Annals of Botany* **44**, 381–3.

Jackson MB (2002) Long distance signalling from roots to shoots assessed: the flooding story. *Journal of Experimental Botany* **53**, 175–81.

Jackson RB, Moore LA, Hoffman WA, Pockman WT, Linder CR (1999) Ecosystem rooting depth determined with caves and DNA. *Proceedings of the National Academy of Sciences of the USA* **96**, 11387–92.

Jackson RD (1982) *Canopy Temperature and Crop Water Stress*. Academic Press, London, New York.

Jackson RD, Idso SB, Reginato RJ, Pinter PJ (1981) Canopy temperature as a crop water-stress indicator. *Water Resources Research* **17**, 1133–8.

Jackson RD, Reginato RJ, Idso SB (1977) Wheat canopy temperature: a practical tool for evaluating water requirements. *Water Resources Research* **13**, 651–6.

Jackson SD (2009) Plant responses to photoperiod. *New Phytologist* **181**, 517–31.

Jacquemoud S, Baret F (1990) PROSPECT: a model of leaf optical properties spectra. *Remote Sensing of Environment* **34**, 75–91.

Jacquemoud S, Ustin SL, Verdebout J *et al.* (1996) Estimating leaf biochemistry using the PROSPECT leaf optical properties model. *Remote Sensing of Environment* **56**, 194–202.

Jaffe MJ (1973) Thigmomorphogenesis: the response of plant growth and development to mechanical stimulation. *Planta* **114**, 143–47.

Jamieson PD, Porter JR, Goudriaan J *et al.* (1998) A comparison of the models AFRCWHEAT2, CERES-Wheat, Sirius, SUCROS2 and SWHEAT with measurements from wheat grown under drought. *Field Crops Research* **55**, 23–44.

Jane FW (1970) *The Structure of Wood*, 2nd edn. Adam & Charles Black, London.

Janott M, Gayloer S, Gessler A *et al.* (2011) A one-dimensional model of water flow in soil-plant systems based on plant architecture. *Plant Soil* **341**, 233–56.

Jarman PD (1974) The diffusion of carbon dioxide and water vapour through stomata. *Journal of Experimental Botany* **25**, 927–36.

Jarvis PG (1976) The interpretation of variations in leaf water potential and stomatal conductance found in canopies in the field. *Philosophical Transactions of the Royal Society of London, Series B* **273**, 593–610.

Jarvis PG (1981) Stomatal conductance, gaseous exchange and transpiration. In *Plants and their Atmospheric Environment* (eds Grace J, Ford ED & Jarvis PG), pp. 175–204. Blackwell, Oxford.

Jarvis PG (1985) Coupling of transpiration to the atmosphere in horticultural crops: the omega factor. *Acta Horticulturae* **171**, 187–205.

Jarvis PG, James GB, Landsberg JJ (1976) Coniferous forest. In *Vegetation and the Atmosphere, Vol. 2: Case Studies* (ed. Monteith JL), pp. 171–240. Academic Press, London.

Jarvis PG, Mansfield TA, eds (1981) *Stomatal Physiology*. Cambridge University Press, Cambridge.

Jarvis PG, McNaughton KG (1986) Stomatal control of transpiration: scaling up from leaf to region. *Advances in Ecological Research* **15**, 1–49.

Jarvis PG, Morison JIL (1981) The control of photosynthesis and transpiration by the stomata. In *Stomatal Physiology* (eds Jarvis PG & Mansfield TA), pp. 248–79. Cambridge University Press, Cambridge.

Jenkins GI (2009) Signal transduction in responses to UV-B radiation. *Annual Review of Plant Biology* **60**, 407–31.

Jensen JR (2007) *Remote Sensing of the Environment: an Earth Resource Perspective*, 2nd edn. Pearson Prentice Hall, Upper Saddle River, NJ.

Jensen M (1985) The aerodynamics of shelter. In *FAO Conservation Guide 10: Sand Dune Stabilization, Shelterbelts and Afforestation in Dry Zones*. FAO, Rome.

Jiang Y, Wang Y (2011) Candidate gene expression involved in drought resistance. In *Handbook of Plant and Crop Stress* (ed. Pessarakli M), pp. 867–76. CRC Press, Boca Raton, FL.

Johnson HB (1975) Plant pubescence: an ecological perspective. *The Botanical Review* **41**, 233–58.

Johnson IR, Thornley JHM, Frantz JM, Bugbee B (2010) A model of canopy photosynthesis incorporating protein distribution through the canopy and its acclimation to light, temperature and CO_2. *Annals of Botany* **106**, 735–49.

Jonckheere I, Fleck S, Nackaerts K *et al.* (2004) Review of methods for in situ leaf area index determination – Part I. Theories, sensors and hemispherical photography. *Agricultural and Forest Meteorology* **121**, 19–35.

Jones HG (1972) *Effects of Water Stress on Photosynthesis*. PhD thesis, Australian National University, Canberra.

Jones HG (1973a) Estimation of plant water status with the beta-gauge. *Agricultural Meteorology* **11**, 345–55.

Jones HG (1973b) Limiting factors in photosynthesis. *New Phytologist* **72**, 1089–94.

Jones HG (1973c) Moderate-term water stresses and associated changes in some photosynthetic parameters in cotton. *New Phytologist* **72**, 1095–105.

Jones HG (1973d) Photosynthesis by thin leaf slices in solution. 2. Osmotic stress and its effects on photosynthesis. *Australian Journal of Biological Sciences* **26**, 25–33.

Jones HG (1976) Crop characteristics and the ratio between assimilation and transpiration. *Journal of Applied Ecology* **13**, 605–22.

Jones HG (1977a) Aspects of the water relations of spring wheat (*Triticum aestivum* L.) in response to induced drought. *Journal of Agricultural Science* **88**, 267–82.

Jones HG (1977b) Transpiration in barley lines with differing stomatal frequencies. *Journal of Experimental Botany* **28**, 162–8.

Jones HG (1978) Modelling diurnal trends of leaf water potential in transpiring wheat. *Journal of Applied Ecology* **15**, 613–26.

Jones HG (1979) Stomatal behavior and breeding for drought resistance. In *Stress Physiology in Crop*

Plants (eds Mussell H & Staples R), pp. 408–28. John Wiley and Sons Inc., New York.

Jones HG (1980) Interaction and integration of adaptive responses to water stress: the implications of an unpredictable environment. In *Adaptation of Plants to Water and High Temperature Stress* (eds Turner NC & Kramer PJ), pp. 353–65. John Wiley & Sons Inc, New York.

Jones HG (1981a) Carbon dioxide exchange of developing apple fruits. *Journal of Experimental Botany* **32**, 1203–10.

Jones HG (1981b) PGRs and plant water relations. In *Aspects and Prospects of Plant Growth Regulators* (ed. Jeffcoat B), pp. 91–100. BPGRG, Lancaster.

Jones HG (1981c) The use of stochastic modelling to study the influence of stomatal behaviour on yield-climate relationships. In *Mathematics and Plant Physiology* (eds Rose DA & Charles-Edwards DA), pp. 231–44. Academic Press, London.

Jones HG (1983) Estimation of an effective soil-water potential at the root surface of transpiring plants. *Plant, Cell and Environment* **6**, 671–4.

Jones HG (1985a) Partitioning stomatal and non-stomatal limitations to photosynthesis. *Plant, Cell & Environment* **8**, 95–104.

Jones HG (1985b) Physiological mechanisms involved in the control of leaf water status: implications for the estimation of tree water status. *Acta Horticulturae* **171**, 291–6.

Jones HG (1987a) Breeding for stomatal characters. In *Stomatal Function* (eds Zeiger E, Farquhar GD & Cowan IR), pp. 431–43. Stanford University Press, Stanford.

Jones HG (1987b) Repeat flowering in apple caused by water stress or defoliation. *Trees – Structure and Function* **1**, 135–8.

Jones HG (1990) Physiological aspects of the control of water status in horticultural crops. *HortScience* **25**, 19–26.

Jones HG (1998) Stomatal control of photosynthesis and transpiration. *Journal of Experimental Botany* **49**, 387–98.

Jones HG (1999) Use of infrared thermometry for estimation of stomatal conductance as a possible aid to irrigation scheduling. *Agricultural and Forest Meteorology* **95**, 139–49.

Jones HG (2004a) Application of thermal imaging and infrared sensing in plant physiology and ecophysiology. *Advances in Botanical Research* **41**, 107–63.

Jones HG (2004b) Irrigation scheduling: advantages and pitfalls of plant-based methods. *Journal of Experimental Botany* **55**, 2427–36.

Jones HG (2007) Monitoring plant and soil water status: established and novel methods revisited and their relevance to studies of drought tolerance. *Journal of Experimental Botany* **58**, 119–30.

Jones HG, Archer N, Rotenberg E, Casa R (2003) Radiation measurement for plant ecophysiology. *Journal of Experimental Botany* **54**, 879–89.

Jones HG, Flowers TJ, Jones MB, eds (1989) *Plants Under Stress*. Cambridge University Press, Cambridge.

Jones HG, Higgs KH (1979) Water potential-water content relationships in apple leaves. *Journal of Experimental Botany* **30**, 965–70.

Jones HG, Higgs KH (1980) Resistance to water loss from the mesophyll cell surface in plant leaves. *Journal of Experimental Botany* **31**, 545–53.

Jones HG, Higgs KH (1989) Empirical models of the conductance of leaves in apple orchards. *Plant, Cell & Environment* **12**, 301–8.

Jones HG, Higgs KH, Hamer PJC (1988) Evaluation of various heat-pulse methods for estimation of sap flow in orchard trees: comparison with micrometeorological estimates of evaporation. *Trees: Structure and Function* **2**, 250–60.

Jones HG, Hillis RM, Gordon SL, Brennan RM (2012) An approach to the determination of winter chill requirements for different *Ribes* cultivars. *Plant Biology* **15**, s1, 18–27.

Jones HG, Luton MT, Higgs KH, Hamer PJC (1983) Experimental control of water status in an apple orchard. *Journal of Horticultural Science* **58**, 301–16.

Jones HG, Osmond CB (1973) Photosynthesis by thin leaf slices in comparison with whole leaves. *Australian Journal of Biological Sciences* **26**, 15–24.

Jones HG, Peña J (1987) Relationships between water stress and ultrasound emission in apple (*Malus* × *domestica* Borkh.). *Journal of Experimental Botany* **37**, 1245–54.

Jones HG, Serraj R, Loveys BR *et al.* (2009) Thermal infrared imaging of crop canopies for the remote diagnosis and quantification of plant responses to water stress in the field. *Functional Plant Biology* **36**, 978–89.

Jones HG, Slatyer RO (1972) Estimation of the transport and carboxylation components of the intracellular limitation to leaf photosynthesis. *Plant Physiology* **50**, 283–8.

Jones HG, Sutherland RA (1991) Stomatal control of xylem embolism. *Plant, Cell & Environment* **6**, 607–12.

Jones HG, Vaughan RA (2010) *Remote Sensing of Vegetation: Principles, Techniques, and Applications.* Oxford University Press, Oxford.

Jones MM, Rawson HR (1979) Influence of rate of development of leaf water deficits upon photosynthesis, leaf conductance, water use efficiency, and osmotic potential in sorghum. *Physiologia Plantarum* **45**, 103–11.

Jordan BR (2011) Effects of UV-B radiation on plants: molecular mechanisms involved in UV-B responses. In *Handbook of Plant and Crop Stress* (ed. Pessarakli M), pp. 565–76. CRC Press, Boca Raton, FL.

Jordan BR, Partis MD, Thomas B (1986) The biology and molecular biology of phytochrome. *Oxford Surveys of Plant Molecular and Cell Biology* **3**, 315–62.

Jordan WR, Ritchie JT (1971) Influences of soil water stress on evaporation, root absorption, and internal water status of cotton. *Plant Physiology* **48**, 783–8.

Joshi MC, Boyer JS, Kramer PJ (1965) CO_2 exchange, transpiration and transpiration ratio of pineapple. *Botanical Gazette* **126**, 174–9.

Kacperska A (2004) Sensor types in signal transduction pathways in plant cells responding to abiotic stressors: do they depend on stress intensity? *Physiologia Plantarum* **122**, 159–68.

Kacser H, Burns JA (1973) The control of flux. *Symposium of the Society for Experimental Biology* **27**, 65–107.

Kaimal C, Finnigan JJ (1994) *Atmospheric Boundary Layer Flows: Their Structure and Measurement.* Oxford University Press, Oxford.

Kaiser W (1987) Effects of water deficits on photosynthetic capacity. *Physiologia Plantarum* **71**, 142–9.

Kaiser WM (1982) Correlations between changes in photosynthetic activity and changes in total protoplast volume in leaf tissue from hygro-, meso-,

and xerophytes under osmotic stress. *Planta* **154**, 538–45.

Kami C, Hersch M, Trevisan M *et al.* (2012) Nuclear phytochrome A signaling promotes phototropism in *Arabidopsis. The Plant Cell* **24**, 566–76.

Kamiya Y (2009) Plant hormones: versatile regulators of plant growth and development. *Annual Review of Plant Biology* **60**, Web compilation, doi: 10.1146/annurev.arplant.1160.031110.100001.

Kanemasu ET, Stone LR, Powers WL (1976) Evapotranspiration model tested for soybean and sorghum. *Agronomy Journal* **68**, 569–72.

Kang MZ, de Reffye P (2007) A mathematical approach estimating source and sink functioning of competing organs. In *Functional–Structural Plant Modelling in Plant Production* (eds Vos J, Marcelis LFM, de Visser PHB, Struik PC & Evers JB), pp. 65–74. Wageningen UR, Wageningen.

Keeley JE, Osmond CB, Raven JA (1984) *Stylites*, a vascular land plant without stomata absorbs CO_2 by its roots. *Nature* **310**, 694–5.

Keller MM, Jaillais Y, Pedmale UV *et al.* (2011) Cryptochrome 1 and phytochrome B control shade-avoidance responses in *Arabidopsis* via partially independent hormonal cascades. *The Plant Journal* **67**, 195–207.

Kendrick RE, Kronenberg GHM, eds (1994) *Photomorphogenesis in Plants*, 2nd edn. Kluwer Academic Publishers, Dordrecht, Netherlands.

Kerr JP, Beardsell MF (1975) Effect of dew on leaf water potentials and crop resistances in a paspalum pasture. *Agronomy Journal* **67**, 596–9.

Khan NA, Nazar R, Iqbal N, Anjum NA, eds (2012) *Phytohormones and Abiotic Stress Tolerance in Plants.* Springer, Heidelberg.

Kim T-H, Bohmer M, Hu H, Nishimura N, Shroeder JL (2010) Guard cell signal transduction network; advances in understanding abscisic acid, CO_2, and Ca^{2+} signaling. *Annual Review of Plant Biology* **61**, 561–91.

King RW, Wardlaw IF, Evans LT (1967) Effect of assimilate utilization on photosynthetic rate in wheat. *Planta* **71**, 261–76.

Kinoshita T, Doi M, Suetsugu N *et al.* (2001) Phot1 and phot2 mediate blue light regulation of stomatal opening. *Nature* **414**, 656–60.

Kirkham MB (2004) *Principles of Soil and Plant Water Relations.* Elsevier Academic Press, Burlington, MA.

Kjelgaard JF, Stockle CO, Black RA, Campbell GS (1997) Measuring sap flow with the heat balance approach using constant and variable heat inputs. *Agricultural and Forest Meteorology* **85**, 239–50.

Klingler JP, Batelli G, Zhu J-K (2010) ABA receptors: the START of a new paradigm in phytohormone signalling. *Journal of Experimental Botany* **61**, 3199–210.

Kluge M, Ting IP (1978) *Crassulacean Acid Metabolism.* Springer-Verlag, Berlin.

Kniemayer O, Buck-Sorlin G, Kurth W (2007) GroImp as a platform for functional–structural modelling of plants. In *Functional–Structural Plant Modelling in Plant Production* (eds Vos J, Marcelis LFM, de Visser PHB, Struik PC & Evers JB), pp. 43–52. Wageningen UR, Wageningen.

Knight MR, Campbell AK, Smith SM, Trewavas AJ (1991) Transgenic plant aequorin reports the effects of touch and coldshock and elicitors on cytoplasmic calcium. *Nature* **352**, 524–6.

Kolber Z, Klimov D, Ananyev G, Rascher U, Berry J, Osmond BA (2005) Measuring photosynthetic parameters at a distance: laser induced fluorescence transient (LIFT) method for remote measurements of photosynthesis in terrestrial vegetation *Photosynthesis Research* **84**, 121–9.

Kolber Z, Prasil O, Falkowski PG (1998) Measurement of variable chlorophyll fluorescence using fast repetition rate techniques: defining methodology and experimental protocols. *Biochimica et Biophysica Acta – Bioenergetics* **1367**, 88–106.

Körner C (1999) *Alpine Plant Life.* Springer-Verlag, Berlin-Heidelberg.

Körner C (2012) *Alpine Treelines: Functional Ecology of the Global High Elevation Tree Limits.* Springer, Basel.

Körner C, Mayr R (1981) Stomatal behaviour in alpine communities between 600 and 2600 metres above sea level. In *Plants and their Atmospheric Environment* (eds Grace J, Ford ED & Jarvis PG), pp. 205–18. Blackwell, Oxford.

Körner C, Scheel JA, Bauer H (1979) Maximum leaf diffusive conductance in higher plants. *Photosynthetica* **13**, 45–82.

Kowal JM, Kassam AH (1973) Water use, energy balance and growth of maize at Samuru, Northern Nigeria. *Agricultural Meteorology* **12**, 391–406.

Kramer PJ, Boyer JS (1995) *Water Relations of Plants and Soils.* Academic Press Inc., London.

Kreith F, Bohn MS, Manglik R (2010) *Principles of Heat Transfer*, 7th edn. Cengage Learning Inc., Mason, OH.

Krömer S (1995) Respiration during photosynthesis. *Annual Review of Plant Physiology and Plant Molecular Biology* **46**, 45–70.

Krul L (1993) Remote sensing in the microwave region. In *Land Observation by Remote Sensing: Theory and Applications* (eds Buiten HJ & Clevers JGPW), pp. 155–74. Gordon and Breach, Yverdon.

Kubota T, Tsuboyama Y (2004) Estimation of evaporation rate from the forest floor using oxygen-18 and deuterium compositions of throughfall and stream water during a non-storm runoff period. *Journal of Forest Research* **9**, 51–9.

Kucharik CJ, Norman JM, Gower ST (1998a) Measurements of branch area and adjusting leaf area index indirect measurements. *Agricultural and Forest Meteorology* **91**, 69–88.

Kucharik CJ, Norman JM, Gower ST (1998b) Measurements of leaf orientation, light distribution and sunlit leaf area in a boreal aspen forest. *Agricultural and Forest Meteorology* **91**, 127–48.

Kumar SV, Wigge PA (2010) H2A.Z-containing nucleosomes mediate the thermosensory response in *Arabidopsis*. *Cell* **140**, 136–47

Kutschera U, Briggs WR (2012) Root phototropism: from dogma to the mechanism of blue light perception. *Planta* **235**, 443–52.

Lakatos M, Obregón A, Büdel B, Bendix J (2011) Midday dew: an overlooked factor enhancing photosynthetic activity of corticolous epiphytes in a wet tropical rain forest. *New Phytologist* **194**, 245–53.

Lakso AN (1979) Seasonal changes in stomatal response to leaf water potential in apple. *Journal of the American Society for Horticultural Science* **104**, 58–60.

Landsberg HE (1961) Solar radiation at the earth's surface. *Solar Energy* **5**, 95–8.

Landsberg JJ, Beadle CL, Biscoe PV *et al.* (1975) Diurnal energy, water and CO_2 exchanges in an apple

(*Malus pumila*) orchard. *Journal of Applied Ecology* 12, 659–84.

Landsberg JJ, Blanchard TW, Warrit B (1976) Studies on the movement of water through apple trees. *Journal of Experimental Botany* 27, 579–96.

Lang ARG (1973) Leaf orientation of a cotton plant. *Agricultural Meteorology* 11, 37–51.

Lang ARG, Evans GN, Ho PY (1974) The influence of local advection on evapotranspiration from irrigated rice in a semi-arid region. *Agricultural Meteorology* 13, 5–13.

Lang M, Kuusk A, Mõtus M, Rautainen M, Nilson T (2010) Canopy gap fraction estimation from digital hemispherical images using sky radiance models and a linear conversion method. *Agricultural and Forest Meteorology* 150, 20–9.

Langsdorf G, Buschmann C, Sowinska M, Banbani F, Mokry F, Timmermann F, Lichtenthaler HK (2000) Multicolour fluorescence imaging of sugar beet leaves with different nitrogen status by flash lamp UV-excitation. *Photosynthetica* 38, 539–51.

Larcher W (1995) *Physiological Plant Ecology*, 3rd edn. Springer, Berlin, Heidelberg, New York.

Larqué-Saavedra A, Wain RL (1974) Abscisic acid levels in relation to drought tolerance in varieties of *Zea mays* L. *Nature* 251, 716–17.

Lauscher F (1976) Weltweiter Typen der Höhen abhängigkeit des Niederschlags. *Wetter und Leben* 28, 80–90.

Lawlor DW (2001) *Photosynthesis*, 3rd edn. Bios Scentific Publishers, Oxford.

Lawlor DW (2002) Limitation to photosynthesis in water-stressed leaves: stomata vs. metabolism and the role of ATP. *Annals of Botany* 89, 871–85.

Lawlor DW, Tezara W (2009) Causes of decreased photosynthetic rate and metabolic capacity in water-deficient leaf cells: a critical evaluation of mechanisms and integration of processes. *Annals of Botany* 103, 561–79.

Le Treut H, Somerville R, Cubasch U *et al.* (2007) Historical overview of climate change. In *Climate Change 2007: The Physical Science Basis. Contribution of Working Group I to the Fourth Assessment Report of the Intergovernmental Panel on Climate Change* (eds Solomon S, Qin D, Manning M *et al.*), pp. 93–127. Cambridge University Press, Cambridge.

Leblanc SG, Chen JM, Fernandes R, Deering DW, Conley A (2005) Methodology comparison for canopy structure parameters extraction from digital hemispherical photography in boreal forests. *Agricultural and Forest Meteorology* 129, 187–207.

Lee RE, Chen C-P, Denlinger DL (1987) A rapid cold-hardening process in insects. *Science* 298, 1415–17.

Lee SH, Singh AP, Chung GC, Kim YS, Komg IB (2002) Chilling root temperature causes rapid ultrastructural changes in cortical cells of cucumber (*Cucumis sativus* L.) root tips. *Journal of Experimental Botany* 53, 2225–37.

Leinonen I, Grant OM, Tagliavia CPP, Chaves MM, Jones HG (2006) Estimating stomatal conductance with thermal imagery. *Plant, Cell & Environment* 29, 1508–18.

Lens F, Sperry JS, Christman MA, Choat B, Rabaey D, Jansen S (2011) Testing hypotheses that link wood anatomy to cavitation resistance and hydraulic conductivity in the genus *Acer*. *New Phytologist* 190, 709–23.

Leuning R (1995) A critical appraisal of a combined stomatal-photosynthesis model for C_3 plants. *Plant, Cell & Environment* 18, 339–55.

Levitt J (1980) *Responses of Plants to Environmental Stresses. Vol. I.*, 2nd edn. Academic Press, New York.

Lewis MC, Callaghan TV (1976) Tundra. In *Vegetation and the Atmosphere. Vol. 2. Cases Studies* (ed. Monteith JL), pp. 399–433. Academic Press, London.

Leyton L (1975) *Fluid Behaviour in Biological Systems.* Clarendon Press, Oxford.

Lhomme J-P, Monteny B, Amadou M (1994) Estimating sensible heat flux from radiometric temperature over sparse millet. *Agricultural and Forest Meteorology* 68, 77–91.

Li QH, Yang HQ (2007) Cryptochrome signaling in plants. *Photochemistry and Photobiology* 83, 94–101.

Li Y, Sperry JS, Taneda H, Bush S, Hacke UG (2008) Evaluation of centrifugal methods for measuring xylem cavitation in conifers, diffuse- and ring-porous angiosperms. *New Phytologist* 177, 558–68.

Li Z, Wakao S, Fischer BB, Niyogi KN (2009) Sensing and reponding to excess light. *Annual Review of Plant Biology* 60, 239–60.

Liang GH, Dayton AD, Chu CC, Casady AJ (1975) Heritability of stomatal density and distribution on leaves of grain sorghum. *Crop Science* **15**, 567–70.

Liang S (2004) *Quantitative Remote Sensing of Land Surfaces.* John Wiley and Sons, Inc., Hoboken, NJ.

Libourel IGL, Sachar-Hill Y (2008) Metabolic flux analysis in plants: from intelligent design to rational engineering. *Annual Review of Plant Biology* **59**, 625–60.

Lin CT, Shalitin D (2003) Cryptochrome structure and signal transduction. *Annual Review of Plant Biology* **54**, 469–96.

Linacre ET (1969) Net radiation to various surfaces. *Journal of Applied Ecology* **6**, 61–75.

Linacre ET (1972) Leaf temperatures, diffusion resistances, and transpiration. *Agricultural Meteorology* **10**, 365–82.

Liu K, Luan S (1998) Voltage dependent K^+ channels as targets of osmosensing in guard cells. *Plant Cell* **10**, 1957–70.

Livingston BE, Brown WH (1912) Relation of the daily march of transpiration to variations in the water content of foliage leaves. *Botanical Gazette* **53**, 309–30.

Lloyd J, Grace J, Miranda AC *et al.* (1995) A simple calibrated model of Amazon rainforest productivity based on leaf biochemical properties. *Plant, Cell & Environment* **18**, 1129–45.

Lobell DB, Cassman KG, Field CB (2009) Crop yield gaps: their importance, magnitudes, and causes. *Annual Review of Environment and Resources* **34**, 179–204.

Lockhart JA (1965) An analysis of irreversible plant cell elongation. *Journal of Theoretical Biology* **8**, 264–75.

Long SP, Ainsworth EA, Rogers A, Ort DR (2004) Rising atmospheric carbon dioxide: plants FACE the future. *Annual Review of Plant Biology* **55**, 591–628.

Long SP, Incoll LD, Woolhouse HW (1975) C_4 photosynthesis in plants from cool temperature regions with particular reference to *Spartina townsendii. Nature* **257**, 622–4.

Long SP, Woodward FI, eds (1988) *Plants and Temperature. Symposia of the Society for Experimental Biology, XLII.* Company of Biologists, Cambridge.

Loreto F, Delfine S, Di Marco G (1999) Estimation of photorespiratory carbon dioxide recycling during photosynthesis. *Australian Journal of Plant Physiology* **26**, 733–6.

Loreto F, Schnitzler J-P (2010) Abiotic stresses and induced BVOCs. *Trends in Plant Science* **15**, 154–66.

Lorimer GH, Andrews TJ (1981) The C_2 chemo- and photorespiratory carbon oxidation cycle. In *The Biochemistry of Plants, Vol. 10: Photosynthesis* (eds Hatch MD & Boardman NK), pp. 329–74. Academic Press, New York.

Luckwill LC (1981) *Growth Regulators in Crop Production.* Edward Arnold, London.

Ludlow MM (1980) Adaptive significance of stomatal responses to water stress. In *Adaptation of Plants to Water and High Temperature Stress* (eds Turner NC & Kramer PJ), pp. 123–38. Wiley, New York.

Ludlow MM, Jarvis PG (1971) Photosynthesis in Sitka spruce (*Picea sitchensis* (Bong.) Carr.). I. General characteristics. *Journal of Applied Ecology* **8**, 925–53.

Ludlow MM, Wilson GL (1971) Photosynthesis of tropical pasture plants. III. Leaf age. *Australian Journal of Biological Sciences* **24**, 1077–87.

Ma BL, Dwyer LM (2001) Maize kernel moisture, carbon and nitrogen concentrations from silking to physiological maturity. *Canadian Journal of Plant Science* **81**, 225–32.

Mackay I, Horwell A, Garner J *et al.* (2011) Reanalyses of the historical series of UK variety trials to quantify the contributions of genetic and environmental factors to trends and variability in yield over time. *Theoretical and Applied Genetics* **122**, 225–38.

MacRobbie EAC (1987) Ionic relations of guard cells. In *Stomatal Function* (eds Zeiger E, Farquhar GD & Cowan IR), pp. 125–62. Stanford University Press, Stanford.

Maes WH, Steppe K (2012) Estimating evapotranspiration and drought stress with ground-based thermal remote sensing in agriculture: a review. *Journal of Experimental Botany* **63**, 4671–712.

Maestre-Valero JF, Martínez-Alvarez V, Baille A, Martín-Górriz B, Gallego-Elvíra B (2011) Comparative analysis of two foil materials for dew harvesting in a semi-arid climate. *Journal of Hydrology* **410**, 84–91.

Malone M, Leight RA, Tomos AD (1989) Extraction and analysis of sap from individual wheat leaf cells: the effect of sampling speed on the osmotic pressure of extracted sap. *Plant, Cell and Environment* 12, 919–26.

Manzoni S, Vico G, Katul GG *et al.* (2011) Optimizing stomatal conductance for maximum carbon gain under water stress: a meta-analysis across plant functional types and climates. *Functional Ecology* 25, 456–67.

Marino G, Aqil M, Shipley B (2010) The leaf economics spectrum and the prediction of photosynthetic light-response curves. *Functional Ecology* 24, 263–72.

Marshall B, Woodward FI, eds (1986) *Instrumentation for Environmental Physiology*. Cambridge University Press, Cambridge.

Martínez-Lozano JA, Tena F, Onrubia JE, De La Rubia J (1984) The historical evaluation of the Ångström formula and its modifications: review and bibliography. *Agricultural and Forest Meteorology* 33, 109–28.

Maskell EJ (1928) Experimental researches on vegetable assimilation and respiration. XVIII. The relation between stomata opening and assimilation: a critical study of assimilation rates and porometer rates of cherry laurel. *Proceedings of the Royal Society of London, Series B* 102, 488–533.

Mathy P, ed. (1988) *Air Pollution and Ecosystems*. Reidel, Dordrecht.

Maurel C, Verdoucq L, Luu D-T, Santoni V (2008) Plant aquaporins: membrane channels with multiple integrated functions. *Annual Review of Plant Biology* 59, 595–624.

Maximov NA (1929) *The Plant in Relation to Water*. George Allen & Unwin, London.

Maxwell K, Johnson GN (2000) Chlorophyll fluorescence: a practical guide. *Journal of Experimental Botany* 51, 659–68.

Mayer H (1987) Wind-induced tree sway. *Trees: Structure and Function* 1, 195–206.

Mayhead GJ (1973) Some drag coefficients for British forest species derived from wind tunnel studies. *Agricultural Meteorology* 12, 123–30.

Mayr S, Rosner S (2010) Cavitation in dehydrating xylem of *Picea abies*: energy properties of ultrasonic emissions reflect tracheid dimensions. *Tree Physiology* 31, 59–67.

McCowan RL, Hammer GL, Hargreaves JNG, Holzworth DP, Freebairn DM (1996) APSIM: a novel software system for model development, model testing and simulation in agricultural systems research. *Agricultural Systems* 50, 255–71.

McCree KJ (1970) An equation for the rate of respiration of white clover plants grown under controlled conditions. In *Prediction and Measurement of Photosynthetic Productivity: Proceedings of IBP/PP Technical Meeting*, pp. 221–9. PUDOC, Wageningen.

McCree KJ (1972a) The action spectrum, absorptance and quantum yield of photosynthesis in crop plants. *Agricultural Meteorology* 9, 191–216.

McCree KJ (1972b) Test of current definitions of photosynthetically active radiation against leaf photosynthesis rate. *Agricultural Meteorology* 10, 443–53.

McCutchan H, Shackel KA (1992) Stem water potential as a sensitive indicator of water stress in prune trees (*Prunus domestica* L. cv. French). *Journal of American Society for Horticultural Science* 117, 607–11.

McElrone AJ, Bichler J, Pockman WT *et al.* (2007) Aquaporin-mediated changes in hydraulic conductivity of deep tree roots accessed via caves. *Plant, Cell and Environment* 30, 1411–21.

McLeod AR, Fackrell JE, Alexander K (1985) Open-air fumigation of field crops: criteria and design for a new experimental system. *Atmospheric Environment* 19, 1639–49.

McNaughton KG (1989) Micrometeorology of shelter belts and forest edges. *Philosophical Transactions of the Royal Society of London, Series B* 324, 351–68.

McNaughton KG, Jarvis PG (1983) Predicting the effects of vegetation changes on transpiration and evaporation. In *Water Deficits and Plant Growth. Vol. 7* (ed. Kozlowski TT), pp. 1–47. Academic Press, New York.

McPherson HG (1969) Photocell-filter combinations for measuring photosynthetically active radiation. *Agricultural Meteorology* 6, 347–56.

Medhurst J, Parsby J, Linder S, Wallin G, Ceschia E, Slaney M (2006) A whole-tree chamber system for examining tree-level physiological responses of field grown trees to environmental variation and climate change. *Plant, Cell & Environment* 29, 1853–69.

Meidner H, Mansfield TA (1968) *Physiology of Stomata*. McGraw-Hill, Maidenhead.

Meijninger WML, De Bruin HAR (2000) The sensible heat fluxes over irrigated areas in western Turkey determined with a large aperture scintillometer. *Journal of Hydrology* 229, 42–9.

Menz KM, Moss DN, Cannell RQ, Brun WA (1969) Screening for photosynthetic efficiency. *Crop Science* 9, 692–5.

Menzel CM (1983) The control of floral initiation in lychee: a review. *Scientia Horticulturae* 21, 201–15.

Mepsted R, Paul ND, Stephen J *et al.* (1996) Effects of enhanced UV-B radiation on pea (*Pisum sativum* L.) grown under field conditions in the United Kingdom. *Global Change Biology* 2, 325–34.

Meroni M, Rossini M, Guanter L *et al.* (2009) Remote sensing of solar-induced chlorophyll fluorescence: review of methods and applications. *Remote Sensing of Environment* 113, 2037–51.

Messinger SM, Buckley TN, Mott KA (2006) Evidence for involvement of photosynthetic processes in the stomatal response to CO_2. *Plant Physiology* 140, 771–8.

Milburn JA (1979) *Water Flow in Plants*. Longmans, London.

Millar AH, Whelan J, Soole KL, Day DA (2011) Organization and regulation of mitochondrial respiration in plants. *Annual Review of Plant Biology* 62, 79–104.

Miller P, Lanier W, Brandt S (2001) *Using Growing Degree Days to Predict Plant Stages*. Montana State University, Extension Service.

Milthorpe FL, Moorby J (1979) *An Introduction to Crop Physiology*, 2nd edn. Cambridge University Press, Cambridge.

Miranda AC, Jarvis PG, Grace J (1984) Transpiration and evaporation from heather moorland. *Boundary Layer Meteorology* 28, 227–43.

Miskin KE, Rasmusson DC (1970) Frequency and distribution of stomata in barley. *Crop Science* 10, 575–8.

Möglich A, Yang X, Ayers RA, Moffat K (2010) Structure and function of plant photoreceptors. *Annual Review of Plant Biology* 61, 21–47.

Molinari HBC, Marur CJ, Daros E *et al.* (2007) Evaluation of the stress-inducible production of proline in transgenic sugarcane (*Saccharum* spp.): osmotic adjustment, chlorophyll fluorescence and oxidative stress. *Physiologia Plantarum* 130, 218–29.

Monsi M, Saeki T (1953) Über den lichtfaktor in den Pflanzengesellschaften und seine Bedeutung für die Stoffproduktion. *Japanese Journal of Botany* 14, 22–52.

Monson RK, Jones RT, Rosenstiel TN, Schnitzler JP (2013) Why only some plants emit isoprene. *Plant, Cell & Environment* 36, 503–16.

Monteith JL (1957) Dew. *Quarterly Journal of the Royal Meteorological Society* 83, 322–41.

Monteith JL (1965) Evaporation and environment. In *Symposia of the Society for Experimental Biology, 19*, pp. 205–34. Cambridge University Press, Cambridge.

Monteith JL (1972) Solar radiation and productivity in tropical ecosystems. *Journal of Applied Ecology* 9, 747–66.

Monteith JL (1973) *Principles of Environmental Physics*. Edward Arnold, London.

Monteith JL, ed. (1975) *Vegetation and the Atmosphere, Vol. 1. Principles*. Academic Press, London.

Monteith JL, ed. (1976) *Vegetation and the Atmosphere, Vol. 2. Case Studies*. Academic Press, London.

Monteith JL (1977) Climate and the efficiency of crop production in Britain. *Proceedings of the Royal Society of London, Series B* 281, 277–94.

Monteith JL (1978) Reassessment of maximum growth rates for C_3 and C_4 crops. *Experimental Agriculture* 14, 1–5.

Monteith JL (1981) Coupling of plants to the atmosphere. In *Plants and their Atmospheric Environment* (eds Grace J, Ford ED & Jarvis PG), pp. 1–29. Blackwell, Oxford.

Monteith JL, Unsworth MH (2008) *Principles of Environmental Physics*, 3rd edn. Academic Press, Burlington, MA.

Montes JM, Melchinger AE, Reif JC (2007) Novel thoughput phenotyping platforms in plant genetic studies. *Trends in Plant Science* 12, 433–6.

Mooney HA, Björkman O, Collatz GJ (1978) Photosynthetic acclimation to temperature in the desert shrub *Larrea divaricata*. I. Carbon exchange characteristics of intact leaves. *Plant Physiology* 61, 406–10.

Mooney HA, Björkman O, Ehleringer JR, Berry J (1976) Photosynthetic capacity of *in situ* Death Valley plants. *Carnegie Institution Year Book* 75, 410–13.

Mooney HA, Ehleringer JR, Björkman O (1977) The energy balance of leaves of the evergreen desert shrub *Atriplex hymenelytra*. *Oecologia* **29**, 301–10.

Mooney HA, Gulmon SL, Ehleringer JR, Rundel PW (1980) Atmospheric water uptake by an Atacama desert shrub. *Science* **209**, 693–4.

Moore AL, Beechey RB, eds (1987) *Plant Mitochondria: Structural, Functional and Physiological Aspects.* Plenum Press, New York.

Morgan JM (1984) Osmoregulation. *Annual Review of Plant Physiology* **35**, 299–319.

Morison JIL (1987) Intercellular CO_2 concentration and stomatal response to CO_2. In *Stomatal Function* (eds Zeiger E, Farquhar GD & Cowan IR), pp. 229–51. Stanford University Press, Stanford.

Morran S, Eini O, Pyvovarenko T *et al.* (2011) Improvement of stress tolerance of wheat and barley by modulation of expression of DREB/CBF factors. *Plant Biotechnology Journal* **9**, 230–49.

Mortensen LM (1987) Review: CO_2 enrichment in greenhouses. *Scientia Horticulturae* **33**, 1–25.

Mott KA, Buckley TN (1998) Stomatal heterogeneity. *Journal of Experimental Botany* **49**, 407–17.

Mott KA, Parkhurst DF (1991) Stomatal responses to humidity in air and helox. *Plant, Cell & Environment* **14**, 509–15.

Mott KA, Peak D (2010) Stomatal responses to humidity and temperature in darkness. *Plant, Cell & Environment* **33**, 1084–90.

Mott KA, Peak D (2011) Alternative perspective on the control of transpiration by radiation. *Proceedings of the National Academy of Sciences of the United States of America* **108**, 19820–3.

Mullen JL, Weinig C, Hangartner RP (2006) Shade avoidance and the regulation of leaf inclination in *Arabidopsis*. *Plant, Cell & Environment* **29**, 1099–106.

Müller J, Diepenbrock W (2006) Measurement and modelling of gas exchange of leaves and pods of oilseed rape. *Agricultural and Forest Meteorology* **139**, 307–22.

Münch E (1930) *Die Stoffbewegungen in der Pflanze.* Verlag von Gustav Fischer, Jena.

Mundy J, Chua N-H (1988) Abscisic acid and water stress induce the expression of a novel rice gene. *EMBO Journal* **7**, 2279–86.

Munné-Bosch S, Alegre L (2004) Die and let live: leaf senescence contributes to plant survival under drought stress. *Functional Plant Biology* **31**, 203–16.

Munns R (2002) Comparative physiology of salt and water stress. *Plant, Cell & Environment* **25**, 239–50.

Munns R, Tester M (2008) Mechanisms of salinity tolerance. *Annual Review of Plant Biology* **59**, 651–81.

Murchie EH, Pinto M, Horton P (2009) Agriculture and the new challenges for photosynthesis research. *New Phytologist* **181**, 532–52.

Nagel OW, Waldron S, Jones HG (2001) An off-line implementation of the stable isotope technique for measurements of alternative respiratory pathway activities. *Plant Physiology* **127**, 1279–86.

Nawrath C (2006) Unraveling the complex network of cuticular structure and function. *Current Opinion in Plant Biology* **9**, 281–7.

Neales TF, Hartney VJ, Patterson AA (1968) Physiological adaptation to drought in the carbon assimilation and water loss of xerophytes. *Nature* **219**, 469–72.

Neales TF, Masia A, Zhang J, Davies WJ (1989) The effects of partially drying part of the root system of *Helianthus annuus* on the abscisic acid content of the roots, xylem sap and leaves. *Journal of Experimental Botany* **40**, 1113–20.

Neel PL, Harris RW (1971) Motion-induced inhibition of elongation and induction of dormancy in liquidambar. *Science* **173**, 58–9.

Nelson N, Yocum CF (2006) Structure and function of photosystems I and II. *Annual Review of Plant Biology* **57**, 521–65.

Newsham KK, McLeod AR, Greenslade PD, Emmett AA (1996) Appropriate controls in outdoor UV-B supplementation experiments. *Global Change Biology* **2**, 319–24.

Ng PAP, Jarvis PG (1980) Hysteresis in the response of stomatal conductance in *Pinus sylvestris* L. needles to light: observations and a hypothesis. *Plant, Cell & Environment* **3**, 207–16.

Nicodemus FE, Richmond JC, Hsia JJ, Ginsberg IW, Limperis T (1977) *Geometric Considerations and Nomenclature for Reflectance.* NBS Monograph 160. National Bureau of Standards, Washington DC.

Niklas KJ (1992) *Plant Biomechanics: an Engineering Approach to Plant Form and Function.* University of Chicago Press, Chicago.

Nilson T, Kuusk A (1989) A reflectance model for the homogeneous plant canopy and its inversion. *Remote Sensing of Environment* **27**, 157–67.

Nobel PS (1988) *Environmental Biology of Agaves and Cacti.* Cambridge University Press, Cambridge.

Nobel PS (2009) *Physicochemical and Environmental Plant Physiology*, 4th edn. Academic Press, Oxford.

Nobel PS, Jordan PW (1983) Transpiration stream of desert species: resistances and capacitances for a C_3, a C_4 and a CAM plant. *Journal of Experimental Botany* **34**, 1379–91.

Nobel PS, Zaragoza LJ, Smith WK (1975) Relation between mesophyll surface area, photosynthetic rate and illumination during development of leaves of *Plectranthus parviflora* Henckel. *Plant Physiology* **55**, 1067–70.

Nobs MA (1976) Hybridizations in *Atriplex. Carnegie Institution Year Book* **75**, 421–3.

Norman JM, Campbell GS (1989) Canopy structure. In *Plant Physiological Ecology: Field Methods and Instrumentation* (eds Pearcy RW, Ehleringer JR, Mooney HA, & Rundel PW), pp. 301–25. Chapman & Hall, London & New York.

Norman JM, Divakarla M, Goel NS (1995) Algorithms for extracting information from remote thermal-IR observations of the Earth's surface. *Remote Sensing of Environment* **51**, 157–68.

Norman JM, Jarvis PG (1974) Photosynthesis in Sitka spruce (*Picea sitchensis* (Bong.) Carr.). III. Measurements of canopy structure and interception of radiation. *Journal of Applied Ecology* **11**, 375–98.

Nowak RS, Ellsworth DS, Smith SD (2004) Functional responses of plants to elevated atmospheric CO_2: do photosynthetic and productivity data from FACE experiments support early predictions? *New Phytologist* **162**, 253–80.

O'Shaughnessy SA, Evett SR (2010) Canopy temperature based system effectively schedules and controls center pivot irrigation of cotton. *Agricultural Water Management* **97**, 1310–16.

Oertli JJ, Lips SH, Agami M (1990) The strength of schlerophyllous cells to resist collapse due to negative turgor pressure. *Acta Oecologia* **11**, 281–9.

Oliver MJ, Tuba Z, Mishler BD (2000) The evolution of vegetative desiccation tolerance in land plants. *Plant Ecology* **151**, 85–100.

Omasa K, Hosoi F, Konishi A (2007) 3D lidar imaging for detecting and understanding plant responses and canopy structure. *Journal of Experimental Botany* **58**, 881–98.

Omasa K, Nouchi I, De Kok LJ, eds (2005) *Plant Responses to Air Pollution and Climate Change.* Springer, Tokyo.

Ong CK (1983) Response to temperature in a stand of pearl millet (*Pennisetum typhoides* S. & H.). I. Vegetative development. *Journal of Experimental Botany* **34**, 322–36.

Oxborough K, Baker NR (1997) An instrument capable of imaging chlorophyll-*a* fluorescence from intact leaves at very low irradiance at cellular and subcellular levels of organisation. *Plant, Cell & Environment* **20**, 1473–83.

Paltridge GW, Denhom JV (1974) Plant yield and the switch from vegetative to reproductive growth. *Journal of Theoretical Biology* **44**, 23–34.

Paoletti E, Grulke NE (2005) Does living in elevated CO_2 ameliorate tree response? A review on stomatal responses. *Environmental Pollution* **137**, 483–93.

Parkhurst DF (1994) Diffusion of CO_2 and other gases inside leaves. *New Phytologist* **126**, 449–79.

Parkinson KJ, Day W (1980) Temperature corrections to measurements made with continuous flow porometers. *Journal of Applied Ecology* **17**, 457–60.

Parkinson KJ, Legg BJ (1972) A continuous flow porometer. *Journal of Applied Ecology* **9**, 669–75.

Parry MAJ, Reynolds M, Salvucci ME *et al.* (2011) Raising yield potential of wheat. II. Increasing photosynthetic capacity and efficiency. *Journal of Experimental Botany* **62**, 453–67.

Parry ML, Canziani OF, Palutikof JP, van der Linden PJ, Hanson CE, eds (2007) *Climate Change 2007: Impacts, Adaptation and Vulnerability. Contribution of Working Group II to the Fourth Assessment Report of the Intergovernmental Panel on Climate Change, 2007.* Cambridge University Press, Cambridge.

Pask A, Pietragalla J, Mullan D, Reynolds M, eds (2011) *Physiological Breeding II: A Field Guide to Wheat Phenotyping.* CIMMYT, Mexico, DF.

Passioura JB (1980) The meaning of matric potential. *Journal of Experimental Botany* **31**, 1161–9.

Passioura JB (1984) Hydraulic resistance of plants. I. Constant or variable? *Australian Journal of Plant Physiology* **11**, 333–9.

Passioura JB (1996) Simulation models: science, snake oil, education or engineering? *Agronomy Journal* **88**, 690–4.

Pastenes C, Porter V, Baginsky C, Horton P, Gonzalez J (2004) Paraheliotropism can protect water-stressed bean (*Phaseolus vulgaris* L.) plants against photoinhibition. *Journal of Plant Physiology* **161**, 1315–23.

Pastori GM, Foyer CH (2002) Common components, networks, and pathways of cross-tolerance to stress. The central role of "redox" and abscisic acid-mediated controls. *Plant Physiology* **129**, 460–8.

Patakas A, Noitsakis B, Chouzouri A (2005) Optimization of irrigation water use in grapevines using the relationship between transpiration and plant water status. *Agriculture, Ecosystems and Environment* **106**, 253–9.

Paw U KT (1992) A discussion of the Penman form equations and comparisons of some equations to estimate latent energy flux density. *Agricultural and Forest Meteorology* **57**, 297–304.

Pearcy RW, Berry JA, Björkman O (1972) Field measurements of the gas exchange capacities of *Phragmites communis* under summer conditions in Death Valley. *Carnegie Institution Year Book* **71**, 161–4.

Pearcy RW, Björkman O, Harrison AT, Mooney HA (1971) Photosynthetic performance of two desert species with C$_4$ photosynthesis in Death Valley, California. *Carnegie Institution Year Book* **70**, 540–50.

Pearcy RW, Ehleringer JR, Mooney HA, Rundel PW, eds (1991) *Plant Physiological Ecology: Field Measurements and Instrumentation*. Chapman & Hall, London & New York.

Pearcy RW, Troughton JH (1975) C$_4$ photosynthesis in tree form *Euphorbia* species from Hawaiian rainforest sites. *Plant Physiology* **55**, 1054–6.

Pedersen O, Rich SM, Colmer TD (2009) Surviving floods: leaf gas films improve O$_2$ and CO$_2$ exchange, root aeration, and growth of completely submerged rice. *The Plant Journal* **58**, 147–56.

Penman HL (1948) Natural evaporation from open water, bare soil and grass. *Proceedings of the Royal Society of London, Series A* **193**, 120–45.

Penman HL (1953) The physical basis of irrigation control. *Report of 13th Horticultural Congress* **2**, 913–14.

Penman HL, Schofield RK (1951) Some physical aspects of assimilation and transpiration. *Symposia of the Society for Experimental Biology* **5**, 115–29.

Penning de Vries FWT, van Laar HH, Chardon MC (1983) Bioenergetics of growth of seeds, fruits and storage organs. In *Potential Productivity of Field Crops under Different Environments*, pp. 37–59. International Rice Research Institute, Los Baños, Philippines.

Pessarakli M, ed. (2011) *Handbook of Plant and Crop Stress*, 3rd edn. CRC Press, Boca Raton, FL.

Pickard WF (2012) Münch without tears: a steady-state Münch-like model of phloem so simplified that it requires only algebra to predict the speed of translocation. *Functional Plant Biology* **39**, 531–7.

Pinty B, Gobron N, Widlowski JL *et al.* (2001) Radiation transfer model intercomparison (RAMI) exercise. *Journal of Geophysical Research – Atmospheres* **106**, 11937–56.

Pockman WT, Sperry JS (2000) Vulnerability to xylem cavitation and the distribution of Sonoran desert vegetation. *American Journal of Botany* **87**, 1287–99.

Pockman WT, Sperry JS, O'Leary JW (1995) Evidence for sustained and significant negative pressure in xylem. *Nature* **378**, 715–16.

Polunin N (1960) *Introduction to Plant Geography*. Longman, London.

Pons TL, Flexas J, von Caemmerer S *et al.* (2009) Estimating mesophyll conductance to CO$_2$: methodology, potential errors, and recommendations. *Journal of Experimental Botany* **60**, 2217–34.

Poorter H, Bühler J, van Dusschoten D, Climent J, Postma JA (2012) Pot size matters: a meta-analysis of rooting volume on plant growth. *Functional Plant Biology*, doi:10.1071/FP12049.

Pope DJ, Lloyd PS (1975) Hemispherical photography, topography and plant distribution. In *Light as an Ecological Factor. Vol II* (eds Evans GC, Bainbridge R & Rackham O), pp. 385–408. Blackwell, Oxford.

Porembski S, Barthlott W (2000) Granitic and gneissic outcrops (inselbergs) as centers of diversity for desiccation-tolerant vascular plants. *Plant Ecology* **151**, 19–28.

Porter JR (1993) AFRCWHEAT2: a model of the growth and development of wheat incorporating responses to water and nitrogen. *European Journal of Agronomy* **2**, 69–82.

Priestley CHB, Taylor RJ (1972) On the assessment of surface heat flux and evaporation using large-scale parameters. *Monthly Weather Review* **100**, 81–92.

Prieto I, Armas C, Pugnaire FI (2012) Water release through plant roots: new insights into its consequences at the plant and ecosystem level. *New Phytologist* **193**, 830–41.

Procházková D, Wilhelmová N (2011) Antioxidant protection during abiotic stress. In *Handbook of Plant and Crop Stress* (ed. Pessarakli M), pp. 139–55. CRC Press, Boca Raton, FL.

Proctor MCF (2000) The bryophyte paradox: tolerance of desiccation, evasion of drought. *Plant Ecology* **151**, 41–9.

Prusinkiewicz P, Lindenmayer A (1990) *The Algorithmic Beauty of Plants*. Springer Verlag., New York.

Puranik S, Sahu PP, Srivastava PS, Prasad M (2012) NAC proteins: regulation and role in stress tolerance. *Trends in Plant Science* **17**, 369–81.

Qi Z, Kishigami A, Nakagawa Y, Iida H, Sokabe M (2004) A mechanosensitive anion channel in *Arabidopsis thaliana* mesophyll cells. *Plant & Cell Physiology* **45**, 1704–8.

Quarrie SA (1991) Implications of genetic differences in ABA accumulation for crop production. In: *Physiology and Biochemistry of Abscisic Acid* (eds Davies WJ & Jones HG), pp. 247–53. Bios, Oxford.

Quarrie SA, Jones HG (1977) Effects of abscisic acid and water stress on development and morphology of wheat. *Journal of Experimental Botany* **28**, 192–203.

Quarrie SA, Jones HG (1979) Genotypic variation in leaf water potential, stomatal conductance and abscisic acid concentration in spring wheat subjected to artificial drought stress. *Annals of Botany* **44**, 323–32.

Quilot-Turion B, Ould-Sidi M-M, Kadrani A *et al.* (2012) Optimization of parameters of the 'Virtual Fruit' model to design peach genotype for sustainable production systems. *European Journal of Agronomy* **42**, 34–48.

Quisenbery JE, Cartwright GB, McMichael BL (1984) Genetic relationship between turgor maintenance and growth in cotton germplasm. *Crop Science* **24**, 479–82.

Rabinowitch EI (1951) *Photosynthesis. Vol. II.* Interscience, New York.

Rains DW (1989) Plant tissue and protoplast culture: applications to stress physiology and biochemistry. In *Plants Under Stress* (eds Jones HG, Flowers TJ & Jones MB), pp. 181–96. Cambridge University Press, Cambridge.

Raschke K (1975) Stomatal action. *Annual Review of Plant Physiology* **1975**, 309–40.

Rauner JL (1976) Deciduous forests. In *Vegetation and the Atmosphere. Vol.II. Case Studies* (ed. Monteith JL), pp. 241–64. Academic Press, London.

Raupach MR (1989a) Applying Lagrangian fluid-mechanics to infer scalar source distributions from concentration profiles in plant canopies. *Agricultural and Forest Meteorology* **47**, 85–108.

Raupach MR (1989b) Stand overstorey processes. *Philosophical Transactions of the Royal Society of London, Series B – Biological Sciences* **324**, 175–90.

Raupach MR, Finnigan JJ (1988) Single-layer models of evaporation from plant canopies are incorrect but useful, whereas multilayer models are correct but useless: discuss. *Australian Journal of Plant Physiology* **15**, 705–16.

Rausenberger J, Hussong A, Kircher S *et al.* (2010) An integrative model for phytochrome B mediated photomorphogenesis: from protein dynamics to physiology. *PLoS ONE* **5**, e10721.

Raven JA (1972) Endogenous inorganic carbon sources in plant photosynthesis. I. Occurrence of dark respiratory pathways in illuminated green cells. *New Phytologist* **71**, 227–47.

Raven JA (2002) Selection pressures on stomatal evolution. *New Phytologist* **153**, 371–86.

Rebetzke GJ, Read JJ, Barbour MM, Condon AG, Rawson HM (2000) A hand-held porometer for rapid assessment of leaf conductance in wheat. *Crop Science* **40**, 277–80.

Rees WG (2001) *Physical Principles of Remote Sensing*, 2nd edn. Cambridge University Press, Cambridge.

Reynolds M, Pask A, Mullen D, eds (2011) *Physiological Breeding I: Interdisciplinary Approaches to Improve Crop Adaptation*. CIMMYT, Mexico, DF.

Reynolds MP, Singh RP, Ibrahim A *et al.* (1998) Evaluating physiological traits to complement empirical selection for wheat in warm environments. *Euphytica* **100**, 85–94.

Ribas-Carbo M, Taylor NL, Giles L *et al.* (2005) Effects of water stress on respiration in soybean leaves. *Plant Physiology* **139**, 466–73.

Richardson EA, Seeley SD, Walker DR (1974) A model for estimating the completion of rest for 'Redhaven' and 'Elberta' peach trees. *HortScience* **9**, 331–2.

Rikin A, Atsmon D, Gitler C (1983) Quantitation of chill-induced release of a tubulin-like factor and its prevention by abscisic acid in *Gossypium hirsutum* L. *Plant Physiology* **71**, 747–8.

Ritchie JT (1972) Model for predicting evaporation from a row of crop with incomplete cover. *Water Resources Research* **8**, 1204–13.

Ritchie JT (1973) Influence of soil water status and meteorological conditions on evaporation of a corn canopy. *Agronomy Journal* **65**, 893–7.

Ritchie JT, Godwin DC, Otter S (1985) *A Simulation Model of Wheat Growth and Development.* Texas A & M University Press, College Station, Texas.

Roberts J (ed.) (2011) Special issue. Exploiting the engine of C_4 photosynthesis. *Journal of Experimental Botany* **62**, 2989–3246.

Roberts SK (2006) Plasma membrane anion channels in higher plants and their putative functions in roots. *New Phytologist* **169**, 647–66.

Robinson MF, Heath J, Mansfield TA (1998) Disturbances in stomatal behaviour caused by air pollution. *Journal of Experimental Botany* **49**, 461–9.

Robson MJ (1981) Respiratory efflux in relation to temperature of simulated swards of perennial ryegrass with contrasting soluble carbohydrate contents. *Annals of Botany* **48**, 269–73.

Rockwell NC, Su YS, Lagarias JC (2006) Phytochrome structure and signaling mechanisms. *Annual Review of Plant Biology* **57**, 837–58.

Roelfsema MRG, Hedrich R (2005) In the light of stomatal opening: new insights into 'the watergate'. *New Phytologist* **167**, 665–91.

Roelfsema MRG, Hedrich R, Geiger D (2012) Anion channels: master switches of stress responses. *Trends in Plant Science* **17**, 221–9.

Rohde A, Bhalerao RP (2007) Plant dormancy in the perennial context. *Trends in Plant Science* **12**, 217–23.

Romero R, Muriel JL, García I, Muñoz de la Peña D (2012) Research on automatic irrigation control: state of the art and recent results. *Agricultural Water Management* **114**, 59–66.

Rorison IH (1981) Plant growth in response to variation in temperature: field and laboratory studies. In *Plants and their Atmospheric Environment* (eds Grace J, Ford ED & Jarvis PG), pp. 313–32. Blackwell, Oxford.

Rose DA, Charles-Edwards DA (1981) *Mathematics and Plant Physiology.* Academic Press, London.

Rosenberg NJ, Blad BL, Verma SB (1983) *Microclimate: the Biological Environment.* Wiley, New York.

Ross J (1975) Radiative transfer in plant communities. In *Vegetation and the Atmosphere. Vol. 1.* (ed. Monteith JL), pp. 13–55. Academic Press, London.

Ross J (1981) *The Radiation Regime and Architecture of Plant Stands.* Dr W Junk, The Hague.

Roujean J-L, Leroy M, Deschamps PY (1992) A bidirectional reflectance model of the Earth's surface for the correction of remote sensing data. *Journal of Geophysical Research* **97**, 20455–68.

Ryan MG, Yoder BJ (1997) Hydrauliuc limits to tree height and tree growth. *Bioscience* **47**, 235–42.

Sachs MM, Ho TDH (1986) Alteration of gene expression during environmental stress in plants. *Annual Review of Plant Physiology* **37**, 363–76.

Sack L, Holbrook NM (2006) Leaf hydraulics. *Annual Review of Plant Physiology* **57**, 361–81.

Sadras VO, Calderini DF, eds (2009) *Crop Physiology: Applications for Genetic Improvement and Agronomy.* Academic Press, Burlington, MA.

Sadras VO, Lawson C, Montoro A (2012) Photosynthetic traits in Australian wheat varieties released between 1958 and 2007. *Field Crops Research* **134**, 19–29.

Sadras VO, Slafer GA (2012) Environmental modulation of yield components in cereals: heritabilities reveal a hierarchy of phenotypic plasticities. *Field Crops Research* **127**, 215–24.

Sage RF (2004) Evolution of C_4 photosynthesis. *New Phytologist* **161**, 341–70.

Sage RF, Christin P-A, Edwards EJ (2011) The C_4 lineages of planet Earth. *Journal of Experimental Botany* **62**, 3155–69.

Sage RF, Sharkey TD, Seemann R (1989) Acclimation of photosynthesis to elevated CO_2 in five C_3 species. *Plant Physiology* **89**, 590–6.

Salisbury FB, Ross CW (1995) *Plant Physiology*, 5th edn. Wadsworth, Belmont.

Salter PJ, Goode JE (1967) *Crop Responses to Water at Different Stages of Growth*. Commonwealth Agricultural Bureaux, Farnham Royal.

Sambatti JBM, Caylor KK (2007) When is breeding for drought tolerance optimal if drought is random? *New Phytologist* **175**, 70–80.

Sandford AP, Grace J (1985) The measurement and interpretation of ultrasound from woody stems. *Journal of Experimental Botany* **36**, 298–311.

Sauer N (2007) Molecular physiology of higher plant sucrose transporters. *FEBS Letters* **581**, 2309–17.

Savage VM, Bentley LP, Enquist BJ *et al.* (2010) Hydraulic trade-offs and space filling enable better predictions of vascular structure and function in plants. *Proceedings of the National Academy of Sciences of the United States of America* **107**, 22722–7.

Scafaro A, von Caemmerer S, Evans JR, Atwell B (2011) Temperature response of mesophyll conductance in cultivated and wild *Oryza* species with contrasting mesophyll cell wall thickness. *Plant, Cell & Environment* **34**, 1999–2008.

Scaife A, Cox EF, Morris GEL (1987) The relationship between shoot weight, plant density and time during the propagation of four vegetable species. *Annals of Botany* **59**, 325–34.

Scandalios JG, Acevedo A, Ruzsa S (2000) Catalase gene expression in response to chronic high temperature stress in maize. *Plant Science* **156**, 103–10.

Scholander PF, Hammel HT, Hemmingsen EA, Bradstreet ED (1964) Hydrostatic pressure and osmotic potential in leaves of mangroves and some other plants. *Proceedings of the National Academy of Sciences of the United States of America* **52**, 119–25.

Schroeder JI, Allen G, Hugouvieux V, Kwak J, Waner D (2001) Guard cell signal transduction. *Annual Reviews in Plant Biology* **52**, 627–58.

Schroeder JI, Hedrich R (1989) Involvement of ion channels and active transport in osmoregulation and signalling of higher plant cells. *Trends in Biochemical Sciences* **14**, 187–92.

Schultz HR (2003) Differences in hydraulic architecture account for near-isohydric and anisohydric behaviour of two field-grown *Vitis vinifera* L. cultivars during drought. *Plant, Cell & Environment* **26**, 1393–405.

Schultze E-D, Čermák J, Matyssek R *et al.* (1985) Canopy transpiration and water fluxes in the xylem of the trunk of *Larix* and *Picea* trees: a comparison of xylem flow, porometer and cuvette. *Oecologia* **66**, 475–83.

Schultze E-D, Hall AE (1981) Short-term and long-term effects of drought on steady state and time-integrated plant processes. In *Physiological Processes Limiting Plant Productivity* (ed. Johnson C), pp. 217–35. Butterworths, London.

Schultze ED (1986) Carbon dioxide and water vapour exchange in response to drought in the atmosphere and soil. *Annual Review of Plant Physiology* **37**, 247–74.

Schulze ED, Lange OL, Buschbom U, Kappen L, Evenari M (1972) Stomatal responses to changes in humidity in plants growing in the desert. *Planta* **108**, 259–70.

Seetharama N, Subba Reddy BV, Peacock JM, Bidinger FR (1982) Sorghum improvement for drought resistance. In *Drought Resistance of Crops with Emphasis on Rice*, pp. 317–38. International Rice Research Institute, Los Banos, Philippines.

Sellers PJ, Dickinson RE, Randall DA *et al.* (1997) Modeling the exchanges of energy, water, and carbon between continents and the atmosphere. *Science* **275**, 502–9.

Sellers WD (1965) *Physical Climatology*. University of Chicago Press, Chicago.

Serbin SP, Dillaway DN, Kruger EL, Townsend PA (2012) Leaf optical properties reflect variation in photosynthetic metabolism and its sensitivity to temperature. *Journal of Experimental Botany* **63**, 489–502.

Šesták Z, Catský J, Jarvis PG, eds (1971) *Plant Photosynthetic Production. Manual of Methods*. Dr W. Junk, The Hague.

Seymour RS, Bartholomew GA, Barnhart MC (1983) Respiration and heat production by the inflorescence of *Philodendron selloum* Koch. *Planta* **157**, 336–43.

Shackel KA, Hall AE (1979) Reversible leaflet movements in relation to drought adaptation of cowpeas. *Australian Journal of Plant Physiology* **6**, 265–76.

Shantz HL, Piemeisel LN (1927) The water requirements of plants at Akron, Colorado. *Journal of Agricultural Research* **34**, 1093–190.

Sharkey TD (1988) Estimation of the rate of photorespiration in leaves. *Physiologia Plantarum* **73**, 147–52.

Sharkey TD, Bernacchi CJ, Farquhar GD, Singsaas EL (2007) Fitting photosynthetic carbon dioxide response curves for C$_3$ leaves. *Plant, Cell & Environment* **30**, 1035–40.

Sharkey TD, Raschke K (1981) Effect of light quality on stomatal opening in leaves of *Xanthium strumarium* L. *Plant Physiology* **68**, 1170–4.

Sharp RE, Davies WJ (1989) Regulation of growth and development of plants growing with a restricted supply of water. In *Plants Under Stress* (eds Jones HG, Flowers TJ & Jones MB), pp. 71–93. Cambridge University Press, Cambridge.

Sharp RE, Poroyko V, Hejlek LG *et al.* (2004) Root growth maintenance during water deficits: physiology to functional genomics. *Journal of Experimental Botany* **55**, 2343–51.

Sharpe PJH, Wu H-I, Spence RD (1987) Stomatal mechanics. In *Stomatal Function* (eds Zeiger E, Farquhar GD & Cowan IR), pp. 91–123. Stanford University Press, Stanford.

Shimazaki K-I, Doi M, Assmann SM, Kinoshita T (2007) Light regulation of stomatal movement. *Annual Review of Plant Biology* **58**, 219–47.

Shinozaki K, Yamaguchi-Shinozaki K (2007) Gene networks involved in drought stress response and tolerance. *Journal of Experimental Botany* **58**, 221–7.

Shuttleworth WJ (1993) Evaporation. In *Handbook of Hydrology* (ed. Maidment DR). McGraw Hill, New York.

Shuttleworth WJ (2007) Putting the 'vap' into evaporation. *Hydrology and Earth System Sciences* **11**, 210–44.

Shuttleworth WJ, Wallace JS (1985) Evaporation from sparse crops: an energy combination theory. *Quarterly Journal of the Royal Meteorological Society* **111**, 839–55.

Siebold M, von Tiedemann A (2012) Application of a robust experimental method to study soil warming effects on oilseed rape. *Agricultural and Forest Meteorology* **164**, 20–8.

Silvey V (1981) The contribution of new wheat, barley and oat varieties to increasing yield in England and Wales 1947–1978. *Journal of the National Institute of Agricultural Botany* **15**, 399–412.

Simmelsgaard SE (1976) Adaptation to water stress in wheat. *Physiologia Plantarum* **37**, 167–74.

Sinclair TR, Seligman NG (1996) Crop modelling: from infancy to maturity. *Agronomy Journal* **88**, 698–704.

Sinoquet H, Thanisawanyangkura S, Mabrouk H, Kasemsap P (1998) Characterization of the light environment in canopies using 3D digitising and image processing. *Annals of Botany* **82**, 203–12.

Slack EM (1964) Studies of stomatal distribution on the leaves of four apple varieties. *Journal of Horticultural Science* **49**, 95–103.

Slatyer RO (1960) Aspects of the tissue water relationships of an important arid zone species (*Acacia aneura* F. Muell) in comparison with two mesophytes. *Bulletin of Research Council of Israel* **8D**, 159–68.

Slatyer RO (1967) *Plant–Water Relationships*. Academic Press, London.

Slatyer RO (1970) Comparative photosynthesis, growth and transpiration of two species of *Atriplex*. *Planta* **93**, 175–89.

Slatyer RO (1977) Altitudinal variation in the photosynthetic characteristics of snow gum, *Eucalyptus pauciflora* Sieb. ex Spreng. III. Temperature response of material grown in contrasting thermal environments. *Australian Journal of Plant Physiology* **4**, 301–12.

Smith H (1973) Light quality and germination. Ecological implications. In *Seed Ecology* (ed. Heydecker W), pp. 219–31. Butterworths, London.

Smith H (1975) *Phytochrome and Photomorphogenesis*. McGraw-Hill, London.

Smith H (1981) Light quality as an ecological factor. In *Plants and their Atmospheric Environment* (eds Grace J, Ford ED & Jarvis PG), pp. 93–110. Blackwell, Oxford.

Smith H (1995) Physiological and ecological function within the phytochrome family. *Annual Review of Plant Physiology and Molecular Biology* **46**, 289–315.

Smith H, Holmes MG (1977) The function of phytochrome in the natural environment. III. *Photochemistry and Photobiology* **25**, 547–50.

Smith JAC, Schulte PJ, Nobel PS (1987) Water flow and water storage in *Agave deserti*: osmotic implications of crassulacean acid metabolism. *Plant, Cell & Environment* **10**, 639–48.

Smith LP (1967) *Potential Transpiration. Technical Bulletin 16.* Ministry of Agriculture, Fisheries and Food, London.

Smith TA (1984) Polyamines. *Annual Review of Plant Physiology* **36**, 117–43.

Smith WK, Geller GN (1979) Plant transpiration at high elevation: theory, field measurements, and comparisons with desert plants. *Oecologia* **41**, 109–22.

Snyder RL, de Melo-Abreu JP (2005a) *Frost Protection: Fundamentals, Practice and Economics. Vol. I.* Food and Agriculture Organization of the United Nations, Rome.

Snyder RL, de Melo-Abreu JP (2005b) *Frost Protection: Fundamentals, Practice and Economics. Vol. II.* Food and Agriculture Organization of the United Nations, Rome.

Snyder RL, Spano D, Paw U KT (1996) Surface renewal analysis for sensible and latent heat flux density. *Boundary-Layer Meteorology* **77**, 249–66.

Sokal RR, Rohlf FJ (2012) *Biometry: the Principles and Practice of Statistics in Biological Research.* W.H. Freeman & Co., New York.

Solárová J (1980) Diffusive conductances of adaxial (upper) and abaxial (lower) epidermes: response to quantum irradiance during development of primary *Phaseolus vulgaris* L. leaves. *Photosynthetica* **14**, 524–31.

Solárová J, Pospišilová J, Slavik B (1981) Gas exchange regulation by changing of epidermal conductance with antitranspirants. *Photosynthetica* **15**, 365–400.

Solomon S, Qin D, Manning M *et al.* eds (2007) *Climate Change 2007: The Physical Science Basis.* Cambridge University Press, Cambridge.

Sperry JS, Donnelly JR, Tyree MT (1988a) A method for measuring hydraulic conductivity and embolism in xylem. *Plant, Cell & Environment* **11**, 35–40.

Sperry JS, Hacke UG, Pitterman J (2006) Size and function in conifer tracheids and angiosperm vessels. *American Journal of Botany* **93**, 1490–500.

Sperry JS, Tyree MT, Donnelly JR (1988b) Vulnerability of xylem to embolism in a mangrove vs an inland species of Rhizophoraceae. *Physiologia Plantarum* **74**, 276–83.

Spitters CJT, Toussaint HAJM, Goudriaan J (1986) Separating the diffuse and direct component of global radiation and its implications for modeling canopy photosynthesis. *Agricultural & Forest Meteorology* **38**, 217–29.

Stålfelt MG (1955) The stomata as a hydrophotic regulator of the water deficit of the plant. *Physiologia Plantarum* **8**, 572–93.

Stanhill G (1981) The size and significance of differences in the radiation balance of plants and plant communities. In *Plants and their Atmospheric Environment* (eds Grace J, Ford ED & Jarvis PG), pp. 57–73. Blackwell, Oxford.

Steppe K, De Pauw DJW, Doody TM, Teskey RO (2010) A comparison of sap flux density using thermal dissipation, heat pulse velocity and heat field deformation methods. *Agricultural and Forest Meteorology* **150**, 1046–56.

Steudle E (2001) The cohesion-tension mechanism and the acquisition of water by plant roots. *Annual Review of Plant Physiology and Plant Molecular Biology* **52**, 847–75.

Stitt M, Sonnewald U (1995) Regulation of metabolism in transgenic plants. *Annual Review of Plant Physiology and Plant Molecular Biology* **46**, 341–68.

Stoll M (2000) *Effects of Partial Rootzone Drying on Grapevine Physiology and Fruit Quality.* PhD thesis, University of Adelaide, Adelaide.

Stoll M, Loveys B, Dry P (2000) Hormonal changes induced by partial rootzone drying of irrigated grapevine. *Journal of Experimental Botany* **51**, 1627–34.

Stroike JE, Johnson VA (1972) *Winter Wheat Cultivar Performance in an International Array of Environments. Bulletin 251.* University of Nebraska Experiment Station, Lincoln, NE.

Sung S, Amasino RM (2005) Remembering winter: toward a molecular understanding of vernalization. *Annual Review of Plant Biology* **56**, 491–508.

Sung YH, Shogo I, Imaizumi T (2010) Similarities in the circadian clock and photoperiodism in plants. *Current Opinion in Plant Biology* **13**, 594–603.

Sunley RJ, Atkinson CJ, Jones HG (2006) Chill unit models and recent historical changes in UK winter chill and spring frost occurrence. *Journal of Horticultural Science and Biotechnology* **81**, 949–58.

Syed NH, Kalyna M, Marquez Y, Barta A, Brown JWS (2012) Alternative splicing in plants: coming of age. *Trends in Plant Science* **17**, 616–23.

Szeicz G (1974) Solar radiation in plant canopies. *Journal of Applied Ecology* **73**, 59–64.

Szeicz G, Monteith JL, dos Santos JM (1964) Tube solarimeter to measure radiation among plants. *Journal of Applied Ecology* **1**, 169–74.

Taiz L, Zeiger E (2010) *Plant Physiology*, 5th edn. Sinauer Associates, Sunderland, MA.

Talbott LD, Zeiger E (1998) The role of sucrose in guard cell osmoregulation. *Journal of Experimental Botany* **49**, 329–37.

Tang A-C, Kawamitsu Y, Kanechi M, Boyer JS (2002) Photosynthetic oxygen evolution at low water potential in leaf discs lacking an epidermis. *Annals of Botany* **89**, 861–70.

Tanner CB (1981) Transpiration efficiency of potato. *Agronomy Journal* **73**, 59–64.

Tardieu F, Simonneau T (1998) Variability among species of stomatal control under fluctuating soil water status and evaporative demand: modelling isohydric and anisohydric behaviours. *Journal of Experimental Botany* **49**, 419–32.

Tazoe Y, von Caemmerer S, Estavillo GM, Evans JR (2011) Using tunable diode laser spectroscopy to measure carbon isotope discrimination and mesophyll conductance to CO_2 diffusion dynamically at different CO_2 concentrations. *Plant, Cell & Environment* **34**, 580–91.

Teeri JA (1980) Adaptation of kinetic properties of enzymes to temperature variability. In *Adaptation of Plants to Water and High Temperature Stress* (eds Turner NC & Kramer PJ), pp. 251–60. Wiley, New York.

Teeri JA, Stowe LG (1976) Climatic patterns and the distribution of C_4 grasses in North America. *Oecologia* **23**, 1–12.

Teeri JA, Stowe LG, Murawski DA (1978) The climatology of two succulent plant families: Cactaceae and Crassulaceae. *Canadian Journal of Botany* **56**, 1750–8.

Teh CBS (2006) *Introduction to Mathematical Modeling of Crop Growth: How the Equations are Derived and Assembled into a Computer Program*. Brown Walker Press, Boca Raton, FL.

Terashima I, Wong SC, Osmond CB, Farquhar GD (1988) Characterisation of non-uniform photosynthesis induced by abscisic acid in leaves having different mesophyll anatomies. *Plant Cell Physiology* **29**, 385–94.

Thames JL (1961) Effects of wax coatings on leaf temperatures and field survival of *Pinus taeda* seedlings. *Plant Physiology* **36**, 180–2.

Thiermann V, Grassl H (1992) The measurement of turbulent surface-layer fluxes by use of bichromatic scintillation. *Boundary Layer Meteorology* **58**, 367–89.

Tholen D, Zhu X-G (2011) The mechanistic basis of internal conductance: a theoretical analysis of mesophyll cell photosynthesis and CO_2 diffusion. *Plant Physiology* **156**, 90–105.

Thom AS (1975) Momentum, mass and heat exchange of plant communities. In *Vegetation and the Atmosphere. 1. Principles* (ed. Monteith JL), pp. 57–109. Academic Press, London, New York and San Francisco.

Thomashow MF (1999) Plant cold acclimation: freezing tolerance genes and regulatory mechanisms. *Annual Review of Plant Physiology and Plant Molecular Biology* **50**, 571–9.

Thompson DA, Matthews RW (1989) *The Storage of Carbon in Trees and Timber. Research Information Note 160*. Forestry Commission, Farnham, UK.

Thompson N, Barrie JA, Ayles M (1982) *The Meteorological Office Rainfall and Evaporation Calculation System:MORECS (July 1981)*. Meteorological Office, Bracknell.

Thornley JHM, France J (2007) *Mathematical Models in Agriculture: Quantitative Methods for the Plant, Animal and Ecological Sciences*, 2nd edn. CABI, Wallingford, UK.

Thornthwaite CW (1944) Report of the committee on transpiration and evaporation. *Transactions of the American Geophysical Union* **29**, 688–93.

Thorpe MR (1974) Radiant heating of apples. *Journal of Applied Ecology* **11**, 755–60.

Tillman JE (1972) The indirect determination of stability, heat and momentum fluxes in the atmospheric boundary layer from simple scalar variables during dry unstable conditions. *Journal of Applied Meteorology* **8**, 783–92.

Ting IP (1985) Crassulacean acid metabolism. *Annual Review of Plant Physiology* **36**, 595–622.

Ting IP, Dugger WM (1968) Factors affecting ozone sensitivity and susceptibility of cotton plants. *Journal of the Air Pollution Control Association* **18**, 810–13.

Tomos AD (1987) Cellular water relations of plants. In *Water Science Reviews, Vol. 3* (ed. Franks F), pp. 186–267. Cambridge University Press, Cambridge.

Tomos AD, Leigh RA (1999) The pressure probe: a versatile tool in plant cell physiology. *Annual Review of Plant Physiology and Plant Molecular Biology* **50**, 447–72.

Trewartha GT (1968) *An Introduction to Climate*, 4th edn. H.H. McGraw, New York.

Trifilò P, Gascó A, Raimondo F, Nardini A, Salleo S (2003) Kinetics of recovery of leaf hydraulic conductance and vein functionality from cavitation-induced embolism in sunflower. *Journal of Experimental Botany* **54**, 2323–30.

Tsuchida-Mayama T, Sakai T, Hanada A *et al.* (2010) Role of the phytochrome and cryptochrome signaling pathways in hypocotyl phototropism. *The Plant Journal* **62**, 653–62.

Tsujimura M, Tanaka T (1998) Evaluation of evaporation rate from forested surface using stable isotopic composition of water in a headwater basin. *Hydrological Processes* **12**, 2093–103.

Tuberosa R, Salvi S (2006) Genomics-based approaches to improve drought tolerance of crops. *Trends in Plant Science* **11**, 405–12.

Turgeon R (2010a) The puzzle of phloem pressure. *Plant Physiology* **154**, 578–81.

Turgeon R (2010b) The role of phloem loading reconsidered. *Plant Physiology* **152**, 1817–23.

Turk KJ, Hall AE (1980) Drought adaptation of cowpea. IV. Influence of drought on water use, and relations with growth and seed yield. *Agronomy Journal* **72**, 434–9.

Turner NC, Begg JE, Tonnet ML (1978) Osmotic adjustment of sorghum and sunflower crops in response to water deficits and its influence on the water potential at which stomata close. *Australian Journal of Plant Physiology* **5**, 597–608.

Turner NC, Jones MM (1980) Turgor maintenance by osmotic adjustment: a review and evaluation. In *Adaptation of Plants to Water and High Temperature Stress* (eds Turner NC & Kramer PJ), pp. 87–103. Wiley, New York.

Tyree MT (1988) A dynamic model for water flow in a single tree: evidence that models must account for hydraulic architecture. *Tree Physiology* **4**, 195–217.

Tyree MT, Dixon MA (1983) Cavitation events in *Thuja occidentalis* L.? Ultrasonic acoustic emissions from the sapwood can be measured. *Plant Physiology* **72**, 1094–9.

Tyree MT, Dixon MA (1986) Water stress induced cavitation and embolism in some woody plants. *Physiologia Plantarum* **66**, 397–405.

Tyree MT, Karamanos AJ (1981) Water stress as an ecological factor. In *Plants and their Atmospheric Environment* (eds Grace J, Ford ED & Jarvis PG), pp. 237–61. Blackwell, Oxford.

Tyree MT, Sperry JS (1988) Do woody plants operate near the point of catastrophic xylem dysfunction caused by dynamic water stress? *Plant Physiology* **88**, 574–80.

Tyree MT, Sperry JS (1989) Vulnerability of xylem to cavitation and embolism. *Annual Review of Plant Physiology* **40**, 19–38.

Tyree MT, Yianoulis P (1980) The site of water evaporation from sub-stomatal cavities, liquid path resistances and hydroactive stomatal closure. *Annals of Botany* **46**, 175–93.

Tyree MT, Zimmermann MH (2002) *Xylem Structure and the Ascent of Sap*, 2nd edn. Springer-Verlag, Berlin, Heidelberg & New York.

UNEP (1997) *The World Atlas of Desertification*. United Nations Environment Programme, London.

Uno Y, Furihata T, Abe H *et al.* (2000) Arabidopsis basic leucine zipper transcription factors involved in an abscisic acid-dependent signal transduction pathway under drought and high-salinity conditions. *Proceedings of the National Academy of Sciences of the USA* **97**, 11632–7.

Unsworth MH, Biscoe PV, Pinckney HR (1972) Stomatal responses to suphur dioxide. *Nature* **239**, 458–9.

Valko P (1984) *Format for Presentation of Data. Subtask D. Task 5. Use of Existing Meteorological Information for Solar Energy Applications*. Sveriges Meteorologiska och Hydrologiska Inst, Stockholm.

Valliyodan B, Nguyen HT (2006) Understanding regulatory networks and engineering for enhanced drought tolerance in plants. *Current Opinion in Plant Biology* **9**, 1–7.

Valluru R, Van den Ende W (2008) Plant fructans in stress environments: emerging concepts and future prospects. *Journal of Experimental Botany* **59**, 2905–16.

Van As H, Scheeren T, Vergeldt FJ (2009) MRI of intact plants. *Photosynthesis Research* **102**, 213–22.

van Bavel CHM, Hillel DI (1976) Calculating potential and actual evaporation from a bare soil surface by simulation of concurrent flow of water and heat. *Agricultural Meteorology* **17**, 453–76.

van den Honert TH (1948) Water transport in plants as a catenary process. *Discussions of the Faraday Society* **3**, 146–53.

van Dongen JT, Gupta KJ, Ramírez-Aguilar SJ *et al.* (2011) Regulation of respiration in plants: a role for alternative metabolic pathways. *Journal of Plant Physiology* **168**, 1434–43.

van Gardingen PR, Jeffree CE, Grace J (1989) Variation in stomatal aperture in leaves of *Avena fatua* L. observed by low-temperature scanning electron microscopy. *Plant, Cell & Environment* **12**, 887–98.

Van Kesteren B, Hartogenesis OK, van Dinther D, Moene AF, De Bruin HAR (2013) Measuring H_2O and CO_2 fluxes at field scales with scintillometry. Part I: introduction and validation of four methods. *Agricultural and Forest Meteorology* **178–9**, 75–87.

Vanlerberghe GC, McIntosh L (1997) Alternative oxidase: from gene to function. *Annual Review of Plant Physiology and Plant Molecular Biology* **48**, 703–34.

Vásquez-Yanes C, Smith H (1982) Phytochrome control of seed germination in the tropical rain forest pioneer trees *Cecropia obtusifolia* and *Piper auritum* and its ecological consequence. *New Phytologist* **92**, 477–85.

Verhoef W, Bach H (2007) Coupled soil–leaf –canopy and atmosphere radiative transfer modeling to simulate hyperspectral multi-angular surface reflectance and TOA radiance data. *Remote Sensing of Environment* **109**, 166–82.

Vermote E, Tanre D, Deuze JL, Herman M, Morcette J-L (1997) Second simulation of the satellite signal in the solar spectrum, 6S: an overview. *IEEE Transactions in Geoscience and Remote Sensing* **35**, 675–86.

Vignola F, Stoffel T, Michalsky JJ (2012) *Solar and Infrared Radiation Measurements*. CRC Press, Boca Raton, FL.

Vilar R, Held AA, Merino J (1995) Dark leaf respiration in light and darkness of an evergreen and deciduous plant species. *Plant Physiology* **107**, 421–7.

Vince-Prue D (1975) *Photoperiodism in Plants*. McGraw-Hill, London.

Vince-Prue D, Cockshull KE, Thomas B, eds (1984) *Light and the Flowering Process*. Academic Press, London.

Visscher GJW (1999) Chapter 72. Humidity and moisture measurement. In *The Measurement, Instrumentation and Sensors Handbook on CD-ROM*. CRC Press LLC, Boca Raton, FL.

Vogel S (1984) The lateral thermal conductivity of leaves. *Canadian Journal of Botany* **62**, 741–4.

von Caemmerer S (2000) *Biochemical Models of Leaf Photosynthesis*. CSIRO Publishing, Collingwood, Victoria, Australia.

von Caemmerer S, Farquhar GD (1981) Some relationships between the biochemistry of photosynthesis and the gas exchange of leaves. *Planta* **153**, 376–87.

von Caemmerer S, Furbank RT (1999) Modelling C_4 photosynthesis. In C_4 *Plant Biology* (eds Sage RF & Monson RK), pp. 173–211. Academic Press, Toronto, Canada.

von Caemmerer S, Quick WP, Furbank RT (2012) The development of C_4 rice: current progress and future challenges. *Science* **336**, 1671–2.

von Mohl H (1856) Welche Ursachen bewirken die Erweiterung und Verengung der Spaltöffnungen. *Botanische Zeiting* **14**, 697–704.

von Willert DJ, Brinckmann E, Scheitler B, Eller BM (1985) Availability of water controls crassulacean acid metabolism in succulents of the Richtersveld (Namib desert, South Africa). *Planta* **164**, 44–55.

Vos J, Marcelis LFM, de Visser PHB, Struik PC, Evers JB (2007) *Functional-Structural Plant Modelling in Plant Production*. Wageningen UR, Wageningen.

Wallace JS, Batchelor CH, Hodnett MG (1981) Crop evaporation and surface conductance calculated using soil moisture data from central India. *Agricultural Meteorology* **25**, 83–96.

Wallace JS, Lloyd CR, Sivakumar MVK (1993) Measurements of soil, plant and total evaporation from millet in Niger. *Agricultural and Forest Meteorology* **63**, 149–69.

Walthall CL, Norman JM, Welles JM, Campbell G, Blad BL (1985) Simple equation to approximate the bidirectional reflectance from vegetative canopies and bare soil surfaces. *Applied Optics* **24**, 383–7.

Wang HX, Zhang WM, Zhou GQ, Yan GJ, Clinton N (2009) Image-based 3D corn reconstruction for retrieval of geometrical structural parameters. *International Journal of Remote Sensing* **30**, 5505–13.

Wang JY (1960) A critique of the heat unit approach to plant response studies. *Ecology* **41**, 785–9.

Wang W, Vinocour B, Altman A (2003) Plant responses to drought, salinity and extreme temperatures: towards genetic engineering for stress tolerance. *Planta* **218**, 1–14.

Wang X, Lewis JD, Tissue DT, Seemann JR, Griffin KL (2001) Effects of elevated atmospheric CO_2 concentration on leaf dark respiration of *Xanthium strumarium* in light and in darkness *Proceedings of the National Academy of Sciences of the USA* **98**, 2479–84.

Wang YP, Jarvis PG (1990) Influence of crown structural properties on PAR absorption, photosynthesis, and transpiration in Sitka spruce: application of a model (MAESTRO). *Tree Physiology* **7**, 297–316.

Warland JS, Thurtell GW (2000) A Lagrangian solution to the relationship between a distributed source and a concentration profile. *Boundary Layer Meteorology* **96**, 453–71.

Warren Wilson J (1957) Observations on the temperatures of arctic plants and their environment. *Journal of Ecology* **45**, 499–531.

Warrit B, Landsberg JJ, Thorpe MR (1980) Responses of apple leaf stomata to environmental factors. *Plant, Cell & Environment* **3**, 13–20.

Watson DJ (1958) The dependence of net assimilation rate on leaf area index. *Annals of Botany* **22**, 37–54.

Weast RC (1969) *Handbook of Chemistry and Physics*, 50th edn. Chemical Rubber Publishing Company, Cleveland, OH.

Wellburn AR (1994) *Air Pollution and Climate Change: the Biological Impact*, 2nd edn. Longman, London.

West AG, Hultine KR, Sperry JS, Bush SE, Ehleringer JR (2008) Transpiration and hydraulic strategies in a piñon-juniper woodland. *Ecological Applications* **18**, 911–27.

West GB, Brown JH, Enquist BJ (1997) A general model for the origin of allometric scaling laws in biology. *Science* **276**, 122–6.

West GB, Brown JH, Enquist BJ (1999) A general model for the structure and allometry of plant vascular systems. *Nature* **400**, 664–7.

West JD, Peak D, Peterson JQ, Mott KA (2005) Dynamics of stomatal patches for a single surface of *Xanthium strumarium* L. leaves observed with fluorescence and thermal images. *Plant, Cell & Environment* **28**, 633–41.

Weyers JDB, Meidner H (1990) *Methods in Stomatal Research*. Longman, London.

Whisler FD, Acock B, Baker NR *et al.* (1986) Crop simulation models in agronomic systems. *Advances in Agronomy* **40**, 141–208.

White JW, Andrade-Sanchez P, Gore MA *et al.* (2012) Field-based phenomics for plant genetics research. *Field Crops Research* **133**, 101–12.

Wikström M, Hummer G (2012) Stoichiometry of proton translocation by respiratory complex I and its mechanistic implications. *Proceedings of the National Academy of Sciences of the USA* **109**, 4431–6.

Wilkinson S, Bacon MA, Davies WJ (2007) Nitrate signalling to stomata and growing leaves: interactions with soil drying, ABA, and xylem sap pH in maize. *Journal of Experimental Botany* **58**, 1705–16.

Wilkinson S, Davies WJ (2002) ABA-based chemical signalling: the co-ordination of responses to stress in plants. *Plant, Cell & Environment* **25**, 195–210.

Willmer CM, Fricker M (1996) *Stomata*. Chapman & Hall, London.

WMO (2008) Chapter 4. Measurement of humidity. In *World Guide to Meteorological Instruments and Methods of Observation*. WMO, Geneva.

Wolfenden J, Mansfield TA (1990) Physiological disturbances in plants caused by air pollutants. *Proceedings of the Royal Society of Edinburgh B* **97**, 117–38.

Wong SC, Cowan IR, Farquhar GD (1979) Stomatal conductance correlates with photosynthetic capacity. *Nature* **282**, 424–6.

Woodhouse IH (2006) *Introduction to Microwave Remote Sensing*. Taylor and Francis, London.

Woodward FI (1988) Temperature and the distribution of plant species. In *Plants and Temperature* (eds Long SP & Woodward FI), pp. 59–75. Company of Biologists, Cambridge.

Woodward FI, Lomas MR, Betts RA (1998) Vegetation-climate feedbacks in a greenhouse world.

Proceedings of the Royal Society of London, Series B **353**, 29–39.

Woodward FI, Pigott CD (1975) The climatic control of the altitudinal distribution of plants with diverse altitudinal ranges. I. Field observations. *New Phytologist* **74**, 323–34.

Wright EA, Lucas PW, Cottam DA, Mansfield TA (1986) Physiological responses of plants to SO_2, NO_x and O_3: implications for drought resistance. In *Direct Effects of Dry and Wet Deposition on Forest Ecosystems: in Particular Canopy Interactions* (ed. Mathy P), pp. 187–200. Commission for the European Communities, Brussels.

Wright IJ, Reich PB, Westoby M *et al.* (2004) The worldwide leaf economics spectrum. *Nature* **428**, 821–7.

Wuenscher JE (1970) The effect of leaf hairs of *Verbascum thapsus* on leaf energy exchange. *New Phytologist* **69**, 65–73.

Wullschleger SD, Gubderson CA, Hanson PJ, Wilson KB, Norby RJ (2002) Sensitivity of stomatal and canopy conductance to elevated CO_2 concentration: interacting variables and perspectives of scale. *New Phytologist* **153**, 485–96.

Xu Y (2010) *Molecular Plant Breeding*. CABI, Wallingford, UK.

Yamaguchi-Shinozaki K, Shinozaki K (1994) A novel cis-acting element in an *Arabidopsis* gene is involved in responsiveness to drought, low-temperature, or high-salt stress. *Plant Cell* **6**, 251–64.

Yang S, Tyree MT (1992) A theoretical model of hydraulic conductivity recovery from embolism with comparison to experimental data on *Acer saccharum*. *Plant, Cell & Environment* **15**, 633–43.

Yang S, Vanderbeld B, Wan J, Huang Y (2010) Narrowing down the targets: towards successful genetic engineering of drought-tolerant crops. *Molecular Plant* **3**, 469–90.

Yates F, Cochrane WG (1938) The analysis of groups of experiments. *Journal of Agricultural Science, Cambridge* **28**, 556–80.

Yin X, Struik PC (2008) Applying modelling experiences from the past to shape crop systems biology: the need to converge crop physiology and functional genomics. *New Phytologist* **179**, 629–42.

Yoshino MM (1975) *Climate in a Small Area*. University of Tokyo Press, Tokyo.

Young JE (1975) Effects of the spectral composition of light sources on the growth of a higher plant. In *Light as an Ecological Factor. II.* (eds Evans GC, Bainbridge R, & Rackham O), pp. 135–60. Blackwell, Oxford.

Zavalloni C, Andresen JA, Flore JA (2006) Phenological models of flower bud stages and fruit growth of 'Montmorency' sour cherry based on growing degree-day accumulation. *Journal of the American Society for Horticultural Science* **131**, 601–7.

Zeiger E, Farquhar GD, Cowan IR, eds (1987) *Stomatal Function*. Stanford University Press, Stanford.

Zelitch I, Schultes NP, Peterson RB, Brown PL, Brutnell TP (2009) High glycolate oxidase activity is required for survival of maize in normal air. *Plant Physiology* **149**, 195–204.

Zhang J, Nguyen HT, Blum A (1999) Genetic analysis of osmotic adjustment in crop plants. *Journal of Experimental Botany* **50**, 291–302.

Zhang W, Fan L-M, Wu W-H (2007) Osmo-sensitive and stretch-activated calcium-permeable channels in *Vicia faba* guard cells are regulated by actin dynamics. *Plant Physiology* **143**, 1140–51.

Zhu J-K (2002) Salt and drought stress signal transduction in plants. *Annual Review of Plant Biology* **53**, 247–73.

Zhu X-G, Long SP, Ort DR (2010) Improving photosynthetic efficiency for greater yield. *Annual Review of Plant Biology* **61**, 235–61.

Zimmermann D, Reuss R, Westhoff M *et al.* (2008) A novel, non-invasive, online-monitoring, versatile and easy plant-based probe for measuring leaf water status. *Journal of Experimental Botany* **59**, 3157–67.

Zimmermann U, Schneider H, Wegner LH, Haase A (2004) Water ascent in tall trees: does evolution of land plants rely on a highly metastable state? *New Phytologist* **162**, 575–615.

INDEX